149 Chambers

867-1944

VIDEO RESEARCH IN THE LEARNING SCIENCES

VIDEO RESEARCH IN THE LEARNING SCIENCES

Edited by

Ricki Goldman
New York University

Roy Pea
Brigid Barron
Stanford University

Sharon J. Denny
University of Wisconsin–Madison

LAWRENCE ERLBAUM ASSOCIATES, PUBLISHERS
2007   Mahwah, New Jersey   London

Lawrence Erlbaum Associates, Inc., Publishers
10 Industrial Avenue
Mahwah, New Jersey 07430
www.erlbaum.com

Cover design by Tomai Maridou

Library of Congress Cataloging-in-Publication Data

Video research in the learning sciences / edited by Ricki Goldman ...
[et al.].
      p. cm.

Includes biographical references and index.

ISBN 978-0-8058-5359-9 — 0-8058-5359-6 (cloth)
ISBN 978-0-8058-5360-5 — 0-8058-5360-X (pbk.)
ISBN 978-1-4106-1619-7 — 1-4106-1619-3 (e book)

1. Television in education. I. Goldman, Ricki.
LB1044.7.V463   2006
371.33'58—dc22   2006038474
                                                    CIP

Printed in the United States of America
10 9 8 7 6 5 4 3 2 1

Contents

PART III

Video Research on Classroom and Teacher Learning

Part III   Cornerstone Chapter

Preface

Video Research in the Learning Sciences aims to deepen our understanding of theory, methods, and practices for using video to add new layers of meaning to scholarship in the learning sciences. Our authors explore the breadth of work underway in the learning sciences that is devoted to fostering the art, science, and practices of video as a way of knowing about and sharing learning, teaching, and educational processes. As a community of authors using video, we seek to advance what takes place when researchers use video to record, annotate, and reflect on their work with teachers and learners in reflexive, epistemological, and hermeneutical collaborative learning endeavors.

Our primary concern is to contribute both to the science of learning through in-depth video studies of human interaction in learning environments—whether in classrooms or other contexts—and the uses of video for creating descriptive, explanatory, or expository accounts of learning and teaching. Our volume is dedicated to taking stock of the growing field and frontiers of video research, both theoretically and methodologically, and to push the technological envelope in designing the next generation of digital video tools.

AUDIENCE

Video Research in the Learning Sciences addresses the key theoretical, methodological, and technical advances concerning uses of digital video-as-data in the learning sciences that will serve as an intellectual guide. It is intended for university professors, graduate students, and anyone interested in how knowledge is expanded using video-based technologies for inquiries about learning and teaching. Although we designed this book to become an important addition within educational research circles, we also hope that scholars in a range of academic fields for whom digital video use is attracting researcher attention will find this book useful (e.g., cultural studies). A significant proportion of the chapters are also of relevance to teacher educators and useful in teacher education courses, either preservice or in-service in nature.

In particular, this book was written to expand the horizons of researchers in the learning sciences conducting research using video-as-data looking for new insights and their colleagues' new developments, researchers in the learning sciences intrigued with doing video-based research who want to learn more about how leading researchers design their theory, methods, and tools, teacher educators who have been exploring or who are considering using video case studies for fostering reflective discussions on teaching practices and learning needs, researchers in Computer Science and Engineering, as well as for-profit education companies, who are interested in learning more about the requirements of and use practices concerning digital video in a real application area of societal importance; and scholars in the human sciences and humanities including anthropology, linguistics, psychology, and cultural studies as well as affiliated scholars of new media who are devoted to understanding sense making of human behavior using video technologies.

We invite our colleagues, especially new video researchers, to engage with us in an analytical effort to address these questions and to find conceptual and thematic commensurability among the various current intellectual threads related to video research in the learning sciences, while also keeping the door open for future advances. We hope that this book will help "usher in" video scholarship and supportive technologies, as well as help mentor video scholars, so that the work of this community will meet its maximum potential to contribute to our growing knowledge base about teaching and learning, and so that video scholarship in the learning sciences grows and thrives as an accepted and respected academic pursuit.

The Content: Understanding the Use of Video in the Learning Sciences

Over the past few decades, the use of video for a range of educational purposes has grown substantially at leading research universities in a range of fields. This book addresses a unique perspective of using video research in the learning sciences. The interdisciplinary field of the learning sciences proclaims as a founding assumption that an understanding of human learning requires integrative inquiries from multiple disciplines including cognitive, developmental, and educational psychology, linguistics, anthropology, education, and computer science. The learning sciences has developed as a distinctive branch of the multidisciplinary cognitive sciences, with distinctive emphases on the problems of education and learning. As a result, research in the learning sciences draws on multiple theoretical perspectives and research paradigms in order to understand the complexities associated with learning, cognition, and development at different levels of analysis.

Learning science researchers are increasingly using video records to advance these inquiries. In research environments, uses of video-as-data are a very different kind of video use than video for telecommunications or distance education. Video is used in a variety of ways depending on the preference of the learning scientist. Some researchers include codings and quantitative analyses of behaviors and interactions in their video records; others use video in support of the purposes of qualitative research and the construction of rich compelling narratives. Some establish coding schemes from the bottom up as they view data, whereas others code hierarchically, establishing higher level descriptors before collecting and analyzing their video data. Some learning science researchers use video to make meaning, involving learners, researchers, and viewers in a community of inquiry by using tools for shared commentary. Others follow experimental approaches, designing tools for coding large databases of video archives.

The thread that connects the variety of approaches discussed in this book is that video records and cases as representations (and re-presentations) of learning and teaching are now more easily than ever being combined with the annotation and communication capabilities to enable unprecedented levels of distributed collaborative research, or *collaboratories*. More than ever, we now need to combine our efforts to understand how to make the most of these distributed affordances.

What are the different dimensions along which video data adds value to learning research? Among the most important that are commonly noted are the prospect of different disciplines and different individuals to "see" exemplars of different theoretical categories in video data versus more reductionistic representations such as written coded data from such records; the recurrent "discoverability" experience in working with video records, in that new phenomena come to light on repeated viewing; and the rhetorical power of viewing video of behaviors and interactions for understanding nuances of social relationships, kinesics, proxemics, prosodics and other situated parameters of human interactions. Do video and textual media forms have to complement each other, or can each one serve different purposes, expanding the results of a research project? Moreover, how do the various issues concerning validity and generalizability of video research presentations relate to comparable issues for traditional paper publications? These questions are just a snapshot of the important issues that authors address in this volume.

Structure

The book is designed around four main sections with a cornerstone chapter in each part written by one of the four editors; each editor has taken the lead role in shaping one of the sections. A cornerstone chapter to each part introduces and synthesizes the cluster of chapters articulating the major themes common to that cornerstone. This design acts as a unifying role for readers to get a handle on a complex subject. The four cornerstones chapters are:

I. Theoretical Frameworks. This cornerstone chapter invites readers to think about how video affects the nature of conducting research when it is used not only to describe, represent, analyze, interpret data on learning and teaching, but also to build learning communities and cultures. In this section, authors discuss a range of theoretical and methodological frames and visions-from ethnography, semiotics, conversational analysis, aesthetics, pleasure, and phenomenology—to conduct and present research on learning and teaching in the learning sciences.

II. Video Research on Peer, Family, and Informal Learning.   This cornerstone chapter illustrates how video-based analysis has been an important tool for advancing theories and understanding of learning that takes place in peer-based or family-based interactions, and for capturing learning processes that arise in informal learning settings such as after-school clubs, museums, and technology learning centers.

III. Video Research on Classroom and Teacher Learning.   This cornerstone chapter focuses on research and theory regarding how and what teachers and students learn through study of complex, real-world practices that are captured and represented through digital video media. Implications for the design of video-based environments and activities for classroom learning and teacher professional development are also addressed.

IV. Video Collaboratories and Technical Futures.   This cornerstone chapter explores and researches the value of new functionalities for representation, reflection, interaction, and collaboration with digital video to support research and communications in the learning sciences.

The chapters within the cornerstones are by researchers who have made what we consider to be lasting contributions to video research. These authors are currently engaged in advanced theoretical, methodological, and design projects, and each has addressed fundamental topics in digital video use in the learning sciences within research studies over the past few decades. Moreover, each has been and is currently engaged in using video and in designing video-based research environments for research purposes.

Affiliated Web Site

An affiliated Web site—http://www.videoresearch.org—is being launched at New York University Steinhardt School of Education (http://www.nyu.edu/education/) by editor Ricki Goldman, a professor in the Program for Educational Communication and Technology at NYU's Steinhardt School of Education. Merlin is the virtual video research lab that is situated within the Consortium for Research and Evaluation of Advanced Technologies in Education (CREATE) that Jan Plass (director) and Goldman (co-director) direct. The mission of CREATE is to advance the cognitive studies foundations of the educational media and technology. Research within the consortium focuses on the design, implementation, and evaluation of emerging advanced technologies. The CREATE laboratory is part of the Program in Educational Communication and Technology in the Department of Administration, Leadership & Technology at New York University Steinhardt School of Education.

ACKNOWLEDGMENTS

This book would not have been possible without the enthusiasm and commitment of the contributing authors whose chapters encapsulate critical issues in this emerging field of study. We would like to thank each and every author for submitting their best and most original work as a vehicle to push the field forward.

An edited volume of this scope requires the close editorial involvement with the publishing team. Heading up our publishing team at Lawrence Erlbaum Associates is Naomi Silverman, whose involvement was felt at each stage of designing and writing this book. We first approached Naomi with this idea at the Annual Meeting of the American Educational Research Association in 2003. She immediately understood the potential significance of this collection of works by leading video researchers in learning sciences and encouraged us to keep our focus on quality and depth. For her constant support and encouragement, we thank her. Other staff members at Erlbaum we would like to acknowledge include Erica Kica and Bonita D'Amil.

—*R. Goldman, R. Pea, B. Barron, & S. Derry*

Part One

Theoretical Frameworks

Video Representations and the Perspectivity Framework: Epistemology, Ethnography, Evaluation, and Ethics

Ricki Goldman
New York University

INTRODUCTION

In the first collected volume of essays on the use of visual representations in the social sciences, *Principles of Visual Anthropology* edited by Paul Hockings in 1973, Margaret Mead remarks that the "hazards of bias, both in those who film from their own particular cultural framework and in those who see their own filmed culture through distorting lenses, could be compensated for ... by the corrective of different culturally based viewpoints" (p. 8). Having pioneered the use of visual anthropology with cyberneticist Gregory Bateson in the late 1930s, her article challenges anthropologists to change their research practice. Her fear was that cultures would disappear with the spread of modernity and that valuable knowledge of cultural performance in worlds still untouched by industrialization would be lost forever:

> Those who have been the loudest in their demand for "scientific" work have been least willing to use instruments that would do for anthropology what instrumentation has done for other sciences—refine and expand the areas of accurate observation. (p. 10)

The practice of anthropology and the study of education have several overlapping similarities—both are grounded in what people say and do when they are in the act of thinking, making, and creating. Both have a history of valuing "outcomes"—artifacts or material representations that demonstrate achievement and advancement. In both practices, there are communities of practice that point out the perils of valuing outcomes more than processes. And, both anthropologists and educators have moved from notions of the grand narrative to a focus on *local, situated* knowledge (Brown, Collins, & Duguid, 1996; Geertz, 1983).

Both disciplines also face an important complexity that Mead could not have foreseen in 1973—how the use of networked visual data, tools, and methods enable researchers not only to reflect more deeply on their observations, but also on the entire research process that has moved away from the solitary researcher to the community with multiple stakeholders. Within the global community, it is no longer possible to describe others as being part of a scientific endeavor without addressing what is commonly referred to as the *crisis of representation*—the dilemma we face when we try to represent others and ourselves as we crisscross boundaries of gender, race, identity, culture, time, and location. We now take a more reflective and ethical look at what is involved in the scientific investigation of human meaning making. As researchers of learning, we now use video, text, and sound as elements that are put together in ways that resemble the collage movement in mid-20th century art. For example, a free web tool called a *wiki* becomes a community space for community members to represent their own voices in ways that were not possible when Mead first studied the Samoans. Video, text, and sounds are now pixels that can be manipulated, sized, shaped, segmented, layered, not streams of moving images from one filmmaker's perspective. Nevertheless, as educational researchers, we still want to know that we have a way to come close to understanding what happens in the learning culture as we observe and participate within it. It is our human nature to make sense of what we experience. And, as researchers of learning, we want to be able to use theories, tools, and methods of investigation that will result in making learning, not only sensible to ourselves, but also, more meaningful for learners. As we learn to participate in the global online community, we might want to consider that the full range of continually emerging media forms can and will be used to describe, interpret, and represent what was happening for the members of the community "by the corrective of different culturally based viewpoints."

FRAMING CHALLENGING QUESTIONS

During these past two decades, I have written, almost exclusively, about the need to embrace diverse *points of viewing* (Goldman-Segall, 1990, 1996, 1998) to prevent the hazards of bias, misrepresentation, and missed-representation, emphasizing the affordances of using advanced video technologies. In the digital commons, knowledge is shared simultaneously, immediately, and sometimes without safeguards. Although this can be somewhat unnerving for educational researchers, the real hazard we now face, 30 years after Mead made her statement, would not only be the failure to use emerging video-based technologies to include diverse perspectives, but also the failure to take the time needed to reflect on how these technologies change every part of the research process from the moment the camera is turned on. Perhaps we should begin the process of reflection from the time a researcher conceptualizes her study of others using video. Perhaps we should continue to examine how these emerging technologies are changing every part of the research process.

From the start of the video research process, educational researchers are confronted with five interwoven questions when using the video camera in research. The first one addresses the importance of understanding the affordances and problematics of using video in the learning sciences. What do we learn during our investigations while videotaping, editing, and analyzing video that we might not be able to learn without having this media form? Are these rich media artifacts a new way of understanding not only those we study, but also, ourselves as researchers as the camera is pointed in a certain direction taping what the camera-person wants to display about these learning cultures? As you read the chapters in this book, you will notice the many affordances that video offers and how these affordances are often met with an equally strong challenge to overcome as researchers attend to the subtleties seen in every frame and in every stream of video. The authors selected for this book have experienced how repeated viewings, for example, are not only an affordance, but also a challenge. When does one stop re-viewing? What is enough viewing for a given study? How can we be sure? Another affordance is the ability to share with others what one sees with colleagues, teachers, and the learners themselves. The challenge is how to manage the rich commentaries and observations that others make on the video data. Another affordance that most every author in this book has had to confront is how the medium of video affects and changes the culture one is studying from the moment the camera is turned on. Can any of us, with real honesty, say that the camera is not affecting our actions? As Barron notes in her study of sixth graders:

> Although it is possible that the video camera may have influenced student behavior, it is difficult to predict in which direction. Being recorded could as easily have been distracting as facilitating with respect to the attention of the student participants. (Barron, 2000, p. 397)

If we can learn not to act in front of a camera, how long does it take to establish that kind of composure? And, do we ever know, even when we seem comfortable with the camera taping us, if we are being true to ourselves? It seems obvious but necessary to state that we should *not* decide to *not* use video because our actions might be affected by the presence of the camera, but rather to accept the performative actions we demonstrate whenever we are being observed.

A second question to ask in our learning science community is whether the use of video in research is only an evidentiary tool or also a media form used to tell a story and convince viewers and readers of emerging texts so that they understand what happened to learners as the research was taking place. Let us use an ethnographic lens to address this question. Both modernist and postmodernist ethnographers have underscored the importance of being there (Geertz, 1973) or being with the community (Heshusius, 1994). The difference is that the postmodern ethnographer understood that convincing the reader that she was there was not the Truth, but a partial truth (Clifford, 1986), a construction of what she experienced and how she interpreted that experience into a textual narrative. Traditionally, ethnographers used field notes to record what was happening as it happened, and, then retreated to an office close to their academic homes to write the compelling story. Often, field notes were gathered while the observing ethnographer sat at the back of a room or village huddled over her notebook. Later, during analysis and inter-pretation, the ethnographer constructed journal articles, chapters, and books. This was the way it was done. That is not to say that anthropologists did not engage in the life of a village. Many had deep relationships with informants and participants, shar-ing in the day-to-day lives of the places they studied.

Research with online digital video with easy access to online environments cre-ates an even more complicated process in spite of the ease of pressing the record but-ton on a camera and then downloading the video onto a computer. Each part of the research process can now be a community activity, with multiple feeds of data, shared video databases, and shared analysis tools. One can also predict in the very near fu-ture collages or blends from one "movie" to another, sharing databases (McWhinney, this volume). Researchers will be able to fluidly work either deductively (top down), inductively (down up), or by using both approaches simultaneously. They may also decide to explore themes as they review digital video records before the focus of the study is defined. Or, they may search video databases to build rich cases from large-scale quantitative or qualitative studies. The point is that video seems to put re-searchers not only in touch with the perspectives of those who design, participate in, and analyze the study; but it also puts them in touch with the multiple methods of conducting a study, a method we often refer to as triangulation. This will, no doubt, become more pervasive within and across research communities of the flattened *world* (Friedman, 2005).

This leads us to the third challenge we face. Each research community using video, even within the learning sciences, has a different epistemological understand-ing about what makes research valid, robust, and reliable. How each community uses and evaluates video will be different. Why? Each community uses video quite dif-ferently. Even within the ethnographic community, some collect video to create more closely grounded stories that include the full range of gestural, auditory, and contex-tual subtlety in the *thick description* of the event (Geertz, 1973). Others map video representations to "locate particular analyses in times and spaces" (Green, Skukauskaite, Dixon, & Córdova, this volume). Others design rich, multilayered sto-ries that convince the reader that the author of the visual representations was "there." Others use video because they find the medium pleasurable and compelling, a better way to tell stories that show readers what they mean (Tobin & Hsueh, this volume; Hayes, this volume.) in short, the exploration of how to use video in the learning sci-ences has just begun.

The fourth problematic area is evaluation. If the use of video in research prac-tices is indeed as diverse as we now know it to be, how do we develop criteria that take into consideration the range of both evaluative measures and e-value-ative qualities for adjudicating the significance of research using video as a research tool?

Maybe we do not need to use video when conducting a study of a learning environment. We should select the appropriate tools needed for a specific study in a particular setting—using whatever combination of media works best for data collection, analysis, and dissemination. What we need to understand is each method, whether qualitative, quantitative, and—what my colleagues (Goldman, Crosby, Swan, & Shea, 2004) and I refer to as—*quisitive methods,* will have a variety of evaluative criteria, sometimes overlapping, and sometimes not.

Over the coming decade, many of the authors in this book will have participated in two symposia sponsored by the National Academies in Washington. The press of the National Research Council produced a report called *The Power of Video Technology in International Comparative Research in Education* (Uleeicz & Beatty, 2001). Certainly the challenge of evaluating video—or what I more commonly refer to as *eliciting* the *value* of video research (e-Value-ation)—will continue to provide us with rich discussions on the changing nature of educational research. From my experience of presenting papers at these sessions (Goldman-Segall, 1999), I am quite certain that many diverse viewpoints of how to evaluate video research will continue to be examined with scrutiny, critical analysis, and vigor. And also with enthusiasm for this emerging methodological tool.

The fifth issue facing video researchers speaks to the main topic running throughout this chapter: How do video representations make us more aware of the ethical stances in our research practices? What are the ethical concerns of using video? Surely it is not just an issue of privacy and confidentiality, although those are important issues. What I propose is that by using video in research, researchers are faced with ethical issues in the research process that they might otherwise overlook. Capturing people on video reminds us of our colonialist past when early explorers collected plants, animals, and people during travels to exotic lands. If we use video representations as disassociated objects to display others, we are indeed repeating past mistakes. As I will discuss later in this chapter, facing the dark legacy of imperialism in educational research (Willinsky, 1998) may serve as a valuable warning to the next generation of video and media researchers in educational settings. As we critique the ventures of early explorers, botanists (Charles Darwin, for example), and collectors of cultural artifacts, we may think differently about how to design and use advanced video and media in the still mostly uncharted territory of learning environments. It is time to change our colonialist past, not only by designing more ethical tools, but also by changing our practices and beliefs about what we are doing when we use video in our research studies.

Frames for Video Research: Chapters in Part 1

What are we doing to address the challenges of using video research in the learning sciences? Although this book offers many diverse uses of video in classroom and informal settings as well as a range of tools used in empirical studies, in Part 1 of the book, authors address how they understand the nature and meaning of video research in the learning sciences. Each chapter takes a different epistemological outlook. Many offer frameworks. One offers a manifesto! And each of the authors has found a unique way of describing how video influences their process of making meaning in and of educational settings and what particular challenges they each have faced in their use of video research. I have arranged these chapters with a narrative flow—a story of coherence within a broad range of diverse *points of viewing*.

Jay Lemke

[W]e cannot understand the epistemology of video as representation unless we also understand the processes by which we make meaning with video when we experience it. I propose that we consider the semiotic uses of video in terms of the ways we meaningfully (and feeling-fully) move across and through immediate and mediated attentional spaces. (Lemke, this volume)

Lemke uses an experiential and phenomenological approach to how video is used to study learning, pointing out the importance of materiality and the felt experience of using this "seductive" medium. He takes us on a journey that starts with the images we experienced inside that magic box called the television when it first arrived into our homes. He notes that the medium takes us beyond our living rooms and beyond ourselves into that other dimension, a dimension that is much more than the 2D images we are watching. He sets the stage for his main critique by warning that if we continue to follow Cartesian duality, we increase the separation between reality and representation rather than see them as intrinsically interconnected. We need to recognize that "all meaningful interactions with realities are also equally mediated by culture-specific interpretative codes, and thus share the feature that seems to distinguish 'representations' as such, but does not." We need to understand how video and other emerging visual media such as electronic games demand (and offer) multiple *attentional worlds* to simultaneously experience.

Lemke suggests a provocative idea—we must begin to attend to the *traversals*, these "temporal–experiential linkings, sequences, and catenations of meaningful elements that deliberately or accidentally, but radically, cross genre boundaries" (Lemke, 2005). This is particularly true when, during our analyses of video records and inscriptions, we are "interacting with and interpreting in very specific ways a very partial (in both senses) record of an activity" (this volume).

François Tochon

François Tochon examines the voyage from video cases to video pedagogy as a path that can focus on the whole situated process of feedback. Viewers can reflect on their own process of learning and teaching by hearing the voice and watching the direction of the gaze. "The signs that organize knowledge construction can be decoded in video feedback" (Tochon, this volume).

In short, Tochon uses a semiotic approach for exploring the constructive meaning-making process. Tochon's expertise is with video study groups. For example, experienced teachers videotape their language teaching methods, and then make video presentations to show to student teachers in the teacher education program. Tochon then creates min-cases for student teachers to reflect on their own cases. Supporting teacher professional growth through the use of video feedback "aims to objectify, conceptualize, and share practices, and then integrate the commonly developed theory into action (Tochon, this volume). However, this process is not about "consumption," but rather about shared exchange.

The frameworks—or thematics—Tochon applies to his video study groups are mastery, pychocognitive, sociocognitive, narrative, critical, and pragmatic. The key to integrating these various frameworks is to understand that readings of video texts must include a range of perspectives to achieve validation of meaning.

What stands out in Tochon's chapter is his discussion of the *third construct*, in addition to the first two, utopian ideals and aesthetics. "The third construct is emerging from visual and systematic feedback for reflective deconstruction and conceptual reconstruction. It uses video to educate in an attempt to integrate social change with aesthetic research" (Tochon, this volume).

Tochon, reviewing earlier research, reminds us of the connection among video research, signs, social effect, and social engagement. "If signs can have social effect, concepts have power. They can be used to transform society." (Tochon, this volume). This stance went underground for a decade with the advent of electronic media, Tochon contends, but now the conceptualization of using video in educational research seems to be taking a more integrative socioconstructivist form based on shared reflections on present and future actions (Tochon, this volume). Video, for Tochon, becomes a mirror for those who are videotaped to reconsider their actions.

Joseph Tobin and Yeh Hsueh

Joseph Tobin and Yeh Hsueh have picked up the mantle of the importance of the pleasurable experience of using video. Tobin's earlier and groundbreaking video ethnography with Davidson and Wu (Tobin et al., 1989) established the importance of video as a form that gives pleasure. And, indeed, Tobin's work over 20 years now demonstrates how video productions of closely grained moments in the lives of young children in early childhood educational settings can be both aesthetically pleasing and effective social science research.

> For videos of classrooms to function as provocation and stimulus, they must be hybrid constructions, blurred genres that are simultaneously social scientific documents and works of art—if they come across as insufficiently systematic, they will be dismissed for lacking rigor; if they feel insufficiently artful, they will be ignored for being boring and visually unappealing. (Tobin & Hsueh, this volume).

Listing ethnographies, documentaries, instructional videos, illustrations of best practices, and tools for critical reflection as specific genres, Tobin and Hsueh focus their lenses on both the ethnographic genre and the critical reflection genre as being essential for their video ethnographies.

As researchers, Tobin and Hsueh struggle with the challenge of coherence, among other challenges and trade-offs. One challenge is how to tell a narrative that is engaging while editing the actual flow of events to a story that conveys meaning. For Tobin and Hsueh compelling narrative forms must have strong protagonists to engage the audience. Another other challenge is *simultaneity of events*. For example, how does one tell a story when events take place in more than one location?

However, what concerns them the most as they take us all step by step through their process of making video ethnographies and sharing their excitement with us through the decision-making process, is that the pleasures involved in making and viewing videos need not be *guilty pleasures*.

Michael T. Hayes

Taking us on a journey into a school building in Northern Idaho, Michael T. Hayes tells us that each time he puts the camera to his eye, he gains "a different and new perspective on the building and the history that surrounds it" (Hayes, this volume). Although Hayes calls his chapter, "Overwhelmed by the Image: The Role of Aesthetics in Ethnographic Filmmaking," he is certainly not overwhelmed! He is simply viewing complex issues from a range of important perspectives, and does so with elegance and sensitivity. He conceptualizes aesthetics as a space where experience and representations of the world meet. For Hayes, experience and representation "overlap, merge, and blend in a shimmering movement."

He focuses on the aesthetic qualities of the image "to (re)present in visual form my physical, intellectual, and emotional encounter with the subject of my research" (Hayes, this volume). And indeed his textual representation not only describes two diverse ethnographic film events—the school in Northern Idaho and Waikiki Beach in Hawaii, but also raises important theoretical and social issues. His most striking concern, and one I discuss later in this chapter, is the problematic nature of recording others—the danger of hegemonic practices when we point our camera and gaze upon others. Hayes notes that:

> [t]his terrain of power turns the persons, groups, or societies represented from the subject of the imagery into the subjected. They are entered into the ongoing legacy of the colonial project that defines the other thus entering them into a relationship of domination and subordination. (Hayes, this volume)

Instead of continuing the legacy of colonialization, Hayes uses the camera to "disrupt the conventions of authoritative voice and uncover the relationships of power that have quietly structured the ethnographic or documentary film style" (Hayes, this volume). His purpose is to enter the *chaotic flow* and to experience both difference and contradiction. In other words, by problematizing the images he represents, he "redirects the viewer's attention away from the content of the film and lodges it with the logic and the technique that guides its production" (Hayes, this volume, paraphrase).

Reminiscent of Marshall McLuhan's popular phrase "the medium is the message," Hayes refers to his technique and logic as the message. Yet, Hayes's intent is not to follow in the footsteps of McLuhan. Far from it. His intent is to disrupt the reader and, through this disruption, connect the scientific and the artistic use of video in re-search.

> My intent is not to degrade or abandon the scientific, because I find the inten-tions and purposes of scientific objectivity to be useful and important. By scien-tific, I am referring to the overarching cultural values of certainty, objectivity, ontology, and veracity …. My intent is to harness the aesthetic qualities of the image and use them to repoint or redirect those aspects that conjure up verac-ity, reality, and objectivity (or perhaps it is to recapture and recreate the aesthet-ics that resides in the scientific). (Hayes, this volume)

Hayes opens the door for future educators to both conduct and represent re-search from an aesthetic perspective—and one must add, from a social conscience per-spective, all the while searching for a new meaning of objectivity.

Shelley Goldman and Ray McDermott

Opening their chapter with the phrase by Goethe, "the hardest thing to see is in front of your eyes," Goldman and McDermott describe a video case study of three boys—Boomer, Hector, and Ricardo—from their middle-school mathematics through applications project. The case study lives up to the words by Goethe that we do not of-ten see what is right in front of us, making the point that what is needed is a close exam-ination of "staying the course" with those we study, rather than giving up on them as learners. Theirs is a poignant discussion of the interactions among the group of boys engaged in learning mathematics and how assessment may miss what the video camera and the researchers do not. One will want to know more about Hector.

The case study is embedded in a behavioral and social interactionist theoretical framework. Their framework addresses the nature of video records and their significance to the research agenda. Goldman and McDermott are unequivocal when dis-cussing video records; they become data only after

> … emergent analytic frames are documented and systematically plied across multiple viewings. Not until the discourse is dissected and aligned with the be-havioral record, one act at a time, and across time, can the opinions and biases of initial viewings give way to more empirically demonstrable accounts. Only then can the word "data" take its place in a research program. (Goldman & McDermott, this volume)

They are similarly certain that this "analytical advance has not resulted in general ways to help schoolchildren." They are outraged that schools are still places of inequity where the promise of democracy is still a promise and not an actuality. Schooling has become a place where more attention is paid to examining students than teaching them or, in the case of research, observing how they learn.

My reading of their chapter is that they hope that applying their analysis approach (situated in conversation analysis and social interaction) of video analysis will lead to taking away the obstacles children face, and will provide new opportunities for their success. Although video will not make analysis easier, they warn us, it "makes communication visible and potentially reveals behavior nested across levels in precarious and contested interactions" (Goldman and McDermott, this volume). Certainly, that was the case with one boy named Hector.

Judith Green, Audra Skukauskaite, Carol Dixon, and Ralph Córdova

Green and her colleagues continue the discussion on video records. However, they focus on *interactional ethnography*, "a theoretically-driven approach that enables us to learn from the social and academic work of class members." In this chapter, they describe Arturo and Alex, two students in a fifth-grade class. In essays written "across times and events," the boys describe their insider/outsider perspectives of being members of the classroom culture while reflecting on their role as ethnographers of that culture. In contrast to Shelley Goldman and Ray McDermott, they do not see ethnography as being "procedurally specific ways inside an emergent and well-documented analytic program" (this volume). In *interactional ethnography*, the participants become researchers. Referring to others' scholarly frameworks, Arturo uses "contrastive analysis to make visible ways in which life in the classroom was socially constructed, local, and often invisible to outsiders who do not share the history, meanings, and language that members have in common." Needless to say, this chapter is close to my heart. It underscores the importance of ethnographic border crossing that we have witnessed as students construct their own portraits and thick descriptions of the cultures within which they participate. I have applied a somewhat similar ethnographic framework—the first was a study of elementary school students from the Hennigan School in the 1980s. These children participated in reflective video-based analyses during and after the filming of *The Growth of a Culture* (Goldman-Segall, 1988). This documentary was about children at the Hennigan School in Boston becoming ethnographers and epistemologists of their learning cultures. The second study at the Bayside Middle School (1998) involved middle-school students using video as their ethnographic research tool, constructing their own media portraits and collaborative analyses of curricular innovations during the early 1990s. As I wrote:

> My audiovisual ethnography is about co-construction; those I view use the same tools to view me and to view themselves. As I elicit meaning from others and from myself, so they elicit meaning from me and from themselves; in this way we build cultural artifacts as a community of inquiry. Images are reflected back and forth as our gazes meet. And, in the end, we are all affected as our relationship grows. (Goldman-Segall, 1998, p. 105)

For Green and colleagues, the use of the interactional ethnographic perspective is not only to involve students as ethnographers studying the growth of their learning cultures, but also to "guide secondary analysis within our ongoing ethnographic corpus in K–20 classrooms (1–12 years of data collection per teacher)." In short, in Green and her colleagues' studies, the teachers become cultural guides enabling the researchers to follow the history of science teaching over a 2-year period. And, indeed, in this chapter, Green and colleagues meticulously describe the identification, collection, and analyses of Ralph Córdova's third- and fourth-grade classes. They focus both on the discourse of the social collective as well as on the individual acceptance or rejection of the resources. Similar to many ethnographic accounts, they examine how the individual and the culture are interconnected. Yet, building on Bakhtin (1986), they weave new life into this discussion by "analyzing chains of (inter)action to construct grounded arguments about intentions speakers and hearers signal to each other" (paraphrase). "From this perspective, what is captured on video records of classroom (and other institutional) life, are intentional actions among members of a sustaining social group" (this volume). Certainly, Green and her colleagues have touched one of the core issues in any expression of ethnography—ethnographic accounts are *thick descriptions*, winks upon winks upon winks, and turtles all the way down, as Clifford Geertz (1973) so vividly notes in his seminal work, *The Interpretation of Cultures*.

Timothy Koschmann, Gerry Stahl, and Alan Zemel

Koschmann, Stahl, and Zemel provoke and shock our sensibilities (and sensitivities) with an erudite chapter they call, "The Video Analyst's Manifesto (or The Implications of Garfinkel's Policies for Studying Instructional Practice in Design-Based Research)." They ask how we should study learning in an already designed context and what vocabulary we should use to express our theories of learning, theories that will help practitioners. To answer their questions, they propose the use of *conversational analysis* within an *ethnomethodological framework*, based on the writings of Howard Garfinkel. They acknowledge that they have chosen to name their chapter with the word, *manifesto*, to be provocative. Noting that other leading researchers have conducted ethnomethodologically informed studies in classrooms, they state that their application of ethnomethodology spearheads a different intention—to improve education through design-based research.

The chapter follows Garfinkel's five policy requirements; indifference; contingently achieved accomplishments; relevance; accountability; and indexicality.

Note that indifference does not mean that the researchers are indifferent to those they study.

> Ethnomethodology's policy of indifference stands in stark contrast to the assumptions underlying conventional experimental research in education. …
> The policy of indifference not only suggests that any instance will do for the purpose of demonstrating some phenomenon of interest, but also that such a demonstration can be based on a single case. (Koschmann, Stahl, & Zemel, this volume)

Indexicality can be defined as expressions of "knowledge of the context within which the expressions were produced," rather than content-free expressions. Locating the utterance of a conversation within a context enables the analyst to get a handle on what that utterance means. (Here, one can see the link to the semiotic lens of video analysts.) In short, these authors suggest that Garfinkel's indexicality policy provides a framework to study actions, not as individual events, but as

> … resources by which actors can produce the sense of prior actions in light of the current action, and make relevant and sensible possible subsequent actions. This clashes with the view of context as a given, as a container within which actors do what they do. Instead, it poses the task for the video analyst of rendering an account of how members in their capacity as "order production staff" (Garfinkel, 2002, p. 102) go about constructing context through their indexical actions. (Koschmann, Stahl, & Zemel, this volume)

Frederick Erickson

Frederick Erickson takes a phenomenological approach to video analysis. He asks what people notice when they watch minimally edited footage. But, let's step back a bit. Minimally edited footage is video that is almost completely unedited; the footage simply captures what is occurring as events unfold, without the usual cinematic techniques of jump cuts, montage, and cutaways layered seamlessly over a soundtrack to simulate continuity. Erickson reminds us that watching minimally edited video is not easy for novice analysts. Viewers are socialized to make meaning of multiple edits that infer the meaning of an event rather than describe the event itself.

Note that Erickson is not advocating the analysis of long video streams. Even a clip of one minute can overload a novice viewer. (What Hayes calls being "overwhelmed by the image."

Referring to Stanley Fish's 1980 book, *Is There a Text in this Class?*, Erickson reminds us that no one person reads a text in the same way. And, therefore, we read the video streams quite differently when we are in the process of analyzing them. To make his point, Erickson recounts how he showed minimally edited video footage of classrooms to his students. The focus of the camera was on a teacher who did not inflect her voice or praise the students. It took inexperienced practitioners a long time to realize that cultural factors were at play. The teacher was not disinterested and indifferent; rather, she was a Canadian First Nations (what is referred to in the United States as Native American) teacher teaching in a Canadian public school for First Nations students. As a First Nations member, a large range of voice inflection and lavishing praise are not how one converses. In other words, what the inexperienced teachers and video viewer needs to know is the context to understand the text of the classroom. And, he argues, that experience is crucial—both as a viewer as well as a teacher.

> However, as we proceed in these design research attempts, further research on the phenomenology of viewing seems warranted, as done by viewers with differing life experiences and differing pedagogical commitments. (this volume)

In short, analysis of video data requires community scaffolding that occurs when minimally edited video is seen and heard phenomenologically.

THE PERSPECTIVITY FRAMEWORK:
EPISTEMOLOGY, ETHNOGRAPHY, EVALUATION, AND ETHICS

To open the door and provide space for new researchers using video and other new emerging technologies, I want to offer an alternative framework for using video research in the social sciences called the *perspectivity framework*. I aim to provide a framework that both embraces conceptual diversity and encourages the use of advanced digital media technologies..

The perspectivity framework encompasses four cornerstones; Epistemology, Postmodern Ethnography, e-Value-ation, and Ethics—each to be discussed in this chapter.

The perspectivity framework illustrates how emerging video technologies become epistemological tools, perhaps better tools than any we have had to date, for researchers, viewers, and those being videotaped to share what they are seeing, making, doing, and thinking while in the process of learning. In short, the focus is also on process and not only on results. Digital video research environments using emerging construction and analysis tools provide a collaborative space for exploring the process of construction (See Cornerstone 4, this volume). Using these tools, we have the opportunity to share our roles as researchers and learners, breaking the hegemonic practices of capturing video records and shooting others. We can share the shooting, editing, and interpretations with those we study. We may even decide to involve the community in the design of the study to ensure that each stakeholder group has an opportunity to represent a range of perspectives.

The perspectivity framework acts as a conceptual scaffold to address the journey from bits and segments (video data in the small) into meaningful stories and valid results (video interpretations in the large). This framework is open, flexible, and inclusive of the diverse theories and methodological approaches that have emerged (and will continue to emerge) as researchers use digital video in their research. In the learning sciences, the theoretical approaches include constructionism (Papert, 1991), situated cognition (Lave & Wenger, 1991), anchored instruction (Bransford, Sherwood, & Hasselbring, 1988); design theory (Kolodner, 1995), and computer-supported collaborative learning (Koschmann, 1996; Stahl, 2006), to name but a few. The methods run the full range from quantitative to qualitative, and to what my colleagues and I have been calling, *quisitive* (Goldman et al., 2004), a form of research that included multimodes and diverse research methods, such as mixed, blended, or complimentary methods.

At its core, the perspectivity framework is a structure that encompasses a range of diverse lenses to view what occurs in a given setting when the camera is turned on, knowing that every stakeholder in front and behind the camera and monitor has a different interpretation of the event, one that may change as the video is later shared, annotated, and put into new configurations. The perspectivity framework also acknowledges that negotiating the meaning of events from multiple points of viewing enables a layering of diversity producing a clearer understanding of the complexity involved in knowing what happened in a given time and place. It also addresses the need for interacting with the artifacts or representations that are continually being created in the process of communication about meaning. As Rowland points out:

> We come to know through interpretation, dialog, and negotiation of meaning with … others, through a conversation with manipulation of the materials of a situation. (2004, p. 43)

One problem with building frames, even open ones such as the perspectivity framework, is best presented by Trinh Minh-Ha (1992) who brings to light the fact that the framer is always framed—entangled and perhaps complicit in the act of framing others. Thus, the framer of others is a *conspirator* unless she continually engages the perspectives of stakeholders in her presentations and interpretations. This commitment to nonhegemonic practices where those in the frame of the camera are not "othered" is at the core of the perspectivity framework. Perspectivity is an open framework, continually transforming itself with each new viewpoint and each new additional layer of interpretation.

The perspectivity framework addresses the problem of surveillance video and its ability to "frame" and "freeze" learners' knowledge production in their continually evolving performance. As those of us who use video in our research practice are aware, video is a very good record of what happened, but, like most media forms, it is always an incomplete story, subject to misrepresentation. Some of these problems can be overcome by including multiple perspectives and by using analysis technologies to build configurationally valid accounts from emergent patterns of convergence (Goldman-Segall, 1995). However, even with the best possible video analysis technologies, video data will always be subject to misinterpretation and bias. The "trick" is to accept and appreciate that, regardless of the method used, data are always subject to some degree of personal framing of what the researcher experiences. Researchers expand the viability and validity of the video records by sharing viewings and interpretations within discourse communities that include the participants who are videotaped. Another vital action is to design video tools that not only shed light on what learners are doing, but also on what learning researchers are doing when using video. That is, researchers need to reflect on the reflections of their framing in a reflexive and, if possible, a critical manner.

As we will discuss in more detail, video-based research artifacts are not just external visual representations, but can also be re-presentations, presentations that can be reviewed, revisited, restructured, and recognized, from multiple viewpoints. They represent the perspectives of all participating members of the community. Video representations seem to be a different kind of re-presentation than textual representations. They display and illustrate a person's expression and experience in the context of a community as an event is taking place. In this shared presentation of the learning experience, video re-presentations are not only evidentiary artifacts, but, more importantly, expressive objects of inspiration, perhaps, as Tobin and Hsueh (this volume) state, they are objects of pleasure. Creators of video texts display the story of people's experiences in a learning setting, stories that are continuously woven and rewoven into material artifacts. Audiences cannot only view what has been created within the research environment, they can participate and interact with the data. Yet, in the new flat world (Friedman, 2005), where participation is now the modus operandi, we can expect even more access to, not only the conclusions of research, but also, the processes of collecting and gathering data, instantly interpreted by communities for their own purposes. YouTube™ is but a small tip of an iceberg that has been growing for over a decade in video research communities.

In short, video re-presentations may never be raw data in the sense that we once understood that phrase, raw; nevertheless, they *are* data to be layered and saturated with interpretation, from the moment the video camera is turned on.

Epistemology and (Video) Representations

What are representations? Are representations *things* that are created or are they, as postmodern semiotician Stuart Hall proposes, processes? Whereas positivists postulate that the world is "out there" to be discovered and categorized, postpositivists have agued that learners construct meaning in their minds and knowledge must be relative. Postpositivists contend that there are no universal truths to discover or uncover, no fixed categories that uniformly describe the world, and no set structures and stages that define how the mind makes meaning, but rather, multiple lenses to apply as the learner interprets and expresses their understanding of events and states being experienced. Given the enormous gap between positivist and postpositivist perspectives in the various discourse communities of educational researchers, how do we reach any agreement about the nature of a representation and its importance in knowledge construction?

To begin this discussion, we will start at the most basic definition: A representation stands for something else. A picture of a boat is not a boat, yet we interpret the picture as a boat when we "read" the picture. The sign and the referrant become linked. With visual signs, the word and the object being referred to are, in most cases, unrelated. The word *boat* does not necessarily look like or sound like a boat. Hall addresses this particular issue, defining representations as "the production of meaning of the concepts in our minds through language" (Hall, 1997, p. 17.) Included in his definition is the link between concept and language that "enables us to refer to either the real world of objects, people, or events, or … to the imaginary worlds of fictional objects, people and events" (p. 17). In other words, without having a representation for objects, people, events, and concepts in our minds, meaning making is impossible. Moreover, Hall's definition of representations includes two related systems, the first being the correspondence between the people, places, things, and ideas and the ever-changing conceptual maps we form as we experience the world, and, the second, the more interesting, being the link between those conceptual maps and the languages (of signs) we speak. The relationship among

... things, concepts, and signs lie at the heart of the production of meaning in language. The process, which links these three elements together, is what we call representation. (Hall, 1997, p. 17)

Note that, according to Hall, representations are not *things*, but rather *processes*. If Hall is correct in his definition, then how do we understand the things, concepts, and signs that get materialized and shaped? How do we interpret what others speak, write, and transmit electronically to us? In other words, can representations be both process and objects? For Hall and other cultural anthropologists and scholars of semiotics, the answer is that being understood means that we agree to participate within the culturally determined codes that are established, not by "what is out there" as an objective reality, but by what we have collectively created within a social code or language as we participate in discourse communities. As a result, a learner cannot interpret meaning as a one-to-one correspondence between object and referrant. Different people will always interpret the same "thing" differently. The key is to reach agreements to try to understand difference and to know how to negotiate difference, keeping the mind open to what may be a best interpretation given the full range of possible explanations.

The Galactic Metaphor

Given emergent technologies and emerging approaches on how to use video in research, how do we discuss the nature of video representations rather than attempting (and probably failing) to codify any particular method, or even a group of selected methods? How do we celebrate the diversity of methods, rather than recommend hard-and-fast rules about what video representations must look like in the learning sciences?

I use the galactic metaphor to describe the nature of visual representations. It provides a handle to enter the dynamic and continually emerging nature of how knowledge is both created and revealed to self and to other. The galactic metaphor applies to the young field of using digital video in educational research.

Positivist theorists discuss representations as either external artifacts or internal mental models. One could graph the bifurcated view of representations as a closed box. This approach divides the epistemological landscape in ways that no longer work in the digital world. (Internal and individual level representations were commonly referred to as mental models, or what Jean Piaget called schema [Piaget, 1930; Piaget & Inhelder, 1956]. When externalized, these representations become artifacts—objects created individually, and whose effect on the external world was considered as emerging from a subjective and personal construction.)

Internal collaborative representations are the kinds of representations that occur from interacting distributed mental models. One could think about this form of mental representation as a computer system with interactive components. When externalized, artifacts continue to be collaboratively produced. In doing so, the subjective and personal elements of the construction are reduced and the artifacts are now more "objectively" produced.

The problem is that our perception and knowledge-producing systems are not so neatly divided. The notion that any outcome can become objective or truthful by removing bias is close to a fantasy, as all representations are constructed through perceptual systems that are unique, personal, and complex, even if similar in structure. Postpositivists do not accept the notion of internal and external representations; when the postpositivist uses the representation, she does not mean ... "as a representation that can be stored and retrieved, but as perpetually constructed patterns of action based on self-organized, every day human interactions" (Rowland, p. 43).

In other words, there is no inside and outside, but rather dynamic, emergent, and spiral systems. It may be helpful to refer to Stacey's theory of learning as the "perpetual construction through the detail of interaction of human bodies in the living present, namely complex responsive processes of relating" (Stacey, as cited in Rowland, 2004). Rowland then adds "These processes are carried out in a similar way individually—as an internal conversation—and socially, so a distinction of individual and group or organizational levels is deemed unimportant" (Rowland, 2004, p. 36, italics added).

Let us now return to the metaphor of a galaxy consisting of constellations and stars, to make this point. From our perspective from any one location, constellations seem to be continually changing locations in the sky as one entity. Yet, they remain intact vis-à-vis each other, from our perspective. Actually, they are clusters interacting within larger dynamic systems. In fact, it is our standpoint that is gradually altering as our planet turns.

This constellation metaphor in the points of viewing theory (Goldman-Segall, 1998) revolves around the idea that each person experiences the world from a standpoint, a viewpoint, or what we might call a situated context emerging from years of perceiving and making meaning of experiences. (For a fuller discussion of this concept and the online digital video analysis tool called Orion, see R. Goldman, this volume; also see Goldman-Segall, 1998). Given that making sense of what is experienced is deeply enhanced by the viewpoints of diverse people, as Margaret Mead (1973) reminded us, this stellar and galactic metaphor provides a way to get a handle on understanding how one event or one particular perspective can also "live" (as a reconstituted entity) in another constellation when viewed from a different "galaxy," knowing that galaxies are also within dynamic and emerging systems. If we were to create an image of this system, we might use a galactic map of the Milky Way as seen from outside the Milky Way. (At http://www.ras.calgary.ca/CGPS/where/plan_basic_big.gif, we can view a schematic image of the Milky Way along with both the sun and the spiral arms of matter. The image was created by the University of Calgary's Canadian Galactic Plane Survey.)

In a recent article published in the *Epoc Times*, David Jones (2005) on the website reports on research by astrophysicist Bryan Gaensler from the Harvard-Smithsonian Centre for Astrophysics. In describing the Large Magellan Cloud, Gaensler relates that the magnetic field is both smooth and ordered in spite of "internal conflict." Describing a galaxy in terms of "having a birthday party ... for a bunch of 4-year-olds, and then finding the house still neat and tidy when they leave," Gaensler anthropomorphizes galactic behavior as human behavior. Gaensler also expands the notion of the *dynamo* as "a process where the overall rotation of a galaxy combines and smoothes the small magnetic fields created by whirls and eddies of gas." He adds: "Stars bursting out at random all over ... strengthen the magnetic field, not mess it up."

It may seem an epistemological leap to compare galaxies, constellations, and stars to conceptualize how representations of learning and conducting research with video data occurs. However, I offer this metaphor as a way to think about thinking, an extremely useful metaphor to use given the interactive and dynamic nature of galaxies. Interacting systems, both large and small, are ordered and patterned within the chaos of random creation. In effect, my comparison may be quite similar to one where Mitchell Resnick (1997) compared computer "creatures" such as dots on a screen called ants or viruses to explain how thinking occurs in more complicated systems, except that I recommend a large system, galaxies, to understand the smallest unit of representation.

Using the galactic metaphor provides a working metaphor for how multiple representations of thinking change our limited perspective. Thinking-in-the-large removes the current obsession with bifurcation into binary dualistic thinking, and replaces it with one that is, at its core, systemic, multiple, interactive, and patterned.

Video and Theories of Knowing

In our continuing discussion on representations, we now turn to the intersection of learning theory and video research. In the early to mid part of the 20th century, behaviorists had an enormous hold on educational research; in fact, its core ideas are so compelling that the paradigm still flourishes in many instructional communities today. The behaviorist approach to learning theory, as derived from J. B. Watson (1913) and B. F. Skinner (1931), is probably the easiest way to measure learner and teacher achievement; it measures the acceptable and expected response to a stimulus rather than the creation of new knowledge representations and creations. It is easy to understand the connection between a theory of learning (such as behaviorism) and representation (such as giving the required answer). The examiner provides a stimulus; the learner responds with the answer that was put into the mind like water pouring into an empty vessel or a blank slate, the *tabula rasa*. For the behaviorist there was, and still is, little need for the creative production of new knowledge in the process of providing the correct answer. And, it is still the easiest way to measure if someone has understood what has been taught. And, for this reason, primarily, behaviorist approaches to learning and even to conducting research have been heavily relied on for almost a century.

When considering the use of video technologies and learning, it is helpful to reflect on the first attempts at designing computing and artificial intelligence systems. These systems build on stimulus/response versions of mentality, using the cause/effect (if X, then Y) way of thinking. This behaviorist influence on the design of learning technologies in education is best exemplified by an early computer-aided instruction (CAI) project at Stanford University called PLATO directed by Patrick Suppes (1966). The project used computers to teach elementary school mathematics and science.

Suppes envisioned computer tutoring on three levels: [T]he simplest is drill-and-practice work, in which the computer administers a question and answer session with the student, judging responses correct or incorrect, and keeping track of data from the sessions. The second level was a more direct instructional approach: [T]he computer would give information to the student, and then quiz the student on the information, possibly allowing for different constructions or expressions of the same information. In this sense, the computer acts much like a textbook. The third level was to be more sophisticated dialogic systems, in which a more traditional tutor–tutee relationship could be emulated. (Goldman-Segall & Maxwell, 2002, p. 6)

The first and second levels are quite obviously connected to a behaviorist paradigm; there is knowledge instruction, not construction. No internal making meaning or interpretation, rather structures designed for input and output. Even in level three, the relationship of the computer to the learner is one of tutor/tutee with no emphasis on the creation of knowledge or invention of ideas and artifacts by learners.

Around this time, the theories of Jean Piaget began to take hold in technology design, offering quite a different way of thinking about the mind, and of mental representations. As we know, Piaget envisioned the mind as the builder of schemata (Piaget, 1930). Schemata are internal storage and interpretation systems for representing ideas and concepts. They provide the building blocks that enable people to solve problems and perform new tasks, yet they are, according to Piaget, limited to age-based stages whose clockwork is internally set. (Seymour Papert, years later, reframed constructivist thinking to what he termed *constructionism.* He rejected fixed age-based developmental stages and replaced stages with *styles* that learners, regardless of age, experience as they make things that represent their thinking about complex problems; Papert, 1988). Papert also avoided the topic of representations, using the term *objects-to-think-with* in numerous lectures and conversations with colleagues.)

In the 1950s, a new way of thinking about the nature of representations was brewing in what became known as the cognitivist revolution—best described in *The Mind's New Science* by Howard Gardner (1985). Gardner asserts, the cognitive scientist "… rests his discipline on the assumption that, for scientific purposes, human cognitive activity must be described in terms of symbols, schemas, images, ideas, and other forms of mental representations " (p. 39).

He goes on to note that with the renewed interest in neurosciences, mental representations may no longer be the core of cognitive psychology, even though among many psychologists, linguists, and computer scientists, a representational explanation of how the mind works is as close to "an article of faith" as one comes (Gardner, p. 40). Linguist Noam Chomsky (1968) added that representations are innate internal mental blueprints or models—similar to a computer program, roughly outlined and filled in through experience over time. (Chomsky himself has changed his thinking offering a derivative of generative grammar with stronger focus on emergence and connection and less on the significance of representation.) Rowland, describing Aadne, von Krogh, and Roos's theory (1996) of representations, also described and then critiqued the cognitivist view, "… [I]n the cognitive view … routines, rules, understandings are defined with respect to objective, universal 'truths' external to the individual and group; that is, knowledge is thought to represent more or less accurately an external pre-given world" (Rowland, 2004, p. 34).

Most cognitive scientists tend to agree that representations are internal mental constructs. For example, Spiro and Jehng's (1990) explain how mental representations work in complex and ill-structured domains. They make a case that the mind, similar to a connectionist machine, processes schema rather than retrieves it. The mind is not only a retrieval machine as earlier behaviorists, and to some degree, the cognitive scientists, had proposed.

By cognitive flexibility, we mean the ability to spontaneously restructure one's knowledge, in many ways, in adaptive response to radically changing situational demands ... This is a function of both the way knowledge is represented (e.g., along multiple rather than single conceptual dimensions) and the processes that operate on those mental representations (e.g., processes of schema assembly rather than intact schema retrieval). (Spiro & Jehng, 1990, p. 165)

For Tzeng and Schwen (2003) mental representation can be broken into models and elicited models:

While the former [models] are constructed representations for mental calculations derived from analyses of collected data (e.g., Gentner & Stevens, 1983), the latter are intended to elicit the constructs of mental calculations and present them as "structural analogues to the world." (Johnson-Laird, 1983, p. 165; text in parenthesis added)

Building on Anderson's theory of *knowledge propositions* (1982; as cited in Tzeng & Schwen, 2003) being the smallest unit of knowledge representation—later called "semantic molecules" by Rickheit and Sichelschmidt (1999)—Tzeng and Schwen (2003) remind us that the essence of mental representations are not internal events, but dynamic "mappings" between the perception of an event and a person's values and background.

This newer approach of mental representations—as mappings between a person's perceptions of an event that are inextricably woven with the experiences of individuals—is not only a departure from previous ways of thinking about representations and how the mind learns, but an even more radical one when thinking about visual representations, such as the way visual sensations interact with our minds. After all, our experience of the world is strongly based on both visual *sense* and cortex, and the *sense* we make of the visual image—video and other moving images being perhaps the most compelling.

A theory that connects our perceptual system with our production of knowledge as a series of dynamic and interactive events trumps the linear, causal, and internal explanation of behaviorists or of cognitivists.

In other words, the discontent with both behaviorism and cognitivism is that they are embedded in thinking about representations as internal "things." In the previous section, I tendered an invitation to consider a more dynamic, galactic metaphor of understanding representations—whether material, textual, visual, or aural—rather than a bifurcated and dualistic approach to knowledge representations. Raising the problematics of considering representations as only internal processes and external artifacts created by individuals or groups of people, I put forward an interpretation that representations are more akin to re-presentations, interactive and ever-changing processes that bridge the internal versus external divide of consciousness and experience.

To explain this idea using a more narrative discourse, let me provide an example from my own practice. As a digital video ethnographer, I struggle to make meaning of the nature of learning as I videotape, edit, and make "re-presentations" from my video selections. Using the points of viewing theory (1998) as a theory to frame an ethnographic approach to representation, I direct my focus to the act of seeing, perceiving, and interpreting my data, using the camera to layer more and more experiences on the subject under investigation during the experience of the event. I understand that my perceptions are based on perspectives gleaned over many years of experiences in learning experiences of all kinds. After all, meaning is made through multiple lenses—using our biological lenses (eyes) and minds or the lenses of the electronic devices, such as cameras, telescopes, and microscopes to amplify, refocus, and expand our sense making. In short, the use of video in research offers a panacea to the discontent with behaviorism and the cognitive revolution, demonstrating that learning is much more than an individual's mental input and output. It is the search for methods to bridge consciousness and experience.

The disillusionment with the cognitive revolution in the late 1970s started after the heyday of individualism. During these intellectually, socially, and culturally turbulent years following the 1960s, educational theorists found themselves on the cusp of a competing approach to understanding representations and how the mind learns—sociocultural. Deeply influenced by Lev Vygotsky (1962, 1978), Jerome Bruner (1990, 1996) and other leading cognitive scientists of the time, the discussion of defining knowledge moved from the individual mind to the collaborative, social, and cultural worlds within which individuals participate in knowledge creation. And, if knowledge construction was somehow linked to a sociocultural context, then what did this mean for the nature of representation?

Along with this sociocultural framework, computational technologies were also undergoing a dramatic change. The mainframe computer (the size of a small room) turns into Alan Kay's vision of the personal computer, a "desktop" with icons, and a Dynabook, a dynamic notebook that we now call a laptop using the Smalltalk software that would enable children and adults to construct knowledge while seamlessly using any media form (1996). Simultaneously, Papert and his research teams during the 1970s and 1980s immersed themselves in the use of a programming language, Logo—the program Papert has spent several decades designing for children to program a computer to get the turtle to act out the program's commands. Suddenly, the object, a small turtle on the computer screen—an object to think with or what we could say a representation of how the computer program was behaving—becomes a thinking tool. But, still most researchers, including Papert, focused on the individual child.

However, a shift, perhaps a paradigm shift, was occurring that offered users of new technologies to collaborate and layer their viewpoints. Video technologies were at the forefront of the changes because, even before the creation of the Internet and computer-supported collaborative communities (CLCL) emerged, video software and hardware enabled multiple viewpoints, repeated viewings, and, most importantly, it enabled the entire stakeholder community to view what was recorded and layer the database with commentary and description (Goldman-Segall et al., 1988, Goldman-Segall et al., 1998). Clifford Geertz's (1973) concept of *thick description* was now within reach of a videographer who could use a computer to control a laser disc player and move around the data chunks adding diverse and insightful comments to specific chunks of video data. Moreover, video technologies provided access, not by one researcher, but by teams of researchers, to explore the situational aspects within which learning was taking place. It was impossible to see the video of a child learning Logo in a classroom without also attending to what the teacher was doing, what the other kids were doing, and mostly, how the young person was gesturing while she spoke about her new creations.

In 1985, I decided to collect my "data" using a camcorder, VHS tapes, and a microphone. As a member of Papert's team, I realized quite soon that video enabled the affordance of collaborative interpretation, what I later called thick interpretation (1998), and then years later called thick communication (2004). Although there was no Internet, networked electronic mail and file transfer programs (FTPs) offered the promise of shared databases and collaborative interpretation. In a videotaped presentation in 1987, I presented my plans to the MIT community and our collaborating teachers from the Hennigan School. I discussed how we would share and annotate the video database of, to my knowledge, the first longitudinal digital video ethnography with a specific computer software application to access video stored on videodisks. By 1988, a working video analysis prototype enabled teachers and students from the participating school, researchers, and visiting scholars had the opportunity to traverse the video database, build clusters from selected video clips (stars) around themes, view each other's comments, and add comments and transcripts. The dissertation was submitted in 1990 (Goldman-Segall, 1990).

During the same time, the sands were shifting, and a community of educational researchers were pushing the boundaries of how computer programs could be specifically designed for what Roy Pea (1985, 1987) called the *amplification of knowledge*, and Marlene Scardamalia and Carl Bereiter (1991) called *knowledge building*.

The appearance and use of online video technologies connecting individuals to each other and expanding knowledge production through collaboration went hand-in-hand with the emergence of theories of learning that embraced the creation of expressive artifacts designed by individuals and groups within situated communities.

One could ask a McLuhan-esque question: Was sharing text and video across distributed computers the cause of our epistemological shift to what Lave and Wenger (1991) termed situated cognition; or, did the shifting theoretical views of learning and cognition create the climate for imagining connected viewpoints using technologies that could explore those boundaries? The question is, of course, flawed, given the nature of a question based in causal dualities. Yet, one does have to consider the iterative impact of the digital revolution with the developing theories of the times.

Postmodernity put the nails into the modernist box internal and external representations.

Educational philosopher James Marshall critiqued linearity, especially any form of mental causality, such a when one first sees a boat and then sees it in the mind, before producing a boat on canvas. Traditionally, educators took the fixed process of a mental model of a boat being "reproduced" and modified on paper as a fixed truth. In other words, there must be a mental representation (idea, schema, or algorithm) that stays in the brain for us to draw on when we want to paint the boat. However, Marshall refutes the simplicity of this kind of causal relationship—from experience to mental model (with or without a blueprint) and then to artifact—stating that the act of writing, for example, creates thoughts, and in the process changes the self, the representation, and the thinking about the topic. For Marshall, the process of creating a representation changes thinking and ideas. In an interview reported by Ghiraldelli (2005), Marshall says:

> I do not know what I think until I see what I write. Foucault would have put it more strongly when he said that if he knew what he would write about he would not have the courage to start writing a book. Writing also changes the person so that the author whilst writing both undergoes and reacts, changing both his thoughts and his self, but also the text. (Marshall, as cited in Ghiraldelli, p. 290)

This idea dovetails with Papert's epistemological approach that tinkering with an object created by a learner changes the learner's thinking (Papert, 1980 & 1991; the author (builder) is changed in the process. To paraphrase Marshall's comment substituting writing for writing of media texts, "Writing [of media texts] also changes the person so that the author whilst writing [of media texts] both undergoes and reacts, changing both his thoughts and his self, but also the [media] text" (Marshall, as cited in Ghiradelli, 2005; text in parentheses added by author).

Obviously, one can easily apply this idea of the creative element of working with video to produce an artifact. Producing (and then reworking with video edits or clips) is not only the result of finding in and out points of the visual stream, but comprises acts of creation experienced by videographers who, through productions, reshape the material into something that has rich meaning and is reshaped in the process. Once again, the galactic metaphor provides us with a way to consider how consciousness and experience are woven together.

Video Ethnography

Quantitative studies often provide researchers with assessments and global predictions, but do not aim to explain the inside story—the meaning that people ascribe to the events they experience in learning environments. Ethnographic accounts tell rich stories that help us to understand the meaning of events. Ethnography is the description, interpretation, and a representation of what researchers experience when they become involved with the day-to-day lives of people carrying out their lives. In short, ethnography is the study of cultures providing us with new ways to view and interpret growth and change within and across cultures.

> The task of the social sciences as is conceived by [anthropologist Charles] Taylor is analogous to Wittgenstein's understanding of the aim of philosophy. In its most general and positive from, this aim is to offer an Ubersicht surview or *perspicuous representation.* (Smeyers, 2005, p. 411; italics added)

Ethnography is best understood as a family of approaches to the study of culture within the knowledge domain of anthropology and, over the past few decades, extending to educational research and other social science research. Ethnography encompasses the full range of traditional descriptive accounts of cultural groups to more socially activist-oriented critical methods whose aim is the empowerment of nonmainstream communities through the study of hegemony and resistance. The incorporation of feminist and postmodern theories to traditional anthropological notions of culture has broadened the intellectual landscape opening new questions and methodological concerns (Clifford, 1986; Lather, 1986; Roman, 1991, among others). Unfortunately, these more progressive and social justice streams of thinking about knowledge, for example—"whose knowledge" is being represented—have not been fully articulated in the learning sciences community. One reason could be that researchers trained in mathematics and science education are too often trained to think of their knowledge domains as representing the truth. If not true, then at least the most truthlike one can have using empirical evidence.

Video research, in particular digital video ethnography, has raised the consciousness of learning scientists who would have once believed that research results should be gleaned only from quantitative data and devoid of personal interpretation. As many of the chapters in this book show, ethnographic video representations in the form of video segments or video streams have provided many learning scientists with the realization that to know and to re-present the process of learning requires a range of emergent tools and techniques.

Video has played a significant role in the learning sciences by demonstrating what constructivists have long contended—that our theories emerge through our deep engagement with what we see by attending closely to the process of learning rather than by only attending to the results of a given treatment on a group of people in an experimental lab-like setting. Video ethnography is personal, close-up, and affected by the views of those who videotape and direct the video camera's lens. As I have noted elsewhere and as should be restated once more, the word theory comes from the Latin word, *theoria*, which means "a viewing."

Currently, the use of video representations, regardless of whether the analysis is frame-by-frame, video case based, or documentary style movies of what is going on in a learning environment, can be shared, discussed, interpreted, and clustered into new mixed by diverse communities of researchers and stakeholders anytime, anywhere. Along with this collaborative approach, learning researchers seem to agree that valid interpretations of video representations are the result of dynamic interactions among, on the one hand, ideas and concepts, and, on the other hand, collaboratively constructed artifacts—texts, videos, software—that emerge within a community of practice. Reed Stevens (this volume), places his emphasis on ideas that people "embody and share within public, material ways": "At least with regard to learning from and with others and most likely also ourselves, the kinds of ideas that we need to attend to are those that people embody and share within public, material ways ...." (Stevens, this volume).

Note Stevens's focus on the creation of shared material constructions when he refers to learning from and with others. Certainly there is a good reason for this change of heart in the learning sciences. First of all, most technology-rich learning and research environments are comprised of a range of people with diverse levels of involvement and contribution ranging from planning large cross-university and interdisciplinary research projects. Learning Scientists regularly collaborate in the collection and analysis of multiple forms of data and the design or modification of tools and web environments. Learning Scientists share ideas, artifacts, tools, and presentations with each other at professional meetings, as do other researchers. Moreover, they can no longer afford to agree with either the cognitivist or behaviorist view of learning, or a bifurcated view of representations being either internal or external. Instead, they need to explore what ethnographers and other social science researchers call the *crisis of representation* in order to reformulate research methods using all forms of media to create participatory communities of practice that are, by their nature, inclusive. Using postmodern ethnographic lenses, learning scientists need not only attend to how they could conduct research of others using methods that could "other" those they study; learning scientists must develop a critical eye on their own practice while making decisions about who they represent and how they represent what they have learned, knowing that any representation of others has ethical implications.

Capturing the Crash of Cultures

A dark side of ethnographic representations of worlds we do not often enough bring to light—especially when discussing the early scientific and anthropological studies of foreign cultures—began with the desires for expansion, trade, and treasures by mostly British, French, and Spanish monarchies. Monarchs commissioned explorers to navigate and chart the distant seas to make future travel easier, ensuring that the appropriation of lands and their resources would become a viable marketplace (Lutz & Collins, 1993). From their journeys to distant lands, explorers brought back cultural artifacts and drawings and, later, photographs of the exotic, frozen for all time in daily rituals of dancing, hunting, and preparing food. It was not uncommon for explorers to entice and coerce the peoples they met. First Nations Peoples along the now-British Columbia coast of Canada became objects of display (Clifford, 1988). Then they were forced to partake in the performance of cultural identity, often displayed in full headdress and costume.

With the rise of the bourgeoisie in Western Europe along with the technological improvements of 18th century industrialization, travel and transportation methods improved and created easier access to far-away lands (Fussel, 1987). Established routes were used to transport enslaved Africans to the vast lands of the Americas to grow crops and mine natural resources, contributing to an ethical and moral devastation from which Western civilizations have yet to recover (Willinsky, 1998).

Once again, people's cultural artifacts were stolen or traded for trinkets; artifacts thought to be valuable or culturally curious were placed in museums or simply destroyed. Missionaries who accompanied the expeditions destroyed artifacts of non-Western worship, especially those that suggested differing sexual values and practices. It is a story we know too well.

In the long term, however, traces of these representational objects and ideas exist in shared memories—sometimes emerging in expressive forms such as art, dance, music, storytelling, medicine, and even in community conversations. Expressions of cultural clashes also continue to be acted out. These artifacts and memories continually morph and blend as experiences and new expressions layer to become part of an intra, inter, and transcultural conversation.

In a global crashing of diverse worldviews is where we find ourselves; views described brilliantly in the 2005 motion picture movie, *Crash,* directed and written by Paul Haggis. This movie depicts the crashing of immigrant cultures in Los Angeles as individuals of all socioethnic backgrounds carry out their day-to-day activities. However, the epistemological crashes we face as learning science researchers are neither local nor only national. Epistemological collisions are global and historical and we cannot pretend that these precedents do not impact our perspectives. And so—as is the theme of the movie—we have to deliberate on how to find antidotes for our research crashes to prevent culture wars. We need to renew opportunities for negotiating shared viewpoints.

Easy-to-upload digital video artifacts—assuming that one has access to a camera and the Internet, knowing that this assumption of access should always be on the forefront of our minds—for collaborative inquiry and viewing, may provide this antidote, or at least a relief, from the practices of misrepresenting others' views and privileging our own. Wherever we are on our planet, we witness a global desire to download and upload images on the web; and this desire is still in its infancy. The framework of shared perspectives has become evident with the advent of the Internet. The Internet as a shared visual and text-based space where people can make public, share, navigate, and use each other's artifacts has generated a new worldview of *mixed communities*, thereby moving from exploitation to placing value on transcultural representations of self, identity, community, and culture.

For educators who study learning, one possible negative consequence of creating ethnographic platforms and databases is that our practice of gathering video could allow unfettered video surveillance. The camera becomes a data collector and an evaluative eye, rather than a tool for researchers and those being videotaped to construct compelling documentary-style stories. At some point, as concerned citizens, we must begin to be more critical of methods and technologies used to trespass the privacy rights of individuals and groups. If we believe that learners need to expand their horizons and see the world with the multiple frameworks required for an open flexible global community, it is time to reform our methods and include diverse research modalities. In doing so, we need to ask ourselves what we are collecting and for what purposes. And, for whose benefit? Are we conducting research that emancipates and empowers learners to become increasingly independent, decent, and collaborative citizens of this planet, or are we conducting yet another kind of imperialism by capturing not only the learning of our students, but learning in other cultures as a means of trying to compete and eventually take over their unique contribution within the global community?

E-Value-ation

As educational researchers, we need to ask ourselves how to be accountable (not counting but being accountable), reliable, and rigorous while using video to build theories of learning. It is easy to believe that the answers lie in the quantity of tapes piled high, the terabytes of storage space used on the hard drive, or the computer tool used to conduct analyses. It is so seductive to be enamored with the resolution of the image, the close-up, the young boy slapping his hand on his knee telling us stories about how you press a key on the keyboard and words pop up on the screen. "Now, how does a computer do that?" he asks. Yet, as a viewer of learning science video data, one has a right to expect compelling signals that what we are viewing represents the best interpretation we can make, given the state of individual perceptual systems that see the world differently. We want to trust and, yet, we also need to be critical of what we see. We know too well the experience of visual technologies as compelling storytelling tools throughout our lives as moviegoers. So, we are left with the question of how we come to *know*. Rowland provides us with a metaphoric explanation of how to we come to know what is valuable.

> We come to know by dancing—by being creative, active, and responsive, and depending on the type of dance, the context, the tempo, the rhythm, and so on, by following predefined patterns in general and taking ad hoc actions in particular. We come to know by making choices about the dance, such as when to be present and where to be present, participate, and follow patterns or not, and by making judgments using wisdom, borne of experience, in the here and now of doing. (Rowland, p. 44)

Any monotheistic viewpoint of conducting research within the *learning sciences* as well as any attempt to reach completely valid conclusions is not only problematic, but also erroneous. Instead, we need to search for ordered patterns within the collaborative conversation among the various research traditions to expose the gaps and the crevices that need to be overcome when searching for new ways to solve emerging phenomena. The most we can have, as James Clifford (1986) reminds us, is partial truths, partial insights, and partial knowledge. The most we have, as astrophysicist Gaensler reminds us is a galaxy "thriving on chaos."

Similarly, any search for a "best video representation" of an event is flawed from the start, given my arguments in this chapter. Instead, when using video representations, we need to provide room for the creation of critical cultures of inquiry where emerging methods, tools, and artifacts are placed on a platform for multiloguing (1998, p. 32), that is, if we are serious about improving our discourse in the (art and) science of learning.

Jerome Bruner, in an invited lecture at Harvard in 1990 (later published in 1991), presented a list of qualities for understanding narrative in educational research. Inspired by his thinking and the work of Clifford Geertz (1973), and Ivan Illich (1973), I have created a list of criteria for e-VALUE-ating (eliciting the value of) video research projects in the digital video ethnography courses I have taught over the last twenty years.

Criteria for Evaluating Media Texts

1. *Wholeness/Particularity*—The video research artifacts provide the reader/viewer with "enough" detail without taking the viewer/reader through the entire body of research. Events are fully presented. Details are meaningful descriptions that bring the reader or viewer "inside" the event. The cases are presented as a microcosm of the range of events and elements that create the exemplary classrooms. These cases do not exist in isolation but in interdependence. (Bruner calls this *hermeneutic sensitivity*.) And yet, the particular case becomes a "token" of a broader and more general view of the culture. Geertz (1973) states that we cannot know God through the details, but we can know the world of people. With online digital video excerpts, researchers may need to be shown how to edit a 1.5-min sequence that captures the essence of a particular event. In my experience with students, they move from 15-min sequences to 1.5 min in a few weeks and produce stirring digital movies.

2. *"Being there/Being with"*—The research products convince viewers that these viewers are "there" and "with" the videographer, or at the very least, "there" in a metaphorical sense of being connected to what is happening in the video. Postmodern ethnographers have contested the notion that we can ever "be there" with others in their situations. Lous Heshusius (1994) states that it is more important for readers to "be with" rather than "be there." Geertz was not to be taken so literally; "being there" is a device used to describe the effort that authors make in composing texts for readers. Just as we cannot find the whole truth of any given situation, although we can search for verisimilitude, so we cannot expect readers to "be there" with the students and teachers. Video researchers, however, can aim for a connection with others in their place. The videographer has a toolkit of devices including using cut-aways, pans, wide-angle shots, and narrative voice-overs.

3. *Perspectivity*—The research products make clear the videographer's point of view or what I call, the multiple *points of viewing* (R. Goldman, 2004, and this volume). The author's point of view is readily apparent to the reader or viewer. The perspectives of the participants being recorded and the treatment of the larger educational context can be easily identified by the reader/viewer, allowing the reader/viewer to locate his or her own viewpoint about the subject being discussed.

4. *Genre consistency/breech*—The "form" of the research products breaks convention but is situated in a genre. In research, as in art, film, or literature, the author of the print or media text uses a particular style or genre to keep the flow consistent. Usually one genre is followed. However, mixing genres can work if there is a reason for the breech. In fact, the breech of genre is what creates and inspires visual media texts. If we only see what we think we are going to see, we stop looking.

5. *Authenticity*—Research video data lead to new interpretations that are grounded in both the rigor of content and the innovation of new connections. They might not be "original" works or completely new approaches. Yet, video researchers create products that shed new light. It may mean that the works produced are artistically based. However, an argument, discussion, or a theorem could as easily represent authentic thinking about a subject.

6. *Chronological Verisimilitude*—The research representations are not an accurate chronological account but are truthlike. The reader or viewer can comprehend an ordering of events suited to the topic being addressed. The product is in sync with the meaning of the events. (Bruner calls this *diachronicity*—the patterns of events are not clock time but human time.)

7. *Conviviality*—The video artifacts are accessible for both public and scholarly consumption. Illich (1973) defines convivial tools as being easy to use, assessable, and beneficial to humanity. Educational research that is convivial should also be created for the public good, serving the larger interests of the community. Often the products assembled for public consumption by researchers are inaccessible. Translating the research for the public good without diluting the content is one aspect. Another is to build tools for real and virtual community building where the public can participate in the debate and the construction of theory.

8. *Resonance*—The research video data are presented in such a way that the reader or viewer of the research is able to make connections to his or her situation (Bateson, 1984). For example, in an annotation by a student named Ashley who accessed my book's online video: "I was really touched and my eyes were opened on the way the young boy spoke of respect and how science teaches you to be a friend. I am in my teens myself and I never noticed how the world and environment taught you about friends and respect. I feel that in the way he used friend and respect that he truly was touched." At http://www.pointsofviewing.com (circa 1997), you can "Participate" and then "Go To" page 225 to read Ashley's annotation of 12/0701998 (11:02:15 AM) after viewing the video.

9. *Immersion* (Murray, 1997)—The research products demonstrate a deep level of engagement and involvement with the topic. Readers/viewers of the products can easily enter into the context that was created by the videographer. Immersion is not to be confused with the mindlessness one might experience when being "lost in space watching a motion picture." Mindfulness (a term

Gavriel Salomon coined in 1979) requires a critical awareness of the process of creation and a deep understanding of cultures studied. Video researchers have the opportunity to work closely with the entire learning culture, paying special attention to the broader learning environment.

10. *Commensurability* (Geertz, 1973)—The research video data provide a toolkit for sharing concerns, beliefs, attitudes, and pedagogical practices. Researchers learn to richly describe a particular culture to make diverse cultural practices understood. The products created with networked media technologies can be shared, but can they also be felt? Can they inspire others to take action and to engage with others whose practices are different from their own and learn? The video artifact created is not a recipe or a formula; it is a snapshot, a context, and a culture for sharing ideas about what happens when the camera is turned on. And, in the best-case scenario, it inspires viewers and participants in the study to create their video representations of what they experience in learning cultures.

In short, video is an epistemological tool, perhaps a better tool than words, for displaying learners' ways of thinking as they engage in learning. This visually based medium called video, which has the power to display nuance and subtlety, provides us not with the traditional split between process and product that emerged after the industrialization of learning, but rather with a postanalog technique to navigate the interactions among individuals and groups as they reflect on their experience and actions in the world, thus creating personalized and shared viewpoints of ideas that they continually transform by the act of interacting with the created works of others engaged in a similar journey.

Ethics

In closing this introductory chapter, let us return to the theme of the voyage and the story of Charles Darwin, the consummate voyager who changed our ways of thinking about science. Our final story is about the collection of representations by naturalists (ethnographic precursors) who were invited to accompany the surveying expeditions in the early 19th century, especially those who came onboard to sketch, collect, and later photograph diverse peoples, species of plant and animal life, classify them, and theorize their origin. Darwin, the most famous of these naturalists, traveled on the *H.M.S. Beagle* to South America and the Pacific including the Galapagos Islands in the late 1820s.

Perhaps influenced both by his father, the physician and naturalist Erasmus Darwin, and by the intellectual milieu of the early 19th century—Darwin is the most celebrated of the traveling naturalists. His theory of evolution was based on *natural selection*, a concept both he and anthropologist Alfred Russel Wallace[1] arrived at simul-

---

[1]For more about Wallace's contributions see http://www.nhm.ac.uk/nature-online/evolution/how-did-evol-theory-develop/evol-wallace/alfred-russel-wallace.html

taneously. Herbert Spencer later applied this conceptual framework to human nature and called it social Darwinism. Social Darwinism uses the theory of evolution to develop the "survival of the fittest"—a theory that Darwin would probably not have approved of as it condoned a hierarchical class structure glorifying the survival of the strong over the weak within a species, rather than simply explaining an evolutionary development of diverse species over thousands of years. That said, Darwin was deeply influenced by his reading of Malthus's *Population,* a treatise calling for the end to overpopulation while basically blaming the lower classes for overpopulations. As Darwin put it:

> In October 1838, that is, 15 months after I had begun my systematic inquiry, I happened to read for amusement *Malthus on Population*, and being well prepared to appreciate the struggle for existence that everywhere goes on from long-continued observation of the habits of animals and plants, it at once struck me that under these circumstances, favorable variations would tend to be preserved, and unfavorable ones to be destroyed. (Darwin, 1876)

Looking back on Darwin and Spencer's theories, one has to reflect on how these theories of continued and sustained development interacted with the technologies of transportation, justifying the potential for increased exploration, exploitation, and cultural imperialism (Fussel, 1987).

In a similar vein, in order to keep a balanced view that tempers our enthusiasm for the "riches" and possibilities that new video technologies offer to our research practices, we also have to ask ourselves how our new online digital technologies not only promote shared perspectives and the appreciation of each others' cultural representations, but also how they create a new kind of cybermarket, exporting and importing virtual cultural artifacts, creating a silicone elite class, and transforming existing cultural mores in alignment with cultural imperialism. Are the same kind of people who viewed foreign lands as markets and resources still laying claim to virtual worlds as they buy and sell virtual property, as, for example, in the game, *Everquest*? Is this a game or a harbinger of a new world?

As travelers and tourists in unknown classrooms, educational researchers are now armed with cameras and handhelds to explore every nook and cranny of the classroom, zooming here and there, observing the real lives of children and teachers in their habitat, not realizing that some element of what we do has a long history based in a worldview that encourages us to shoot, capture, dissect, and organize the bits and pieces of embodied chunks in systemic and 'objective' practices—to build one best Truth bespoke by the gathered evidence. Being trapped in a culture of evidence-based research, we, too often, cling to an interpretation of evidence that entails reproduction, as if we can simply learn the best practices in one situation and scale and reproduce it in another. We forget what we know about cultures and communities, that they are situated contextual organisms. As much as we would like to control and change what we put into the classroom Petri dish so that the outcome is replicable and scaleable, we know that the real world of learning is more akin to complex biological systems that can adapt morph, reconfigure, and interact with events and the experience of those events, over time. Akin to the ever-changing galactic metaphor.

Educational researchers have, throughout the years, critiqued how media tools can privilege certain mainstream perspectives (Bryson & de Castell, 1994; de Castell, Bryson, & Jenson, 2001). This occurs even when our research is conducted with the honorable intention of improving the lives of "others" in our classroom. The problem is that colonialism and the material benefits gained from this exploitation too often permeate the educational system, treating the underprivileged as end-users of products rather than as creators of knowledge that will provide them with the tools to be free from the oppression of generations of poverty, lack of educational and health benefits, and lack of opportunity.

When we videotape our subjects over and over again in classrooms and informal learning environments, we face an ethical challenge to not repeat with these powerful visually based electronic tools what the tall ships of yesteryear did to others—capture, collect, dissect, categorize, and construct hierarchies.

Our history as learning researchers is one that has aimed to include full participation and diverse viewpoints. And, yes, we do collect material artifacts, including still and moving images of others not like us. The point is that we do our best to use these material shared re-presentations to build strong bridges between what we know and what others know to make changes across the borders and boundaries of diverse educational and social systems. Bridges that will provide learners with greater understanding of each other's cultures. Still, there is much more to be done. Perhaps it is time to ask how we can use digital video and postmodern ethnographic sensitivities and sensibilities to create the kinds of stories, cases, and examples of learning that will produce a new generation of learning scientists who, in their future studies, will not use video as a tool for capturing others for personal gain, but rather for the purpose of building convivial learning communities, as Ivan Illich (1973) would propose.

The legacy of our video research in the learning sciences should not be a repeat of colonial survival of the (fittest and) best practices that can be marketed to other nations, but rather to rethink the role of technologies as a force for greater equity and opportunity. Video research should comprise the histories that live in shared memories and experiences of all the participants of the learning community. Video research in the learning sciences should become a method that integrates the art and science of creating meaning. Video research should provide us with an approach where we can always question our methods, tools, and theories; where we consider how our actions as designers and members of a research community affect those we study. Postmodern and postcolonial ethnographic frameworks—including the perspectivity framework—offer important clues as to how, using video research, we represent the points of view of others, and, in that process, how such knowledge is best re-presented to each other.

ACKNOWLEDGMENTS

During the course of writing on this chapter, conversations with colleagues and friends, John Willinsky and Mary Bryson, and Jay Lemke inspired me to move toward a more critical discussion. These discussions led me to think about video research less as a method and more as a context. They led me to consider that video research is intertwined with the larger political realities of the times in which we live. I use the word, conversation, to include my interactions with the many authors in this book whose chapters influenced me while I read and reviewed Each chapter author, especially those by co-editors Roy Pea, Brigid Barron, and Sharon Derry, had a deep effect on my thinking since I first conceived of this book in 2001. I thank them all for their many contributions.

REFERENCES

Aadne, J. H., von Krogh, G., & Roos, J. H. (1996). Representationism. In G. Von Krogh. & J. H. Roos (Eds.), *Managing knowledge: Perspectives on cooperation and competition* (pp. 9–31). London, England: Sage.

Anderson, J. R. (1982). Acquisition of a cognitive skill. *Psychological Review 89*, 369–406.

Bakhtin, M. (1986). *Speech genres and other late essays* (V. McGee, Trans.). Austin: University of Texas Press.

Barron, B. (2000). Problem solving in video-based microworlds: Collaborative and individual outcomes of high achieving sixth grade students. *Journal of Educational Psychology, 92*(2), 391–398.

Bateson, M. C. (1989). *Composing a life*. New York: Grove Press.

Bransford, J., Sherwood, R., & Hasselbring, T. (1988). The video revolution and its effects on development: Some initial thoughts. In G. Forman & P. B. Pufall (Eds.), *Constructivism in the computer age* (pp. 173–201). Hillsdale, NJ: Lawrence Erlbaum Associates.

Brown, J. S., Collins, A., & Duguid, P. (1996/1989). Situated cognition and the culture of learning. In H. McLellan (Ed.), *Situated learning perspectives* (pp. 19–44). Englewood Cliffs, NJ: Educational Technology Publications.

Bruner, J. (1990). *Acts of meaning*. Cambridge, MA: Harvard University Press.

Bruner, J. (1996). *The culture of education*. Cambridge, MA: Harvard University Press.

Bryson, M., & de Castell, S. (1994). Telling tales out of school: Modernist, critical, and postmodern "true stories" about educational technologies. *Journal of Educational Computing Research, 10*(3), 199–221.

Chomsky, N. (1968). *Language and mind*. New York: Harcourt Brace & World.

Clifford, J. (1986). Introduction: Partial truths. In J. Clifford & G. Marcus (Eds.) *Writing culture: The poetics and politics of ethnography*. Berkeley: University of California Press.

de Castell, S., Bryson, M., & Jenson, J. (2002). Object lessons: Towards an educational theory of technology. Retrieved March, 2005 from http://www.firstmonday.org/issues/issue7_1/castell/index.html

Darwin, D. (1876). Autobiography. In Francis Darwin (Ed.), *The Life and Letters of Charles Darwin*, Vol I. London: John Murray.

Friedman, T. L. (2005). *The world Is flat: A brief history of the twenty-first century*. New York: Farrar, Straus & Giroux.

Fussell, P. (Ed.) (1987). *Norton book of travel*. New York: Norton.

Gardner, H. (1985). *The mind's new science: A history of the cognitive revolution*. New York: Basic Books.

Geertz, C. (1973). *The interpretation of cultures*. New York: Basic Books.

Geertz, C. (1983). *Local knowledge: Further essays in interpretive anthropology*. New York: Basic Books.

Gentner, D., & Stevens, A. (Eds.). (1983). *Mental models*. Hillsdale, NJ: Lawrence Erlbaum Associates.

Ghiraldelli, J. (2005). Interview with James Marshall. *Educational Philosophy and Theory, 37*(3), 285–423.

Goldman, R., Crosby, M., Swan, K., & Shea, P. (2004). Introducing quisitive research: Expanding qualitative methods for describing learning in ALN. In R. Starr Hiltz & R. Goldman (Eds), *Learning together online: Research on asynchronous learning networks* (pp. 103–120). Mahwah, NJ: Lawrence Erlbaum Associates.

Goldman-Segall, R., & Maxwell, J. W. (2002). Computers, the Internet, and new media for learning. In W. M. Reynolds & G. E. Miller (Eds.), *Handbook of psychology. Volume 7: Educational psychology* (pp. 393–427). New York: John Wiley & Sons.

Goldman-Segall, R. (November, 1999). *Using video to support professional development and improve practice.* Paper presented to the Board on International Comparative Studies in Education (BICSE) Invitational Consortium on Uses of Video in International Studies, Washington, DC.

Goldman-Segall, R. (1998). *Points of viewing children's thinking: A digital ethnographer's journey.* Mahwah, NJ: Lawrence Erlbaum Associates. Accompanying video cases retrieved November 10, 2005 from http:www.pointsofviewing.com/

Goldman-Segall, R. (1996). Looking through layers: Reflecting upon digital ethnography. *JCT: An Interdisciplinary Journal for Curriculum Studies, 13*(1), 10–17.

Goldman-Segall, R. (1995). Configurational validity: A proposal for analyzing ethnographic multimedia narratives. *Journal of Educational Multimedia and Hypermedia, 4*(2/3), 163–182.

Goldman-Segall, R. (1990). *Learning constellations: A multimedia ethnographic research environment using video technology to explore children's thinking.* Unpublished doctoral dissertation, MIT, Cambridge, MA.

Goldman-Segall, R. (1988). *The growth of a culture.* [Film]. MIT Council of the Arts Award. Cambridge, MA: MIT.

Goldman-Segall, R., & Riecken, T. (1989). Thick description: A tool for designing ethnographic interactive videodisks. *SIGCHI Bulletin, 21*(2), 118–122.

Hall, S. (1997). The work of representation. In S. Hall (Ed.), *Representation: Cultural representations and signifying practices* (pp. 13–74). London, England: Sage.

Heshusius, L. (1994). Freeing ourselves from objectivity: managing subjectivity or turning toward a participatory mode of consciousness? *Educational Researcher, 23*(3), 15–22.

Hockings, P. (Ed.). (1975). *Principals of Visual Anthropology*, The Hague: Mouton Publishers.

Illich, I. (1973). *Deschooling society.* Harmondsworth: Penguin.

Johnson-Laird, P. N. (1983). *Mental models. Towards a cognitive science of language, inference, and consciousness.* Cambridge, MA: Harvard University Press.

Jones, D. (April, 2005). Unknown forces keep galaxies in order. *Epoc Times.* Retrieved November 15, 2005 from http://english.epochtimes.com/news/5–4–2/27535.html

Kay, A. C. (1996). The early history of SmallTalk. In J. Thomas, J. Bergin, J. Richard, & G. Gibson (Eds.), *History of programming languages—II* (pp. 511–578). New York: ACM Press / Addison-Wesley.

Kolodner, J. L., & the EduTech Design Education Team. (1995, June). Design education across the disciplines. *Proceedings ASCE Specialty Conference: 2nd Congress on Computing in Civil Engineering, 318–333.*

Koschmann, T. (Ed.). (1996). *CSCL: Theory and practice.* Hillsdale, NJ: Lawrence Erlbaum Associates.

Lather, P. (1986). Issues of validity in openly ideological research: Between a rock and a soft place. *Interchange, 17,* 63–84.

Lave, J., & Wenger, E. (1991). *Situated learning: Legitimate peripheral participation.* Cambridge University Press.

Lemke, J. (2005). Toward a theory of traversals. Retrieved on November _DATE_, 2005 at http://www-personal.umich.edu/~jaylemke/papers/traversals/traversal-theory.htm

Mead, M. (1973). Visual anthropolgy in a discipline of words. In P. Hockings (Ed.), *Principles of visual anthropology* (pp. 3–10). The Hague, The Netherlands: Mouton.

Murray, J. (1997). *Hamlet on the holodeck: The future of narrative in cyberspace.* New York: Free Press.

Papert, S. (1980). *Mindstorms: Children, computers, and powerful ideas.* New York: Basic Books.

Papert, S. (1988). The conservation of Piaget: The computer as grist to the constructivist mill. In G. Forman, & P. B. Pufall (Eds.), *Constructivism in the computer age* (pp. 3–13). Hillsdale, NJ: Lawrence Erlbaum. Associates

Papert, S. (1988). The conservation of Piaget: The computer as grist to the constructionist mill. *Constructivism in the computer age* (pp. 75–96). Hillsdale, NJ: Lawrence Erlbaum Associates.

Papert, S. (1991). Situating constructionism. In I. Harel & S. Papert (Eds.), *Constructionism*. Norwood, NJ: Ablex.

Pea, R. (1985). Beyond amplification: Using the computer to re-organize mental functioning. *Educational Researcher, 20*(4), 167–182.

Pea, R. (1987). Integrating human and computer intelligence. In R. Pea & K. Sheingold (Eds.), *Mirrors of minds* (pp. 128–146). Norwood, NJ: Ablex.

Piaget, J. (1930). *The child's conception of the world.* London: Harcourt, Brace.

Piaget, J., & Inhelder, B. (1956). *The child's conception of space.* London: Routledge & Kegan Paul.

Rickheit, G., & Sichelschmidt, L. (1999). Mental models – some answers, some questions, some suggestions. In G. Rickheit & C. Habel (Eds.), *Mental models in discourse processing and reasoning*(pp. 9–40). Amsterdam: North-Holland.

Roman, L. G. (1991). The political significance of other ways of narrating ethnography: A feminist materialist approach. In M. LeCompte, W. Millroy, & J. Preissle Goetz (Eds.), *The handbook of qualitative research in education* (pp. 556–594). San Diego, CA: Academic Press.

Roschelle, J., Pea, R. D., & Trigg, R. (1990). *VideoNoter: A tool for exploratory video analysis* (Tech. Rep. No. 17). Palo Alto, CA: Institute for Research on Learning.

Rowland, J. (2004). Shall we dance? A design epistemology for organizing learning and performance. *Educational Technology, Research, and Development, 52*(1), 33–48.

Scardamalia, M., & Bereiter, C. (1991). Higher levels of agency for children in knowledge building: A challenge for the design of new knowledge media. *Journal of the Learning Sciences, 1*(1), 37–68.

Skinner, B. F. (1931). The concept of the reflex in the description of behavior. *Journal of General Psychology, 5*, 427–458.

Smeyers, P. (2005). The laboring sleepwalker: Evocation and expression as modes of educational research. *Educational Philosophy and Theory, 37*(3), 407–423.

Spiro, R. J., & Jehng, J. C. (1990). Cognitive flexibility and hypertext: Theory and technology for the nonlinear and multidimensional traversal of complex subject matter. In D. Nix & R. J. Spiro (Eds.), *Cognition, education, and multimedia: Exploring ideas in high technology* (pp. 163–205). Hillsdale, NJ: Lawrence Erlbaum Associates.

Stahl, G. (2006). *Group cognition: Computer support for building collaborative knowledge.* Cambridge, MA: MIT Press.

Suppes, P. (1966). The uses of computers in education. *Scientific American, 215*(3), 206–220.

Tobin, J., Wu, D., & Davidson, D. (1989). *Preschool in three cultures: Japan, China, and the United States.* New Haven, CT: Yale University Press.

Tzeng, J. Y. & Schwen, T. (2003). Mental representation-based task analysis for analyzing value-laden performance. *Educational Technology Research and Development, 51*(3), 5–23.

Trinh, M. H. (1992). *Framer framed.* New York: Routledge.

Ulcicz, M., & Beatty, A. (2001). *The power of video technology in international comparative research in education.* Washington, DC: National Academy Press.

Vygotsky, L. S. (1962). *Thought and language* (E. Hanfmann & G. Vakar, Trans.). Cambridge, MA: MIT Press.

Vygotsky, L. S. (1978). *Mind in society: The development of higher psychological processes.* Cambridge, MA: Harvard University Press.

Watson, J. B. (1913). Psychology as the behaviorist views it. *Psychological Review, 20*, 158–177.

Willinsky, J. (1998). *Learning to divide the world: Education at empire's end.* Minneapolis, MN: University of Minnesota Press.

Video Epistemology In- and Outside the Box: Traversing Attentional Spaces

Jay Lemke
University of Michigan

EXPERIENCING VIDEO MEDIA

In its earliest form, we encountered video images inside the magic box of early broadcast television. To some observers it might well have seemed that inside that box there was an animated dollhouse inhabited by miniaturized people, caught in some eldritch sorcery. On the tiny stage, classic theatrical dramas were being performed. The bulky boxes held a magical space, a space that in turn magically held our attention. A space that was rescaled relative to that of our living rooms, but that also seemed a spatial extension of them into another dimension. Some of these boxes had wooden doors, to be opened like any other cabinet or tiny closet. As children, we pressed our noses against the glass, wondering if we could somehow get inside this ante-room that opened up to elsewhere.

Early television programs played with the fascination of the boundary between living rooms and the living space of the video world. Children wrote on transparent screens (or naively on the glass itself) and the video characters would complete their pictures or give them the magic codes of their games; for example, *Winky Dink and You* (Prichett & Wyckoff, 1953–1957) or *Captain Video* (Brock, 1949–1955). This was the first interactive video. The recent film *Pleasantville* (Ross, 1998) revived the fantasy of television as a portal into a 1950s black and white world, as Woody Allen's 1985 *Purple Rose of Cairo* similarly imagined a 1930s movie screen as an interface between the worlds of a movie and of the theatre where it was playing. We never perceive cinematic and televisual worlds as two-dimensional surfaces, but always as three-dimensional spaces that are worlds in their own right and, at the same time spatial extensions of our own local places.

I want to consider here issues of epistemology and verisimilitude for video both as a medium we experience and as a medium through which we record data for research in the learning sciences. These are linked because we make use of video data by viewing the video record, and we cannot understand the epistemology of video as representation unless we also understand the processes by which we make meaning with video when we experience it. I propose that we consider the semiotic uses of video in terms of the ways we meaningfully (and feeling fully) move across and through immediate and mediated attentional spaces. I will develop the notion of multiple "attentional spaces" to stand for both the telemediated worlds to which we pay attention(real or fictional) as well as those more immediately navigable by our bodies.

I am starting from an experiential or phenomenological perspective because, regardless of the underlying technology, whether film or videotape or digital display, there is a quality of experience that remains constant. I will certainly also be considering this experience as a semiotic construction of meaning and as a social practice characteristic of a community and its culture. But I do not want to lose sight of materiality, or of the bodily and felt qualities of the experience. Feeling as well as meaning matters, and indeed, feeling matters to meaning, and vice versa. I am not limiting materiality here to the proprioceptive or visceral body. Contrary to myth, video is not a passive medium; we *do* act and move in using the technology and in response to its images and sounds. Embedded in interactive multimedia, as it increasingly is, video becomes an even more active medium. We make meaning with video and its extensions (video games, interactive video, interactive multimedia and hypermedia) in real space and in real time. Video genres have chronotopes (Bakhtin, 1981); typical patterns of movement among sites (*topoi*) and of pacing action in time (*chronos*), and there are chronotopes also for the kinds of activities in which we use video, including those for research.

I want to examine the diversity of video genres—commercial, ethnographic, and research records and the commonality and variety of ways in which we respond to them—as virtual realities, as representations, and as data. Before doing so, I need to sort out first some issues of the nature of the medium and then some of the ways in which we experience it as part of larger activities in which we traverse across different sites and other media. Afterwards, I want to sketch one possible future direction for the use and study of this medium in our work in the learning sciences.

VIDEO AND KINDRED MEDIA

First, a few preliminaries. My primary object of concern here is the experience of video as a space of visible and audible dynamic activity: We see and hear something happening in a virtual space and time that has some indeterminate relationship to the immediate space and time that our bodies inhabit. In this sense, experienced video may not be significantly different for research purposes from experienced film or cinema, although there are certainly differences in "feel" arising from the conventionally greater resolution of film images (a temporary difference, already disappearing) and the conventionally larger spatial scale of the cinema screen. By "video medium," I will mean primarily the sensory-attentional features of the medium to which we respond in making meanings and in experiencing feelings, rather than the material technologies that record and present these features to our senses. We can call "video technology" the assemblage of operable artifacts, and the conventional ways of using them, that enable us to record and play back video data streams. There are also, of course, "video texts" that we experience as having some coherence or cohesion of meaning and feeling over their run time of play back, and "video genres" that are typical and recognizable cultural forms with reference to which we make sense of what is happening in most video texts. These "product" genres are the result of "process" or activity genres: The conventional ways in which video texts of various genres are made or produced (including scripted, unscripted, and/or directed action, camera-work, editing, etc.).

With the definitions given, both video and film experience share substantially the same audio-visual semiotic; the same interpretative conventions for their salient sensory features. They have more or less different technologies, but very similar or identical genres, and there are many texts that exist in both media. We all know that video technologies are used increasingly in cinema production, along with digital technologies, and there is an increasing convergence among these technologies and also among the media. Watching a film on a big screen in a theatre is still a somewhat different social and phenomenological experience from watching substantially the same text (perhaps with different aspect ratio, another temporary difference) at home on even a large television monitor, much less on a smaller flat screen digital video monitor. The extreme scale of the IMAX technology (a screen that fills peripheral vision, tens of meters high and wide, or in older similar technologies wrapping around a curved screen) does produce unique phenomenological effects of immersion and telepresence, similar to those of Virtual Reality (VR) technologies (screen-based caves or goggle-based optics). We get strong visceral and vestibular (proprioceptive) sensations of motion, and so a more vivid sense of a three-dimensional spatial environment.

In between video and VR is the increasingly popular medium of video games. Whether presented on televisions or computer monitor screens (or in arcades on dedicated systems), video games animate a dynamic, interactive, apparently three-dimensional world, but one in which the realism is generally reduced compared to most film and video texts by the more cartoonlike actors, scenery, and in-world artifacts. Video games however, do include motion-capture sequences made with live actors, cut-scene video, and other elements that overlap with those of the recorded video medium. In the case of co-produced films and video games such as *The Matrix Reloaded* and *Enter the Matrix* (Shiny Entertainment, 2003; Wachowski & Wachowski, 2003) or *The Lord of the Rings: The Return of the King* (Electronic Arts, 2003; Jackson, 2003) not only is there great overlap in features, meanings, feelings, and so forth, but the media themselves are mixed and hybridized. The film, video, and videogame texts are also marketed in relation to one another. The greatest difference is that the videogame medium is highly interactive in ways that the video medium itself is not: We more frequently engage in physical contact and motor movement interactions with the technological artifacts that provide our interface to the medium (keyboards and game controllers vs. television or VCR remotes and buttons), and we more regularly see and hear changes in the videogame features that we interpret as direct responses to our actions, just as we do in the world of immediate experience. I emphasize this new member of the family of videolike media because of its prospective importance for research in the learning sciences, to which I will return later.

The video medium of course has long included genres of reduced realism, beginning with animated cartoon films, and has also experimented with hybrids of visually realistic genres (the feature film) and cartoon animations, such as in *Who Framed Roger Rabbit* (Zemeckis, 1988). On this analytical dimension as well, there is a quasi-continuum of media and genres linking film/video and video games.

In the cases of video games and IMAX movies we are most vividly presented with the paradox of virtual spaces whose spatial and temporal relationships to our immediate surroundings are indeterminate. (For immersive VR media, there may be no visual or auditory input from our unmediated surroundings, although there is still a doubled tactile–kinesthetic world.) We may focus on the interface and try to imagine the screen as a portal between two worlds, or two places in the same world (as is the case with video-conferencing experiences), or we may yield to the illusion of virtual presence in the video world, whether as observer or as participant. We can imagine that we could, or already have, passed through the portal, or that the interface technology exists in some sense in both worlds, and that in video games, by touching the controls in this world, we are performing actions in the other world. Video-system (TV, VCR) remote controls can already be viewed as precursors of this interface metaphor, and in fact, are readily integrated as artifacts with videogame controller units.

ATTENTIONAL WORLDS AND TRAVERSALS

In the cases just described, we may either feel experientially that we are living in two worlds at once, or that we are effectively immersed "in" the virtual world. I want to speak of both the immediate and the mediated worlds in the same terms semiotically, affectively, and insofar as possible, bodily and materially (including spatiotemporally). It seems useful to construct a notion of "attentional worlds" to indicate that experientially, we attend to the happenings, sights, sounds, meanings, and feelings of a particular space or place, even if it is a virtual one (i.e., one with which we do not have all the same material interactional affordances as we do with what we take to be the world mediated by our unaided senses and physical contacts). We are capable of attending simultaneously to, or at least cycling rapidly among multiple attentional worlds. We have the experience of moving from one attentional world to another in many aspects of our more and more technologically mediated lives. We attend to the meeting we are in, to the space inhabited by the person we are talking to on our cell phone (or which we co-habit), to the relatively flat (i.e., feature-poor) spaces of text windows on our laptop, and perhaps also to the video presentation on a room screen or on our laptop screen.

Even without mediating attentional technologies, we are accustomed to shifting the primary focus of our attention rapidly and on many time scales, from the differing second-to-second salience of features in the sensory environment, to the slower shifts between different conversations and activities in a complex setting (a crowded room at a party, shopping at a mall with a loose-knit group of friends). Over longer time scales we also make meaning along the *traversals* (Lemke, 2002a, 2002b, 2003) of our channel surfing and web surfing, or the multiple scenes of our daily life. In the complexity of daily life, and increasingly with the affordances of new technologies, we are offered choices as to how we shall take up these affordances and what habits of engaging with multiple attentional worlds we each will prefer.

What kinds of meanings do we make, and what kinds of feelings do we experience, as we attend to simultaneous and sequential attentional worlds, with a constructed sense of spatiality, interactivity, and dynamic temporality within and across them? How do we integrate or reconcile the different pacings and apparent flows of time, the interruptions and resumptions of events or activities, the disjointed topological spaces of our multiple attentional worlds? It seems to me that these questions are fundamental to any reflexive and critical use of the video medium in our research in the learning sciences.

THE GENRES AND USES OF VIDEO IN RESEARCH

The kind of research with which most of us are concerned is the study of human learning and behavior in natural settings. We may also wish to study human meaning-making practices, affective responses, and the production, use, and interpretation of artifacts, whether more toollike or more textlike. I am going to propose that we also need to study human learning and behavior within, between, and across virtual attentional worlds. In the tradition of this kind of research, the use of video has multiple ancestry. It descends from the use of photography, of audio recording, and of ethnographic film making. I will be concerned less with the lineage of the medium and more with the lineage of the practices of using these materials in research.

Two early uses of photography in the human sciences were time lapse and stroboscopic photography in the study of human movement (e.g., Edgerton & Killian, 1939) and ethnographic photography (Collier & Collier, 1986) and film making (e.g., Mead & Bateson, 1952) in cultural anthropology. The first made use of the temporal dimensions of the medium, a forerunner of slow-motion and fast-forward video, and of cinematic animation. It enabled us to study the detailed articulations and transitions from one microtimed moment of a movement to the next. It points us to the importance of temporality in our uses of video. The second made us realize how little of what can be seen in any scene we actually attend to: Re-viewing a photographic record with a different "set," a different purpose or interest, the original photographers, and others, could see far more than was remembered or noted on the occasion preserved in the photograph. Photographs were re-usable and almost endlessly rich as a data type. This points us to the role of the viewer in our use of research video.

One can also note that photographs can be juxtaposed and grouped and linked in a large number of possible ways, each of which can potentially inspire a particular story or document a particular insight, provided only that some interpretable basis of commonality or contrast is present (e.g., that all were taken in the same village in the same year, or in two different villages being compared, etc.). That the same is true of video episodes from a unified or coherent corpus (say, many short videos made by different students in a class each exploring some aspect of the same town) was recognized by Goldman-Segall (1998), who created *Constellations* (and more recently *Orion*) as hypertextual environments for grouping, linking, and annotating video clips, precisely to allow us to aggregate multiple "points of viewing" in our research in the learning sciences.

Audio recording was employed by dialectologists eager to capture authentic samples of local speech and folklorists seeking to preserve local story traditions. Even more than time-lapse and stroboscopic photographs, audio recordings allowed us to play with time (e.g., slowing or looping speech to better transcribe it), and they too were susceptible to re-analysis, such as with acoustic oscilloscopes, or when Hymes (1981) advanced a theory of the role of prosodies in traditional story telling.

Ethnographic video descends most directly from ethnographic film and the pioneering work of Gregory Bateson in setting up a camera in his own home to record daily life. Now the issue of temporality comes very much to the fore in the process of production (i.e., recording) as well as in the use of the recorded product. If we leave the camera running all day and night, we miss nothing, but we then need 24 hours to view the recording. With fast forward, we can inspect the recording more quickly, but very imperfectly (especially for speech). If we want to do a detailed analysis on days or months worth of tapes, it could take years, perhaps more years than mere mortals have. This is still a primary problem of video research.

Ethnographic video is an even richer source of data for re-analysis than are photographs or audiotape. Not only does it include all the information that could be obtained from these two sources separately, but it provides information on the temporal relationships of speech and sounds to visually depicted actions and events. One of the first distinctive results from such video research was the discovery of interactional synchrony, both between speech and unconscious movements for a single individual, and between both of these for two different individuals in conversation or extended social interaction (Condon & Ogsten, 1967; Kendon, 1973). Moreover, it provides us with data on pacing and rates of speech and action, and on their potentially significant modulation and variation (accelerations and decelerations, contrasts of rapid and hesitant, interruptions and resumptions, breaking off, self-correction, pauses, etc.) These are perhaps even more evident in abstract animated diagrams, but the temporal dimension of video also provides the perceptual basis for many of our inferences of cause–effect relationships.

As researchers, our uses of the video medium have primarily been to record behavior and social interaction in their ordinary contexts and settings, and to make these recordings available for later and more leisurely inspection. Once again, the temporal affordances of the medium have been critical for us. Human social activity goes by too quickly for us to remember its details. With video, we can either slow it down or replay it again and again. We can then transcribe it and link the transcription (of speech, action, and events) to the relevant segments of the video, available for replay and further analysis (as well as for linking to other segments). We have been led by this tradition of using video to a very *microscopic* approach to the study of human learning and behavior. We have learned a great deal from it, but the approach does fundamentally distort and limit our view of human learning and social interaction. How so?

First, we typically examine behavior mainly on very brief time scales to the neglect of longer time scales. There are thousands of excellent analyses of 5 min episodes in school classrooms, very few of whole 40 min lessons, and almost none of either whole school days for individual pupils or teachers or a whole week (much less a whole year) in the life of one class. Similarly so for learning under artificial laboratory conditions. We do not even have a means of describing in brief the salient patterns in learning or social activity over longer time scales. Time-lapse video might give us one kind of start toward that, but by itself, it will not necessarily reveal salient patterns unless we already know what we are looking for.

Second, by putting brief episodes under the analytic microscope, we magnify small details and minor events out of all proportion to the flow of activity on a longer time scale. A hesitation here, a mis-speaking there, a self-correction or a mis-statement that may loom large in an utterance-by-utterance analysis may in fact be missed by most of the participants and have no consequences for the rest of a lesson. If a teacher uses a term 30 times in a lesson and mispronounces it once, will students care or pay much attention to the error? Some such examples are obvious, but more generally, how can we judge the relative significance of short-term events for longer term developments if we do not even have a descriptive apparatus for the longer time scales?

Meaning is not just made moment by moment. It is also made across longer time scales. So also with learning and development, with the performance of identity, with invention and discovery, with the emergence of social relationships, with the conduct of projects and tasks, the solution of problems, the creation of artifacts and texts. We need to evolve meso and macroscale uses of video to balance our inherited tradition of microscale uses.

Finally, video texts as research data are "unedited" in the sense that no cuts and joins have been made, no time periods left out, and generally there is a view from only a single camera, not intercut with other spatial–perspectival viewpoints. We also have a tradition of edited ethnographic film and video as interpretative presentations, in which there *are* juxtapositions of temporally separated scenes, recorded on different occasions, and with commentary added. Epistemological alternatives and their supporting technologies such as Goldman-Segall's *Constellations* or *Orion* allow us in effect to make diverse interpretative complexes of video segments from a shared video database, to view other people's, and to link and annotate these new wholes as well as the original segments. We also have available to us now picture-in-picture and picture-with-picture video that allows us to see (almost) simultaneously the synchronized video streams from different cameras viewing the same scene. A new technology (Pea, 2003) allows 360-degree video recording and a play back in which we can zoom in on the part of the scene on which we wish to focus, thus reducing the limitations of re-usability of data that come from the viewpoint selections of the original directional camera. Although audio is less directional, it is still true that there are multiple speech and sound sources in many scenes and we will want to be able to attend differentially to these without being limited by the placement of a single recording microphone. Selection among many video and audio data streams in an integrated, synchronized software environment is already possible in modest-cost research systems (Hay, 2003).

RESEARCHERS INTERACTING WITH VIDEO MEDIA

Video is a seductive medium, especially when its visual realism and audio fidelity are high and the quality of sensory experience comes to resemble (or even exceed) that of ordinary perception of the immediate environment. It makes us believe that we are simply seeing what is there, rather than interacting with and interpreting in very specific ways a very partial (in both senses) record of an activity. This is particularly true when we view the recording at a rate of playback that seems natural for human action and speech. If we slow it down, we gain some analytic and interpretative distance and reflexivity, and even more so when we denaturalize it by speeding it up. Once we have viewed and reviewed the same scene many, many times it does begin to seem to us more like an artifact and less like a moment of life. Something of the same effect can occur if we zoom in on a detail or zoom out, or change our angle of view. The more we interact with and become conscious of deliberately affecting the visual experience presented by the medium, the less likely it is to seem merely a natural unfolding of recorded events.

But this critical perspective on our use of video is itself limited to a very specific time scale in the process of analysis, namely the minute-to-minute time scale of inspecting or replaying a part of the tape. What about the longer time scales of our uses of video; the whole longer activity of which the minute-to-minute re-viewings of segments are directly constituent parts? The activity in which we select which tape to view and what we are going to do with it, or look for in it? The activity of creating the record itself? The larger project of which the use of the video is one part? How can we get a critical and reflexive perspective, and some prophylactic practices to avoid unwarranted interpretation and claims?

This same problem has already been analyzed in the work of Bruno Latour and other researchers in the social studies of science and technology (Latour, 1987, 1999; Lynch & Woolgar, 1990), who look at scientists and what they are actually doing in moving from data collection and analysis to published results and claims. For our purposes here, a key analytic conceptualization from that body of work is the notion of "chains of translations of inscriptions." In this view, what is happening over longer time scales is that one inscription (for us, say, a social phenomenon perceived in real time) is being "translated" by technical practices and technological mediations into another inscription (in a different medium, here, say, a video recording the phenomenon), and then into another (a transcription), and yet another (a running description or summary, a commentary, an article for publication). In many cases, the later inscriptions are embedded in artifacts (charts, tables, texts, "constellations") in which they are integrated with inscriptions having a different prior chain (i.e., starting from different phenomena or events). Both these integrations and the procedures for translating an inscription to a new medium or form must insure faithful or defensible connections; connections that preserve the chain of logic and consequence that supports the eventual argument of the researcher.

We have a tendency as researchers to overclaim a reversibility for these chains of translations of inscriptions that we have constructed. Our positivist traditions in the learning sciences want to make claims about the reality we assume to lie at the head of these chains. We have followed Cartesian dualist logic too far in dichotomizing realities and representations. All the links in a chain of translations are equally real; indeed their reality is attested precisely by our ability to include them in a chain, or more generally in networks of interlocking chains. All meaningful interactions with realities are also equally mediated by culture-specific interpretative codes, and thus share the feature that seems to distinguish "representations" as such, but does not.

In this view, the video, the researcher, the camera, the play-back apparatus, the transcript, the drafts of the article, and so forth, are all interdependent parts of a network, tied together by their roles in producing a sustainable chain. This is saying more than simply that the researcher plays an inescapable role in interpreting the video. It is saying that it is only in relation to their roles in the chain (and in other chains) that any of these entities or "actants" (including ourselves) has meaning (or even a determinable existence). Latour's ontology does not allow a view from outside, even for the purposes of taking a picture of our own activity in order to be reflexive about it. All the more reason for us to complement it, nonetheless, with a more phenomenological perspective.

What are we doing when we come to view a video? In one sense we are making a traversal, and experiencing along it some meanings and feelings, across at least two distinct attentional spaces: that of our office or lab and its activities, and that of the events in the virtual world in the video. We are, in Latour's sense, trying to make a translation between these two worlds, to forge an enduring connection between them that will ultimately support the claims or interpretations we will want to make about what the video "shows." We are doing this in real time and in real space, in our own material bodies, with meanings and feelings, sensations and actions. But unlike what we might be doing in the absence of video or other virtual attentional environments, we will necessarily find ourselves trying to forge connections between worlds where time may be flowing at different rates, where space can have different relative scales, where we can move backward and forward in virtual time, where we can interrupt action and resume it without consequence, where we can move our viewpoint at arbitrary virtual speeds or not at all, and where we can view multiple attentional spaces almost simultaneously, side by side, in temporal synchrony or not.

We are not just making meaning "within" the virtual world of the video as if we were an interpreting observer present at the original event, we are also and crucially making meaning *across* the world of our research work and of the world of video events. Experientially and phenomenologically this has to feel strange to us, and in many cases even disorienting. What kinds of manipulations of the video and cyclings back and forth between the two attentional worlds, in and out of the video "illusion," are we comfortable with or deeply disturbed by? How does this bias or limit the kinds of uses of video we make in our research?

Imagine viewing a presentation of four video windows in each of which virtual time is passing at a significantly different rate (and so showing us what changing features are salient on each different time scale), and all different from the rate of time passing in our nonvideo environment. Imagine that each window zooms in rapid sequence to fill most of your visual field, with a pause between zooms to see all the windows at the same size (but with the videos still running apace in each). Imagine now that you can intervene in the automatic cycle and hold a window at maximum zoom, and then jump from window to window, while also attending to a graphic display that shows where the action in each window is on a timeline in relation to what is being shown in all the other windows? Maybe you would never want to do this (or maybe we will all be doing it in 10 years), but imagine how it might *feel*. How does it feel to do the kinds of work with video that we do now? What role do such feelings play in how we do our video research? What can reflection on these feelings tell us about how we are using the video medium and about the directions we are more or less likely to take in developing the medium and our uses of it in the future?

There are obviously limits to our sensory and cognitive capacities to process multiple video streams, to live with our attention rapidly cycling among radically different attentional worlds. But there are relatively few limits to our ability to create technological aids to enable us to ratchet down the cognitive and sensory demands of such uses to a viable level. Our uses of video media are themselves becoming, and likely in the future to become even more, mediated. The audiotape transcriber machine is the forerunner of a vast technology to enable us to traverse multiple video worlds on multiple space and especially time scales.

One More Future for Video Research

The future of video research is only partly about getting more complete and multiply reusable kinds of video data. And it is only partly about devising new technologies to allow us to manipulate and connect multiple video texts and multiple views of the same video data. It is most importantly, I think, about the evolution of the medium itself and about the kinds of research projects in which we will want to enroll video data.

I believe that video games point the way to one likely future for research video. A generation of social researchers that has grown up playing video games is not going to rest until we have a technology in which we can visit recorded real-world scenes with all the flexibility of movement in a three-dimensional world that video games now provide for designed simulation worlds. We are not likely for a while yet to recreate the "holodeck" of Star Trek imagination (Murray, 1998) or to have full-scale immersive virtual reality, at least for underfunded social research. But we can certainly have something equivalent in its affordances to a video game or an animated VRML scene. What will be most significant, in terms of the points I have been making here about such a development, will be the increased interactivity for the researcher user. I think we can extrapolate from experience in other advanced media (full presence VR and to a lesser degree, high-end video games) to predict that the sense of immersion and telepresence for the researcher will increase with interactivity.

I need to be a little more precise here about what I mean by interactivity. I mean a sense of agency in the virtual world, so that it is not simply an observed attentional world, but a world in which there is the full feedback cycle of efference and afference (action and perceived response), which we humans interpret as causal efficacy. On the opposite side, I believe that other, more "external" kinds of manipulation (starting and stopping the scene, replaying it, rewinding it, fast forwarding, etc.) will reduce the sense of presence and immersion. Why does this matter? Partly because of the delicate balance researchers must maintain between the immersion that leads to insight and the distance that allows critical reflection. But more generally because video is already a powerfully affective medium. Watching a video, even on a small screen, even in black and white, even with bad audio fidelity, can make you cry. It can make you feel fear or at least apprehension and dread. It can be sexually arousing. It can make you feel good, even elated. That power increases with all the qualities that promote a sense of attentional engagement and immersive presence in the virtual world. Researchers are going to have to learn how to deal with our own affective responses to video data.

As we should already be doing. For all our commitment to reflexivity in research, where is the research on how researchers engage with video as a medium? What do we know about the complex processes of making meaning and experiencing feelings not just when attentionally engaged in the video world, but also when moving in and out of it? What do we know, systematically, about the affective elements of our engagement with video and how they play a role in the larger agendas of our research? And in our choices among research agendas and procedures in the learning sciences?

I would like to close with a modest proposal. If the arguments I have given so far make sense, then there is a larger research enterprise in which we may wish to engage. We may want to learn more generally about how people make meaning and experience feelings in engaging with attentional media and in moving across the interface between the immediate and mediated attentional worlds. And we may want to find out how people make meaning and experience feeling when they move within and among multiple virtual attentional spaces, as they do in video games, or in jumping among windows on a conventional computer display.

There is increasing interest today in ethnographic research on spatiality and temporality and on the mediation of meaning, feeling, and action by artifacts and structures in the material environment, as well as by social interaction, and how all these interact with one another. One promising conceptual approach uses Bakhtin's (1981) notion of the "chronotope": a typical pattern of timings and pacings of action interconnected with movements in and among particular settings. Bakhtin saw chronotopes as important organizing characteristics of literary narrative genres. Very likely they are also organizing patterns in life activity and in the design of video games. We could use this concept to help us combine research on engagement with attentional media and research on the role of space, time, and artifacts in the mediation of culturally meaningful activity by examining these latter phenomena in the context of video games.

Video games foreground affect. They create simplified spatial environments and genre-typical chronotopes of action. They make it easy to observe artifact and social mediation of action. They afford a range of time scales and variation in pacing of action. Some are even beginning to permit manipulation of time within the game world, for example *Blinx: The TimeSweeper* (Artoon, 2002) and *Prince of Persia: The Sands of Time* (Ubisoft, 2003). In playing most video games, there are frequent opportunities to pause the action of the game to consider strategy or resources, and so there is often a cycling between the attentional world of the game play and various supplementary "screens" where other kinds of actions are performed, relevant to the game world but experientially outside it and distinct from it as attentional spaces (while still being within the overall game environment as designed). Players also move between their immediate nongame environment and the attentional space of the game play, and sometimes these worlds interact with one another in various ways. We might find a group of players on a local-area network (e.g., students in a school computer lab during lunch hour) who are talking to each other in the conventional world, while communicating "in character" to one another in the game world, and acting collaboratively or competitively in both worlds (but not necessarily in the same combinations or alliances). We might also find players on wide-area networks, where they have both in-game interaction with other players-as-characters and CHAT or IM (Instant Messaging) interactions with the same players in real time, outside the game and the game system windows, but still within the attentional space of the computer monitor display.

A deep ethnography of computer gaming could have much to tell us not only about this new video medium, and basic issues of meaning making, affect, spatiality, temporality, and artifact mediation, but also about the dimensions of our own interactions as researchers with the video media of our present and our future. We need to frame our video epistemologies both in and outside the box, and most especially along our own traversals into, through, and out of their multiple attentional spaces.

REFERENCES

Artoon. (2002). *Blinx the Timesweeper* [Digital Game]. USA: Microsoft Game Studios.

Bakhtin, M. M. (1981). Forms of time and of the chronotope in the novel. In M. Holquist (Ed.), *The dialogic imagination* (pp. 84–258). Austin, TX: University of Texas Press.

Brock, M. C. (1949–1955). *Captain Video and his Video Rangers* [Television Program]. J. L. Caddigan (Producer). New York: DuMont Television Network.

Collier, J., & Collier, M. (1986). *Visual anthropology: Photography as a research method*. Albuquerque, NM: University of New Mexico Press.

Condon, W. S., & Ogsten, W. D. (1967). A segmentation of behavior. *Journal of Psychiatric Research, 5*, 221–235.

Edgerton, H. E., & Killian, J. R. (1939). *Flash! Seeing the unseen by ultra high speed photography*. Boston, MA: Hale, Cushman & Flint.

Electronic Arts. (2003). *The Lord of the Rings: The Return of the King* [Digital Game]. USA: Electronic Arts.

Goldman-Segall, R. (1998). *Points of viewing children's thinking: A digital ethnographer's journey*. Mahwah, NJ: Lawrence Erlbaum Associates.

Hay, K. (2003). *Integrated temporal multimedia data (ITMD) research system* Retrieved on December 11, 2006 from http://crlt.indiana.edu/research/vrpd.html

Hymes, D. (1981). *In vain I tried to tell you*. Philadelphia: University of Pennsylvania Press.

Jackson, P. [Director]. (2003). *The lord of the rings: The return of the king* [Motion picture]. New York: New Line Cinema.

Kendon, A. (1973). The role of visible behavior in the organization of social interaction. In M. von Cranach & I. Vine (Eds.), *Social communication and movement* (pp. 29–74). New York: Academic Press.

Latour, B. (1987). *Science in action*. Cambridge, MA: Harvard University Press.

Latour, B. (1999). *Pandora's hope: Essays on the reality of science studies*. Cambridge, MA: Harvard University Press.

Lemke, J. L. (2002a). Discursive technologies and the social organization of meaning. *Folia Linguistica, 35*(1–2), 79–96.

Lemke, J. L. (2002b). Travels in hypermodality. *Visual Communication, 1*(3), 299–325.

Lemke, J. L. (2003). The role of texts in the technologies of social organization. In R. Wodak & G. Weiss (Eds.), *Theory and interdisciplinarity in critical discourse analysis* (pp. 130–149). London: Macmillan/Palgrave.

Lynch, M., & Woolgar, S. (1990). *Representations in scientific practice*. Cambridge, MA: MIT Press.

Mead, M., & Bateson, G. (1952). Trance and dance in Bali [Film]. *Character Formation in Different Cultures* [Series]. USA: Penn State Media.

Murray, J. H. (1998). *Hamlet on the holodeck*. Cambridge, MA: MIT Press.

Pea, R. (2003). *DIVER: Point-of-view authoring of panoramic video tours for learning, education and other purposes*. Santa Clara, CA: EOE Foundation.

Prichett, H. W., & Wyckoff, E. B. [Producers]. (1953–1957). *Winky Dink and you* [Television broadcast]. Barry/Enright/Friendly Production. New York: CBS-TV.

Ross, G. [Director]. (1998). *Pleasantville* [Film]. New York: New Line Cinema.

Shiny Entertainment. (2003). *Enter the matrix* [Digital Game]. USA: Atari.

Ubisoft. (2003). *Prince of Persia: The sands of time* [Digital Game]. USA: Ubisoft.

Wachowski, A., & Wachowski, L. [Directors]. (2003). *The matrix reloaded*. Hollywood, CA: Warner Bros.

Zemeckis, R. [Director]. (1988). *Who framed Roger Rabbit?* [Film/Animation]. Hollywood, CA: Buena Vista Pictures.

From Video Cases to Video Pedagogy: A Framework for Video Feedback and Reflection in Pedagogical Research Praxis

François V. Tochon
University of Wisconsin-Madison

Since the time when video recording made its first appearance, video technology has made huge advances in both ease of handling and extent of general use. Today, video equipment is reasonably priced and available in every school. When we consider the limitations faced by the first educators who worked with video cameras, we can fairly say that, technologically speaking, it is now easy to record in the workplace, and we can ask teachers and students to tape themselves in the school context. The essential advantage of video as opposed to audio recording is that it recreates both the voice and the behavior, the physical context, the direction of the gaze. It allows for "situated research" (Tochon, 1999a), referring to lived experience for the purpose of understanding and reflection. Video education has thus freed itself from an exclusively laboratory-based approach and has broadened to include self-viewing and other viewing in reflection groups on the image. Video pedagogy is a worthwhile approach at all levels of education and in various other professional sectors. It emerged as a process-oriented response to criticisms that the sequential acquisition of skills, centered on performance, is inadequate for developing professional standards. Standards cannot, in the last analysis, be dissected. Video feedback can focus on the whole, situated process.

Much more is revealed when human interaction is studied through images as well as through discourse. How video can help build meaning in the social sciences, and especially in education, is the main topic of this chapter. The signs that organize knowledge construction can be decoded in video feedback. Video feedback offers one means of contributing to the constructive, meaning-making process. Used as such, it constitutes one of the loci of exchange and validation of knowledge.

Meaning-making processes can be studied within an applied semiotics framework. Applied semiotics is the study of sign-based, meaningful interactions and enacted meaning-making processes. Simply said, applied semiotics may be defined as the study of how meaning is construed in situations through specific types of signs. Thus, video feedback can be seen as an applied, semiotic process through which learners and teachers make sense of their actions. The resulting attention has implications for how teaching and learning are conceived and studied at various levels.

LINKING VIDEO CASES TO VIDEO STUDY GROUPS

I try to encapsulate experiential knowledge in world language education. I had years of experiences with video study groups for teacher education (Tochon, 2002, 2003). I help student teachers create materials for their web-based portfolios in the teacher education program, and I ask experienced teachers to make demonstrations of language teaching methods, for which we videotape lessons and edit video materials. These videotaped experiences are sometimes presented in an evening show, where they comment their own cases in front of an audience of student teachers. In that way, I have built video cases that were used in supervisory conversations for teacher education, sorts of minivideo study groups to help student teachers reflect on their own cases.

There are different stages of video capture: In the teacher's or student teacher's classroom when she uses the targeted method. The teachers have to require the necessary authorizations for videotaping an activity in their class before this event; and we also record the presentation of selected video cases to our preservice teachers. Field evenings are organized as part of the World Language Education program. For a couple of years we had one or two teachers with one student teacher voluntarily organize the evening from 5:00 to 7:00 p.m. The presentation is made in front of preservice teachers in our program, and some inservice teachers who can receive clock hours from the Department of Public Instruction for their participation. When tapes concern student teachers, we can try to compress some excerpts and put them in their web-based portfolio.

Thus there are three steps in the development of each methodological video case:

1. An activity is videotaped in a class, and the video is edited to illustrate a teaching case related to one specific teaching method; the approach to cases is not systematic and the results are not generalizable: They are meant to help people reflect.
2. The movie is presented with methodological explanations by practitioners during a pedagogical evening.
3. The explanations and methodological discussion also are videotaped for further editing. Video editing is based on problem cases.

Following Croué (1997), a typology of video cases could be raised: (a) Cases that dealt with the nature of a problem were related to decision making, to assessment or feedback; (b) cases that dealt with the range of the problem presented a problem or showed the interconnection between various problems; (c) cases that took a story format presented a story line, with episodes, props, and portraits.

How Personal Cases Can be Shared in Reflective Video Study Groups

What principles can be recommended for organizing video case-based shared reflection? I should like to summarize here, and further explore, some approaches I developed to support professional growth and reflection through the use of group video feedback. A more detailed account can be found in *Video Study Groups for Education, Professional Development, and Change* (Tochon, 1999b). I received positive feedback from a teachers' network in New York state who used this book to create their own video study groups with much success. Video study groups are designed to share professional reflections among cooperative practitioners doing research on their own actions and their professional constraints.[1] Different types of video study groups exist, each with specific working principles. The idea of the video study group is based on a flexible educational model founded on reflective exchange in response to videotaping of activities conducted by the participants. An educational video study group aims to objectify, conceptualize, and share practices, and then integrate the resulting communally developed theory into action. The work done becomes the basis for planning. Thus, the video study group provides a versatile educational model with which we can learn to learn, to develop, and to change.

The point is not just to consume the video image. Rather, it must be integrated into shared practice, and that sharing becomes the site of exchange.

The educational video study group may include pedagogical support in order to ensure that participants derive as much information as possible from what has been videotaped. Nevertheless, those running the study group must bear in mind that over-organizing other people's knowledge tends to demotivate everyone but themselves. Feeling at ease in our chosen professions takes time, energy, and research. Little by little, through trial and error, we learn what professional attitudes are convenient and what modes of action and principles bring us personal satisfaction.

The video study group is specifically intended to facilitate this process by enabling participants to benefit from the advice of their peers. In the video study group, reflection can occur in two stages:

1. Understanding of a specific standard within a professional act. Here, an effort is made to capture the image of the situated competency as displayed by professionals in action.

---

[1]For the concept of reflective teaching, see Zeichner and Liston (1996); video was also used to elicit practical arguments among inservice teachers (Fenstermacher & Richardson, 1994).

2. The participant's implementation of the targeted standard in a given situation. Here, a process of defining the situated competencies is undertaken with the mentors. Then, certain delicate analytical tools are set up, and a process of reflection involving a dialectic between theory and practice is launched.

Every week, the teachers record, in their classrooms, 20 mins of a key lesson on a subject and then evaluate their progress. They serve as consultants to each other and reflect on ways of improving their approach. Change relates to the practice of the individual participant.

The reflective process is supported by video pedagogy, which is, above all, a new method of growth. It may be defined as a method of interaction and feedback through dialogue during encounters of reflective practice in which the essential instrument of implementation is feedback on practice. Video pedagogy involves the art of choosing the right framework for shared reflection. This constitutes a basic principle of video study groups: The framework of reflection specifies the point of impact of video feedback. The way video is used is closely dependent on the framework chosen by the group. It must be concerted and well defined.

What frameworks are available for video study groups? Several have been found especially helpful to shared reflection. The mastery framework orients activities in light of the desired results of the task to be performed. The psychocognitive framework emphasizes the conceptual structures of information and strategies for learning. In the sociocognitive framework, the process of bringing thought to awareness is linked to authentic experiences that challenge widely held notions. The narrative framework is the foundation for autobiographical or personal approaches to human experience. The critical framework functions in a social and participatory perspective. It proposes empowerment over the act of learning and an examination of those aspects of interaction that turn education into a process of either oppression or liberation. The pragmatic framework is focused on the language of practice. It explicates practical discursive arguments and intentions related to situations.

How to Avoid Classroom Fictions?

Pedagogical content knowledge is empirical, nonsystematic, and probably not generalizable. It has to become personal. It is a clinical knowledge that belongs to what Russell and Munby (1992) named the authority of experience. Different questions can arise when choosing a good case from the video rushes: Do we choose a case related to the contents, to the methods, or to the classroom management and strategy? Experiences show that all depend on the footage and the type of event. Thus, a bottom-up process seems an easier fit than do top-down, prior prescriptions.

How do we organize a case on video? Time is reprocessed within a case to give a mythic extension to it through increased indexation. The reading of the case has to be fairly obvious. This gives rise to a validity problem: By reinforcing one interpretation of the case in the editing process one reduces the complexity of the field data. It may become classroom fiction. Each time I could, I tried to avoid that reductionism in choosing excerpts that were in rupture vis-à-vis the culture of niceness we are used to in the world of teacher education. We should allow for different readings of a video case, otherwise a procedural, pedagogical indoctrination may replace the declarative indoctrination of contents. We should allow different interpretive voices in the process of case analysis, not focus on one specific rule as if the answer was a closed one. Ricki Goldman as well as Judith Green (in this volume) emphasize the importance of open perspectives in the configuration of validity, curricula, research, and discourse. Indeed, solutions may be different for different individuals. Reading and interpretation of the case probably depends on experiential knowledge. It is a reading process. Education tends to be increasingly perceived in terms of personal inquiry. Constructive learning occurs through research and documentation. In this context, the use of video appears particularly relevant. Video editing is a research process: It involves a search for authentic data, a selection of data, and the construction of an interpretation that has to be shaped into a scenario. The technology of selecting pictures and transitions and choosing colors and shapes, pace and meaning, sequences, and nonlinear frames and subframes underpins the new mode of communication: It comprises the way instruction is communicated and the way in which educational forms convey their meaning. There is a risk of creating classroom fictions. With digital video the classroom walls become translucent: Worldwide visual communication frames the new boundaries of classroom research. As a metacommunication, video constitutes the literacy of our age. The way to avoid possible delusive effects is probably to keep up with interpersonal research goals.

What the Literature Has to Teach us About Video Feedback

Two constructs have emerged in the history of videography: The first one, driven by utopian ideals, presents a reflective message for social change; the second, driven by abstract aesthetics, views video as a medium-centered art (Tochon, 2001a; 2004). I am not discussing here the field of arts per se. Rather, I am exploring how these constructs are currently being transferred in the educational field. Indeed, a third construct is emerging from visual and systematized feedback for reflective deconstruction and conceptual reconstruction. The third construct is driven by attention to both the medium and the message. It uses video to educate in an attempt to integrate social change with aesthetic research (Derrida, 2004).

Semiotic theory rests on the proposition that we cannot know the world as it truly is, but only through signs. From a semiotic viewpoint, video is becoming a language. Its signs create realities. In this light, the study of videographic signs addresses issues related to the creation of reality. There is a close link between video, research, and the study of signs. This leads further to theories of engagement related to how signs can have a social effect. If signs can have a social effect, concepts have power. They can be used to transform society. Such a perspective is *conceptualist* in the sense this word had in the early days of video technology: "Conceptualism was intended as a liberation from the shackles of the object ... encouraging for some a potential for greater social engagement" (Hall, 1996, pp. 2–3). From video's beginnings, video creators explored the perceptual and conceptual processes in order to deconstruct experience. Authors dealt with issues of feedback and social presence.

There is a physicality to knowledge: Knowledge comes from engaging in communication with a cultural community (Streibel, 1998). In the semiotic, interpretive framework that supports video feedback, meaning-making processes are shared within communities of learners. In this frame, it may no longer be adequate to speak of "teaching" and "training." Education is not conceived of independently of those educated. At the very most, one may speak of a pedagogy of sharing, which is cooperative in nature. Adjustment to situations allows for the gradual development of balanced action. Thus, video feedback provides a mode of action for cooperative self-development. Because it reveals identity in action, it can provide an integrated approach to identity formation.

The concepts behind the use of video have power, a power for feedback and a power for social change. This was the conceptualist stance: Feedback obtained through video could be provocative. Conceptualism is opposed to pure, detached aesthetics. Conceptualism and aesthetics constitute two constructs, two ways of contemplating what video should be. In each construct, the function of video research is different. The object is the focus for formal aesthetics; the social subject is the focus in conceptualism. Where conceptualism is social, aesthetics is individualist. Exploration of video space as a medium for expressing abstractness introduced the second construct of video technology. For a while, aesthetics became the prime motive of video research. The medium had to be explored per se. That is just what video creators of the aesthetic school did. They considered the medium as the message. Very soon, a specific body of video works explored the issue of abstraction as a means for defining their medium. Abstraction and not social change became the key term. Video entered the electronic game and for a time lost sight of its social goals. In this second construct, space won and history failed. The scenario was lost for a generation. The message had gone underground.

After a first shift from social change to abstract aesthetics, video production seems to be becoming more integrative: Form and meaning can be united in its educational function. Today, we are witnessing the resurgence and development of educational conceptualism in a new, socioconstructivist form. This emerging, postaesthetic construct brings video very close to the highest educational goals, as feedback becomes the basis for self-actualization and human growth. The 21st century may become the era of education. We are currently witnessing a stunning development: New trends in video production tend to meet educational goals, just as structuralism evolved toward poststructuralism and its social constructivist expressions. The goals of conceptualism tend to merge into a larger constructivist perspective. When constructivism is interpreted as an ongoing process of meaning making in context, involving both mind and body and with implications for the developing self, then it enters the semiotic mainstream (Smith, 2001). Constructivism has had an impact on both aspects of the coin; on the definition of aesthetics and the way concepts are understood. Mainstream social constructivism is about meaning making and sharing processes; it constitutes an integrative framework. Even the most abstract aesthetics seem to be undergoing a process of reappropriation in terms of knowledge rupture, resistance, and social consciousness raising. As education comes to be defined as constructive social change, it meets the goals of prior concepts like conceptualism. For instance, colorful video recording may be a way of unthinking whiteness. Thus, the integrative framework that is emerging seems to match aesthetic goals as well as conceptual leitmotifs. It defines a practical eclecticism; a method that derives from a wide range of historic styles.

Since the days of microteaching, most practices involving video feedback have been grouped under the heading of stimulated recall (Wilson, Rodrigues, & L'Anson, 2003; see also, but not exclusively, much recent work published in the journal *Teaching and Teacher Education*). Stimulated recall consists of various techniques used to record and remind students or teachers of their previous episodic thoughts. The recorded indices stimulate memory and allow for more accurate verbalizations. This label is perhaps not wholly appropriate in cases where the purpose of video feedback is to stimulate verbalization of thinking concurrent with viewing of the video, or to stimulate shared reflection on practice. In Tochon (2001b), I have proposed that some clarity be introduced into this area of educational practice and research praxis through the use of three different terms corresponding to the specific time orientation of the various kinds of feedback; (a) stimulated recall for the recovery of past thinking; (b) clinical objectification for work on present, emerging metacognitions (Flavell, 1976); and (c) shared reflection when feedback is done in video study groups.

Stimulated Recall, or Remembering Past Thoughts

Stimulated recall is the traditional term used to speak about video-based studies that focus on interactive teacher and learner thinking. Tapes are made of learning or teaching interactions and, because it is unfeasible to interview people about their thinking while they are engaged in action, the interview time is postponed to the moment when they are able to view their own actions on a monitor. At that time, a series of questions is asked that may involve different methods of questioning that vary in time and regularity. An additional issue is whether the participant or the researcher uses the remote control to interrupt the tape and reflect on past thought.

Stimulated recall is by definition directed toward the past. Viewing past actions is a way to remember one's past thoughts with greater validity than recall done without the benefit of video feedback stimulation. Numerous research studies have been conducted using stimulated recall (Benjamin Bloom presented the first study in 1953), but few major reviews of these have been done and none recently but in the field of second language research (Gass & Mackey, 2000). Nevertheless, video technology is receiving increasing use in educational settings around the world, and it would seem indeed that approaches using stimulated recall have diversified in ways that no longer respect the original assumptions of the technique.

Clinical Objectification, or Present Building Upon the Past

Clearly, by definition, stimulated recall supports the elicitation of past thinking. Many researchers, however, came to use it to prompt metacognitions on the viewing process itself. The use of video email is no exception (Inglis, 1998). Increasingly, researchers tried to put subjects in the mood for expressing their representations and beliefs while verbalizing thoughts in the process of arising, that is, thoughts that arose as they watched the video. In this way, as Yinger (1986) had anticipated, the process of inquiry shifted from remembering and studying past thinking to objectifying the thoughts emerging from direct viewing and eliciting metacognitions about them.

Objectification consists of systematic questioning, with the goal of raising tacit mental processes to conscious awareness. It involves a systematic feedback loop that increases knowledge of, and control over, oneself and one's actions. Thus, researchers had come to work on thinking concurrent with the watching of videos of one's actions, seeking to develop valid knowledge while keeping within a cognitive and metacognitive framework. Some tried to analyze practice in an ethnographic fashion but, under this approach, teachers remained "others."

For years, no real cooperation existed between researchers and practitioners in this approach to research. The inquiry process constituted a goal in itself. The isolated production of knowledge by the researcher for the research community was the norm. By the end of the 1980s, however, some research studies were presenting breakthroughs such that participant-based research became more of the rule by the beginning of the 1990s. At that time, the goal of engaging in video feedback changed to sharing one's reflections in a group. In this way, then, video-based cognitive research joined other, older educational trends that had evolved from microteaching to video feedback related to student teaching and inservice practice.

Shared Reflections, or Reflecting the Future

In the last editions of the APA style manual, the subjects of research are termed "research participants." This move accords with the transition from the dualism of the researcher–practitioner split to a more cooperative view of school/education-faculty partnerships. It is also characteristic of the change from objectification per se to intersubjectivity, from other viewing by the researcher to a sharing of experience for a practical purpose.

This is the way video study groups tend to help practitioners share both their practices and their increases in awareness. Objectification is not an end in itself; it is a springboard for planning and change. The slide from the cognitive perspective to a presumed reflection-oriented subjectivity went unnoticed in the literature on stimulated recall. The term "stimulated recall" remained in use, however, obscuring the fact that it now referred to very different practices, ones whose orientation had shifted from research to practice itself, from factual objectification to the construction of new professional knowledge. Today, it appears more appropriate to abandon the term stimulated recall in talking about concurrent thinking (what I propose to call "clinical objectification") or in dealing with video-based prospective knowledge construction (what I propose to call "video-based shared reflection"). These seem at present to be the strongest trends in the use of video feedback.

To sum up, the emphasis in video feedback has shifted from a focus on reconstructing past thinking, to a focus on postactive metacognitions, toward a focus on constructive, shared reflections on present and future actions. Due to methodological confusion, those three focuses may have been embedded in each other in many past studies, but their goals are clearly different. This methodological clarification will help in understanding the promising characteristics of current practitioner research based on video feedback. One of the earliest discoveries about video's usefulness by videographers appears to have been related to instantaneous feedback. Video provides information on oneself: It is both witness and analytical tool. Two major intellectual currents strengthened the use of video for social change; deconstruction and reflexivity (or feedback). Video can be understood as a complex mirror. Video reflexivity is a component of conceptualization and social action and permits a collaborative dialogue. Videotaping oneself raises awareness of the action and allows the actors to understand their practice and sometimes to improve their next moves. The video camera's shots are taken to build a sense of identity and connectedness that holds a strong appeal for the viewer.

Cases to Share and Cases to Educate

Video-based shared reflection has sometimes been framed within reflective analysis of practice (Schön, 1987). Video feedback stimulates reflection and semiosis; the control and cognition inherent in the comprehension and production of signs. Elicitation of semiosis is the locus of a quest; semiosis is the basis of a personal search (Merrell, 2000). The concept of education research arises out of the social constructivist approach within which research is not separated from education. Video plays an active role in this research. Through its ease of use and practicality, digital video allows for fast turnaround in editing to support constructive reflection on practice and educational change within the groups involved. A major leap was made when shots and excerpts from video feedback sessions started to be selected, reconstructed, and edited by educators and teachers for the purpose of teaching experiential knowledge through video cases. Following that leap, another important methodological clarification needs to be made. Analyzing practice for oneself or one's team—for instance in a video study group—is crucially different from analyzing it for others. In each case, the knowledge representation is different in nature. For instance, when the framework is preformatted by the researcher as in Sherin's video clubs (in this volume), then the logic is very different.

Cases for oneself or one's group of peers are directly linked to personal stories and intimacy. They are unique. They can be shared among peers but they are not meant to be publicized. The crucial difference represents the public/private clash. Personal knowledge is idiosyncratic, it cannot be replicated (it might be mimicked). Think about journal keeping practice as an illustration of this discrepancy: You will write differently if nobody will read you, or maybe just somebody you love very much. Personal cases are subtle and complex; everything cannot be really expressed, part of it is still at an implicit level. They have an etic dimension. Getting public would lead you to shed a very different light on each experiential case.

In contrast, cases for others are built with generalizable knowledge. They have an emic dimension. These cases are often claimed to be situated when they use video associated with narratives of experience. So-called situatedness may be delusive. The way of construing these types of cases is to infer generalities from particular experiences. The knowledge they provide might even enter the realm of what Jean Piaget would have named 'quasi-nomothetical'—that is classificatory and typological. They have the appearance of ideography. However their apparent situatedness hides an attempt at generalizing educational knowledge. These two dimensions—cases to share and cases to educate—overlap the etic/emic distinction, and are crucial to defining video cases.

What a Video Case Should Be: Toward an "Etic" Definition

Video cases 'for others' could be the products of sound ethnographic research. Yet, in my view, their best use should be process oriented. My claim is that the emic, normative value in the way they are used—as if they were finished, stand-alone products—can be counterproductive to the meaning-making process. Could video cases get an etic, truly situated dimension? I will assume that this can only be if they are genuinely shared mirrors of emergent processes in peer study groups. My experience thus leads me to define the video case's educational power in the process and eticity rather than the product and emicity.

This stand is somewhat different from most current ways of thinking in the field of educational video. Video cases used in education explore personal knowledge to illustrate general principles, rules, and laws and help students solve generic problems with the illusion that they are idiosyncratic. The difference with cases for oneself (or one's peers) is that the context is partly lost and the teacher is not there to explain it. That being said, it does not totally invalidate emic video cases as useful instruments for education, however the search for criteria defining good video cases is a risky compromise with normative methods. For instance the emic video case addresses the criteria of common problem cases (Muchielli, 1992):

1. The criterion of authenticity: The video case deals with a concrete situation of day-to-day professional life. Nonetheless, as earlier mentioned, authenticity is always a reconstruction as those cases have an emic dimension.
2. The criterion of situated emergency: The video case deals with an urgent problem requiring complex diagnostics and decision making. Of course, this is reconstructed situatedness: The users have to identify themselves to the case, which helps dissolve the distance vis-à-vis the other who lived the case.
3. The criterion of pedagogical relevance: The video case cannot be solved without specialized knowledge and training. Ironically, declarative knowledge would be the leading way to solve problems rather than procedural and situated knowledge.
4. The criterion of wholeness: All the necessary and relevant information is included in the video case. Ecological pragmatics, situated meaningfulness, and interpretive diversity are considered inexistent.

The criteria of this definition are obviously limited by the conceptual framework used and its strategic orientation. Moreover, the postulated self-sufficiency of the case in no way takes into account the contextual complexities. There is no such thing as a stand-alone life segment: It is always intricated into situations from which one builds interpretations.

What could be the criteria for an etical approach, then? For quite a few years now, I have been using video clips in an "education-research," processual orientation (Tochon, 2001b). Following Feyerabend's (1975) stand that running science according to fixed and universal rules is unrealistic and pernicious, and Schön's (1987) case against technical rationality, I shared with Allwright (1993) a certain disillusionment with technicist approaches and built a case against methods, looking for epistemological solutions that would make Verständnis a priority (in philosophy, German word for understanding) rather than Erklärung (or explanation). I was looking for a way of bridging research and practice. In that move, science had to elicit knowledge as its epistemic as well as etymological root, reasserting a sense of research that would be life supportive rather than ecologically destructive. It had to sustain communities of practice (Tochon & Hanson, 2003). Situated meaning had to prevail over form. For instance, in spring 2004, 17 teaching assistants and secondary teachers of different languages gathered in an online seminar that I organized to help them share video clips of their teaching practices on the university website, and to discuss their own local uses of different forms of video study groups. In this process, intuition and friendship played as much a role as analytic rigor and technology. The video cases were not built up to demonstrate a preconceived theory. Rather, they were the expressions of attempts at new, creative, and emerging theorizing. The emphasis was on perfectible, temporary, and evanescent processes, not on perfect and final products.

To sum up, in my view, best practice with emergent, educational video cases implies an etic, process-oriented approach to each lived case. One condition for this to happen is that video cases proceed from genuinely shared experiences in peer study groups.

The Paradoxical Lack of Match of Video Cases With Their Optimal Definition

A vast majority of video cases do not match this requirement. Cases may be used in a web environment that limits the use of the video itself while the use of explanatory texts takes the fore. Students sometimes express that educational websites look like frustration tests. The reason why this is the case is partly due to our current limitations in terms of broadband and speed in sending and receiving video data, and to the inheritance of a textbook tradition that we tend to project onto the websites that serve as a threshold for working on video cases. Most of the relevant information provided to solve the cases is in text format, educational videos are not self-contained, and the resources are not in video format.

What is left in video format is often a minimalist and hypercompressed editing of some excerpts that may only give a rough idea of the context in which the real case occurred. The case context is often to be found in the accompanying, descriptive text, as a challenge for the students who have the patience to read the uneasy electronic textbook that is being proposed. Websites are still mainly about texts, not about video.

The potential of video cases is to bring a rich source of content for a community to use and change. Nonetheless, it seems to currently benefit the industry more than learning, and to propel neoliberal consuming in knowledge factories rather than community feel and shared experiences. In this light, the production of video cases could be considered as a downscaling and even a corruption of the best results of ethnography and video study groups. In other words, the risk is that video cases transform knowledge into products whereas it should remain a process. Confusing the product and the process might look like misleading constructivism.

In this chapter, I propose a semiotic reading of what we do in terms of reflective video and video pedagogy. I discuss what the literature has to tell us about the history of video and video feedback. I trace the use of video as tool for social change, aesthetic object, and educational tool. The most recent construct is related to shared reflection on practice. Educational technology cannot stay in the hand of the researcher only! New ways of using video feedback are proposed to illustrate the new approach. We need the viewpoint of the teachers—the learners in the case of ethnographic learning research, as demonstrated by Goldman in this volume. Namely, reflection can be shared within video study groups and video can support shared reflection on practice.

The search for efficient learning through modem connections transforms the wealth of video experience into minute aspects of professional interactions that are accompanied by flood of textual information and categories with too many options to catch their nature. Thus,what coating helps compacting the video pill? Certainly not text. Strategic research tends to indicate that the most indigestible way to compile knowledge for learning is through declarative texts. Our current direction toward standardized products may be going the wrong way.

ACKNOWLEDGMENTS

This author's video study groups were supported by grants from the Social Sciences and Humanities Research Council of Canada, and by the Graduate School of the University of Wisconsin-Madison.

REFERENCES

Allwright, D. (1993). Integrating 'Research' and 'pedagogy': Appropriate criteria and practical possibilities. In J. Edge & K. Richards (Eds.), *Teachers develop teachers' research* (pp. 125–135). Oxford, UK: Heineman.

Croué, C. (1997). *Introduction à la méthode des cas* [An Introduction to the case method]. Paris: Gaëtan Morin Europe.

Derrida, J. (2004). *For what tomorrow: A dialogue.* Stanford, CA: Stanford University Press.

Fenstermacher, G., & Richardson, V. (1994). L'explicitation et la reconstruction des arguments pratiques dans l'enseignement [The elicitation and reconstruction of practical arguments in reading]. *Cahiers de la recherche en éducation, 1*(1), 3–6.

Feyerabend, P. (1975). *Against method.* New York: Verso.

Flavell, J. (1976). Metacognitive aspects of problem solving. In L. Resnick, (Ed.), *The nature of intelligence* (pp. 231–235). Hillsdale, NJ: Lawrence Erlbaum Associates.

Gass, S. M., & Mackey, A. (2000). *Stimulated recall methodology in second language research.* Mahwah, NJ: Lawrence Erlbaum Associates.

Hall, D. (1996). Early video art: A look at a controversial history. In J. Knight (Ed.), *Diverse practices: A critical reader on British video art* (pp. 34–49). London: Arts Council of England / John Libbey.

Inglis, A. (1998). Video email: A method of speeding up assignment feedback for visual arts subjects in distance education. *British Journal of Educational Technology, 29*(4), 343–354.

Merrell, F. (2000). *Change through signs of body, mind, and language.* Prospect Heights, IL: Waveland Press.

Mucchielli, R. (1992). *La méthode des cas* [The Case Method]. Paris: ESF.

Russell, T., & Munby, H. (1992). Teacher knowledge and teacher development. London: Falmer.

Schön, D. A. (1987). *Educating the reflective practitioner.* San Francisco: Jossey-Bass.

Smith, H. (2001). *Psychosemiotics.* New York: Peter Lang.

Streibel, M. J. (1998). Information technology and physicality in community, place, and presence. *Theory Into Practice, 37*(1), 31–37.

Tochon, F. V. (1999a). The situated researcher and the narrative reference to lived experience. *International Journal of Applied Semiotics, 1*, 103–114.

Tochon, F. V. (1999b). *Video study groups for education, development and change.* Madison, WI: Atwood Publishing.

Tochon, F. V. (2001a). Video art as a new literacy or the coming of semiotics in education. *Arts and Learning Research, 17*(1), 105–131.

Tochon, F.V. (2001b). 'Education-research': New avenues for digital video pedagogy and feedback in teacher education. *International Journal of Applied Semiotics, 2*(1–2), 9–28.

Tochon, F. V. (2002). *L'analyse de pratique assistée par vidéo* [Video assisted analysis of practice]. Sherbrooke, Quebec: Éditions du CRP, Université de Sherbrooke.

Tochon, F. V. (2003). *El análisis de la práctica con ayuda del video* [Video assisted analysis of practice]. Rio Cuarto, Argentina: Editorial de la Fundación Universidad Nacional de Rio Cuarto.

Tochon, F. V. (2004). Video education as a new literacy. In M. Pereira (Ed.), *New literacies after the visual turn in educational research and teacher education.* Granada, Spain: University Consortium for Open/Distant Education of Andalusia-Consorcio Fernando De Los Rios.

Tochon, F. V., & Hanson, D. (2003). *The deep approach: World language teaching for community building.* Madison, WI: Atwood Publishing.

Wilson, G., Rodrigues, S., & L'Anson, J. (2003). Mirrors, reflections and refractions: The contribution of microteaching to reflective practice. *European Journal of Teacher Education, 26*(2), 189–199.

Yinger, R. J. (1986). Examining thought in action: A theoretical and methodological critique of research in interactive teaching. *Teaching and Teacher Education, 2*(3), 263–268.

Zeichner, K. M., & Liston, D. P. (1996). *Reflective teaching: An introduction.* Hillsdale, NJ: Lawrence Erlbaum Associates.

Overwhelmed by the Image: The Role of Aesthetics in Ethnographic Filmmaking

Michael T. Hayes
Washington State University

The building was quite cold that day. The sun was shining bright but February in northern Idaho is cold no matter the conditions. I was working with a colleague on a film about an Indian Boarding school on the Couer d'Alene reservation. To the best of our knowledge, the building was erected around 1901 and had operated as a boarding school until 1985. Since the closing of the school, the tribe had been using the building as their education offices, but had just vacated the premises for new space in a town about 20 mins north. Although the building operated as a boarding school for over a century, it did not match the image of a boarding school we expected to see. We felt that the building and the Jesuit-run education that occurred on this site might provide a unique insight into the history of Indian boarding schools. Except for my presence and a few abandoned or stored items; the building was empty. My purpose this day was to experiment with using the confines of the camera lens to make sense of the space that the building simultaneously occupies and creates.

The building is remarkable for its architectural design and for the way it occupies space on the landscape. It is a three-story red brick building reminiscent of 19th-century Georgian architecture. The exterior is dominated by windows that bring enormous amounts of light to the interior space. Inside, the space of each floor is cordoned into a long hallway that bisects offices on either side. A stairwell spirals upward to connect the three floors. The highly polished wood floors creek beneath your feet and each of the office entryways is elegantly framed in crafted dark woods. Throughout the interior space, a slatted dark wood wainscoting adorns the lower portion of each wall. This day there is no need to turn on the lights because the sun shone crisp through the windows.

The building is perched atop a hill overlooking the small town of Desmet; named for the Jesuit priest who founded the mission in 1847. Its sheer size, location, and architectural history make it the dominant and most conspicuous building in the town. Geographically, it is located in northern Idaho's panhandle in a valley of rolling wheat fields surrounded by pine-covered mountains. In the spring, the valley is dotted with fields of bright yellow rape seed and the tribally significant purple Camas. Because of this location the building is conspicuous on the landscape and visible from miles around.

I put the camera to my eye and walked through the building, continually readjusting my perspective as it was mediated by the camera. Immediately I find that I must hesitate as I walk though each door. Light streaming through the window overwhelmed the camera and it takes a moment for the aperture to adjust. As I walk back out into the hallway, I must again hesitate for the moment it takes the camera to adjust to the shadow. By mediating my experience with the building, the camera brought the qualities of light and shadow to the foreground. I found that as I walked toward a window, the bright light momentarily obscured my vision, then instantly, I was presented with a clear view to the outside and I could see for miles around. This became my experience of the building. Light poured in from the many windows creating geometries of light and shadow that played upon the interior spaces. I experienced a constant movement from inside to outside and back again. It was possible to see outside from any position in the building, which meant that it was possible to see inside from many different vantage points. The building observed the land and the land gazed back. Light and shadow, inside and outside, to see and be seen became aesthetic metaphors around which the project was organized.

As a researcher who has recently begun to experiment with film as a way of conducting and representing my research, I have become increasingly interested in aesthetics. Because the camera directly records the world in the form of visual imagery and the film is constructed on the flow of visual elements, aesthetics comes naturally to the foreground. I conceptualize aesthetics as the intersection of experience with the world, particularly as it is mediated through a camera, and how it is organized into the represented structure of a film. Experience and representation overlap, merge, and blend in a shimmering movement. Aesthetics, according to Deleuze and Guatarri (1994) embraces and harnesses the chaotic flow of experience and representation such that the queer, the unforeseen, and the unthinkable emerge as central organizing principles.

My experiences with filmmaking have drawn these concerns front and center. The experience of looking through the camera and working with images has forced me to think about the conduct and representation of ethnographic research in different ways. It is as if the process of filming and mediating my experience through a camera requires a consideration of the aesthetic dimensions of life in the world. The medium of film demands that I foreground the aesthetic qualities that enter my research and inflect the processes of participation, observation, and representation. These aesthetic qualities are not "other than" or in contradiction to the scientific values of rational analysis. In fact, the aesthetic dimension found in film expands, gives extra added meaning to, and even fulfills the promise of the scientific.

The one experience with which I started this chapter is a small opening that I use to begin theorizing the potential of the aesthetic in ethnographic filmmaking. My study issues from the nature of the image in contemporary society. Although it is within film that I am offering my interpretations, it is the image that forms the basis of film. Film is what some film theorists refer to as the moving image in which a sequence of still images is flashed onto a screen or computer monitor at the standard rate of 29.97 frames per sec. As images are presented for this brief moment, and if they are linked by small differences, the illusion of movement is created. It is, in fact, the image as mechanically produced and reproducible photograph that makes film possible. My position is that there are certain tensions extant in the image and its production that trouble traditional standards of scientific rationality, thus opening the possibility for conducting and representing research in ways that foreground aesthetic elements.

THE PROBLEMS AND POSSIBILITIES OF FILM IN AN ETHNOGRAPHIC RESEARCH TRADITION

I begin with the anthropological uses of film because I conceptualize my film work as operating within the anthropological tradition of documentary filmmaking. Some of the very first uses of both photography and film were in the service of observational sciences such as anthropology. Anthropologists quickly grasped the value of "recording" the events they observed and experienced, which then evolved into the documentary style. For the anthropologist and the audience of anthropological work, this imagery played a clear scientific role in verifying or validating the experiences of the anthropologist. Drawing on the social and cultural sense of photography and film as a realistic representation, or even as a direct and unmediated recording of events, anthropologists pressed these image-making technologies into action to validate their traditional ethnographic research methods (Morphy & Banks, 1999). The general feeling in these early historical moments was that the recorded imagery of photography and film added an element of scientific objectivity. Photographs became standard additions to research books and articles, and the documentary film style gained wide acceptance as a way of "showing" the lives of "other" people. Photography and film, however, were unable to untangle the field from the ongoing problem of what it means to understand and represent the other.

Images of "exotic" people, places, and things had a profound impact on late 19th-century European cultural life. Travel books written from an explorer's adventures or a traveler's journey were quite lucrative for the writer and represented some of the most popular selling and most read books of the era (Browne, 1996). Also wildly popular at this time were zoological gardens and national expos that displayed photographic and filmed images of exotic locales, as well as displays of the animals, plants, human artifacts, and human beings collected on the colonial voyages (MacKay, 1986). These events drew huge crowds and were often at the center of social life in European countries. In conjunction with the scientific aims, such imagery served as real, concrete verification of the progress European countries obtained in building a worldwide colonial empire.

Recent reflection on the role of film and photographic imagery in anthropological research has cast doubt and suspicion on the realistic and objective representational strategies purported to film and photography (McDougall, 1999). The anthropological documentary form, like any genre of filmmaking, is filtered through the imagination of the filmmaker. At all judgment points, including where to point the camera, how to frame the context, and what is left on the editing floor (or in digital filmmaking the hard drive) the filmmaker makes socially and academically domesticated and regulated decisions about what is included in and what is excluded from the film. Thus, the eye, the apparatus, the gaze, and the image constitute a terrain of power through which the subject of the film is defined (Minh-ha, 1991). In the case of ethnographic research, this has raised disconcerting questions about the colonizing intents and purposes that surround the use of such imagery, both in their historical context and in their contemporary use. This terrain of power turns the persons, groups, or societies represented from the subject of the imagery into the subjected. They are entered into the ongoing legacy of the colonial project that defines the other thus entering them into a relationship of domination and subordination.

A number of researchers who use images in an ethnographic approach to conducting research have begun experimenting with photography and film in ways that trouble and disrupt the image's inherent problem of veracity. Historically, film has been used as a transparent medium for delivering recorded information mediating experience like a window; offering the viewer a clear entrance into a portion of the life-world of those on the screen. This kind of realism was made possible by making the filmmaking process invisible to the viewer and by concocting an impersonal, internally coherent and authoritative narrative. Recent experiments have attempted to disrupt the transparency of film and to use imagery to challenge the coherent authoritative narrative. In talking about the new directions that ethnographic or documentary filmmakers are taking, Arine Kirstein (2002) suggests that the goal of such experiments "is to call attention to the filmed world and to unsettle and disturb our need to understand what we see" (p. 211). Trinh T. Min-Ha (1991) issues a much stronger challenge to ethnographic filmmaking. She draws on the logic of poetry to argue that the ethnographic film only provides an illusion of realism because, "the nature of poetry is to offer meaning in such a way that it can never end with what is said or what is shown, destabilizing thereby the speaking subject and exposing the fiction of all rationalization (p. 216). Minh-Ha offers the ambiguous and unstable imagery of poetry for its potential to confront the authoritative discourse of ethnographic film. The purpose of these challenges to documentary films is to suggest a genre of experimental documentary filmmaking that challenges the very problematic nature of the image as a direct, objective, and real representation, and calls on aesthetic qualities to fill the gap.

The film *Cannibal Tours* (O'Rourke, 1998) attempts such an aesthetic dimension. The filmmaker followed a group of German tourists on a boat tour to an aboriginal village in Papua New Guinea. He edited the film so that images and sounds were ambiguously and sometimes oddly linked, consciously favoring image over narrative. The purpose was not to create a story that explained the lives of the film's subject, but to disrupt this sequencing of events and throw explanation into disarray. There is no disembodied narration to guide the viewer; just the images and the sounds of the moments that were recorded. The editing is intentionally choppy, reverberating back and forth in time, juxtaposing incongruous images and situations making it difficult to imagine a chronological time.

The camera is unsteady and intrusive, acting as just another participant in the conversation or event. It is clear from the movement and instability of the frame that the camera is being held by someone. Intentionally destabilizing the camera draws attention to the camera as a recording device. The filmmaker and the filmmaking process is revealed in an effort to place the viewer's attention with these elements of the film. The camera, the editing machine, and the gaze become an intentional and visible part of the film. O'Rourke is astute in that he is not using the film production process to attempt the futile political goal of returning subjectivity to the subject. Instead, he is harnessing the aesthetic potential of film to subtly disrupt the conventions of authoritative voice and uncover the relationships of power that have quietly structured the ethnographic or documentary film style.

The intent of the ethnographic film practice and theory I have described here is to recast the relationship between subject and world to show that, for the filmmaker or the audience, film does not offer the possibility of an unmediated window into the world of another. The relationship can in no way be so direct and explanatory, that is, causal. This does not disrupt the desire for a relationship of causality between the film and the viewer because the nature of a film is to cause, to make happen a particular kind of resonance between the viewer and the film. But it is causal, not in the predictable manner relevant to science. Such a mechanical relationship is replaced with a poetic and playful space in which possibility and potentiality are allowed. It is not a space that can be controlled and manipulated in any direct fashion, but it can be borrowed and harnessed for particular reasons. The space of allowance, particularly as it is practiced in the image, is equally important for the scientist and the artist because each requires the ability to detect or fashion novelty and possibility (Galison, 2002). Scientists use terms such as hunch and hypothesis to denote a space that exists outside or beyond their current realm of knowledge and experience. The artist uses parallel vocabulary, such as intuition and vision, to describe the space in which the world is dissimulated into the creative product. It is the space that exists part and parcel to the cognitive and the concrete, but that allows or offers the possibility of something other than the cognitive, the certain, and the verifiable. It is the space of imagination and the imaginary and it is intimately connected to and even woven into what is known and experienced. In fact, this space of allowance gives a full, expanded, and deeply embodied meaning to experience that is understood at a conscious, rational level.

Before I go too far afield, I will bring this idea back to the matter at hand, the image. The photographic image manifests numerous aesthetic tensions that have long troubled scientists, artists, critics, and society in general (Latour & Weibel, 2002). The image reveals and conceals, destroys, and creates at exactly the same moment. As a society, we are comfortable talking about images in terms of what is in them or what they convey to the viewer at the everyday level. For example, we might pull out a photograph and say to a visitor to our house "here is uncle Harold on vacation wearing his funny hat." Attention is accorded to what is revealed in the image and it relies on the scientific argument of ontology or veracity or reality. Yet, I argue that the import of the image lies in the tension between what is revealed and what is concealed. Concealment does not necessarily mean to hide or cover up but can also mean to suppress, not include, excise, or erase. Although there are many cases in which the image was used in a conscious manner to hide and deceive, I am referring more to the pragmatic reality that an image, any image, must distort the world of experience that it purports to represent. As a matter of fact, the image may be more appropriately defined by what it excludes from the frame than what it includes (Bergson, 1988). The camera frames the world by simultaneously including a small portion of what is before the viewer and hiding or excising much larger portions. The image, because it is an image, conceals in at least two ways. First, it renders onto a two-dimensional space a three-dimensional experience. This is one of the major critiques of the contemporary image in art; because it is instantly imposed on a flat surface, it is superficial and lacks the depth and power of historical visual arts such as painting and sculpture. Second, the image excises a very specific and particular moment in history. What we perceive in normal everyday experience as the flow of time is disrupted and recreated as moment. Depth is concealed to reveal surface and the flow of time is disrupted to reveal moment. What is revealed in the image is a direct function of and only possible because of what is concealed.

These are aesthetic and poetic qualities of the image because they are at play with the essential attributes of experience, space, and time. These particular aesthetic qualities of the image reside in the tensions between absence and presence, visibility and invisibility, known and unknown. In an aesthetic, they are not tensions that usher from the pull of incommensurable opposites, but they are intimately linked and interwoven in the image; one creating the possibility for the other. Aesthetics of the image is a function of these tensions. What is hidden and what is revealed exists in an ongoing coordination of complimentarity and conflict. Because the image articulates these tensions, it is never at rest, never complete, and never certain. The image is the result of constant struggles with movement, fragmentation, and hesitation. Although the written text manifests similar tensions, I would suggest that these are urgent, even compelling components of the image and the viewer's experience of it.

For those seeking in the image a relationship of veracity with the experienced world, these are problems to be addressed and eliminated, but for myself, they are potentials to be explored and harnessed. In my work with film as a method of conducting and representing research, I draw a certain amount of inspiration from the aesthetic and political experiments in contemporary anthropological and ethnographic visual culture. Rather than attempting to represent the subject of my research in an objective causal or scientific manner, I trade on the aesthetic qualities of the image and its production as a way of moving in another direction. My intent is not to degrade or abandon the scientific, because I find the intentions and purposes of scientific objectivity to be useful and important. By scientific, I am referring to the overarching cultural values of certainty, objectivity, ontology, and veracity, always understanding that how these become articulated in a range of scientific practices is quite complex. My intent is to harness the aesthetic qualities of the image and use them to repoint or redirect those aspects that conjure up veracity, reality, and objectivity (or perhaps it is to recapture and recreate the aesthetics that reside in the scientific). I wish to infuse a space of allow-ance into my film projects that proposes an element of playfulness with these scientific values. It is the aesthetic space of allowance that gives to the certain, the absolute, and the objective a pliability that imparts the possibility of change, rearticulation, and, I might add, progress.

I will now focus on a film project I am in the process of conducting. The project has moved forward in fits and starts and at this point is incomplete. In conducting re-search and creating films, the aesthetic potential of the image allows for certain ideas to come forward and to be represented while others recede to the background. The sci-entific is always there in some sense; always available for the viewer; it just does not form the core of this project.

THE DESIRING MACHINE: A FILM ON THE PEDAGOGY OF WAIKIKI

As an educator, I have recently started thinking of pedagogy, not as an act or ac-tion that takes place in schools and classrooms or the less formal pedagogical situation of the museum, but as a generalizable principle that can be applied to examine and un-derstand most any social situation. I started on this venture during my first academic position at a university in Hawaii. When I first moved to the island of Oahu, I con-sciously stayed away from Waikiki because of the crowds and its crass commercializa-tion of all things "Hawaiian." Slowly, the space of Waikiki drew me in exactly for these reasons. I began taking my undergraduate and graduate students to Waikiki to examine this space as a certain kind of pedagogical text requiring analysis and deconstruction. I began to examine the spatial construction of Waikiki for the way in which this space and its attendant images drew the attention of the visitor for the purpose of delivering a message or proposing an argument. Waikiki is a pedagogy.

As a researcher, my attention was drawn to the overwhelming quantity of images that formed the pedagogical structure of Waikiki. This space operated much like a school history or social studies textbook exploits U.S. history for the Texas and Califor-nia marketplace. Available imagery incessantly appropriated elements of Hawaiian cul-ture and unspoiled nature and turned them back out to the visitor to Waikiki as a commodity. Hawaii and Waikiki were being recreated through the appropriation of all things about Hawaii that might be desirable to a tourist, turn it into a visible commod-ity, then stuff it into every single space and plane, nook and cranny. It became over-whelmingly absurd: The image of a bird of paradise flower meticulously pressed into the sand of an ashtray, an image of a "hula girl" embossed translucent on the dinner menu of a beachside hotel, and a youthful Elvis adorned with Aloha shirt and flower lei. These images were powerful for their ubiquity and subtlety, yet disturbing in their pedagogical intentions.

To make sense of this pedagogy seemed to require special attention to the images and how they articulated this space. I started taking my video camera to Waikiki and capturing images. After my first excursions, I returned feeling as if I were overly regulat-ing my gaze through my ethnographic training. I was being too discriminate and hunt-ing for imagery that appeared to have clear and direct meaning to the project. Each time I returned from filming, I was surprised at how few images I had actually captured. This troubled me because my experience of the place was of being bombarded by them on each visit. One Saturday morning, I resolved to be much more open and less discriminating in my filming and capture anything that drew my attention. Because the frame of the camera mediated my gaze, I experienced an overwhelming rush of images. The larger context of the place was lost in the disconnected and disorienting fragments that streamed through the camera. After about an hour of doing this, I developed a headache and became almost ill. I found a dark quiet spot and spent a few minutes recovering from the assault on my senses. Quickly I realized that this was my experience of the place; very different from the peaceful images of Waikiki that are sold through vacation ads, coffee mugs, beach towels and so forth.

The mass of images, both as I experienced them and as captured on tape, was overwhelming. My initial move was to impose a rational organization on them. My training and experience in ethnographic research methods required me to create an order by defining and supporting clear precise categories. But I found that the realism I gained by imposing such categories entailed that I would lose the chaotic experience of being overwhelmed, which was the very essence of the place I sought to represent. Still, I had to create some organization; I had to take the images in my camera and place them into an arrangement that somehow made sense to the viewer. But the arrangement I sought was one that enhanced the overwhelming experience of a chaotic flow of imagery rather than reducing it to an order.

As I was putting images, sounds, and narration together for the film, I endeavored to capture this constant incessant bombardment of images. To achieve my goal, I strung together about 2 minutes of photographs, none appearing on screen longer 1/2 sec. This technique instantiated in the film a frantically paced movement from one image to the next. Over these photographs I floated live footage of Waikiki sped up three times natural speed. Short snippets of Hawaiian music, sounds of waves, cars on the street, and people on the beach were layered with narration that consisted of reading labels from products purchased on Waikiki. The intended effect was a confused mass of visual and auditory imagery. The representational strategy, the logic of this portion of the film, was not of presenting *an* image or *the* image for inspection by the viewer, that is, holding it in stasis for examination and contemplation (as would be the case for painting or a photograph). My purpose was not to correspond an image or a string of images to a distinct research generated category, concept, or idea. Although each image is clear and distinct, it is not viewable long enough to be rigorously observed or contemplated; it is taken in the moment.

To represent my experience of Waikiki, I played on the image as surface and as moment. Because the image is pure surface, it is perceived and taken in its entirety in the instant of perception. Each moment, however, contradictory or incommensurate with the previous one, is brought together in an aesthetics of montage. Many different, even seemingly unrelated images are formed into a whole. The aesthetic strategy of montage allows for a coherency that trades on confusion, fragmentation, and dispersion but a coherency nonetheless. It is not the image in its individuality that is at stake in the film but the flow of imagery; how one image leads the viewer to the next one and at the same time, references back to a previous one. My intent is for the viewer to see and apprehend each image but most importantly to sever or disrupt the potential for contemplation. I want the viewer to experience a chaotic flow that is composed of differences and contradictions.

By enhancing and foregrounding the aesthetic qualities of the image, I am very carefully considering the audience who will view the film. Aesthetics, as a space of allowance and play, is an invitation to engage with me and my experience of the place represented in the film. I do not see the purpose of the film as dictating an explanation that is intended to strictly regulate the viewer's experience and interpretation of the recorded imagery; instead, I have willfully pointed them in a particular direction. I think of it as an invitation to engage and to focus their attention in a particular way while at the same time expecting and embracing the fact that their attention will be drawn elsewhere or directed in ways I did not intend

My representational sense is to favor image over narrative and to draw on the various troubling and problematic features of the image to construct the logic of the film. This means redirecting the viewer's attention away from the content of the film and lodging it with the logic and the technique that guides its production. It is a reflexive move. Rather than attempting to make the film or its construction a transparent vehicle for the delivery of content, the film organization and logic are brought forward as part of the content, as part of the message. In some respects, it is the technique and its logic of organization that is the message.

Since my initial experience with a camera in the boarding school building, I have returned numerous times. Each time I put the camera to my eye, I gain a different and new perspective on the building and the history that surrounds it. For instance, on another visit, the camera directed my attention to the shadows cast by half-open doors and concealed corners. Amalgamating the various and different visions together into a coherent flow is a process that merges the expanded and exploded meaning of the artist with the normalizing work of the scientist. These different and contradictory values hold the imagery together in an uneasy tension. It is the peculiar quality of the mechanically reproduced imagery and the device that records them that offers such possibilities. I have embraced them as possibilities and put them to work in my films. This attitude takes my filmmaking in a direction that doesn't so much challenge historical notions of film as an accurate recording of experience as much as it redirects and expands these qualities. It's not the object to which I target my attention as much as the way it reflects light and directs shadows. My purpose in paying attention to the aesthetic qualities of the image is to (re)present in visual form my physical, intellectual, and emotional encounter with the subject of my research.

REFERENCES

Bergson, H. (1988). *Matter and memory* (N. M. Paul & W. S. Palmer, Trans.). New York: Zone Press.

Browne, J. (1996). *Charles Darwin: A biography*. Princeton, NJ: Princeton University Press.

Deleuze, G., & Guattari. F. (1994) *What is philosophy?* (H. Tomlinson & G. Burchell, Trans.). New York: Columbia University Press.

Galison, P. (1997). *Image and logic*. Chicago: University of Chicago Press.

Kirstein, A. (2001). Documentary realism in the age of changed subjectivities. In A. Jerslev (Ed.), *Realism and 'reality' in film and media* (pp. 204–218). Copenhagen: Museum Tusculanem, University of Cophenhagen.

Latour, B., & Weibel, P. (2002). *Iconoclash: Beyond the image wars in science religion and art*. London: MIT Press.

MacKay, D. (1986). *In the wake of captain Cook: Exploration, science, and empire, 1780–1801*. Wellington, NZ: Victoria University Press.

McDougall, D. (1999). The visual in anthropology. In M. Banks & H. Morphy (Eds.), *Rethinking visual anthropology* (pp. 76–95). New Haven, CT: Yale University Press.

Minh-Ha, T. T. (1991). *Framer framed*. London: Routledge.

Morphy, H., & Banks, M. (1999). Introduction: Rethinking visual anthropology. In M. Banks & H. Morphy (Eds.), *Rethinking visual anthropology* (pp. 1–35). New Haven, CT: Yale University Press.

O'Rourke, D. (Director/Producer). (1998). *Cannibal tours* [Film]. (Available from Dennis O'Rourke, GPO Box 199, Canberra 2601, Australia)

The Poetics and Pleasures
of Video Ethnography of Education

Joseph Tobin
Arizona State University

Ych Hsueh
University of Memphis

The central argument of this chapter is that in thinking about the uses of video in educational research, we should break free of educational videos' roots in instructional films and observational analysis, and add to the goals of documenting and informing the goals of provoking self-reflection, challenging assumptions, creating things of beauty, entertaining, and giving pleasure. When we enter into video making with this expanded set of goals in mind, we end up with very different sorts of videos than if we begin with only the first set. And these more aesthetically pleasing, entertaining, compelling videos are not just pleasing and entertaining—they also make for more effective social science.

THE NEW "PRESCHOOL IN THREE CULTURES" STUDY

We are in the midst of conducting a major study, "Continuity and Change in Preschools of Three Cultures." This study is a sequel to *Preschool in Three Cultures: Japan, China, and the U.S.* (Tobin, Wu, & Davidson, 1989). In the new study, we are using basically the same method that was used in the original. In this method, which we sometimes call "video-cued multivocal ethnography," the videos function primarily neither as data nor as description but instead as rich nonverbal cues designed to stimulate critical reflection. In developing this method, we were heavily influenced by the ethnographic film *Jero on Jero: A Balinese Trance Séance Observed* (1981), in which the filmmakers Timothy and Patsy Asch and the anthropologist Linda Connor first filmed a trance séance and then returned to the field to show the footage of herself in a trance and to ask her to comment on her actions (Connor, Asch, & Asch, 1986). Our particular methodological contribution was to combine this use of video as a tool for feedback (Rouch, 1995) and as a "mnemonic device" (Asch & Asch, 1995) with James Clifford's 1983 call for ethnographies to be multivocal texts, Jay Ruby's (1982) admonition that ethnographic films be considered not objective data but reflexive mirrors, and Mikhail Bakhtin's writings on heteroglossia, dialogism, and answerability (1981, 1986, 1990). The result is a method in which videotape is used to provoke reflection not just from the teachers videotaped, but also from their colleagues, their supervisors, and from their counterparts in other cities and other countries.

The steps of the method are straightforward. We (1) videotape a day in a preschool; (2) edit the tape down to 20 mins; (3) show the edited tape to the classroom teacher, and ask her to comment and offer explanations; (4) hold focus-group discussions of the tape with other staff at the preschool; (5) hold focus-group discussions with staff of other preschools around the country (to address the question of typicality); and (6) hold focus-group discussions with staff of preschools in the two other countries in the study.

In the preschool in three cultures method, as originally conceived, the goal was to produce as a final product not a video, but a book, based on an interweaving of the voices of the staff of preschools in three countries, explaining and evaluating videotapes of days in their own and each others' preschools. But once the study was completed and the book published, we realized that the footage had value beyond its use as a research tool. Although we didn't intend for the tapes to have a life of their own, independent of the book, this is just what's happened. We've distributed about 1,000 copies of the videotape from the first study, most of which have been bought by universities for use in classes, which suggests that the tapes have been seen by many more people than those who have read the book.

The knowledge that the videos we are making for the new study will be not only tools of research but also one of our final products is daunting—we were trained primarily as scholars, which means as writers, not as filmmakers. Making videos that function as research tools is one thing; making videos that can stand on their own is quite another. In the original study, the videos turned out to be far better than we planned or could have hoped, considering that we had very little filmmaking experience and virtually no budget—the original video was shot on borrowed, consumer grade camcorders. For the new study, we have a grant from the Spencer Foundation that has allowed us to purchase "prosumer" level equipment and to edit not, as in the first project, from one VCR deck to another, but in Final Cut Pro on a Macintosh computer. This top of the line equipment, combined with our years of experience and the higher level of technical expertise gathered on our new team should make for a better set of videos. There is no guarantee this will be the case (or, for that matter, that the new book will be better than the old just because the first author is older and wiser). Video making and book writing are tough, uncertain enterprises.

For videos of classrooms to function effectively as provocations and stimuli, they must be hybrid constructions, blurred genres that are simultaneously social scientific documents and works of art—if they come across as insufficiently systematic, they will be dismissed for lacking rigor; if they feel insufficiently artful, they will be ignored for being boring and visually unappealing. This is a daunting challenge to educational researchers considering using video because it calls on skills most of us weren't taught in graduate school and on abilities many of us fear we don't have, abilities we associate more with artists than with scholars.

In this chapter, using as an example a videotape we shot in 2002 in a preschool in Kyoto, Japan, we foreground the importance of character, plot, narrative, drama, and aesthetics in the construction of classroom video ethnography. We also discuss the tensions between art and science and between aesthetics and drama that are inherent in our project, and more generally, in the field of educational videography.

Shooting Goals.   We go to preschools with some predetermined criteria for what to shoot:

Routines.   Days in preschools are predictable in structure, across as well as within cultures. In order to facilitate cross-cultural comparison, we shoot and edit our videos to include a set of routines found in the all three countries' preschools; arrival; free play; bathroom; lesson; lunch; nap; snack; departure. By organizing the videos around these routines, we provide a context for comparison. It is in the ways preschools across cultures manage the same set of routines that cultural differences emerge most clearly

Key Issues.   Each scene in our videos function as a nonverbal question. This is based on the idea of Henry Murray (the father of the Thematic Apperception Test) that each TAT card in the collection would elicit information on a different psychological issue. For example, Card 1, a drawing of a boy sitting in front of a violin, was included to get at conflicts about achievement motivation. Key issues we videotape include separation problems (scenes of children and parents having trouble saying goodbye in the morning); fighting (including not just the behavior of the fighting children but also the reactions of their classmates and teachers); misbehavior (for example, a child refusing to follow directions or share); mixed-aged play; and intimacy between teachers and children (for example, a teacher comforting a crying child).

Provocative Issues.   As we shoot and edit our videos, we include scenes that we anticipate will be provocative to viewers. For example, in our new Komatsudani videotape, one of the most provocative and controversial scenes is one that shows the school's bus driver and aide drying off the children after swimming. What makes this scene particularly provocative is not just that it shows a man physically caring for young children, but also that the children are naked and the man shirtless and heavily muscled. Provocation, of course, is in the eye of the beholder. This scene elicited little or no reaction when we showed it to the staff at Komatsudani and in other Japanese preschools, but it invariably provokes a strong reaction in American viewers, for whom

Figure 5.1.   A provocative shot: An aide dries children after their swim.

it stimulates discussions of males working in the field of early childhood education and more generally of the moral panics swirling in the contemporary United States around questions of sexuality, touch, and young children.

These first three concerns that guide our videotaping—routines, key issues, and provocative issues—reflect our perspective as social scientists. But for our method to work, we have to think beyond these concerns, to think not just like social scientists but also like artists. This means we have to be concerned that our videos be compelling, gripping, engaging, and beautiful. This leads us to shoot and edit with a second set of concerns in mind:

Protagonists.   We need strong central characters. This is one area where most videos made by educational researchers fall short. A classroom has too many characters to fit into one coherent storyline. A good teacher appreciates the unique personalities and concerns of each of her students, but a viewer of a 20-min video cannot be expected to pay attention to or care about all of the students in a class. Many educational videos suffer from a dominance of wide shots and a shortage of close-ups, which makes for an absence of compelling characters. Video ethnographers are storytellers and storytellers need to focus on key characters. Before taping in each preschool, we identify four or five children on whom we will focus our cameras' attention. Because our videos are not data but interviewing cues, we select these key children not randomly, but instead based on those we anticipate are most likely to be at the center of social interactions in the classroom and to be appealing on camera. Because we are shooting spontaneous action and not a scripted play, on the day we tape we inevitably have to modify our list of focal children as other children present more compelling storylines.

In our original Preschool in Three Cultures study, there was one clear star, Hiroki, a naughty 4-year-old boy who interrupted lessons by singing songs from popular cartoons; holding a black crayon up to his crotch and announcing, to the delight of his classmates, that he had a black penis; and stepping on the hand of Satoshi, leading him to burst out in tears. In the new study, the new Hiroki is Nao, the youngest and least mature of the 4-year-olds at Komatsudani, who holds on to her mother's leg in the morning at the school gate; refuses to share a stuffed bear; and engages in a prolonged verbal and physical battle over a stuffed bear.

Dramatic Tension.   This is another area where educational videos come up short. We should think of our videos as not (just) data or lessons but also and perhaps primarily as narratives. Our videos need a narrative structure more compelling than the steps of a lesson plan or the movement through a series of daily routines. Every day in every preschool there are lots of good stories. In addition to the teacher, each student is potentially the protagonist of his/her own story, a story with hopes and dreams,

Figure 5.2.   Nao holding her Mother's leg.

antagonists and allies, and twists and turns as the day plays out and for each child, reaches its denouement. All of these stories are potentially worth telling, but in each of our 20-min videos, we have to limit ourselves to telling just one or two. In our new Komatsudani videotape, the core storyline is Mao's complicated relationship with the older girls in her class, who alternately taunt, correct, and comfort her.

Visually (and Auditorly) Compelling, Attractive, and Inviting.   In educational video making, we often talk about quality, but rarely about aesthetics. Framing shots, remembering to set the white balance, and using a tripod are necessary but not sufficient to produce a video that is visually attractive and inviting and able to hold an audience's attention and give pleasure. The aesthetic quality of an ethnographic video should be thought of not as a luxury but as inherent to the purposes of the project, purposes that should include not just teaching audiences a lesson but also giving them pleasure. Aesthetic quality is a worthwhile goal in its own right and also because its absence carries the possibility of interfering with the audience's flow of attention. Amateurish shots can interrupt audience engagement, but so, too, can slickness. Shots that are out of focus will be irritating to almost every audience, but so are shots that call attention to their artfulness and thus to the video's constructedness, as, for example, shots from odd camera angles or dissolves that break our attention as we picture an editor playing around with an effects generator. An effective ethnographic video is constructed in a way that draws audience attention away from its constructedness.

Figure 5.3.   Nao and other girls wrestling on the floor over a teddy bear.

Figure 5.4. Negotiations: "Keep this promise or swallow 1,000 needles."

Figure 5.5. Saika comforts Nao.

Finding the right level of production values for an educational video is a delicate process, necessitating balancing considerations of genre, settings, budget, and audience expectation. Eighteen years ago, when we made the videos for our original study, audiences were forgiving of our amateurish zooms and framing and of the lack in most scenes of audible children's voices. Now that most families show up at school events with their own video cameras, expectations have risen and aesthetic shortcomings are more likely to work against the state of free flowing engagement that we are after. Moreover, attractive images and good sound better reflect the look and feel of the worlds we are trying to represent. Moments in our video that capture a child choking back a tear when being corrected by a teacher or that record a whispered conspiratorial comment to an ally do not just make our videos aesthetically pleasing but also contribute to the sense of verisimilitude that is a central goal of our filmmaking.

Coherence.   Coherence combines shot and scene continuity with good storytelling to provide the audience with the sense that events are unfolding logically. This sense of coherence is in part an illusion of the editing process. Unlike in making a fictional movie in which scenes can be shot in any order and later assembled to produce a linear structure, we order our shots in the edited version just as they played out in real life (with one exception, as explained later). In reducing an 8-hour day recorded on two cameras to 20 minutes, we must cut out many more shots and scenes than we can keep. This cutting process strips from the day the connections that tied one event to the next. We must assemble the shots and scenes in such a way as to create a sense of coherence that approximates the coherence of the original events and that presents the illusion of a whole day passing by in under 20 minutes.

In choosing the next clip to add to our running sequence on the time line, we sometimes must choose between what happened next in the school day and what happened next, emotionally, to a key child in our story. (See Sorenson & Jablonko, 1995, for a discussion of the trade-offs between approaches to choosing what to film, which they call "opportunistic sampling," "programmed sampling," and "digressive search.") Another challenge we face is simultaneity of events. There were times when one teacher stayed inside with some children while the other teacher took children out to the playground. We are experimenting with such cinematic tropes as montage and with dissolves and wipes and other transition effects to suggest simultaneous events taking place in two locations, but none of these solutions seems quite right.

The collapsing of time presents a more daunting editing challenge than tying together changes in location. Many young children attend preschool from early in the morning until the early evening. Our videos, which are titled "A Day at (Blank) Preschool," must convey to viewers the sense of a full day. Typically, our first shot is of a child arriving at school before dawn, our last shot a child (sometimes the same child) departing in the twilight. It's tough to reduce 12 or more hours of videotape to fewer than 20 minutes. Our first round of editing generally produces a video of about one hour, which we cut to 30 minutes the second time through, and then, after several more drafts, down to under 20 minutes. A danger of this drastic compression is a loss of the sense of the pace of a typical school day. Using a similar number of shots in each of our edited videotapes may give the erroneous impression that the preschools in each country have a similar pace or tempo. A willingness to allow for "wait time" and for periods of silence are hallmarks of the classroom tempo of talented teachers. But to devote 3 minutes of one of our videos to a teacher waiting for a child to respond to a question would be to give inordinate coverage to one of many interesting events in the day and to risk driving the audience to a degree of ennui that will break their state of attentive viewing. We generally deal with this problem by using transition effects to suggest the passage of time. We admire the work of the documentary filmmaker Nicolas Philibert, who in *Être et Avoir* (2002) captures the extraordinary patience of a teacher of young children by presenting long, unedited sequences in which very little seems to happen.

Trade-Offs

Our project contains inherent tensions necessitating trade-offs, tensions both within and between our social-scientific and aesthetic goals:

The Interests of Teachers, Researchers, and Audiences. At the end of the week we've spent visiting and videotaping, we show "rushes" to the classroom teacher to get her explanations and reactions while they are still fresh as well as to get some quick feedback on what she would like to have included and dropped from the edited videotape. We use this input to make a rough draft of the video, which we then show to the classroom teacher and director. Based on their reactions, we re-edit the tape, adding scenes they feel are missing and deleting scenes they would rather not have included. There are times when the school staff and we researchers disagree about these edits. There are events recorded in our videos that the teacher or director think are interesting but that we believe will be of little interest or even boring to our audience; events the teacher thinks are mundane that we believe will be of interest to our audience; and events the teacher is embarrassed by or fears will be misleading but that we believe will be among the most interesting and compelling scenes. In these situations, we give teachers and schools the veto to have scenes cut, but we try to negotiate this, to balance their wishes with our research goals and with what we take to be the wishes of our audience. We are aware that although this is a negotiation and that the teachers have the right to veto scenes, that as researchers and filmmakers, we have professional status and experience that gives us greater control than the teachers we videotape over the content of the final product (Tobin & Davidson, 1991).

Audiences of Insiders and Outsiders. This negotiation between the teachers and we videographers over which scenes to include is related to another concern familiar to ethnographers of both the video and conventional variety; the need to balance the points of view, worldviews, and understandings of insiders and outsiders to the culture being studied. Most ethnographies are stories of the cultural beliefs and practices of a group of insiders written by outsiders for other outsiders to consume. The people studied and videotaped by ethnographers are rarely considered an important audience for the research. There are, happily, exceptions to this rule. Margaret Mead stresses that the primary value of filming cultural practices that are in danger of disappearing is to provide members of a culture with an archive that can aid cultural preservation (1972/1995). Linda Smith's *Decolonizing Methodology* (1999) offers insights into how ethnography can be restructured so as to serve the interests of those who traditionally have been the objects but not subjects or audience of anthropology. Making a video that can be understood, enjoyed and found both believable and provocative by insiders as well as outsiders to the culture is a tough task. The outsiders we consider as potential audiences for our videos and book are not just literally outsiders in the sense of being foreigners but also metaphorically outsiders, in the sense of being outsiders to the world of early childhood education. We want and intend that our videos and book be watched and read by an audience that includes not just early childhood educators but also anthropologists, sociologists, area specialists, and psychologists.

Genre. Is "video shot for educational research purposes" a genre? We would argue that there are several genres of educational video including ethnographies, documentaries, instructional videos, illustrations of best practice, and tools for critical reflection. Before shooting, it is crucial to be clear about what genre of video one is making and to stay true to that genre throughout the shooting and editing. Audiences cannot absorb, enjoy, or learn much from videos that are not of a recognizable genre. Even mixed genre and experimental videos must pay attention to genre questions, for in order to be seen as mixed or experimental they require an audience's knowledge and expectations of generic conventions and restraints. Our videos are a mixture of ethnography and projective tools to stimulate critical reflection. Although they are sometimes used in inservice and preservice teacher education settings, our videos are not documentaries or teacher training films. We have learned that if we include too much explanation in subtitles (as, for example, by adding a title that explains that the older children at Komatsudani routinely care for babies), the genre of our film subtly shifts from stimulus to didactics, and viewers respond less by providing their judgments of what they are viewing ("Wow-I can't imagine letting our older children feed babies. It's too dangerous!") and more by asking informational questions ("Do all of the older children care for babies?"). Once viewers decide that what they are watching is an educational documentary, they adopt a more passive, student-like stance and drop the critical perspective we want them to bring to their engagement with our tapes.

Aesthetics Versus Content. To achieve the highest possible production values we could have hired a professional crew. But we did our own videotaping because we decided that our experience in early childhood classrooms combined with our understanding of the goals of our project allowed us to make decisions about where to aim our cameras not perfectly, but better than a crew we would hire for a week. The decision whether anthropologists should shoot their own films or collaborate with professional filmmakers depends on many factors including concerns about intrusiveness, the project budget, the technical and aesthetic skills of the anthropologist, the genre of the study, and the anticipated audience. Timothy and Patsy Asch recommend that "ethnographers should certainly be encouraged to become filmmakers themselves" because "they should be able to predict more accurately and quickly what is about to occur. And they will know what aspects of a sequence should be filmed in order to get the data they want" (1995, p. 345). Jean Rouch declares that he "is violently opposed to film crews" because "the ethnologist alone … is the one who knows when, where, and how to film" (1995, p. 87). Margaret Mead argues "the best work is done when filmmaker and ethnographer are combined in the same person" (1995, p. 7).

Having decided for these reasons to do our own taping, we purchased the best equipment we could afford. We later discovered that higher end equipment has its pluses and minuses. It is impressive to schools to see us arrive with professional-looking gear, but it also can be intrusive and off putting. Our prosumer equipment—big cameras, big tripods, headphones, boom mounted shotgun mics, photog vests, and so forth—ended up causing suspicion in Japan among some parents, who questioned the preschool's director about our purposes in taping at the school. Even though the parents knew that foreign researchers would be taping in classrooms that week, the gestalt we projected turned out to be a bit too much like a TV crew and not enough like academic researchers.

Technical Trade-Offs. Adding a second camera doubles your chances of getting good footage, but it also makes shooting and editing far more difficult. Movies and TV have perfected the grammar of the two-camera edit (establishing wide shot, over shoulder shot, reaction shot, and so forth), a grammar we borrow, not only because we lack imagination to shoot and edit otherwise but also because we would be foolish to not follow a convention our audience knows so well how to read. But because we shoot in real time, without the luxury of retakes, because we are always shooting "on location" in less than ideal sound and lighting conditions, and because we are not professional camera operators, it often turns out that the footage from the two cameras we would like to shuffle into one sequence cannot be easily combined. We tried to anticipate this challenge by sketching out camera locations for each activity of the day, with both cameras on the same side of the plane of action, so the protagonists in a scene will not suddenly jump from the right to left side as we cut from one camera to the other. But we are nevertheless forced at the time of editing to make excruciating choices between the shots from each camera. Our most excruciating two-camera editing dilemma so far comes at the denouement of our most dramatic scene, the girl's fighting over the stuffed bear. As Nao is led away in tears from her battle over the bear by Saika, who offers consoling words and puts her arm around Mao's shoulder, a careful viewer can see the second camera and camera crew in the mirror in the rear of the image. Here, because Saika's empathetic response in the aftermath of the altercation is crucial to appreciating one of the potential benefits of a teacher not intervening in a dispute among children, we have chosen to risk seeming amateurish by leaving in a clumsy shot.

As we edit the videos, we struggle with the question of how long shots need to be in order to avoid a sense of choppiness viewers will find annoying and distracting. What counts as choppy has changed dramatically in the post-MTV era. We notice that even for our mature eyes, shots that would have seemed too short 18 years ago now seem fine. Another technical issue is the balance of wide shots and close-ups. In the first Preschool in Three Cultures study, we planned what we would shoot and where we would locate the camera, but we neglected to discuss how much we would zoom in

Figure 5.6.   Camera visible in the background.

on individual children. When it was time to edit the tapes, we discovered that the footage in China, shot by David Wu, who was raised in a more group-oriented society, has mostly wide shots of the whole class while the footage Joseph Tobin shot in Japan and the United States featured far more close-ups of individual children, reflecting the American cultural valorizing of individuality. In the new study, our whole research team videotaped in all three countries, leading to a more uniform mixture of wide, medium, and close-up shots.

We continue to wonder how much time we should spend in the classroom familiarizing ourselves and letting the kids and teacher get used to us before we begin videotaping. The answer might seem to be, "as much as possible." But this turned out not to be the case. At each school, we spent Monday in the classroom without the cameras, Tuesday with the cameras set up but not actually videotaping, and then Wednesday videotaping. Being around for a couple of days before taping worked well to get the children to stop looking at the camera. But it also produced fatigue, especially for the teachers, who seemed by the third day ready to get the taping over with and return to normal life. The fatigue is caused largely by the energy it takes to perform a version of themselves for the cameras. This is not to say that in our videos that the teachers behave artificially or insincerely, but rather that teaching in front of the cameras is inherently challenging. No matter how long we hang out in the classroom before taping, we cannot claim that our presence made no difference. We anthropologists know that even when we do our research without cameras, our presence must be taken into account. With cameras, our presence is more potentially intrusive and likely to cause changes in routines. We instructed the teachers (not always successfully) to do what they would normally have done had we not been present videotaping. But even if the same lessons are taught and routines followed, the mood, comportment, and behaviors of the teacher and children inevitably change. How could it be otherwise? In our method, this is not as big a problem as it is for most ethnographers and filmmakers. When we show the tapes to the teacher, the first question we ask her is, "Does this look like a typical day?" How did our presence change you and the kids? Were you nervous? Do the kids look different?

Providing Context Through Subtitling and Narration. We add no narration to the stimulus tapes, for to do so would be to cue viewers to which aspects of the tapes we think are most important and to provide explanations for teacher's practices, two tasks that our method dictates must be left up to the viewer. Like projective tests, our research tapes are intentionally ambiguous. But ambiguity is not the same thing as confusion or incoherence. As in the construction of a TAT card, a research video has to be optimally ambiguous—the question to be answered should be clear, only the answer ambiguous. For example, in the scene in which Nao fights over the bear, what is going on should be clear to the audience; all that should be unclear is whether the teacher is wrong not to intervene. Our decision was to provide subtitles of what's said by the teachers and students, but no narration. The absence of narration in our stimulus tapes behooves us to provide as much clarifying context as possible through the strategic use of images. In the case of the fight over the doll, images we use include close-ups of little hands gripping a doll and of children pushing and shoving as the teacher's body moves through the frame, to show that she was in the vicinity during the altercation, and thus could have intervened if she had chosen to.

The ambiguity we build into our research tapes to make them an effective tool for interviewing we reduce when we re-edit the tapes into a final product. As discussed earlier, this is a change of genres, from stimulus tape to ethnographic video. The biggest change we make here is the addition of a narration track that provides the classroom teachers' explanations. The goal of our final products is to present audiences with concrete examples of differences across cultures in praxis, and not just of differences in behavior. A video that presents only cultural variations in teacher and children behavior is what we call "Mondo Kane" video ethnography, in reference to the series of films in the 1960s that presented weird, bizarre, and horrifying cultural practices from around the world. Such scenes can titillate and shock, but not provoke deep reflection unless viewers are helped and pushed to consider the meanings behind the strange behaviors. It is surprising and even disturbing to American viewers that Japanese teachers are slow to intervene in children's fights. Japanese teachers' explanations that they do so to communicate to children that dealing with disputes is a responsibility of the whole class and because they believe that when teachers intervene too quickly, children miss out on opportunities to develop important social skills is deeply challenging to most American viewers, provoking not just surprise and disapproval but also self-reflection, self-critique, and, in some cases, change in practice.

CONCLUSION: THE MISSING DISCOURSE OF PLEASURE AND DESIRE IN ETHNOGRAPHIC VIDEOS

In "Sexuality, Schooling, and Adolescent Females: The Missing Discourse of Desire," (1988) Michele Fine points out that in sex education class, you can talk about sex in terms of reproduction and as disease, but not in terms of pleasure. This left some of the high school girls in the classes Fine observed wondering if their teacher was a knave or a fool—either she knew sex was pleasurable, and was trying to con them into thinking it wasn't, or she didn't know as much as they did about the pleasures of sex, in which case she had no business teaching the class.

We are suggesting that something similar tends to go on in discussions of using video in education. Video is often pushed as a research tool in terms of its virtues of being an efficient way of telling a story ("a picture is worth a thousand words") and its utility as a data-recording tool (providing opportunities for multiple coding, slowing down action, etc.). These claims are true, but they leave something out: video can be very pleasurable, both to make and to watch. These need not be guilty pleasures. Why not name and acknowledge them?

We have found that despite protestations of camera shyness, teachers are more likely to agree to participate in a study when it involves the use of video. Video carries a sense of glamour and of immediacy that makes it more attractive as a research method, for both the researcher and researched, than a study based on just observation, questionnaires, or interviewing. Many of the teachers in our studies report not just that they enjoyed being part of a film production, but also that have enjoyed watching the finished video with their students, their colleagues, and their friends and family. On the other hand, rising concerns about video voyeurism on the Internet are making parents more hesitant to give consent for their young children to be videotaped and the fact that the identities of the students and teachers studied cannot be disguised as easily in video ethnographies as they can be in most other forms of research are often raised by university human subjects committees. It is ironic that, in an age when video cameras are increasingly ubiquitous in everyday life, it is increasingly difficult to get institutional approval to use video in educational research.

Shooting and editing videos has given us an opportunity to be artistic, an opportunity lacking in our other professional work. Video editing, like writing, can be tedious, but it also can be very pleasurable, in the way that arts and crafts are pleasurable and writing usually is not, perhaps because it's what we do for a living, but we suspect for other, deeper reasons as well—the written word both as something we create and consume is not as visceral or immediate as something we listen to or watch.

Our use of video as a tool to stimulate reflection and to provoke a questioning of taken-for-granted assumptions carries with it the potential for another kind of pleasure, a pleasure akin to the pleasure of doing well on an intellectual task. We ask a lot of informants in asking them to be videotaped and then to explain their behaviors as captured in our videos. This process can sometimes produce moments of awkwardness and discomfort, including feelings of "camera shyness," the "tape recorder effect" (being surprised and disappointed to see how one looks and sounds to outsiders—"Is that what I look/sound like?"), to more deeply unsettling feelings of self-critique. But in most cases or even perhaps in every case so far, for teachers in our studies, the experience in the end has been one of self-discovery and learning about others that is not only intellectually and professionally rewarding but also pleasurable, the pleasure coming in the joy of catching a glimpse of oneself in the midst of practice and of getting to relive and ponder fleeting moments from the daily life of the preschool classroom.

When we show a videotape for the first time to a preschool teacher, she tends to be nervous and intense and anxious about both whether the tape portrays her and the children accurately and about how well she will be able to respond to our queries. But when we return to the school to show the re-edited tape to the teacher and her colleagues, there are always smiles and laughter and good-natured teasing. When we show the videotapes to the children, they not only point and laugh but sing along with the video versions of themselves and each other—invariably, at the moment in one of our videos when the teacher leads the class in song, a prelunch prayer, or a counting exercise, the children watching the video sing and chant along, the joy on their faces apparent in performing alongside themselves.

In Shanghai, at the end of a week of visiting and videotaping in a preschool classroom, we held a screening for the children in the class. At the end we asked them if they enjoyed our videotaping. All but one raised their hands. When we asked the one girl who had not raised her hand what she didn't like about the experience she told us, "it was kind of irritating and I wasn't in it enough." Such are the perils and pleasures of our method.

REFERENCES

Asch, T., & Asch, P. (1995). Film in ethnographic research. In P. Hocking (Ed.), *Principles of visual anthropology* (pp. 335–362). New York: Mouton de Gruyter.

Asch, T., Connor, L., & Asch, P. (1981). *Jero on Jero: A Balinese trance séance observed* [Film]. Watertown, MA: Documentary Educational Resources.

Bakhtin, M. (1981). Discourse in the novel (M. Holquist & C. Emerson, Trans.). In M. Holquist (Ed.), *Dialogic Imagination: Four essays* (pp. 259–422). Austin: University of Texas Press.

Bakhtin, M. (1986). *Speech genres and other late essays* (V. McGee, Trans.). Austin: University of Texas Press.

Bakhtin, M. (1990). *Art and answerability* (M Holquist & V. Liapunov, Trans.). Austin: University of Texas Press.

Clifford, J. (1983). On ethnographic authority. *Representations, 1*(2), 118–146.

Connor, L., Asch, T., & Asch, P. (1986). *Jero Tapakan: A Balinese healer.* Cambridge, England: Cambridge University Press.

Être et Avoir, 1982, Nicolas Philiburt [documentary film].

Fine, M. (1988). Sexuality, schooling, and adolescent females: The missing discourse of desire. *Harvard Educational Review, 58*(1), 29–53.

Mead, M. (1995). Visual anthropology in a discipline of words. In P. Hocking (Ed.), *Principles of visual anthropology* (pp. 3–12). New York: Mouton de Gruyter. (Original work published 1972)

Rouch, J. (1995). The camera and the man. In P. Hocking (Ed.), *Principles of visual anthro-pology* (pp. 79–98). New York: Mouton de Gruyter.

Ruby, J. (1982). Introduction. In J. Ruby (Ed.), *A crack in the mirror* (pp. 1–20). University of Pennsylvania Press.

Smith, L. (1973). *Decolonizing methodologies.* London: Zed Books.

Sorenson, E. R., & Jablonko, A. (1995). Research filming of naturally occuring phenom-ena: Basic strategies. In P. Hocking (Ed.), *Principles of visual anthropology* (pp. 147–162). New York: Mouton de Gruyter.

Tobin, J., & Davidson, D. (1991). Multivocal ethnographies of schools: Empowering vs. textualizing children and teachers. *International Journal of Qualitative Research in Education, 3*(3), 271–283.

Tobin, J., Wu, D., & Davidson, D. (1989). *Preschool in three cultures.* New Haven, CT: Yale University Press.

Reflections on a Post-Gutenerg Epistemology for Video Use in Ill-Structured Domains: Fostering Complex Learning and Cognitive Flexibility

Rand J. Spiro
Michigan State University

Brian P. Collins
Michigan State University

Aparna Ramchandran
Michigan State University

This is a chapter about the underlying epistemic goals and assumptions that structure our approach to using digital video, and the kinds of "video moves" employed in our systems as a result of assuming that epistemic stance. Why we say this epistemology is "Post-Gutenberg" is addressed at the end.

Our purpose in using video is to promote deep learning that results in conceptual mastery and preparation for practice (knowledge application) in complex and ill-structured domains. An ill-structured domain is a conceptual arena in which the instances of knowledge application are both individually complex and in irregular relationship to each other—that is, instances that might be called by the same name vary considerably from one to the next. As a result, we argue, it is not possible in such a domain to have a "prepackaged prescription" for how to think and act that covers a wide range of circumstances.[1] Instead, we offer an alternative based on cognitive flexibility theory.

To introduce our chapter, we note that our work with video should be situated in the context of the video research of Ricki Goldman (this volume; Goldman-Segall, 1998), Sharon Derry (this volume), and Maggie Lampert and Deborah Ball (Ball & Lampert, 1998; Lampert, 2001). In many ways, Goldman is the founder of this field of digital video research in education. Her focus on perspective, points of viewing, layering, the constellations metaphor, juxtaposition, children's thinking attitudes as seen in online video cases, and configurational validity (Goldman-Segall, 1998), as well as her development of more than five experimental video analysis tools over the past two decades (this volume), continues to lead the way, and certainly finds numerous points of affinity with our own approach.

FEATURES OF COMPLEX LEARNING IN COGNITIVE FLEXIBILITY THEORY AND ASSOCIATED VIDEO STRATEGIES

What more can be done for learning with digital video cases to promote deeper and more applicable knowledge? This chapter has a particular and very limited focus; the special kind of learning that video affords if one pushes beyond showing video, cataloguing it, and talking about it. We offer a concise statement of our philosophy of teaching for complex learning with digital video. That philosophy has two parts; (1) an underlying ontology of ill-structuredness (i.e., domains of knowledge and practice that are in their very nature, as they occur in the world, characterized by indeterminacy, change, resistance to global generalizations, and so on); and (2) a theory of learning, instruction, and mental representation, cognitive flexibility theory (CFT), whose features are determined by the ontology of ill structuredness and that in turn guides the development of hypermedia learning environments (e.g., Spiro, Collins, Thota, & Feltovich, 2003; Spiro, Coulson, Feltovich, & Anderson, 1988; Spiro & Jehng, 1990). How this ontology determines the features of CFT that in turn determine the kinds of video moves we make in our learning systems is the focus of discussion in the sections that follow. It is a truism, we think, that what you do with video is not as important as why you do it. This chapter is about our "why."

Why Video?

If the goal is presenting complexity as it naturally occurs in order that learners may acquire knowledge of that complexity, then video is obviously a big step forward. Can video present full complexity? Obviously not. Can a fuller presentation of com-

---

[1]To see examples of classroom video uses illustrating all of the points made in this chapter, write to rspiro@msu.edu <mailto:rspiro@msu.edu> for URLs and passwords (which can't be publicly posted because of human subjects requirements). For a nonprotected site that shows the interface we use in our classroom video work and *some* of the features of the latter, see EASE History at <http://www.casehistory.org>, which has as its focus 20th centure American history, with an emphasis on the history of presidential campaign ads (Collins, Spiro, Ramchandran, & Ruggiero, 2007).

plexity be approximated? Yes. Is one camera angle too limiting? Add another. Still too limiting? Add more. Provide auxiliary material not captured on video. Add commentaries from different perspectives. And so on. Will you have fully portrayed the complex reality? Again, no—but you will have gotten a lot closer. Shoot for better approximations to the fullness you need for future knowledge application, and don't worry too much that you'll never get all the way there. It's still a lot more than we had before. (By the way, it should be understood throughout this chapter that we always intend for our video systems to be used in the context of some ecologically valid learning purpose or task.)

Changing Underlying Habits of Mind: Prefiguring Complexity and Opening Perception

We have identified over the years a tendency toward oversimplification we have called the *reductive worldview* (Feltovich, Coulson, & Spiro, 2001; Feltovich, Spiro, & Coulson, 1989; Spiro, Feltovich, & Coulson, 1996). Video affords many ways to combat these habits of mind and replace them with an alternative more suitable to dealing with complexity. Learners need to be predisposed to say "It depends," and "It's not that simple." In our video presentations, we do things like literally having the screen startlingly disintegrate into what look like glass shards at points where we know there is a tendency to think what is being viewed is straightforward, obvious, predictable. A loud voice booms something like "It's not that simple, look again," at which point scenes are re-viewed with special-effects overlays that highlight what was missed on the first viewing. It is not long before learners start to automatically question their pat assumptions and begin to habitually look harder, look again, expect to see more. Habits of mind are hard to change. Video affords ways to catch people's eye and to call attention to the often unconscious assumptions they are making.

Opening Knowledge Structures: Conceptual Variability Demonstrations

The goal in CFT systems is to produce open and flexible knowledge structures to think within context, not closed structures that tell you what to think across contexts. Thus, one of the first things we do with video examples that have been conceptually categorized is to show many variants from the same category. The purpose is to demonstrate that category members in ill-structured domains bear only a family resemblance to each other, kind of similar and kind of different. There are no core defining characteristics, so the meaning is in the patterns of real-world use (Wittgenstein, 1953). Learners with our systems quickly see variability in conceptual application across different video clips as basic to understanding those ill-structured concepts.

So, for example, in one CFT-based video system (Palincsar et al., in press) to teach reading comprehension strategies, the thematic concept of *scaffolding* would be taught by first having the learners look at a large number of examples of scaffolding in instruction. They then come to see that scaffolding is a very complex concept for which they cannot prepackage a definition that would adequately guide use. And they also see demonstrations of the rich variety of contextual features that affect how the concept is applied.

One of our interface features, the "weave" mode (which allows four video clips to be compared in simultaneously appearing quadrants—this feature can be seen in EASE History, at the URL provided earlier), permits a certain kind of exercise that is useful for helping people to understand the ill-structured character of such conceptual families. We ask people to use the weave interface to set up four clips that belong to the same conceptual category and then to identify surprising similarities (clips that don't appear similar on the surface but on closer inspection can be seen to be instances of the same concept) and surprising differences (aspects of clips that seem similar on the surface but that are different in interesting ways when viewed more closely). Learners are quickly dissuaded from reductive notions of meaning and concept use; and a richer sense of meaning for the particular concepts is provided.

Generally, we find that four-way comparisons using the weave quadrant interface have many features that permit subtleties of similarity and difference to emerge that would not in standard two-way comparisons. ("In what ways are A and B similar, but different from C, while A and D have similarities that B and D don't ...."—although clips A, B, C, and D are all similarly categorized.) See Spiro, Collins, and Ramchandran (in press) for a discussion of other ways that CFT-based video learning environments promote openness in conceptual representations.

Learn From Cases, See Lots of Cases, and See Cases Multiple Times In Different Contexts: Revisiting Is Not Repeating

In an ill-structured domain, wide-scope generalization is not possible—in principle (that's what makes it an ill-structured domain). In the absence of general principles or schemas to guide knowledge application, transfer depends on having a rich store of experiences that capture the variety of ways in which both events in a domain happen and concepts of a domain combine. Rather than acquiring knowledge from examples (as in well-structured domains), the knowledge is in the examples.

So we don't just show one video as an exemplar case, we always show *a lot* of video cases. And we don't just show each case once, we show it more than once, in different contexts, so as to bring out alternative facets of its complexity. When one criss-crosses landscapes of knowledge in many directions (the main instructional metaphor of CFT, drawn from; Spiro et al., 1988; Wittgenstein, 1953), a revisiting is not a repeating. The result is knowledge representations whose strength is determined not by a single conceptual thread running through all or most parts of the domain's representation, but rather from the overlapping of many shorter conceptual "fibers" (Wittgenstein, 1953), as befits an ill-structured domain. (This reuse of complex video clips will come up again later, in the section on Experience Acceleration.)

Conceptual Combinations and Knowledge Assembly

Of course, our systems have thematic coding of clips so that they may be retrieved in various ways. However, one important feature of our multiple codings of individual clips is that multiple theme searches are possible. This permits a kind of combinatorial idea play. The further you go in developing hypotheses about a domain, the richer the kinds of search for video evidence you can undertake, allowing the testing against video data of increasingly complex conceptual hypotheses.

This use of conceptual combination searches also instills the idea that a clip is not just an example of one kind of thing. Rather, concepts combine in context in ways they often do not do in concept-based organizations of textbooks. And, of course, learning the patterns of concept combination in context, and how each concept in turn serves as a context for each other, is an important lesson about ill-structured domains.

Crossroads Cases: Clip Selection to Maximize Transfer

If the main metaphor in CFT is "criss-crossing a landscape" of knowledge, one of the best ways to engage that process is by investing at first in cases that have many lessons to teach. Such rich cases, at the "crossroads" of the landscape teach many lessons with relatively small cognitive investment (because the clips are short, usually anywhere from a half-minute to a few minutes in length). In a sense they are partial microcosms of the landscape as a whole. A major goal of CFT is to teach complexity, but to do it in a cognitively tractable way. These crossroads microcases are "bite-size chunks" of manageable processing size that don't strip away the features that make for complex ways of thinking. For example, one can promote seeing multiplicity with a small, manageable multiple (which is still enough to discourage the reductive bias of seeing a single "answer" in things, a single best way of looking or thinking).

Thus, by using these dense, representative short videos, these crossroads cases, the often antagonistic goals of presenting complexity and making learning cognitively manageable are simultaneously achieved. This mode of instruction is synecdochal, in that it shows the whole in the part. It allows a view of a "world in a grain of sand."

Accelerating Experience

It takes way too long to become an expert practitioner of anything that matters, whether it's being a teacher, a doctor, or an engineer. We've all heard of the "Ten Year Rule" for attaining expertise. One reason so much experience is needed is because instances of knowledge application take so many forms in ill-structured domains of real-world practice. A fundamental goal of video use in CFT-based systems is to accelerate the process of familiarizing learners with many of these forms (and the connections among them) in a much shorter amount of time, within the span of instructional contact.

This experience acceleration happens in a variety of ways in our systems. One way is by capitalizing on the previously discussed principle of providing repeated viewings of complex clips. This would be important to do in any case because you could not see their full complexity in a single viewing from a single contextual perspective. However, there is the additional benefit that once you have seen a clip that is a few minutes long two or three times, you become quite familiar with the particulars of that event. Once a learner has made that investment in a rich, crossroads case, no more than a few seconds of it needs to be seen in order to be reminded of the rest.

Once this stage is reached, we can have people making dozens of informative comparisons and contrasts in an interface that permits a large number of clips to be bounced off of each other in a short span of time. It would take more than 7 mins to contrast a 4-min and a 3-min clip. In our approach, by comparing overlearned, highly familiar clips presented in seconds, one can make a huge number of comparisons in that same 7-min amount of time. The learner is reminded of the rest of the familiar clip by the brief "distinctive highlight," and the nonpresented content "comes along for the ride" in connecting to nonpresented content in other abbreviated clips. And the more connections you make, the more cognitive momentum you build up, further accelerating the experience acquisition process. The learner becomes like a conductor, orchestrating the rapid bouncing of clip off of clip, permitting the noticing of connections that might not have been noticed otherwise, and rapidly building a supported and sustainable web of overlapping representations that in turn forms a basis for situation-sensitive knowledge assembly in the future. And preparation for knowledge assembly is the key if your knowledge cannot be precompiled for use (see also Bransford & Schwartz, 2000; Hatano & Inagaki, 1986). In CFT, we prepare people to put together a "schema of the moment" out of fragments of the past to suit the needs of the present. At its epistemological core, *everything* in CFT revolves around this goal.

POST-GUTENBERG?

We have highlighted some of the main epistemological underpinnings of learning systems based on cognitive flexibility theory and the kinds of video moves that are associated with those foundations. So why did we call this a "post-Gutenberg epistemology"? Because these new media are digital. And that makes a kind of nonlinearity and multidimensionality possible that could not be achieved with traditional linear media, refiguring thought from the ground up (Spiro, 2006a, 2006b).

Let's look at an example. We just wrote of learners being conductors (or jazz improvisers), rapidly bouncing excerpts from rich video clips off of each other. Those of us who are a bit older feel a little strange contemplating this (although we have found that adults have no trouble learning to do it). But there is a younger generation, raised on MTV, video games, television ads, and the Web, who have been doing it their whole lives. And for them, it's more natural than a lecture or a book. So why not capitalize on their affinity for this mode of "quick-cutting" across dense images (cf. Stephens, 1998)—and their accustomedness to nonlinear processing generally—for teaching, and to promote more complex and flexible learning? Pack video with important content as it naturally occurs in real-world events. Build in overlays to help manage conceptual and perceptual complexity, and to adjust habits of mind, re-present the video in new contexts to teach the richness of events, but also to make those clips very familiar. And then capitalize on that familiarity to criss-cross between many video excerpts to speed up and deepen the process of building interconnected knowledge from experience (again, always in the context of some authentic task or purpose).

This example does not look like the kind of learning and teaching that goes on from any traditional epistemological base. And I would argue that the same goes for the other aspects of CFT that infuse our use of video, as we have presented in this chapter.

That is because this kind of instruction could not happen without digital media and their random access capabilities. The new media are making possible a new kind of nonlinear and multidimensional learning.

And that is why we say this epistemology is a post-Gutenberg one. It is an entirely different mindset from the ground up, prefiguring thought (and knowledge) in an entirely new way. Just as language and story changed the species, and then Gutenberg's printing press and the widespread availability of text fundamentally changed how we think, digital epistemologies (perhaps like the one offered here to undergird video use and other kinds of learning) will alter the very nature of what it means to think, to learn, to teach, to see (Spiro, 2006a, 2006b).

The world is what we are trying to understand, and there is very little that is linear about the world. This kind of technology-based, post-Gutenberg nonlinearity allows learning to follow the contours of the world—and, paraphrasing McLuhan, the medium to embody a new cognitive message.

Language, story, text—the successive achievements of the species—won't be left behind. But they will be assimilated into a new way of thinking that digital technologies are making possible —and that our ever more complex and rapidly changing world of life, work, and study so urgently needs.

REFERENCES

Ball, D. L., & Lampert, M. (1998). Multiples of evidence, time, and perspective: Revising the study of teaching and learning. In E. C. Lagemann & L. S. Shulman (Eds.), *Issues in education research: Problems and possibilities* (pp. 371–398). San Francisco: Jossey-Bass.

Bransford, J. D., & Schwartz, D. L. (2000). Rethinking transfer: A simple proposal with multiple implications. *Review of Research in Education, 24*, 61–100.

Collins, B. P., Spiro, R. J., Ramchandran, A. R., & Ruggiero, C. C. (2007). *EASE history: Using new media and learning theory to promote deep understanding of history.* Unpublished manuscript.

Feltovich, P. J., Coulson, R. L., & Spiro, R. J. (2001). Learners' understanding of important and difficult concepts: A challenge to smart machines in education. In P. J. Feltovich & K. Forbus (Eds.), *Smart machines in education* (pp. 349–376). Cambridge, MA: MIT Press.

Feltovich, P. J., Spiro, R. J., & Coulson, R. L. (1989). The nature of conceptual understanding in biomedicine: The deep structure of complex ideas and the development of misconceptions. In D. Evans & V. Patel (Eds.), *The cognitive sciences in medicine* (pp. 113–172). Cambridge, MA: MIT Press.

Goldman-Segall, R. (1998). *Points of viewing children's thinking: A digital ethnographer's journey.* Mahwah, NJ: Lawrence Erlbaum Associates.

Hatano, G., & Inagaki, K. (1986). Two courses of expertise. In H. Stevenson, H. Azuma, & K. Hakuta (Eds.), *Child development and education in Japan* (pp. 262–272). NY: Freeman.

Lampert, M. (2001). *Teaching problems and the problems of teaching.* New Haven, CT: Yale University Press.

Palincsar, A. P., Spiro, R. J., Kucan, L., Magnusson, S. J., Collins, B. P., Hapgood, S., Ramchandran, A., & DeFrance, N. (in press). Research to practice: Designing a hypermedia environment to support elementary teachers' learning of robust comprehension instruction. In D. McNamara (Ed.), *Reading comprehension strategies: Theory, interventions, and technologies.* Mahwah, NJ: Lawrence Erlbaum Associates.

Spiro, R. J. (2006a). The "new Gutenberg revolution": Radical new learning, thinking, teaching, and training with technology. *Educational Technology, 46*(1), 3–4.

Spiro, R. J. (2006b). The post-Gutenberg world of the mind: The shape of the new learning. *Educational Technology, 46*(2), 3–4.

Spiro, R. J., Collins, B. P., & Ramchandran, A. R. (in press). Modes of openness and flexibility in "cognitive flexibility hypertext" learning environments. In B. Khan (Ed.), *Flexible learning*. Englewood Cliffs, NJ: Educational Technology Publications.

Spiro, R. J., Collins, B. P., Thota, J. J., & Feltovich, P. J. (2003). Cognitive flexibility theory: Hypermedia for complex learning, adaptive knowledge application, and experience acceleration. *Educational technology, 44*(5), 5–10. [Reprinted in A. Kovalchick & K. Dawson (Eds.), *Education and technology: An encyclopedia* (pp. 108–117). Santa Barbara, CA: ABC: CLIO.

Spiro, R. J., Coulson, R. L., Feltovich, P. J., & Anderson, D. (1988). Cognitive flexibility theory: Advanced knowledge acquisition in ill-structured domains. *Tenth Annual Conference of the Cognitive Science Society* (pp.375–383). Hillsdale, NJ: Lawrence Erlbaum Associates.

Spiro, R. J., Feltovich, P. J., & Coulson, R. L. (1996). Two epistemic world views: Prefigurative schemas and learning in complex domains. *Applied Cognitive Psychology, 10,* 52–61.

Spiro, R. J., & Jehng, J. C. (1990). Cognitive flexibility and hypertext: Theory and technology for the nonlinear and multidimensional traversal of complex subject matter. In D. Nix & R. J. Spiro (Eds.), *Cognition, education, and multimedia: Explorations in high technology* (pp. 163–205). Hillsdale, NJ: Lawrence Erlbaum Associates.

Stephens, M. (1998). The rise of the image, the fall of the word. Oxford, UK: Oxford University Press.

Wittgenstein, L. (1953). *Philosophical investigations*. New York: Macmillan.

Staying the Course With Video Analysis

Shelley Goldman
Stanford University

Ray McDermott
Stanford University

The hardest thing to see is what is in front of your eyes.

—*Johann Wolfgang von Goethe*

For 35 years, video records have supplied data for careful analyses and precise conclusions in educational research. Research proposals often include a methods section calling for video records and discourse analysis, although just how behavior is to be recorded, transcribed, analyzed, and why are often left unspecified. The promise is so powerful, its mention can carry a proposal, as if video could speak for itself, as if discourse had only to be heard to make a case. This chapter tells a different story. The power of video records is not in what they make easily clear, but in what they challenge and disrupt in the initial assumptions of an analysis. They are a starting point for understanding the reflexive, patterned ways interactions develop, and often must develop, inside the structures and interpretations with which kids, teachers, and researchers establish their work.

Video records in real time, like life, go by too quickly to allow more than a confirmation of opinions and biases, but unlike life, unlike paper and pencil note taking, video allows a slow down and multiple viewings. Reexamination invites new methods of analysis, new ways of looking and listening that can reveal both the complexity of participants and the poverty of language available for describing them. Video records allow analysts to keep track of how observations develop systematically; they become data after emergent analytic frames are documented and systematically plied across multiple viewings. Not until the discourse is dissected and aligned with the behavioral record, one act at a time, and across time, can the opinions and biases of initial viewings give way to more empirically demonstrable accounts. Only then can the word "data" take its place in a research program. Video records taken at face value encourage mundane opinions and biases, what we call "of course" analyses. Many video analyses stay on a preset course, with behavior viewed at full speed and reality reported by a simple restatement of what has been transcribed, as if what was known before analysis needs only to be confirmed. Analyses leaving undisturbed the problematic categories that have organized the troubles we research eventually lead everyone "off course."

Behavior captured on video becomes data only when operated on in procedurally specific ways inside an emergent and well-documented analytic program. This commitment does not limit the range of analytic activities that can be initiated. The world is complex, even on video, and so too the people watching the video and the institutions the analysis describes. Such complexity is not an excuse for sloppy work, but an opportunity for diverse ways of being thorough and rigorous.

We offer three ways of organizing careful approaches to and outcomes of video analysis in education. We name them reform, interactional, and historical approaches; great work would use all three as simultaneously as possible. We develop the first with a long example, because reform is both crucial to learning research and the least carefully articulated approach to video analysis. We use the same example in a minimalist discussion of interactional and historical approaches. Interactional approaches are the best formulated, and we recommend their findings as a guide to precise and well-constrained conclusions from video analysis. We recommend an historical analysis as a worthy and necessary goal for educational research.

A REFORM ANALYSIS

Video analyses directed at educational reform stay the course long enough to redefine an initial problem in a way that suggests a response not considered before analysis began. Usually, educational problems are understood simply in terms of individual learning. Video analysis can deliver a more complex account in which the analytic unit goes from someone who does not know something to the patterned social forces that construct educational contexts with differential access to positive learning outcomes. A reform approach to video develops this complex view and invites new versions of intervention.

To illustrate our point, we tell a story of how research questions and findings changed as we examined a corpus of tapes from a mathematics classroom. Across multiple viewings, the analysis changed our perceptions of students, teachers, and classrooms as environments for math learning. It also redirected our work on curriculum development. An analysis of a few students working together brought us four revelations:

1. We saw a supposed "unengaged" student working hard, engaging in tasks and activities, and mastering new mathematical concepts and operations.

2. We saw a supposed "promising" student performing smartness for the teacher and building his resume in inverse relation to, and often at the expense of, the "unengaged" student.

3. We saw the newly noticed hard working student avoid all official and public assessment opportunities, thus keeping his label as "unengaged," even as he successfully completed new tasks and assignments.

4. Although our curriculum materials enabled traditionally less successful students to tackle school math and perform competently inside work groups, we realized they would be ultimately ineffective until we could integrate assessments that could cast a wider net on student learning.

We tell our story in three stages showing; (1) initial understandings that had us off course, (2) viewings that added layers of complexity, and (3) new understandings we were forced to develop.

Observations and Initial Understandings

We undertook field tests of a middle school mathematics curriculum we were developing across multiple iterations in conjunction with teachers, curriculum writers, and researchers (Goldman, Knudsen, & Latvala, 1998; Greeno et al., 1999).[1] We began observing and writing field notes a few weeks prior to the curriculum field test to obtain a base line on classroom organization, teaching styles, and learning practices. We identified two groups of students marked by an ethnic-racial mix and good attendance records.

We observed and videotaped every day from the start of a curriculum unit. We often had four researchers present; two working cameras and two taking field notes. We used two cameras with each group, one focused on students and their materials, the other on the computer screen. Within 2 days, the research team typed field notes and created content logs that recorded an index to the archived tape, a summary of the lesson, and notes on segments for transcription, for example, introductory and closing sequences, question and answer exchanges, explanations and presentations by students, teacher interactions with the group, and specific activities attended to—math happening here, but not there, both as expected, and not.

We display the evolution of our analysis in the story of one work group of three boys. From the start, there were stories attached to the boys as math learners. Boomer was known as a good student. Teachers had identified him as a "star" upon entry into middle school, until he developed an "attitude" problem in his first year. They agreed to work with his "issues" and talked about making an "investment" in him. In class, Boomer alternated quickly between challenging teachers by yelling, cursing, and calling them names and enthusiastically answering lesson-related, teacher-posed questions. He raised his hand, shouted out, and, misbehavior aside, gave reasonable and

---

[1]We report on the Middle-school Mathematics through Applications Project (MMAP), an NSF-funded curriculum reform project we developed and researched. For more information see http://mmap.wested.org. We express our opinions, not those of the NSF.

helpful answers. His teacher kept him in class by ignoring her own behavior-related rules and by chastising him outside of class.

Hector stayed out of the way of teachers, and the teacher had little to say about him. She thought he was "a nice kid" but in trouble academically. He never raised his hand, never gave an answer, and spent whatever time he could sharpening his pencils. He did not hand in assignments, or keep work papers, and often sat idly through class. He was social with other students and expert at teenage banter.

The third boy, Ricardo, was quiet, attentive, and a listener. He was transferred to another school after the first 3 weeks of the field test because he had sent a death threat to a teacher. Before leaving, he socialized with other students when time and situation allowed and made occasional contributions to the math work of the group. All three boys were popular with peers, and teachers labeled them all "at-risk."

Adding Layers of Complexity

Accounts of the three learners were constantly questioned and, as justified, verified, altered, or rejected. Teachers may have called the boys "at risk," but our job was to find out what "at risk" meant in relation to the boys in situ, in the actual unfolding of their behavior in relation to other kids, teachers, and curriculum. Boomer was forgiven for his misbehavior, Ricardo was transferred for his, and Hector was allowed to slide by unattended. What is going on here? Did the accounts hold up to detailed analysis? Did the accounts coordinate with the math the boys had to learn?

We videotaped the boys for 6 weeks as they completed a math unit that required students to design, heat, and pay for a research station for a team of scientists in Antarctica. Despite at-risk rumors, Hector was on task and productive in the Antarctica project. He did everything he was asked to do and then some. By the end, Hector was expert at using the modeling software, and he created the floor plan on which he and Boomer did their math. We saw this engagement on the first day of videotaping, and the inconsistency of engagement with Hector's "story" led us to more detailed analyses of the video records. We came to see him as a social kid who worked steadily but shied away from evaluation.

We offer descriptions of three classroom events that span the beginning, middle, and end of our video records. Together they show how, with detailed analyses, we came to understand our students, their access to learning, our materials, the pressures of assessment demands, and our best guesses about what to do next.

Event I: Reconciling Real-World and Scaled Representations.   The boys have to compare a scale of one meter on a computer screen to one meter in the real world. They also have to draw with dots a small plan for a living space, this to mimic the layout of the computer modeling environment they would use next. Hector decides to figure the size of a real 6x6 m room. He leaves his group to get a meter stick. The teacher catches his eye from across the room. She starts to get him a smaller ruler before realizing and acknowledging why he wants a meter stick. She puts the stick on the wall, flipping it over lengthwise to model the measurement of meters. When she runs out of wall space, she suggests he place the stick on the floor. Hector turns to his group mates and says, "I just wanted to play with it." He moves his chair away and puts the stick on the ground. He stands, bends, stands, takes out a pencil, bends again, and puts marks on the floor to designate each meter until he measures 6 lengths. Meanwhile his group mates work on floor plans at their desks. Boomer looks up to see what Hector is doing and asks, "Is that 6 meters?"

Hector: Here! [stands up]. Six!

Boomer: That's big enough.

Hector: From that thing [points to where he started at door].

[Both boys use their right hands and point to the perimeter of what could be a 6 × 6 m room.]

Hector: That's BIG! [returns to his desk]

Hector: That's bigger than my whole house! [laughs].

[Over the next minute, Hector returns to his desk, looks at his paper, counts off dots, picks up the assignment, reads it, and writes on his dot paper. Ricardo looks at what Hector is writing.]

Hector: They're going to sleep in an area 6 × 6. That's big! [Gestures "big" with his hands.]

Ricardo asks about how many rooms are modeled in their developing floor plans. Hector joins the conversation, but is now concerned that their rooms are too big.

Hector: This was a living area. No mas una. Pueda grande.

Boomer: [Boomer looks over] Everybody should have their own room.

Hector: Eh ... so I made a big room. [The group talks quietly among themselves].

Hector: 4 and a half ... [Boomer is looking at his own work]

Hector: I'm going to leave it like that [puts his pen down, looks at his hand, plays with a pen on his desk, and looks at Ricardo's paper].

Hector: Pues, una grande, ja!

Adult: We need to start cleaning up.

Hector: This time I'm gonna make little small rooms. [picks up a towel to erase the board and works his design].

Hector: That's good. Look [turns to Ricardo with his paper then to the camera. Ricardo doesn't look at his paper and Hector moves his arms closer to get his attention. He puts his picture down on the desk and closes his pen].

Hector: Now we clean this up. We erase everything we did to this, right [to Adult behind camera]?

Adult: No, no, no, no. [He leaves the design on the desk].

Hector compares meters on his paper with meters in the real world and the living space in his home. He worries about room size and its scaled representation and eventually convinces his group to make rooms smaller than 6 × 6 m. Across segments from other class days, we see he becomes somewhat expert on measurements and scale translations. He also masters the software and inputs all floor plan requests. Modeling and revising the research station takes six classroom periods, and much of the group's work is based on Hector's model. He builds on his first day of measuring activity throughout the project. Hector engaged the modeling and the mathematics of scale and proportion, but we missed the importance of his first-day question about erasing his work. Only after we tracked tasks requiring math engagement over time did we discover the relationship between Hector, school math, and assessment.

Event II: Graphing Data.   The curriculum asks students to balance trade-offs in the research center design across three variables; building costs, heating costs over 20 years, and the insulation value of the research station. Based on data charts from a prior investigation with the same variables, the teacher required each group to create a series of graphs for these relationships.

After the teacher's lecture about how to create graphs and plot data points, Hector and Boomer work on graphs from their data chart (see Fig. 7.1). Two significant events take place. First, Hector is away from the desk when Boomer calls for the teacher's help. He returns to see Boomer asking for confirmation on how to plot data points. He watches Boomer plot the first point and the teacher remarks how that point was easy to locate. Hector, looking at the graph, says, "Where is it?" The teacher shows him where the value of 114,500 was placed. Hector begins to plot the point on his own graph. Hector and Boomer discuss rounding off their number and plotting the point at 115,000, the nearest unit on the *X*-axis. The teacher asks Hector, "Do you know why you're putting it there?" When he doesn't answer, she asks, "Where did he get the 14,5 [114,500]?"[2] Hector looks around, then points to the 114,500 on the data chart, saying, "From right here." The teacher says, "but that says 14,4 [114,400]."

Hector: But he went to the [eyes and hands pointing up], I don't know, to the five.

Boomer: I rounded it.

Hector: Yeah, rounded it.

Teacher: OK … Ok, so you should have an easy time putting these on. [Hector breaks into a very big smile].

Hector: All right [leaning in with Boomer to look at their data table].

---

[2] Participants discuss values from the chart using short cuts in their talk. The bracketed numbers indicate the actual number being discussed.

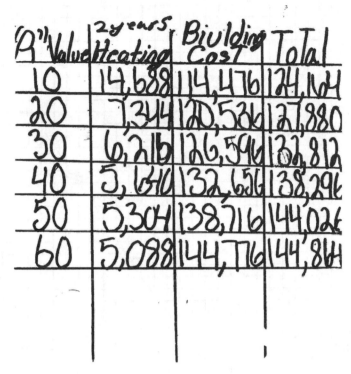

Figure 7.1.   Group's chart of insulation, heating, and building costs.

The two boys and the teacher turn their attention to the next data point. Hector suggests where to plot the point on the graph (at 121,000), and the teacher questions him. She asks where he got the idea for his suggested plot point. Hector answers the teacher's question by pointing to 120,536 on the data table and smiles when she confirms his answer. He is in the flow of the interaction and stays with the graphing.

The analysis documents a challenge to Hector's contribution. When the teacher asks where the "14,5" [114,500] point came from and what the relation is between "14,5" [114,500] and "14,4" [114,400], she faces Boomer while she talks, looks at and touches only Boomer's papers, and makes Hector lean across the desk to see where she is pointing. Hector leans over the desk and answers the second challenge with help from Boomer who gives him "rounding" to express the move of considering 114,476 as 114,500. Again he smiles and points to the next data point to be graphed. Small victories with the teacher encourage his interacting further with the tasks and data representations at hand.

The second significant interaction follows the first graphing problem. Hector and Boomer turn to their data chart to graph the next point. Hector seems unsure and asks Boomer to show him. Boomer shows where he plotted the point. Hector puts a point on his paper and then says, "Let me see if I did it right." Together, they plot four more data points (see Fig. 7.2). They are in a dance: (1) looking for the number on their data chart; (2) looking for the closest marked point on their graph; and (3) deciding whether to round up or down, identify a point, and place a mark on their graphs. On

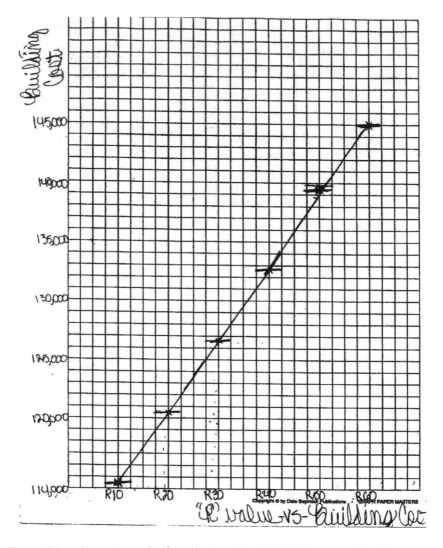

Figure 7.2.   Group's graph of insulation value ("R"value)–versus–building cost.

the fifth point, they dispute whether the value should be rounded up or down. Boomer calls for the teacher, but Hector says he knows where the point should be and twice begs Boomer to listen to him. He mumbles that his answer is "right." During this exchange, the bell rings. Hector makes one last plea for where to plot the point, looks up, and says, "The bell already?" Hector asks Boomer to keep his papers for him because he doesn't have a notebook. Hector slips them into Boomer's notebook, and Boomer says, "It's gonna be all messed up."

Hector interrogated his data and learned to make graphs of the relations between building costs, heating costs, and insulation values. He did not at first know how to graph and turned to Boomer. We see them practice together, and Hector developed enough competence to argue for his solution on graphing a specific point. He became so engrossed in this new activity, that the bell surprised him. Hector engaged assigned activities, handled a challenge from his teacher, grappled with graphing and translating numerical data, and even asserted his sense of competence.

Event III: Public Assessments.   Hector seemed to be learning a great deal, but did not display it publicly. Twice he was asked to report on his groups' work to the class, and twice he failed to show his accomplishments. The first time Hector participated in a classwide design review, he had to give classmates a tour through his group's research station—his specialty. We were eager to witness his performance, especially because Ricardo and Boomer were absent due to behavior problems. Hector made jokes. He got the class laughing hysterically and then sat down. He pointed at the large monitor displaying their research station, but barely gave any information and even claimed to not know much about it.

A second presentation came at the end of the project. Boomer did most of the talking, and he called Hector "stupid" a few times. When a few classmates noticed a mistake in the size of a bed, Hector left in midpresentation and went to the computer to fix the bed size. The computer was hooked to the large classroom monitor, and it crashed. The presentation disintegrated. The teacher tried to salvage it, but the bell rang and everyone started packing up backpacks. Boomer was mad and verbally abusive, Hector was trying to deal with the computer, and accomplishments of the prior 6 weeks seemed small. Boomer's papers, charts and graphs were the only project materials handed in. The teacher had no sense of Hector's accomplishments; his work was invisible to all but us researchers.

Forced to New Understandings

By current classroom accountability systems, Hector looks like a failure even when he is learning. The failure slot is ever ready to acquire children as a contrast set to those who succeed (Mchan 1996; Varenne & McDermott 1998). Hector did not put his learning on display for teachers or fellow students and used display opportunities to perform as a class clown or to take a one-down position to his classmate. Without an intervention, Hector would quietly remain under the achievement radar in math and other academic subjects. Our student-centered curriculum worked at one level—we had Hector engaged and learning mathematics—but failed at the institutionally crucial assessment level. The teacher needed to see Hector's learning without a research team analyzing his every move. To accomplish reform, curriculum must enable even the most system-recalcitrant students to have their learning noticed.

Hector's story had a nice ending. His teacher watched videotapes with us and was excited to see his work. Before seeing the video, she had assumed Boomer did the group's work, but after the video, she gave Hector an "A" for the project—the first unit he had passed that year. The teacher reported his good work to his mother and began to experiment with new ways to notice and document the efforts of Hector and others.

Video analysis led us beyond our original research questions and categories. One major consequence of the analysis was to implicate the structures inviting teachers and students to code classroom events exclusively in terms of competence and incompetence and without attention to engagement patterns and practices. Our analyses also forced us to consider how social interaction situates math learning. We saw Hector alternating between competence and noncompetence displays depending on social and institutional demands. The conflict between social and math contexts was most apparent in whole-class public assessments, when Hector, if on his own, joked and entertained or when, with Boomer, mistakes were revealed and the social fabric of the classroom degenerated.

With video records, we expanded our research questions and foci. Analysis took more time, but it allowed us to capture patterns of engagement not visible to participant observers. We were also forced to confront biases in an attempt to build a curriculum designed to attract alienated students. We learned that mere engagement in math work was not a cure-all; we would have to dig deeper to provide educational environments disruptive of patterns of school failure. Our initial questions were, "What mathematical practices are here? Are the children engaged mathematically at sustained levels?" As we examined classroom tapes in greater detail, our questions expanded: "When are mathematical practices?" "How do we see mathematical practices in language and actions?" "How do schools, classrooms, teachers, students, curriculum, and assessments form a structure for creating differential student performance?" and "What structures are needed to change and what resources will have to be in place for more engaged, equitable mathematical work and assessment?"

INTERACTIONAL APPROACH

A profound development in the social sciences of the last half-century is the relentless description of the interaction order. Deep below the level of anything we can say about each other, below the easy application of terms for consciousness and intention, there is a complex web of connections between people and, in one mix and match or another, they are available for use among participants in social interaction. In a videotape analysis directed at documenting the interaction order, the goal is to stay the course long enough to define the social and behavioral mechanisms people use to coordinate their activities across persons. Almost always, a detailed analysis can deliver accounts of the many ways people are involved in the organization of each other's behavior—even when the behavior is exactly the behavior they would love to extinguish. It is no accident the analysis of body movement as communication (Birdwhistell, 1970; Kendon, 1990, 2004; Scheflen, 1974) developed first in the study of schizophrenia, nor that conversational analysis (Goodwin & Goodwin, 1992; Sacks, 1990; Schegloff, 1992) developed first in the study of suicidal calls for help. The world is both a terrifyingly mysterious and wonderfully well-organized place, and that we do not always understand it does not mean there are no regularities to what happens, how we contribute to what happens, or how others interpret what happens. The analysis of social interaction is foundational to the use of videotape in learning research, because it has consistently delivered on a promise: that if looked at carefully enough, where carefully means an hour of transcription and analysis per every second of videotape, every behavior, no matter how bizarre, or seemingly nonresponsive, is part of a coherent string of behaviors developed across persons and across time, by some mix of agreements, contrasts, and contradictions in tune with the demands of the moment. This means that everyone—every child learning or not, misbehaving or not—can be shown to make sense and likely in ways that reveal additional structures in play.

Analyses directed at educational reform often stop before analyses directed at a full description of the give and take between people in interaction. The one seeks ways to make change, the other, ways to find new phenomena. This difference should not obscure the importance of interaction analysis for a reform agenda. We were often moved to give up on Hector, but we stayed the course for two reasons: We mistrusted the ease with which everyone else had given up on him, and we knew on the basis of interaction analysis that if we continued looking, his every move would make sense given what others around him were doing. Eventually, we saw the patterns; the teacher's inattention, Boomer's constant insults and put-downs, and the insistence by all adults and most kids that public verbal displays of math competence were the coinage of the classroom economy. This was enough to change our research questions and plans for new math units. The message was clear: The more we looked, the more we understood the competence of the students and the more we had to worry about the baggage we, their adults, brought to school.

If interaction analysis urges a more careful look at behavioral detail, our next approach urges a wider look. These are not separate activities, not in the long run. Both are crucial to making any learning agenda less embedded in the status quo.

HISTORICAL APPROACH

In the early 1970s, the rush to videotape classrooms was driven by an immediate purpose and a long-term vision. The immediate purpose was to defend children, particularly minority children, against the capacities of schools to label and disable their relation to schoolwork. Analytically, the purpose was achieved, and we now have a tradition showing that kids are more interesting and quick to achievement than is allowed by the institutionalized language of formal evaluation (from a wide range of perspectives, see Chaiklin & Lave, 1993; Cicourel et al., 1974; Erickson, 2004; Gilmore & Glatthorn, 1982; Goldman, 1996; M. Goodwin, 1990; Greeno & Goldman, 1998; Mehan, 1979; Wortham & Rymes, 2002). With the analytic good news comes the political bad news. The immediate purpose of classroom video has not been achieved institutionally; the analytic advance has not resulted in general ways to help schoolchildren. As the economic gap between social classes has widened, the promise of democratic schooling has dwindled, and the possibility of achieving even a minimal functionalist ideal—that is, if not educating to make the most of every life, then at least getting the best people in the right jobs—has evaporated in an evaluation system pitting all against all on arbitrary tasks. Schools deliver success in accord with the established order. As a political institution, education is a dependent variable with little room for analytic accomplishments showing the potential and productive power of discarded students.

If classroom video analyses served their immediate purpose with both success and disappointment, the underlying long-term vision has disappeared. More than a century before video, Ralph Waldo Emerson read the human situation in the body, in the movements of persons: "The whole economy of nature is bent on expression. The telltale body is all tongues. Men are Geneva watches with crystal faces which expose the whole movement" (1860, p. 156). For Emerson, the mannered social body holds in its every move the secrets of the person's connections with history, nature, and possibility. By this vision, we can use video to defend those left out and to identify forces to which schools are responsive. We could see, for example, not only the problems of children, but their possibilities; not only the constraints on teachers, but their sensitivities; not only the mismeasures of institutional evaluators and researchers, but their dreams for helping; and all that as institutionally and historically shaped and potentially reshaped by our analyses and reform activities. In the forces moving kids about is the history of the world, its antiquity and forms, its economies and social structures. Classrooms are the staging ground for large-scale politics and social structure, places where a competitive capitalism, with apparent sensitivity to individual differences, sorts the supposed best from the disjointed rest. Video ethnographies of classrooms show the moment-to-moment lives of tiny selves in exploration and confrontation with social structure.

Most classroom video projects do not privilege this vision and instead ignore the worlds in which children and teachers conduct business intertwined with larger social forces. Theories of ability, disability, cognition, learning, mind, self, and language have developed without enough attention to forces organizing classrooms literally accountable to the demands of social structure and political economy. Born in Mexico, how did Hector come to a California classroom? If schools actively suppress his skills in Spanish, why would it be strange for him to hide his skills in math? If gangs offer Hector, Boomer, and Ricardo a vibrant enough alternative to school to exist in identical form across California and the Southwest, do they also emerge in math classrooms? Why is the prison budget of California outstripping the education budget? And how did mathematics, a powerful tool in the construction of modernity, come to classrooms without connection to being useful? We built our curriculum with these questions, and from the analysis of kids and teachers, we gained insight into, and played back and forth between the wider system and moments in the classroom. We brought new questions to the analysis and received, in exchange, better new questions. This is the dynamic of video research tied to a historically specified program of social reform.

CONCLUSION

Video records do not make analysis easier. They provide no short cut. Analysts should use video because it makes communication visible and potentially reveals behavior nested across levels in precarious and contested interactions. Video records capture what individuals seemingly attend to, talk about, and do with what is at hand, and they allow, more crucially, an analysis of how all this is arranged with the most locally demanding and collectively constructed constraints of time and space. In video analysis, time is transformed from a simple matter of then and now to a more reticular and reflexive then in anticipation of a now and now in response to a then, both occurring simultaneously. Every moment is a retrospective and prospective advance into the future; as George Herbert Mead ([1938]1972:65) confided, "The unit of existence is the act, not the moment." Space is similarly transformed analytically from a simple matter of here and there to a more embodied place for the negotiation of person and social relations. This is conceptually difficult. We set out to teach forgotten children school math and ran into a world—our world—of roadblocks recursively organized to keep the social production of knowledge, ignorance, and hierarchy hidden from view. By looking more closely, we made students disappear as problem learners in their own right and put in their stead, the problems of the immediate and future worlds they inhabit.

REFERENCES

Birdwhistell, R. (1970). *Kinesics and Context*. Philadelphia: University of Pennsylvania Press.

Chaiklin, S., & Lave, J. (Eds.). (1993). *Understanding practice*. Cambridge, England: Cambridge University Press.

Cicourel, A., Jennings, K., Jennings, S., Leiter, K., Mackay, R., Mehan, H., et al. (1974). *Language use and school performance*. New York: Academic Press.

Emerson, R. W. (1860). *The conduct of life*. New York: A. L. Burt.

Erickson, F. (2004). *Talk and social theory*. London: Polity Press.

Gilmore, P., & Glatthorn, A. (Eds.). (1982). *Children in and out of school*. Washington, DC: Center for Applied Linguistics.

Goldman, S. V. (1996). Mediating microworlds. In T. Koschmann (Ed.), *CSCL: Theory and practice of an emerging discipline* (pp. 45–81). Hillsdale, NJ: Lawrence Erlbaum Associates.

Goldman, S., Knudsen, J., & Latvala, M. (1998). *Engaging middle schoolers in and through real-world mathematics*. In L. Leutzinger (Ed.), *Mathematics in the middle* (pp. 129–140). Reston, VA: National Council of Teachers of Mathematics.

Goodwin, C., & Goodwin, M. (1992). Assessments and the construction of context. In A. Duranti & C. Goodwin (Eds.), *Rethinking context* (pp. 147–190). Cambridge, England: Cambridge University Press.

Goodwin, M. (1990). *He said she said*. Bloomington: Indiana University Press.

Greeno, J. G., & Goldman, S. (Eds.). (1998). *Thinking practices in mathematics and science learning*. Mahwah, NJ: Lawrence Erlbaum Associates.

Greeno, J., McDermott, R., Cole, K., Engle, R., Goldman, S., Knudsen, J., et al. (1999). Research, reform, and the aims of education. In E. Lagemann & L Shulman (Eds.), *Issues in education research* (pp. 299–335). San Francisco: Jossey-Bass.

Kendon, A. (1990). *Conducting interaction*. Cambridge, England: Cambridge University Press.

Kendon, A. (2004). *Gesture: Visible action as utterance*. Cambridge, England: Cambridge University Press.

Mead, G. H. ([1938]1972). *The philosophy of the act* (p. 65). Chicago: University of Chicago Press.

Mehan, H. (1979). *Learning lessons*. Cambridge, MA: Harvard University Press.

Mehan, H. (1996). The construction of an LD student. In M. Silverstein & G. Urban (Eds.), *Natural histories of discourse* (pp. 253–276). Chicago: University of Chicago Press.

Sacks, H. (1990). *Lectures on conversation*. Oxford, England: Blackwell.

Scheflen, A. E. (1974). *How behavior means*. New York: Anchor.

Schegloff, E. (1992). In another context. In A. Duranti & C. Goodwin (Eds.), *Rethinking context* (pp. 191–228). Cambridge, England: Cambridge University Press.

Varenne, H., & McDermott, R. (1998). *Successful failure*. Boulder, CO: Westview Press.

Wortham, S., & Rymes, B. (Eds.). (2002). *Linguistic anthropology of education*. New York: Praeger.

Epistemological Issues in the Analysis of Video Records: Interactional Ethnography as a Logic of Inquiry

Judith Green
University of California, Santa Barbara

Audra Skukauskaite
Texas A&M University, Corpus Christie

Carol Dixon
University of California, Santa Barbara

Ralph Córdova
Southern Illinois University, Edwardsville

Community

In our Tower community, we have our own language as well as the languages we bring from outside (like Spanish and English) which helped us make our own language. So, for example, someone that is not from our classroom community would not understand what insider, outsider, think twice, notetaking/notemaking, literature log and learning log mean. If Ms. Yeager says we are going to "make a sandwich," the people from another class or room would think that we were going to make a sandwich to eat. Of course we aren't, but that is part of our common language.

To be an insider, which means a person from the class, you also need to know our Bill of Rights and Responsibilities which was made by the members of the Tower community. And if Ms. Yeager said, "Leave your H.R.L. on your desk," people would not understand unless someone from the Tower community told him/her and even if we told him/her that H.R.L. stands for 'Home Reading Log,' they still would not understand what it is and what you write in it. If we told a new student, "It's time for SSL and ESL," he would not understand.

These words are all part of the common Tower community language and if someone new were to come in, we would have to explain how we got them and what they mean. We also would tell them that we got this language by reports, information, investigations, and what we do and learn in our Tower community.

—Arturo Zaragoza, 5th grade student, 1995

We open this chapter on interactional ethnography and the analysis of video records with an essay from Arturo, a fifth-grade student, engaging in a common practice in his bilingual class—writing a community essay at the end of the year. In this essay, Arturo, drawing on ethnographic language and concepts he learned across the year as he became an ethnographer in his class (Yeager, Floriani, & Green, 1998), inscribes a set of methodological and conceptual principles guiding the work of ethnographers seeking to study classrooms (and other social/institutional settings) as cultures (Collins & Green, 1992). He writes about insider/outsider (emic/etic) perspectives, part–whole relationships, ways of making extraordinary the ordinary, and taking a point of view (Green, Dixon, & Zaharlick, 2003), key elements of an interactional ethnographic approach. Throughout his essay, he uses contrastive analyses to make visible ways in which life in his classroom was socially constructed, local, and often invisible to outsiders who do not share the history, meanings, and language that members have in common (Edwards & Mercer, 1987; Lin, 1993).

Arturo, speaking as both an ethnographer and as a member of the class, makes visible the challenges facing ethnographers and others, who are seeking to enter and participate in an ongoing social group, or to interpret the patterns of life of a group recorded on video or other artifacts. Both need to uncover what members need to know, understand, produce, and predict (Heath, 1982) within and across times and events, and how this knowledge is local, situated, and constructed by members through their actions and interactions across times and events (Gee & Green, 1998; Santa Barbara Classroom Discourse Group, 1992a, 1992b). Arturo makes visible the challenges that we and other ethnographers face, when engaging in ethnographic studies of the social construction of knowledge, identity, disciplinary knowledge, literate practices, and social/academic access in classrooms (areas of interest to our research community). These challenges, like those facing outsiders or newcomers to the class, include identifying and understanding what counts as ways of communicating, knowing, being, and doing in the class or group within the class.

Although Arturo did not directly speak of how life in his class changes over time, other students wrote about the changing nature of classroom life. For example, in 1991, Alex, who entered sixth grade in the middle of the school year, stated that

> Our community has a lot to do over the year. Sometimes our community gets different during the year. What I mean is like the first day I walked in the door, I was new and nervous, just me thinking who am I, trying to make friends. I came in the door. Other students explained how to do the Writer's Workshop. I didn't understand the three logs. Other kids and the teacher explained. Now I'm just part of everyone else. (Green & Dixon, 1993, p. 235)

In this essay, Alex, like Arturo, speaks about how particular events and artifacts, constituting life in the classroom, are formulated and reformulated by the teacher and students. He also claims that to be just part of everyone else, the person entering needs to know what something is and how to take action in the events of classroom life. In other words, the ethnographer and the outsider need to learn from and with the members of the class.

In this chapter, we draw on theoretical concepts that Arturo and Alex make visible, to present the logic of inquiry guiding interactional ethnography as a theoretically driven approach that enables us to learn from the social and academic work of class members. Although our general approach to ethnographic research in classrooms is to work collaboratively with teachers and students over time (at least 1 year), we also use this approach to analyze the work of students and their teachers on video records (Castanheira, Crawford, Green, & Dixon, 2001) that are not part of an ethnographic corpus. In both types of studies, overtime ethnographic research and videobased studies, we use the ethnographic perspective as an orienting theory that enables us to learn from the work of actors inscribed on video records collected by other researchers (Green & Bloome, 1997). We also use an ethnographic perspective to guide secondary analysis within our ongoing ethnographic corpus in K–20 classrooms (1–12 years of data collection per teacher).

To illustrate the epistemological stance underlying this approach, we present steps taken in identifying, collecting, and analyzing video records from a 2-year study in Ralph Córdova's third- and fourth-grade classes. The specific problem we sought to understand occurred during a presentation that Mr. Córdova and his students in Santa Barbara made to a class in San Diego during a videoconference to share their science research projects. During this conference in which three of the authors of this chapter were part of the local audience, Mr. Córdova, in responding to the San Diego teacher's suggestion that they work together across time in the next year, agreed to her suggestion and also stated that working together for 2 years makes a big difference.

This statement puzzled us, because this was his first year in fourth grade and the first time that the two classes had met. A discussion with Mr. Córdova, who was also an ethnographer in his own class (Reveles, Córdova, & Kelly, 2004), led to an understanding that three of his students, who were participating in this videoconference, had been in his third- and fourth-grade classes and had worked on science projects both years. The new knowledge led to the study that we undertook in collaboration with Mr. Córdova to explore what counted as science to students in his class. Our goal was to

uncover the opportunities they had for learning science and to locate specific moments in which Mr. Córdova made visible to the students what it meant to become a scientist. We used the video record of the presentation as a way of triangulating the
opportunities afforded in the classroom with the view of science presented to the
community.

Mr. Córdova acted as a co-researcher and a cultural guide, making available artifacts that would enable us to trace the history of science teaching across the 2 years—for
example, the index of his data, his teacher's plan book for fourth grade, and his teaching
partner's plan books for third grade (he taught one day per week both years). His participation also provided a basis for ongoing conversations about the practices being uncovered and provided contextual information, enabling us to locate cycles of activity, and
intertextually tied events central to the onset of science in each year.

INTERACTIONAL ETHNOGRAPHY AS AN ORIENTING THEORY
AND A SET OF RESEARCH PRACTICES

Interactional ethnography is the approach developed by members of the Santa
Barbara Classroom Discourse Group over the past 15 years (Rex, 2006). This approach
integrates practice-centered theories of culture (e.g., Ortner, 1984) with discourse
analyses to examine how, over times and events, members of a social group (a class or a
group within the class) construct local knowledge (Geertz, 1983) and patterned ways
of communicating, knowing, being, and doing (e.g., Goodenough, 1981; Spradley,
1980) through the moment-by-moment interactions (Green & Dixon, 1993; Santa
Barbara Classroom Discourse Group, 1992a, 1992b). This approach involves two interrelated angles of analysis—one focusing on the discourse(s), social actions, accomplishments and outcomes at the level of the collective, and one focusing on individuals
within the collective, how they take up (or not) what is constructed at the collective
level, and how they use these material resources in subsequent events. Each can be the
primary focus; however, the two angles of analysis are complementary, each contributing to a part–whole relationship needed to obtain a fuller understanding of the interre-
lationship of collective and individual learning and development within classrooms
(Souza Lima, 1995).

From this perspective, what is captured on video records are the actors, their
words and actions within a developing cultural context, as well as visual texts related to
the physical spaces, objects, and graphic artifacts of the classroom. Theoretically, we
understand the actors to be texts for each other, not merely for the ethnographer
(Erickson & Shultz, 1981; McDermott, 1976). As such, they discursively and socially
signal to each other (and to us) what their actions mean, what counts as appropriate
and/or expected actions, and how these observed actions tie to prior and future activity
and knowledge (e.g., as indicated through verb tenses and/or direct references). They
also signal roles and relationships in the ways that members orient to and position with
each other, creating particular opportunities for identity formulations and take-up
(Heras, 1993) as well as access to academic knowledge and social participation. By examining chains of interactions about a particular topic or sequence of activity, we also
explore the agency of individual members as they read, interpret, and act on these

graphic, written, visual, and/or oral texts constructed at the collective level (Putney, Green, Dixon, Duran, & Yeager, 1999).

Building on Bakhtin (1986), we understand that the discourse among members makes visible speaker–hearer relationships central to our analysis. He argues that speakers speak with an implicated hearer and hearers listen with an implicated speaker. In other words, speakers take the audience into consideration when choosing what to say to whom, where and when, how, in what ways, and for what purpose(s). The hearer, whether the conversational partner, or an overhearing audience (Larson, 1995), that is, another group member or an interactional ethnographer, also takes into consideration what they know about the history of the event, their relationship with the speaker, and the topic under consideration, to interpret the meaning of the speaker's message as well as their possible intent. By analyzing chains of (inter)action and what is accomplished, we construct grounded arguments about intentions speakers and hearers signal to each other. From this perspective, what is captured on video records of classroom (and other institutional) life, are intentional actions among members of a sustaining social group.

Given this set of theoretical assumptions about the discursive construction of everyday life, to understand what is represented on video records (and in other artifacts—field notes, pictures, objects), interactional ethnographers engage in analyses at different levels of scale. Each level focuses on particular sets of actions, making it possible to uncover layers of co-occurring constructions by members—such as events, phases of activities, topically related sequences of interaction, turns, and actions (Green & Wallat, 1979, 1981). Analyses of these varying units provide a basis for identifying local constructions of identities for students and teacher(s), patterns of access to particular disciplinary and cultural knowledge and practices, and patterns of impact of policy decisions on the opportunities for learning afforded members of the class, among others (Dixon, Green, Yeager, Baker, & Franquiz, 2000). This analysis focuses on the developing collective level of scale, and does not make visible the full range of individual actions or take-up across time and events.

Analyses at the group or collective level provide a sketch map of the world members jointly construct (cf. Frake, as cited in Spradley, 1980), a broad picture of what members accomplish discursively across times and events. Depending on the level of scale of detail represented on the sketch map (e.g., event level through descriptions of sequences of actions), interactional ethnographers are able to uncover varying layers of structures and structuring practices intertextually produced and used by members to accomplish collective life in their class.

To explore the contributions of individual members and/or the impact of the collective actions on individual's within the collective's opportunities for learning, identity development, and other social and academic work, we shift the angle of analysis from the focus on the developing collective to a focus on individuals within the collective. This shift in angle of analysis provides a basis for tracing individuals' actions and discourse to identify what they take up and how they read, interpret, and use (or not) material resources and cultural practices of the class (Putney et al, 1999). To accomplish this microanalysis, we draw on a complementary set of theories of discourse; interactional sociolinguistics (Gumperz, 1986; Gumperz & Levinson, 1996), critical discourse analysis (Fairclough, 1995; Gee & Green, 1998; Ivanic, 1994) and microethnography (Bloome, Carter, Christian, Otto, & Shuart-Faris, 2004; Erickson, 1986). These discourse theories and approaches provide systematic and theoretically complementary ways of transcribing, interpreting, and representing the discursive work among members and what they accomplish.

Each theory entails a particular level of analysis, object of study, and potential for knowledge construction. Taken together, they provide theoretically driven ways of identifying how members discursively construct and take up the social, communicative, and referential systems and practices of life within a social group (Bloome et al., 2005; Green & Wallat, 1979, 1981). Central to this approach is the identification of a key speech (Gumperz, 1986) or academic event to anchor a series of contrastive analyses of the discourse and actions in prior or subsequent events. This form of contrastive analysis involves backward and/or forward mapping from a key event or anchor point (Putney et al., 1999). Using this approach, we trace the roots or routes of particular texts, topics, actions, concepts, and roles and relationships, among others, to construct a grounded interpretation of intertextual relationships and what members need to know, understand, produce, and predict to act as insiders of a social group.

These theories also enable us to shift focus from analyses of the intertextual nature of classroom life at a collective level to the analyses to how individual members take up and use language in the processes of learning, identity construction, and other social and academic events of classroom life. Together, the two angles of analyses—the group construction of the discourse and the individual discourse use and take up, construct a picture of where, when, and how knowledge was made available, what counts as knowledge, and how common knowledge and access to knowledge are socially constructed (e.g., Edwards & Mercer, 1987; Heap, 1980; 1991) both in the moment and over time. Juxtaposing the two levels of scale provides a way of identifying and analyzing how situated academic identities and academic access are constructed, and who can and does take up this access and these identities, for what purpose(s), under what conditions, and with what outcomes.

Analyses at Multiple Levels Scale: An Illustrative Case

As indicated previously, the question guiding the research process and the logic of inquiry we present in the remaining sections of this chapter was part of a larger study examining what counted as science in Mr. Córdova's third and fourth grade classrooms. The specific questions driving the process of identifying, collecting, and analyzing video records and related artifacts were: What are the opportunities for learning science afforded students across the 2 years? What data help us understand how Mr. Córdova's students might have interpreted his statement that "working together for 2 years makes a big difference?" Specifically, to clarify potential sources of our confusion, we sought to identify what occurred during the 2-year time period mentioned by Mr. Córdova, and who was part of these 2 years. Further, we explored how this time period was intertextually tied to the presentations the students were giving in the videoconference.

In framing these questions, we sought to turn the confusion (frame clash) we were experiencing into what Agar (1994) called a *rich point*, a place where culture happens. A rich point is both a physical (a point in time) and a discursive place where a person has an opportunity to learn about the others' viewpoint or cultural practices, and a place to learn through contrasting personal expectations with observed actions of others. Thus, we sought places on video records where Mr. Córdova made visible to students cultural practices and formulations of science projects. We view our research, data collection, and analyses as opportunities to learn from the discourse between Mr. Córdova and his students, the ways in which being a scientist was talked into being across time and events. Our approach to identifying potential records, from which we would construct data to be analyzed, entailed a series of interactive and responsive steps, similar to those used in a comprehensive ethnography.

We located the onset of the science projects through an exploration of both the index of the events of each day recorded on video (and other artifacts-fieldnotes) and the teacher's plan books for days on which an ethnographer was not present. We then reviewed the video records of the events identified to locate intertextual references across events. Using these purposefully and theoretically sampled events, we then engaged in a series of analyses at different levels of scale. These levels of scale are represented in Figure 8.1 by three types of event maps, each map representing a different time scale (2 years, one day, 4 months as well as subscales of months, minutes, and weeks; Castanheira, Crawford, Green, & Dixon, 2001; Green & Meyer, 1991). Guided by the questions—On what did members spend time? When was science?—we constructed these three event maps to represent different information that enabled us to locate part—whole relationships. Although these analyses often overlapped in time, we present them as a set of progressive disclosures.

The first map on Figure 8.1 represents the period of time in months and years constituting the science work referenced in the videoconference (2 years). On this map, we also locate the period of time of the actual projects for the videoconference, April through June 2003, creating one perspective on whole—part and part—whole relationships. The second map represents the events of April 21st and introduces a series of focused explorations, moving closer in time to the moments of discursive construction of the texts of events, and providing a representation of events constructed on that day. This map represents two different time scales, the day and the parts of the day. Thus, what counts as an event depends on the time scale being used, and its size is a theoretical decision within a particular level of analysis. An event, therefore, is a bounded series of actions, accomplished through a coordinated set of interactions, with coherence of content leading to the construction of a particular topic and/or purpose. An event can only be identified post hoc by observing changes in activity signaled referentially and accomplished by members (e.g., we are now doing social science, not mathematics); it is the product of textualizing work of members (Bloome et al., 2005). One way of viewing the relationship between the first two maps is that the map of April 21st was designed as a *swing out map*, one that, in photographic terms, enlarges the image in order to explore particular moments in more detail while maintaining its intertextual relationship to a larger time scale.

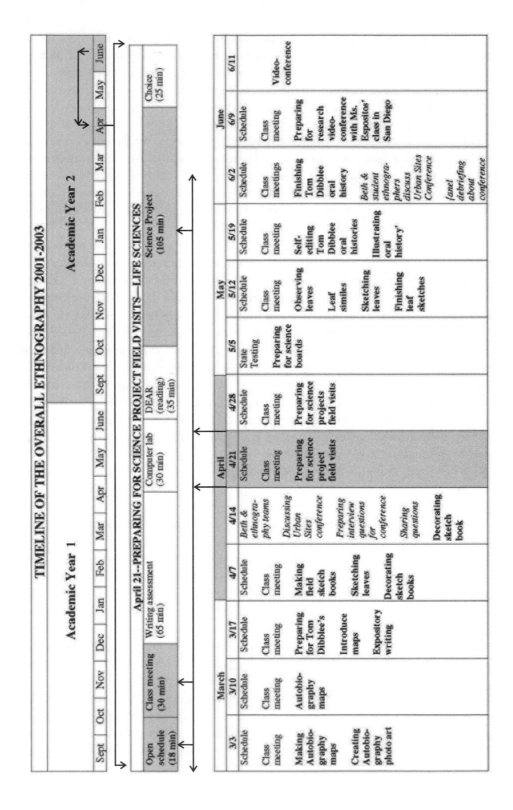

Figure 8.1. Time line situating frame clashes and analyses of the two years of science (Grades 3 and 4).

The third map represents still another time scale—the 4 months in which we identified direct references to the science content and/or practices related to the science projects that the students presented to the San Diego class. This map represents work in two areas of science and records the transition from Earth Science (3/3 through 4/14) to Life Science (4/21–6/11), both of which were represented in the research meeting of the two classes. This event map focuses on those events of the day in which students were engaged in actions (represented as verbs) associated with being scientists and learning science. The level of scale in this map serves to make visible resources members brought to, referenced, and used on April 21st and in the video-conference.

The arrows between the three maps indicate intertextual and intercontextual ties across levels of time scale. The use of maps at multiple levels of time scale is central to our work, because we seek to maintain levels of context and to show historical contexts (past, present, and future) that constitute an intertextual and intercontextual web of consequential progressions (Putney et al., 1999) of classroom life. Without the video records, the time scales would be lost, since written records are often inexact in regards to time and boundaries of events. Such boundaries are rarely recorded and require post hoc analysis to show the ways in which events are constructed through differentiated activity (Green & Wallat, 1979, 1981). Video records also provide a basis for moving across levels of scale to focus in on particular moments without losing the larger context on a given record or expanding levels of context across records. Figure 8.1, therefore, represents intertextual relationships at different levels of scale and how we use mapping to represent and locate particular analyses in times and spaces.

The logic of inquiry of the multiple levels of scale presented earlier suggests that the types of claims possible about knowledge construction, access, identity, and other outcomes of socially constructed dimensions of classroom life are limited when one analyzes a single event of classroom life and/or a single moment captured on video. What is needed to make larger claims about these phenomena are records collected over time and events that permit an exploration of intertextual relationships that members signal as socially significant.

Transcribing as a Basis for Grounded Interpretations: Mapping Discursive Work

The next series of steps involved transcribing the work of the teacher and students to construct local, situated (re)presentations of what counted as being a scientist in the Life Science projects. We view transcription as a form of mapping, one that represents particular relationships among speakers and units of discourse. Building on Ochs (1979), we view transcription as theory that represents our ontological view of classroom life as discursively constructed in the moment-by-moment and over-time interactions among members. The ways in which we represent the discursive work of members shapes the text that we then read and interpret at another level of scale.

The basic unit of transcribing that we use is the message unit. A message unit is a minimal burst of talk that provides social information (Green & Wallat, 1981).

Central to the identification of this unit is Gumperz' (1982) argument that prosodic cues (e.g., pitch, stress, pause, juncture, intonation contours, eye gaze, proxemics and kinesics) provide contextual information about the meaning and intentions of speakers. Message units, like Lego blocks, are bounded units, which when combined together create larger patterns of interaction and units of work action, turns at talk, interaction units, sequences of tied interactions about a single topic, phases of activity accomplished through tied sequence units, and events themselves (e.g., Bloome et al., 2005; Green & Wallat, 1979, 1981; Putney, 1997). This approach uses contextualization cues as a means of making visible and representing multiple levels of speech, text construction, and activity members construct as they read, interpret, and respond (or not respond in the moment) to what is being proposed by a speaker. This discourse analysis approach requires the transcriber to take a listener/hearer's perspective.

The ethnographer/discourse analyst identifies the ebb and flow of the talk and the ways in which the members, through this talk, draw on cultural practices of the group (e.g., discourse and social processes) to contribute to the text being constructed. By exploring how members are constructing these different levels of units, the discourse analyst/interactional ethnographer seeks to identify the work among members to construct a collective text. As Fairclough (1992) argues,

> … discourse constitutes the social. Three dimensions of the social are distinguished—knowledge, social relations, and social identity—and these correspond respectively to three major functions of language. Discourse is shaped by and shapes relations of power, and is invested with ideologies, both historical and constructed through local texts. (p. 8)

To illustrate the social nature of discourse and the constructed nature of these texts as well as their interpretation, we progressively disclose a segment of transcript from the Science event on the map of events on April 21 (Fig. 8.1). Through this progressive unfolding, we illustrate the layers of intertextual work the students and teacher undertook at the collective level, and how individuals contributed to this, creating a particular ideology of science practices and a local set of roles and relationships.

Scene: T shifts orientation from talking to a coach who had entered to take a group of students out of the class for testing to talking to the whole class sitting at their desks.

1. okay
2. so you are right
3. when you talked about asking questions
4. about la paloma
5. cuz you've been there
6. you have a little bit of history
7. about that place

This transcript begins with two chains of tied message units in which the teacher addressed the whole class (lines 1–5 and lines 6–7). These chains constitute action units through which the teacher reoriented students and reestablished the topic under discussion prior to the interruption. These units (lines 6–7) also provided new information that signaled to students an intertextual link between the question being asked and their prior experience, obtained from a visit to one of the three field-research sites.

The next chain of actions of the teacher took was also initiated by the use of okay, marking another transition point. Whereas the first shift was a physical one, leading to a reformulation of the topic under discussion, the second shift was a conceptual one. In both instances, the use of okay served to signal a shift. In lines 9–12, he proposed actions he wanted the students to take—to develop some themes, some questions, some general themes.

8. okay
9. i want you to develop
10. some themes
11. for some questions
12. some general theme
13. what's an example of a theme
14. that you might want to think about having
15. when we go to these places
16. we're all going to have the same themes
17. i want you to know that

Lines 13–15 elicited examples of themes, tying them to the places to be studied. Lines 16–17 specified conditions under which students would work—all students will have the same themes. He then went on to let them know what he expected from them as they worked in their small groups (one for each site to be studied).

18. so i want you all to participate
19. in coming up with the themes
20. what theme
21. that you want us
22. to keep in mind
23. when we go to one of those places
24. to research it

In this sequence, he restated and elaborated the actions that they needed to take and how they were to think about themes that they, the team, wanted to research at one of the sites. At this point, the students self-selected a site to study, creating three interest groups, one for each site. The teacher's actions and discourse made visible the common actions that each self-selected group needed to take—identify themes that will be common and then think about how the theme will be studied at the selected site.

At this point, individual students nominated potential themes, without direct question or designation of speakers by the teacher.

25.   Brady to T (class as overhearing audience): what kind of plants do you have
26.   T: okay
27.   so one of them is plant life
28.   that touches this concept
29.   so that's wonderful
30.   so how can we ask
31.   well
32.   we'll come up with questions later
33.   so plant life at that particular habitat
34.   right brady
35.   (inaudible)

In this chain of actions, the teacher accepted Brady's question, signaling a student's right to propose ideas. The idea was acknowledged (lines 25–26), and then used as a basis for elaboration, creating an intertextual tie between a graphic chart of concepts that students identified from the state standards in an earlier discussion (lines 27–29 and lines 33–35). Between these two actions focusing on theme–concept–standards relationships, the teacher began to ask a question but aborted this course of action, indicating that questions will come later, and that the current focus was on developing general themes.

His next action signaled continuity in task, while summarizing and reformulating what had been proposed and accepted. He then requested another "one" (theme) and called on Kari, whose hand was raised.

36.   so far we have plants
37.   that's good
38.   what's another one
39.   (St raises hand)
40.   Kari

In this brief chain of discursive work, less than 1 minute of classroom life, we identified a range of levels of work by the teacher: to reorient students to a common interactional space after an interruption, to re-establish the topic on the floor, to make visible to the group how an individual's proposal fit within the topic being constructed, and how his own talk about questions was not the current focus but would be the topic of actions later.

In this segment, we were also able to see how two different students took up the proposed activity without the teacher needing to distribute turns. The first student, Brady, read the implied invitation to contribute and proposed a question, when the teacher ceased speaking. That his contribution was given at an appropriate time and in an appropriate way was indicated by the teacher's take up (Collins, 1987) of the student's proposed topic. The teacher's take up however, while accepting the student's right to contribute, reformulated the contribution so that it met his intention to generate themes, not questions at this point in time. The second student, Kari, can be seen as acting in a consistent manner, when she raised her hand in response to the teacher's request for another theme. In this brief segment, we demonstrate the join construction of task, content, and social relationships. Further, the analysis shows that the students were not mere receivers of information; rather, their actions show that they read the social, academic, and communicative expectations and demands and responded in an appropriate manner.

Table 8.1 presents a representation of the chains of actions on April 21st taken by the teacher and students in the three events in which science practices and actions were the focus of discursive work. As indicated in this table, each event had a particular focus and each subsequent event assumed the knowledge and actions of the previous ones, creating a horizontal (sequential) and vertical (nonsequential) set of intertextual relationships that together constituted what counted as science on this day. Our over-time analysis represented in Figure 8.1 shows a similar pattern in which earlier events form a part of a consequential progression across times in the classroom. These progressions can be seen as building expanding understandings and uses of science practices and content.

Interactional Ethnography as a Descriptive Language: Some Final Comments

The dynamic relationships identified in this study make visible how opportunities for learning, identity formulation, and take up, and access to academic knowledge are not located at one point in time nor are they the purview of a single individual. Rather, the collective and individuals within the collective construct them over time and over events. In presenting the different levels of scale and the rationale for each, we made visible how an interactional ethnographic perspective leads to the construction of a particular logic of inquiry that is grounded in understanding and uncovering the actions of members of a social group as they jointly construct the everyday events of life within the group.

The description of the steps constituting the logic of inquiry created for this set of analyses demonstrates how discourse analyses, grounded in an ethnographic perspective and data corpus, enable the researcher to learn from the recorded actions of members of a social group. These steps illustrate how the juxtaposition of analyses at different levels of scale made visible intertextual relationships constituting classroom life. Such relationships are proposed, socially accomplished, recognized and acknowledged as socially significant by members of a group (Bloome & Egan-Robertson, 1993)

TABLE 8.1
Science in Three Parts of the Day: An Example of Vertical Intertextuality

Overviewing Schedule 18 minutes (8:20–8:38) students sitting in desk groups	*Conducting class meeting 30 minutes (8:38–9:09) students in oval on floor*	*Preparing for field research 105 minutes (1:25–2:30) students in site groups*
indicating sequence of activities in class	**eliciting** ideas from students on ways of organizing groups	**establishing** boundary of event to come
contextualizing an action to follow within the day	**foreshadowing** ways of organizing into groups	**inscribing** sketchbook as artifact for science project
Telling students importance of science project	**discussing** responsibility for project	**contextualizing** project in moment & over time
requesting students listen to something important	**Inscribing/discussing** three places	**reformulating** project as complex
giving date for future events—open house & science fair	**inscribing** a list of work involved in the fieldtrips	**reformulating** project as multi-layered
placing today's work in sequence of work over time	**discussing** doubts/questions adult leaders in each place have about student ability, maturity & preparation	**establishing** listening & attending as conditions for learning complex ideas
formulating the need to know what is expected	**signaling** actions as discipline specific, not site specific	**inviting** to student to remember three places
reminding class mrs. B already mentioned science projects	**discussing** how site determines questions to be asked	**listing** three places
presenting district requirement: every 4th grader must do a science project	**tying** actions inscribed to the spaces for the actions to student identities as researchers	**reformulating** 3 places as "out in community"
linking 4th grade requirement intercontextually to project undertaken with mr. C & mrs. H in 3rd grade		**asking** student for attention (first time individual singled out to take up expected action)

engaging students in thinking about prior work in "earth science projects"

contextualizing the notion of "concept" through prior work (earth science)

introducing life science concepts by reviewing earth science concepts

sharing 4th grade district requirements for learning science

linking what students know to requirements

inscribing purposes and expect outcomes for life science projects

guiding class in brainstorming questions & themes to investigate in three places

engaging students in small group in identifying potential questions & themes

inviting groups to share questions/themes with class

---

invoking student expertise

reformulating intercontextual and intertextual rues between 4th grade requirement & science in 3rd grade

reformulating request to "listen" and "don't interrupt" as mr.c introduces the science project

defining what is meant by "it's much bigger" (earlier reference)

inscribing three places graphically on the board

naming students as "researchers"

specifying what researchers do and prepare to do

reformulating "three places" as "three different places"

specifying science projects as "field research"

defining field research as involving particular activities: interviewing people, collecting data, collecting plant samples.

defining area of field research—are plants indigenous?

stating an outcome: learning about native plants

and are consequential for constructing local, situated practices, and content of classroom life within and across moments, events as well as across events.

The analyses at different levels of scale underlying the interactional ethnographic perspective to data collection, analysis, interpretation, and representation, also made visible that the challenges facing ethnographers who seek to understand the actions, interactions, and their outcomes in classrooms as cultures, are not unlike the challenges facing students. Like Arturo and Alex, ethnographers face daily challenges of gaining access to and interpreting the insider discourse and actions through which academic and social texts are constructed by the group as well as by individuals. As the essays and analyses in this chapter demonstrate, interactional ethnography, with its theoretically driven principles of practice, provides a descriptive language and a systematic approach to examining social construction of life in classrooms. Viewed in this way, interactional ethnography is a resource for students, teachers, ethnographers, and other researchers, that makes visible the often invisible work, accomplishments, and consequences of classroom life for both students and teachers.

REFERENCES

Agar, M. (1994). *Language shock: Understanding the culture of conversation*. New York: Morrow.

Bakhtin, M. M. (1986). *Speech genres and other late essays*. Austin: University of Texas Press.

Bloome, D., Carter, S. P., Christian, B. M., Otto, S., & Shuart-Faris, N. (2005). *Discourse analysis and the study of classroom language and literacy events: A microethnographic perspective*. Mahwah, NJ: Lawrence Erlbaum Associates.

Bloome, D., & Egan-Robertson, A. (1993). The social construction of intertextuality in classroom reading and writing lessons. *Reading Research Quarterly, 28*(4), 304–334.

Castanheira, M., Crawford, T., Dixon, C., & Green, J. (2001). Interactional ethnography: An approach to studying the social construction of literate practices. *Linguistics and education: Analyzing the discourse demands of the curriculum, 11*(4), 353–400.

Collins, E., & Green, J. L. (1992). Learning in classroom settings: Making or breaking a culture. In H. Marshall (Ed.), *Redefining student learning: Roots of educational restructuring* (pp. 59–85). Norwood, NJ: Ablex.

Collins, J. (1987). Using cohesion analysis to understand access to knowledge. In D. Bloome (Ed.), *Literacy and schooling* (pp. 67–97). Norwood, NJ: Ablex.

Dixon, C., Green, J., Yeager, B., Baker, D., & Franquiz, M. (2000). "I used to know that": What happens when reform gets through the classroom door. *Bilingual Education Research Journal, 24*(1&2), 113–126.

Edwards, D., & Mercer, N. (1987). *Common knowledge: The development of understanding in the classroom.* New York: Falmer.

Erickson, F. (1986). Qualitative research. In M. Wittrock (Ed.), *The handbook of research on teaching* (3rd ed., pp. 119–161). New York: Macmillan.

Erickson, F., & Shultz, J. (1981). When is a context? Some issues and methods in the analysis of social competence. In J. L. Green & C. Wallat (Eds.), *Ethnography and language in educational settings* (pp. 147–150). Norwood, NJ: Ablex.

Fairclough, N. (1992). *Discourse and social change.* Cambridge, England: Polity Press.

Fairclough, N. (1995). *Critical discourse analysis: The critical study of language.* London: Longman.

Gee, J., & Green, J. (1998). Discourse analysis, learning, and social practice: A methodological study. *Review of Research in Education, 23,* 119–169.

Geertz, C. (1983). *Local knowledge: Further essays in interpretive anthropology.* New York: Basic Books.

Goodenough, W. (1981). *Culture, language, and society*. Menlo Park, CA: Cummings.

Green, J., & Bloome, D. (1997). Ethnography and ethnographers of and in education: A situated perspective. In J. Flood, S. Heath, & D. Lapp (Eds.), *Handbook of research on teaching literacy through the communicative and visual arts* (pp. 181–202). New York: Simon & Schuster Macmillan.

Green, J., & Dixon, C. (1993). Introduction to special issue, "Talking knowledge into being: Discursive and social practices in classrooms." *Linguistics and Education, 5*(3&4), 231–239.

Green, J. L., Dixon, C. N., & Zaharlick, A. (2003). Ethnography as a logic of inquiry. In J. Flood, D. Lapp, J. R. Squire, & J. M. Jensen (Eds.), *Handbook of research on the teaching of the English language arts* (2nd ed., pp. 201–224). Mahwah, NJ: Lawrence Erlbaum Associates.

Green, J. L., & Meyer, L. A. (1991). The embeddedness of reading in classroom life. In C. Baker & A. Luke (Eds.), *Towards a critical sociology of reading pedagogy* (pp. 141–160). Philadelphia: John Benjamins.

Green, J. L., & Wallat, C. (Eds.). (1981). *Ethnography and language in educational settings*. Norwood, NJ: Ablex.

Green, J. L., & Wallat, C. (1979). What is an instructional context? An exploratory analysis of conversational shifts across time. In O. Garnica & M. King (Eds.), *Language, children, and society* (pp. 159–174). New York: Pergamon.

Gumperz, J. (1982). *Discourse strategies*. Cambridge, England: Cambridge University Press.

Gumperz, J. (1986). Interactive sociolinguistics on the study of schooling. In J. Cook-Gumperz (Ed.), *The social construction of literacy* (pp. 45–68). New York: Cambridge University Press.

Gumperz, J., & Levinson, S. (1996). *Rethinking linguistic relativity*. Cambridge, England: Cambridge University Press.

Heap, J. L. (1991). A situated perspective of what counts as reading. In C. D. Raver & A. Luke (Eds.), *Towards a critical sociology of reading pedagogy* (pp. 103–139). Philadelphia: John Benjamin.

Heap, J. L. (1980) What counts as reading: limits to certainty in assessment. *Curriculum Inquiry, 10*(3), 265 292.

Heath, S. B. (1982). Ethnography in education: Defining the essentials. In P. Gillmore & A. Glatthorn (Eds.), *Children in and out of school: Ethnography and education* (pp. 33–55). Washington, DC. Center for Applied Linguistics.

Heras, A. I. (1993). The construction of understanding in a sixth grade bilingual classroom. *Linguistics and Education, 5*(3&4), 275–299.

Ivanic, R. (1994). I is for interpersonal: Discoursal construction of writer identities and the teaching of writing. *Linguistics and Education, 6,* 3–15.

Larson, J. (1995). Talk matters: The role of pivot in the distribution of literacy knowledge among novice writers. *Linguistics and Education, 7*(4), 277–302.

Lin, L. (1993). Language of and in the classroom: Constructing the patterns of social life. *Linguistics and Education, 5*(3&4), 367–409.

McDermott, R. P. (1976). *Kids made sense: An ethnographic account of the interactional management of success and failure in one first-grade classroom*. Unpublished doctoral dissertation, Stanford University, Stanford, CA.

Ochs, E. (1979). Transcription as theory. In E. Ochs &. B. B. Schefflin (Eds.), *Developmental pragmatics* (pp. 43–72). New York: Academic Press.

Ortner, S. B. (1984). Theory in anthropology since the sixties. *Society for Comparative Study of Society and History, 26,* 126–166.

Putney, L. (1997). *Collective-individual development in a fifth grade bilingual classroom: An interactional ethnographic analysis of historicity and consequentiality*. Unpublished dissertation, University of California at Santa Barbara.

Putney, L. G., Green, J. L., Dixon, C., Duran, R., & Yeager, B. (1999). Consequential progressions: Exploring collective–individual development in a bilingual classroom. In C. D. Lee, P. Smagorinsky, R. Pea, J. S. Brown, & C. Heath (Eds.), *Vygotskian perspectives on literacy research: Constructing meaning through collaborative inquiry* (pp. 86–126). New York: Cambridge University Press.

Reveles, R., Córdova, R., & Kelly, G. (in press). Science literacy and academic identity formulation. *Journal of Research on Science Teaching, 41*(10), 1111–1144..

Rex, L. A. (Ed.). (2004). *Discourse of opportunity: How talk in learning situations creates and constrains—Interactional ethnographic studies in teaching and learning.* Cresskill, NJ: Hampton Press.

Santa Barbara Classroom Discourse Group. (1992a). Constructing literacy in classrooms: Literate action as social accomplishment. In H. Marshall (Ed.), *Redefining learning: Roots of educational restructuring* (pp. 119–150). Norwood, NJ: Ablex.

Santa Barbara Classroom Discourse Group. (1992b). Do you see what we see? The referential and intertextual nature of classroom life. *Journal of Classroom Interaction, 27*(2), 29–36.

Souza Lima, E. (1995). Culture revisited: Vygotsky's ideas in Brazil. *Anthropology & Education Quarterly, 26*(4), 443–457.

Spradley, J. (1980). *Participant observation.* New York: Holt, Rinehart & Winston.

Yeager, B., Floriani, A., & Green, J. (1998). Learning to see learning in the classroom: Developing an ethnographic perspective. In D. Bloome & A. Egan-Robertson (Eds.), *Students as inquirers of language and culture in their classrooms* (pp. 115–139). Cresskill, NJ: Hampton Press.

9

The Video Analyst's Manifesto (or The Implications of Garfinkel's Policies for Studying Instructional Practice in Design-Based Research)

Timothy Koschmann
Southern Illinois University

Gerry Stahl
Drexel University

Alan Zemel
Drexel University

STUDIES OF PRACTICE AND DESIGN-BASED RESEARCH

Over the past 10 years, there has been a turn away from conventional approaches to designing innovative curricula toward what has been described as "design experiments" (Brown, 1992; Cobb, Confrey, diSessa, Lehrer, & Schauble, 2003) or "design-based research" (Barab & Squire, 2004; Fishman, Marx, Blumenfeld, Krajcik, & Soloway 2004; Design-Based Research Collective, 2003). Cobb et al. (2003) described this shift in these terms:

Prototypically, design experiments entail both "engineering" particular forms of learning and systematically studying those forms of learning within the context defined by the means of supporting them. This designed context is subject to test and revision, and the successive iterations that result play a role similar to that of systematic variation in experiment. (p. 9)

Design-based research has been characterized as exhibiting the following five features:

> First, the central goals of designing learning environments and developing theories or "prototheories" of learning are intertwined. Second, developments and research take place through continuous cycles of design, enactment, analysis, and redesign …. Third, research on designs must lead to sharable theories that help communicate relevant implications to practitioners and other educational designers …. Fourth, research must account for how designs function in authentic settings. It must not only document success or failure but also focus on interactions that refine our understanding of the learning issues involved. Fifth, the development of such accounts relies on methods that can document and connect processes of enactment to outcomes of interest. (Design-Based Research Collective, 2003, p. 5)

However, how do we go about studying forms of learning within a "designed context?" How do we account for "how designs function in authentic settings?" What are the methods by which we might "document and connect processes of enactment to outcomes of interest?" And, finally, but most crucially, where do we turn for a vocabulary that will allow us to express our "prototheories of learning" in terms that are useful both to "practitioners and other educational designers?" All of these questions point to the need for conducting fine-grained studies of instructional practice and this is the role that video-analytic research is presumably designed to fill. But this still leaves unanswered the biggest question of all: How do we go about systematically and rigorously studying practice?

For an answer, we must turn to the disciplines in the social sciences, namely sociology, anthropology, communication studies, that explicitly take up practice as a legitimate object of study. One particularly promising line of inquiry originating within sociology is the tradition known as Ethnomethodology (EM). EM is centrally concerned with practical reasoning and the procedures (i.e., "methods") that participants (i.e., "members") routinely employ when making sense of their own actions and the actions of others, that is in producing what could be called "local rationality" (Heap, 1995). EM's concern with sense making makes it a natural framework for undertaking a study of instructional practice. Instruction and instructability have in fact been featured topics in ethnomethodological research from its earliest days.[1] Conversation Analysis (CA), the largest and best-known of ethnomethodology's various programs of study, has focused on how participants produce sense and order within talk-in-interaction. It has generated a large and well-articulated body of literature (Heritage, 1984, 1995; Sacks, 1992;). Research in CA employs specialized transcription conventions

---

[1]The second half of Garfinkel's (2002) recent book on EM is devoted to an extended discussion of "instructed action."

and a logic of "analytic induction" (Heritage, 1995) that could be productively applied to the study of instructional practice.[2]

There are many ways of approaching the task of analyzing video materials. Because of its central concern with practical reasoning and meaning-making practices, EM would seem to present a useful disciplinary foundation for building a systematic and rigorous program of video-analytic research related to design experiments. This chapter, therefore, explores what it might mean to do ethnomethodologically informed video analytic research. To do so, we turn to some of the early writings of the founder of the field, Harold Garfinkel. We have labeled this chapter a manifesto, partly to be provocative, but also because that precisely describes what the chapter strives to achieve (i.e., "a public declaration of policy and aims," *Oxford Modern English Dictionary,* 2nd ed., p. 606).

GARFINKEL'S POLICIES

Garfinkel (1967) defined ethnomethodological research in terms of five policy statements. For expository purposes, we have attached a name (e.g., indifference, relevance, indexicality) to each policy statement, summarizing what we took to be a defining feature of the policy. Garfinkel's policies are densely worded and, although presented as five independent items, are complexly interconnected and overlap considerably in their scope. Indeed, some of his policies (e.g., indexicality, contingently achieved accomplishment) have a fractal character in that they can be read to encapsulate or summarize some or all of the other policies. The policies are less practical specifications for doing video-analytic work than prescriptive or normative guides for how to document a practice in a valid way. For each policy statement, we have endeavored to explain the implications of the policy for video-analytic research.

Policy 1: Indifference

An indefinitely large domain of appropriate settings can be located if one uses a search policy that any occasion whatsoever be examined for the feature that "choice" among alternatives of sense, of facticity, of objectivity, of cause, of explanation, of communality *of practical actions* is a project of members' actions. Such a policy provides that inquiries of every imaginable kind, from divination to theoretical physics, claim our interest as socially organized artful practices. (Garfinkel, 1967, p. 32)

This aspect of EM research is sometimes described as *ethnomethodological indifference*. This is not meant to suggest a lack of care or interest on the part of EM researchers, but rather to observe that any instance of social action is as good as any other for the purposes of understanding how social action is organized. EM is concerned with the practices people engage in to make sense of their own and other's activities.

---

[2]We do not provide a description of the specifics of this methodology here but instead refer the interested to reader to any of the several available introductory guides (e.g., ten Have, 1999; Hutchby & Woolfit, 1998; Psathas, 1995).

Because human interaction always constructs meaningful order, the EM researcher can analyze almost any interaction and discover interesting processes of meaning construction and order negotiation. An EM researcher has great latitude, therefore, in selecting settings in which to do analysis. In particular, any circumstance, situation or activity that participants treat as one in which instruction and learning is occurring can be investigated for how instruction and learning are being produced by and among participants.

Ethnomethodology's policy of indifference stands in stark contrast to the assumptions underlying conventional experimental research in education. Experimental designs based on statistical models depend on acquiring a sufficiently large sample of instances before any valid inferences can be drawn. The policy of indifference not only suggests that any instance will do for the purpose of demonstrating some phenomenon of interest, but also that such a demonstration can be based on a single case.

Policy 2: Contingently Achieved Accomplishment

> Members to an organized arrangement are continually engaged in having to decide, recognize, persuade, or make evident the rational ... character of such activities of their inquiries as counting, graphing, interrogation, sampling, recording, reporting, planning, decision-making, and the rest .... "[A]dequate demonstration," "adequate reporting," "sufficient evidence," "plain talk," "making too much of the record," "necessary inference," "frame of restricted alternative," in short, every topic of "logic" and "methodology," including these two titles as well, are glosses for organizational phenomena. These phenomena are contingent achievements of organizations of common practices, and as contingent achievements they are variously available to members as norms, tasks, troubles. (Garfinkel, 1967, pp. 32–33)

Garfinkel elaborated that by *rational character,* he meant, "coherent, or consistent, or chosen, or planful, or methodical, or knowledgeable" (p. 32). The imputed sense or meaning of an action or of a sequence of actions is not determinate, however, but is instead endlessly open to new interpretation. Actions produce their own sense since they are designed in their achievement to be recognizable as what they are. The "interpretation" or "recovery" of that sense rests on co-actors' abilities to induce or infer their sense from actions themselves as they are performed/achieved locally in the circumstances of their production.

Garfinkel described this notion of inferring sense from actions as the "documentary method of interpretation" (p. 78). Garfinkel credits the expression to Mannheim, although his usage goes quite beyond that developed by Mannheim (1968, pp. 53–63). In using this expression, Mannheim was addressing the problem of how to interpret the "motivated character" (Garfinkel, 1967, p. 95) of observed action. As Garfinkel described,

> The method consists of treating an actual appearance as "the document of," as "pointing to," as "standing on behalf of" a pre-supposed underlying pattern. Not only is the underlying pattern derived from its individual documentary evi-

dences, but the individual documentary evidences, in their turn, are interpreted on the basis of "what is known" about the underlying pattern. Each is used to elaborate the other. (p. 78)

For Garfinkel, the documentary method is "a convenient gloss for the work of local, retrospective–prespective, proactively evolving ordered phenomenal details of seriality, sequence, repetition, comparison, generality, and other structures" (Garfinkel, 2002, p. 113). It is a ubiquitous and unavoidable feature of all social interaction. Actions must provide for their own recognizability as instances of some broader category of behavior.

Another implication of the documentary method of interpretation is that actors are selective in what they treat as relevant. As Heritage (1984) explained, "The task of fellow-actors … is necessarily one of *inferring* from a fragment of the other's conduct and its context what the other's project is, or is likely to be" (p. 60). In other words, actors are faced with what others are doing and must select which fragments of the other's conduct and its context to consider in inferring the sense of the actions under consideration. The only requirement that actors themselves place on their sense making is that it be adequate for the purposes at hand. Meaning, therefore, is "a contingent accomplishment of socially organized practices" (Garfinkel, 1967, p. 33).

Instruction is, to use Durkheim's term, a "social fact." It is an organized arrangement constructed in all its details to be recognizable as what it is. Its recognizability is a contingently achieved accomplishment of the members to the activity, that is the teacher and her/his students. This aspect of instruction is sometimes overlooked in educational research that treats a curricular innovation as something that has an existence outside of what actually takes place in classrooms. Instead, it is something that is produced through the moment-to-moment interactions of the members in the setting. This is not to deny that teachers do have teaching plans or to suggest that there cannot be abstract models of effective instruction. When we speak of the "designed context" (Cobb et al., 2003, p. 9) of instruction, however, we must be ever mindful of its status as a contingent achievement (see discussion under Policy 5 of how action is "doubly contextual"). The task before the analyst remains one of adequately accounting how participants actually *do* the teaching plan or *do* the instructional model as ongoing interactional achievements.

This policy highlights the need for sequential analysis of interaction (cf. Sacks, 1992), that is an analysis of how instruction is produced within an unfolding stream of interaction. Such an analysis would constitute a description of the unfolding determinate sense of the situation that members construct through their actions. Video technology is a valuable tool for performing a sequential analysis of a contingently achieved activity. It enables the analyst to inspect the interaction at a much higher level of detail than could ever be achieved through direct participant observation. Video also provides for repeated inspectability of the recorded materials. As Schegloff (as quoted in Prevignano & Thibault, 2003) argued, "These days, only such work as is grounded in tape (video tape where the parties are visually accessible to one another) or other repeatably (and intersubjectively) examinable media can be subjected to serious comparative and competitive analysis" (p. 27f). The use of video-based materials also provides a means for satisfying McDermott, Gospodinoff, and Aron's (1978) criterion for "ethnographically adequate description," namely that such a description "be presented in a way that readers can decide for themselves whether or not to believe the ethnographer's account of what it is that a particular group of people is doing at any given time" (p. 245).

Policy 3: Relevance

A leading policy is to refuse serious consideration to the prevailing proposal that efficiency, efficacy, effectiveness, intelligibility, consistency, planfulness, typicality, uniformity, reproducibility of activities—i.e., that rational properties of practical activities-be assessed, recognized, categorized, described by using a rule or a standard obtained outside actual settings within which such properties are recognized, used, produced, and talked about by settings' members. (Garfinkel, 1967, p. 33)

In developing Policy 2, Garfinkel wrote,

It is not satisfactory to describe how actual investigative procedures, as constituent features of members' ordinary and organized affairs, are accomplished by members as recognizably rational actions *in actual occasions* of organization circumstances by saying that members invoke some rule by which to define the coherent or consistent or playful, i.e., rational, character of their actual activities." (author's italics, pp. 32–33)

He went on, "Nor is it satisfactory to propose that the rational properties of members' inquiries are produced by members' compliance to rules of inquiry" (p. 33). He was railing against the practice in formal sociological analysis of treating the analytic subject as a "judgmental dope" (p. 70) acting in ways determined by the subject's status as a student, teacher, a gendered person, learning disabled, low achieving, language impaired, and so forth. For Garfinkel, it is not acceptable to offer descriptions that depend on analyst-imposed categories as accounts for what participants do or don't do.

Wittgenstein (1953) had already demonstrated the incoherence of treating social practices as a matter of following culturally defined rules. Tacit practices and group negotiations are necessary at some level to put rules into practice, if only because the idea of rules for implementing rules involves an impossible recourse. Although there is certainly order in social interactions of which people are not explicitly aware but that can be uncovered through microanalysis, this order is an interactive accomplishment of the people participating in the interactions. While the order has aspects of rationality and meaning, it is not the result of simply invoking or complying with a determinate rule. Consider, for instance, the orderliness of traffic flows at stop signs. The smooth functioning in accordance with traffic laws is continuously negotiated with glances, false starts and various signals. Although we do not usually explicitly focus on how this is accomplished unless we take on an analyst's perspective (because explicit awareness is not usually necessary for achieving the practical ends), the signs that are exchanged are necessarily visible to the participants and accordingly accessible to a researcher with appropriate means of data capture.

Methodologically, Policy 3 calls for "bracketing out" our pre-existing theories and interpretations while constructing our analyses. The introduction of categories to account for behaviors should only take place when we can empirically demonstrate their "relevance" as evidenced by the talk and activities of the participants. Schegloff (Prevignano & Thibault, 2003) specified, "The most important consideration, theoretically speaking, is (and ought to be) that whatever seems to animate, to preoccupy, to shape the interaction *for the participants in the interaction* mandates how we do our work, and what work we have to do" (p. 25). Schegloff (1991) described the ethnomethodological notion of relevance when he wrote:

> There is still the problem of showing from the details of the talk or other conduct in the materials that we are analyzing that those aspects of the scene are what the parties are oriented to. For that is to show how the parties are embodying for one another the relevancies of the interaction and are thereby producing the social structure. (Schegloff, 1991, p. 51)

The important point for conducting an ethnomethodologically informed analysis (video-based or otherwise) is that it is up to the members themselves to work out through their interaction what is to be treated as relevant and it is the task of the analyst to *discover* what these relevancies might be. This has clear implications for the study of instructional practices in the context of design experiments. Whether or not a situation is an instance of learning and instruction or of successful innovation is not a matter for curricular designers or program evaluators to judge a priori, but for video analysts to demonstrate in their empirical analysis of what the participants took their own activities to be. Meaning, therefore, is not a matter for the participants to address in post hoc surveys, interviews, or focus groups either; for retrospective rationalizations are not the same as the sense making that is produced in situ. It is up to the video analysis to locate and document what counts as learning and instruction within "actual occasions of organizational circumstances." Garfinkel's policy of relevance also calls into question the commonly used practice in discourse studies of coding talk using externally defined coding categories.

Policy 4: Accountability

> The policy is recommended that any social setting be viewed as self-organizing with respect to the intelligible character of its own appearances as either representations of or as evidences of a social order. Any setting organizes its activities to make its properties as an organized environment of practical activities detectable, countable, recordable, reportable, tell-a-story-aboutable, analyzable—in short, *accountable*. (Garfinkel, 1967, p. 33)

Actors organize their activities in ways that provide for their intelligibility as reportable and inspectable. To be more specific, we assume that people do things in ways that are inherently designed to make sense. This is a powerful assumption because it allows us to say that actions and the sense associated with them are sequential in nature and that this sequential organization produces, sustains and is informed by members' shared sense of the local social order. This allows members to recognize prospectively and retrospectively that they are engaged in the work of instruction and learning as they engage in that work. Garfinkel's policy of accountability highlights that actors are capable of making choices and they have a shared, if provisional and defeasible, sense of propriety with respect to what they both can and cannot do and what they should and should not do. While this sense of propriety may or may not be something actors can account for, it is evident in what they do and the way they do it. When Garfinkel refers to behavior as being accountable, the word can be understood in two senses. First, members can be (and are) responsible for their actions and are accountable to their interlocutors for utterances and actions that may appear to be without reason or rationale. Second, and more obliquely, Garfinkel is contending that all behavior is designed in ways to give an account of the action as an instance of something or the other (see the description of the documentary method of interpretation under Policy 1). It is the work of the video analyst to document how this is accomplished.

EM begins from the assumption that members are doing something competently and that our job is to figure out what it might be. When we say that members are competent, we mean that they are qualified to recognize (and assess) the competence of their own actions and those of others. It is this very competence that provides for their status as membered participants and that provides the basis for accountability. This competency is referred to by ethnomethodologists as the "unique adequacy requirement."[3] Garfinkel (2002) stipulated that the ability to recognize competence is "staff-specific, work-site-specific, discipline-specific" (p. 113). The unique adequacy suggests that it is the analyst's job to document what the participants *are* doing, rather than what they *should be* doing based on some set of a priori expectations. In this way, it prohibits assessment by analysts of members' achievements.

Policy 5: Indexicality

> The demonstrably rational properties of indexical expressions and indexical actions is an ongoing achievement of the organized activities of everyday life. (Garfinkel, 1967, p. 34)

Indexical expressions are those whose sense depends crucially upon knowledge of the context within which the expressions were produced (as opposed to "objective" or context-free expressions). The most obvious examples are expressions that contain deictic terms such as here, there, I, you, we, now, then, and so forth. To make sense of an utterance containing such terms, it will generally be necessary to know who is the speaker, who is the audience, where the speaker and audience are located, when the utterance was produced, and so forth. Any sentence containing such elements will have different interpretations or meanings depending on the circumstances in which it

---

[3]See Garfinkel, 2002, pp. 175–176 for an elaborated discussion of this requirement in its weak and strong forms.

is produced. Because of this, deictics are sometimes referred to as "shifters" by linguists. Logicians and linguists "have encountered indexical expressions as troublesome sources of resistance to the formal analysis of language and of reasoning practices" (Heritage, 1984, p. 142).

One of Garfinkel's contributions was to note that deictic terms are not the only ones that have indexical properties. Heritage (1984) provides the example of the assessment, "That's a nice one," offered while the speaker and the listener are attending to a particular photograph. What qualifies the picture as nice (e.g., its composition, color rendering, content, etc.) is not made evident by the utterance and must somehow be worked out by the listener by inspecting the object in question. In this way, nondeictic terms such as *nice* are also indexical in use. Not only expressions, but also socially organized actions can have indexical properties. Imagine two people standing face to face and one participant reaching out and touching the other. The meaning of this act, however, as a warning, provocation, greeting, demonstration, empathetic gesture, act of belligerence, and so forth depends crucially on context, on the nature of the interaction that immediately preceded and immediately follows the action.

The fact that the meaning of indexical expressions and actions cannot be determined isolated from the circumstances within which they were produced does not usually present a problem for participants. For starters, participants inhabit the situations within which the expressions and actions are produced and, as a result, are naturally supplied with many resources for resolving their meaning for present purposes. Further, participants have the opportunity to dispel any residual ambiguity through additional sense negotiation. Ultimately, however, all indexical expressions and actions are always contingent and to some degree indeterminate to a degree that is deemed acceptable to actors themselves. For Garfinkel, the question of how this indeterminacy is managed in the nonce on a routine basis was at the heart of all ethnomethodological inquiry.

Members' talk and action has a reflexive character, which is to say that it is simultaneously "context-shaped" and "context-shaping" (Heritage, 1984). The meaning of any action depends crucially on the context within which it is performed. At the same time, the action itself re-shapes the context in ways that will inform the understandability of other actions that follow. Heritage (1984) referred to this as being "doubly contextual" (p. 242). To study instruction and learning as a form of practice, therefore, we need to examine how particular actions provide for their own understandability *as* instruction and learning. Said another way, we need to study observed actions as resources by which actors can produce the sense of prior actions in light of the current action, and make relevant and sensible possible subsequent actions. This clashes with the view of context as a given, as a container within which actors do what they do. Instead, it poses the task for the video analyst of rendering an account of how members in their capacity as "order production staff" (Garfinkel, 2002, p. 102) go about constructing context through their indexical actions.

And so sayeth the book of Harold.

EM AND DESIGN-BASED RESEARCH

Garfinkel's policies stipulate that there is more to studying instructional practice than simply mining examples of "good instruction" or "bad instruction" from compiled recordings. It is the responsibility of the analyst to discover within the recorded materials what the members are actually accomplishing (Policies 2 and 4) and are making relevant (Policy 3) through their interaction. The requirement to "bracket out" pre-existent interpretations (Policy 3) might suggest the need for a division of labor within design-based research teams. In particular, it makes explicit that the tasks of design and analysis must be treated as distinct, at least logically, if not in terms of specific team member responsibilities. This does not imply that they are independent, however. Instead, their relationship is a symbiotic, one-design must be informed by analysis, but analysis also depends on design in its orientation to the analytic object.

We are not the first to propose applied ethnomethodological research (cf. Heap, 1990); nor are we the first to undertake ethnomethodologically informed studies within classrooms and other sites of instruction (cf. Ford, 1999; Koshik, 2002; Macbeth, 2000; Mehan, 1979). What is novel, however, is the proposal to use ethnomethodologically informed findings as one component of a larger program to improve education, specifically through design-based research. At the moment, however, the initiative outlined in this chapter has only the status of a proposal in that we know of no research team that has actually attempted to employ video analysis in precisely the way that we have described here.

As a discipline focusing on members' methods for practical reasoning, EM provides a useful foundation for research into the practices of learning. Garfinkel's policies for ethnomethodological studies, therefore, provide a reasonable starting point for building a rigorous program of video-based analysis connected to design-based research. Much remains, however, to bring such a program to life. This chapter is not intended to substitute for a careful reading of Garfinkel's own development of the five policies summarized here. We recommend that interested readers study not only the full text of the policies themselves (Garfinkel, 1967, pp. 31–34), but also Garfinkel's three advisories (Garfinkel, 2002, pp. 112–114), and his description of the prominent objects of study in ethnomethodological inquiry (Garfinkel, 1967, pp. 4–10).

ACKNOWLEDGMENTS

Support for the writing of this chapter came from grants from the National Science Foundation to the first and second authors. Any opinions, findings, and recommendations expressed are those of the authors and do not necessarily reflect the views of the funding agency.

REFERENCES

Barab, S., & Squire, K. (2004). Design-based research: Putting a stake in the ground. *Journal of the Learning Sciences, 13*, 1–14.

Brown, A. (1992). Design experiments: Theoretical and methodological challenges in creating complex interventions in classroom settings. *Journal of the Learning Sciences, 2*, 141–178.

Cobb, P., Confrey, J., diSessa, A., Lehrer, R., & Schauble, L. (2003). Design experiments in educational research. *Educational Researcher, 32*(1), 9–13.

Design-Based Research Collective. (2003). Design-based research: An emerging paradigm for educational inquiry. *Educational Researcher, 32*(1), 5–8.

Fishman, B., Marx, R., Blumenfeld, P., Krajcik, J., & Soloway, E. (2004). Creating a framework for research on systemic technology innovations. *Journal of the Learning Sciences, 13*, 43–76.

Ford, C. E. (1999). Collaborative construction of task activity: Coordinating multiple resources in a high school physics lab. *Research on Language and Social Interaction, 32*, 369–408.

Garfinkel, H. (1967). *Studies in ethnomethodology.* Englewood Cliffs, NJ: Prentice-Hall.

Garfinkel, H. (2002). Ethnomethodology's program: Working out Durkheim's aphorism. Lanham, MD: Rowman & Littlefield.

Heap, J. (1990). Applied ethnomethodology: Looking for the local rationality of reading activities. *Human Studies, 13*, 39–72.

Heritage, J. (1984). *Garfinkel and ethnomethodology.* Cambridge, England: Polity Press.

Heritage, J. (1995). Conversation analysis: Methodological aspects. In U. Quasthoff (Ed.), *Aspects of oral communication* (pp. 391–418). Berlin: Walter de Gruyter.

Hutchby, I., & Woolfitt, R. (1998). *Conversation analysis.* Cambridge, England: Polity Press.

Koshik, I. (2002). Designedly incomplete utterances: A pedagogical practice for eliciting knowledge displays in error correction sequences. *Research on Language and Social Interaction, 35*, 277–309.

Macbeth, D. (2000). Classrooms as installations: Direct instruction in the early grades. In S. Hester & D. Francis (Eds.), *Local education order: Ethnomethodological studies of knowledge in action* (pp. 21–72). Philadelphia, PA: John Benjamins.

Mannheim, K. (1968). On the interpretation of *Weltanschauung.* In P. Kecskemeti (Ed.), *Essays on the sociology of knowledge* (3rd ed., pp. 33–83). London: Routledge & Kegan Paul.

McDermott, R. P., Gospodinoff, K., & Aron, J. (1978). Criteria for an ethnographically adequate description of concerted activities and their contexts. *Semiotica, 24*, 245–275.

Mehan, H. (1979). "What time is it, Denise?": Asking known information questions in classroom discourse. *Theory into Practice, 18*, 285–294.

Oxford Modern Dictionary, The. (2nd ed.). D. Thompson (Ed.). (1996). New York: Oxford University Press.

Prevignano, C., & Thibault, P. (Eds.). (2003). *Discussing conversation analysis: The work of Emanuel A. Schegloff.* Amsterdam, Netherlands: John Benjamins.

Psathas, G. (1995). *Conversation analysis: The study of talk-in-interaction.* Thousand Oaks, CA: Sage.

Sacks, H. (1992). *Lectures on conversation.* Oxford, England: Blackwell.

Schegloff, E. (1991). Reflections on talk and social structure. In E. Boden & D. Zimmerman (Eds.), *Talk and social structure: Studies in ethnomethodology and conversation analysis* (pp. 44–70). Berkeley, CA: University of California Press.

ten Have, P. (1999). *Doing conversation analysis: A practical guide.* Thousand Oaks, CA: Sage.

Wittgenstein, L. (1953). *Philosophical investigations.* New York: Macmillan.

Ways of Seeing Video: Toward a Phenomenology of Viewing Minimally Edited Footage

Frederick Erickson
University of California, Los Angeles

What do people notice as they view audiovisual recordings of naturally occurring social interaction in everyday life—what and how do they see and hear in engaging minimally edited footage? We know very little about this.

How fictional cinema communicates is somewhat better understood. There is a vast literature on the semiotics of professionally produced fictional film and video. But films made to tell a story in themselves are constructed very differently from most research footage. In fictional cinema, the human interaction that is photographed is scripted and rehearsed. The film is shot typically in short "takes," often with multiple cameras. The takes are then joined together in the editing process as sequences of strips (not necessarily in the order in which the takes were originally photographed), cutting back and forth between the views of different cameras. In their edited form the individual strips derived from camera takes may be as short as 3 seconds and are often no longer than 10 seconds. The editing of sets of strips together—"montage"—is done in such a way as to provide a coherent narrative—what the viewer's attention is directed toward, in and across moments of viewing, is highly scaffolded by the montage as well as by the plot line of the script.

In documentary or ethnographic film, the human action is, for the most part, naturally occurring (although filmed interview is an exception here) and the camera's takes are usually longer than those in fictional cinema. In subsequent editing after the initial shooting, stringing the camera takes together, as well as shortening portions of the takes when necessary, tends to keep closer to the original sequence of the events that were recorded (this is especially true for "ethnographic" documentary film). In ethnographic film, there is typically one cameraperson rather than multiple persons recording simultaneously. Yet, in the *genres* of documentary and ethnographic film, editing still produces a strong narrative frame within the film itself and this scaffolding guides the viewer's moment by moment attention and overall narrative interpretation.

With minimally edited research footage, the situation is quite different. There is very little camera editing. The takes are often very long—often a single take lasts the entire duration of the event that is being recorded. There is also very little moving of the camera from side to side ("panning") or alternating between close-up and wide angle views ("zooming"). In editing, a strip of audiovisual recorded material is chosen to present to viewers as an example of some phenomena that are of research interest (e.g., conversational turn taking, a way of emphasizing a substantively important point, helping a student to articulate his or her thinking more clearly, shifting from routine action to high interest/high involvement action, using gesture or other visual displays to accompany speech as a way of communicating meaning).

The strip, a video clip with minimal camera editing, may be less than a minute long, or in the interest of showing more complex sequential chains of action, a strip may be as long as 2 ½ minutes. Usually, edited examples are no longer than that. This is because clips of even a minute in duration are very difficult for viewers to watch without getting overloaded with informational detail. Experts can stay with a strip of unedited footage for a while, bringing to their viewing explicit and focused substantive interests. But novice viewers as they try to view minimally edited footage tend to "zone out" fairly quickly, as they are flooded by many more information bits than they can process cognitively. Socialized by the visual and narrative conventions of fictional cinema and television, with their strong scaffolding for focus of attention and interpretation, novice viewers of minimally edited video find themselves at sea, as it were, in a stream of continuous detail they don't know how to parse during the course of their real-time viewing in order to make sense of it. They see many trees but little forest, and gradually, in the absence of a sense of dimensions of analytic contrast, all the trees begin to look alike.

One way that presenters of unedited footage can address this problem—especially when video clips are presented in multimedia arrays—is to simplify the visual image—tightening the visual frame to a close-up. One can also shorten the illustrative clip to less than a minute. These are means of reducing extraneous detail for the viewer. But these simplifications sacrifice portrayal of a more comprehensive, ecological sense of human social interaction. Such interaction is not just a succession of talking heads, as in the "talk show" shooting of commercial television. Rather, as speakers are speaking, listeners are influencing them through listening behavior (Erickson, 1986; Erickson & Shultz, 1982; Goodwin, 1981; Hall, 1964) and this cannot be shown without presenting both speaker and listener together in the same visual frame. Also, social action doesn't just happen as a succession of short, discrete, bursts of activity—the nonverbal plus verbal equivalents of "sound bites"—but in longer, more complex sequences that require viewing time in order to comprehend.

If simplifying the picture and shortening the clip are inappropriate means of handling the problems of information overload and incoherence that minimally edited footage present to the novice viewer, what are more appropriate ways of dealing with this? First we need to recognize that the difficulties novice viewers have with minimally edited footage are a result of their use of their normal processes of visual perception and sense making.

SOME INTELLECTUAL BACKGROUND

The cognitive psychology developed over the last 30 years (e.g., Newell & Simon 1972; Schank & Abelson 1977; Simon, 1972) has shown persuasively that there are severe limits on humans' capacity of information processing in real time—limits on attention and short-term memory. A continuous recording of audio and video presents us with far more information than we can take in and make sense of. Further, Gestalt psychology and constructivist psychology more generally has shown that *seeing* is not simply a passive matter of the reception of sensory stimuli but that it is an active process—in everyday visual perception, we construct what we see (Gibson, 1979) just as we do in viewing and making sense of visual art (Arnheim, 1969).

The philosopher Wittgenstein's example of the outline of a picture of a duck that can also be seen as a rabbit (Wittgenstein, 1953, Part II, Sect. 11, p. 194e) as well as myriads of other examples of visual illusions further show that to see is to construct a coherent whole. Wittgenstein extends this insight in considering how we understand words and phrases in ordinary talk. The psychologist Bruner has recently claimed that most human understanding is narrative in character—we produce storylike models of events and of sequences of events (Bruner, 1991, 2002; see also Bartlett 1932; Schank & Abelson 1977). The founder of the school of "ethnomethodology" in sociology, Garfinkel, argues along similar lines as those of Bruner. Garfinkel claims (1967) that all "common sense" reasoning involves constructing storylike frames within which particular details of observable behavior make sense (calling this the "documentary method"). For example, we are walking along a street and we see a man lying in the gutter shaking his head slightly. Has he just had a heart attack— or is he just waking up from a drunken stupor—or is he a homeless person who habitually sleeps there? The storylike schemata we posit as a frame for the behavior we can observe powerfully influences what we do, in social action, as a response to that behavior. (Further, suggestive evidence comes from the experience of teachers of the reading of literature that has led to "reader-response theory." For example, the literary theorist Stanley Fish in a classic monograph entitled "Is there a text in this class?" (Fish, 1980) argues that all readers construct the texts they read, bringing differing life experience to the text, and influenced by various communities of interpretive practice, so that one person's or one interpretive community's reading of a particular text (say, a sonnet of Shakespeare) is not the same reading, phenomenologically, as that of another reader or community.)

Sheer familiarity or unfamiliarity with certain phenomena also influences how we see and hear them in cinema and video. When I was in graduate school, one of my teachers, the anthropologist Edward T. Hall, told the following story in a course on culture and communication. He had taught intercultural communication for the United States Foreign Service and heard this anecdote from a United States Information Service (USIS) worker. The USIS had been showing a film about life in America in numerous villages in rural east Africa. The film showed iconic representations of American life—city skylines and street scenes, manufacturing in factories, mountains, seacoasts, and crop fields cultivated by machines. USIS was supposed to "evaluate" its attempts to communicate about American life, and so after each film screening, the villagers were asked what parts of the film they liked best. From village to village, the response was the same, and it was given immediately, without pondering. "We liked the chicken," they said. "That was our favorite part." The USIS worker was confused because he couldn't remember anything about a chicken in the film. When he returned from the field, he watched the film very carefully, shot by shot. Sure enough, in one brief shot of a farm scene, a chicken ran across the edge of the screen. The villagers had never seen New York skyscrapers or Mt. Rushmore. But they knew chickens!

Similarly, I remember being confused at about age 6 at Christmas time, as I looked at pictures of the Nativity and found something missing. There was the stable, with Mary holding her baby and Joseph at her side, with animals and shepherds looking on. But where was Round John Virgin? I was conflating the words of the Christmas hymn "Silent Night" with Robert Louis Stevenson's story *Treasure Island* and thus, my narrative expectations led me to look for a character in the Nativity pictures that was never there.

My general point here is that what we see as present or as absent in audiovisual footage—and as intelligible and expectable or as unintelligible and bizarre—is not simply a matter of what is contained in the audiovisual record itself. We are indeed constructivists—we make sense in our seeing and hearing. And thus, for us, minimally edited video footage of human social action is ambiguous and equivocal—it does not speak clearly for itself. Consider the reaction of the trial jury in the Rodney King police brutality criminal trial. Jurors were shown video by the prosecution—that footage showed policemen repeatedly beating King with their nightsticks as he was lying on the ground. The defense attorney, using slow motion analysis, persuaded enough jurors to "see" the footage alternatively—King's movements on the ground as he was being beaten could possibly be his attempts to get up and assault the policemen. The jurors were unwilling to convict the officers. The minimally edited video footage did not show what had happened (narratively) "beyond reasonable doubt."

Watching Classroom Examples

Since 1977, I have been using in my teaching at the graduate level a 2½ minute video clip of a Native American first-grade teacher from a reserve in Northern Ontario, Canada, as she instructs a classroom of students who are all from that reserve (for a description of the study, see Erickson & Mohatt, 1982). I have used this example in social foundations classes and in qualitative research methods classes. The video clip—from a single take that lasted an hour—began just before the official school day had begun. (The teacher was wearing a wireless microphone so we could hear every word she said, and the teacher was always central in the visual frame.) She asked a child to take the attendance slip to the principal's office and then began the morning routine; recitation of the Rosary, locating today's weekday and date on the calendar, identifying a "sight" vocabulary word written on the chalkboard, and then the teacher handing out reading workbooks and beginning a reading lesson.

What I do in showing this clip is to say: "Watch carefully what is being said and done, paying close attention to what things look and sound like. I will play the tape once through so you can get the feel of it. Then I will play it again, and, this time, write notes. Try to get in your notes as much as you can of what you are seeing and hearing." I play the tape twice. After that I say, "What did you notice, especially? What did things sound like and look like?" Invariably, the viewers begin to comment on the teacher—she seems bored to them, or tired, or going through her pedagogical motions routinely. Some viewers notice that the teacher did not vary pitch or volume much as she talked.

In the clip, one can see that the teacher, a grandmother in the reserve community, moved deliberately from place to place at the front of the room, without accelerating or decelerating as she walked. She spoke with relatively low voice volume and did not vary pitch much from syllable to syllable (in ordinary parlance this is "talking in a monotone"). In the 2½ minutes of the clip she neither praised nor reprimanded any child verbally, and she directed content questions only to the entire classroom group, not to any individual children. She did not change facial expression much during the clip. As she walked from one station to the next, between successive events, she occasionally exhaled audibly and rather sharply, in what could be heard as a sigh. (In the winter of 1979, I showed this clip at a colloquium for staff at the National Institute of Education [NIE] in Washington. I did my usual lead-in [as described earlier], showed the clip twice, and then used my standard prompt to begin discussion. Immediately, a man at the back of the room said, "Well it's obvious—this woman is clinically depressed." I said, "Oh, that's interesting. What do you do here?" The man said that he was a clinical psychologist who was visiting NIE for a year as a consultant on research program development.)

I then began to provide some interpretive scaffolding for the viewing (see especially Erickson & Mohatt, 1982, and Philips, 1983). I explained that in this community, it was culturally inappropriate to vary volume and pitch while speaking—that was only done when people were furious and about to resort to violence. In everyday life, individuals were not singled out for direct questions or for praise or blame in front of others because such "spotlighting" of public attention was considered coercive and rude. I explained that the teacher had a cardiopulmonary condition (a precursor of congestive heart failure, from which the teacher eventually died) in which water vapor builds up inside the lungs and is then expelled periodically in an involuntary outbreath.

Every first-time viewer hears the teacher's outbreaths as sighs (I did too). Almost every viewer, unless they are familiar with what Philips (1983) calls "invisible culture" patterns in Native American speech style, hears the absence of variation in speech pitch and volume as behavioral evidence of at least boredom—and possibly of depression—on the part of the teacher. Similar attributions are made about the motion quality of the teacher's walking around the room—she must be bored, or tired.

North American Anglo viewers seem to bring with them a scriptlike or schemalike set of expectations for what they will see in a first-grade classroom—a vocally and kin-esthetically animated teacher who smiles broadly and moves rapidly from place to place, praising individual children for what they have just done right and reprimanding children when they do wrong. The Native American teacher did none of this. Yet, she was judged as a very effective teacher by the Native American principal of the school and by her Native American fellow teachers. Her students scored at or above the Canadian national average (when Canadian Native American students typically scored far below that average). Viewers, in using their previously established narrative understandings of a "normal" first-grade teacher and classroom, seemed to be attending as much to what was not there to be seen in the clip as they were attending to what was there. (And what about the recitation of the Rosary? Were you wondering about that? As you read about, it did you ask whether this was a public school or a parochial religious school? Watchers of that video clip often ask that question. The school was part of the Canadian public school system for Native Americans, but because in schooling, Church and State are not separated as much in Canada as they are in the United States, it was permissible for this teacher to begin her school day with the recitation of one verse of the Rosary.)

From 1981–1984, as a senior researcher in the federally funded Institute for Research on Teaching at Michigan State University, I directed a study entitled "Teachers' Practical Ways of Seeing and Making Sense." Our main activity was to observe, video-tape, and interview five second-grade teachers, each for an entire school year. We wanted to identify what teachers noticed—what they attended to visually and auditorially while they were in the midst of teaching, and then see what pedagogical use they made of what they had noticed. In addition, we showed video clips of other teachers to panels of experienced teachers and we also showed the same video clips to novices to teaching—teacher education students who had not yet had field placement as student teachers.

The first of the video clips that I showed, in separate viewing sessions, to a panel of experienced teachers and to one of inexperienced teachers, came from a kindergarten–first grade classroom in an elementary school located in a White working-class neighborhood of a Boston suburb. (This classroom is further described in Erickson, 1996, 2004; Florio & Walsh, 1981.) The clip shows the beginning of an early afternoon reading lesson held in the first days of school in September 1974. The teacher, Ms. Walsh, was seated in front of a u-shaped table and the students (a small group of the first graders) faced her, sitting around the outside of the u. The kindergartners had gone home after lunchtime and the rest of the first graders were engaged in free reading, writing, and playing with puzzles that involved letters and words (i.e., they were not doing the more usual "seatwork" to keep them occupied while the teacher was working with a small reading group—they were not sitting at their desks filling out ditto sheets containing simple literacy exercises).

The viewing group of experienced teachers consisted of those in whose classrooms I and my research assistants had previously observed, videotaped, and interviewed for one or more entire school years, teaching in inner city and suburban schools in a small mid-western city. We had watched them teach—now we were asking them to watch someone else teach, on video. The inexperienced teachers I had not met before.

In the video viewing sessions, each set of viewers was asked to write notes on what they had noticed and then to discuss what they had seen in the tape. The discussions had the character of focus group interviews. As the teacher education students watched the video clip, they wrote quite detailed behaviorally descriptive notes, but the notes were fragmentary, without a coherent pedagogical story line. Their comments on what they noticed also had a fragmentary quality, although some of the inexperienced teachers expressed concern about the level of ambient noise in the classroom, and the amount of student movement around the room. In contrast, the notes written by the experienced teachers were much briefer, but they were more schematic. The notes were either globally descriptive (suggesting some kind of pedagogical story line or a tactical "move" within such a storyline) or the notes raised questions concerning background information that was not visible on the tape, such as "How much preschool have these children had?" (Like the global descriptive notes, such questions indexed a narrative understanding of teaching as a process that develops over time.)

Three of the four teachers in the viewing group were very concerned about noise and movement in the classroom. They saw the classroom as disorderly, and our previous fieldwork and interviewing with two of those three teachers had shown that they believed firmly in the proposition that "classroom order necessarily precedes student learning." Their own rooms had a much greater surface appearance of order than did the one shown on the video clip. After viewing the clip, the teachers expressed doubts about the competence of the teacher in the clip. (An exception here was the fourth teacher, who the previous summer had taken a course at the local university in which she watched a variety of tapes of Ms. Walsh and the students in her classroom. Through this study, the fourth teacher in the viewing group had become persuaded that Ms. Walsh was highly skilled and effective.)

The concern of the three other experienced teachers about what they perceived as a lack of overall classroom order was so strong that they overlooked some pedagogically relevant information that was seeable and hearable in the video clip—the first-grade students (who had been kindergartners with Ms. Walsh the previous year) were, on a day in the first week of the current school year, reading aloud fluently from the first-grade books. They were visible on the video screen already knowing how to read. Whatever the customary level of ambient noise and kinesthetic activity in Ms. Walsh's classroom might have been in the previous school year, her classroom had been a place in which at least some kindergartners had learned to read quite well, as evidenced by what those children, now beginning first grade, were doing on the video clip. The fourth experienced teacher (who had studied tapes of Ms. Walsh in a previous summer school course) gently pointed this out to the other teachers in the viewing group. But the other teachers were not entirely convinced by their colleague's observation. Could so disorderly a room be one in which student learning could take place? Three of the four teachers continued to wonder about that. Their pedagogical commitments (to the proposition that classroom order precedes students learning) seemed to be driving what they could see in the tape.

The discussion then turned to questions about background information—going beyond what could be heard or seen on the video screen. These questions included the previously written question about how much preschool the students had had. The teachers began to speculate about what the classroom would look like later in the school year. We agreed to meet again and view some footage from Ms. Walsh's classroom later in the year. (When we held the second meeting, it was apparent that that year's kindergartners were learning to read and that the first graders were continuing to develop on the foundation that had been established the previous year when they were kindergartners. Three of the four teachers still wondered if Ms. Walsh had adequate skills in "classroom control." And once when I showed a video clip from Ms. Walsh's classroom to the Native American teachers from the northern Ontario school where we had videotaped the first-grade teacher who was a grandmother, that set of teachers watched the clip and then said "We don't think we could teach those children—they're so loud.")

The inexperienced teachers, during their viewing session, had not asked for background information "beyond the screen." But such requests were frequent for the experienced teachers (and I have found that this is usually the case in viewing sessions with experienced teachers). One kind of background information had to do with a strong sense of the "yearliness" of early grades teaching in a self-contained classroom. The experienced teachers wanted to know about the previous school year and they asked to see footage from later in the current year. Another kind of background information the experienced teachers sought concerned matters of institutional context. One of the teachers asked about the reading materials—had they been chosen by the district, or by the local school faculty and the principal, or by the classroom teacher herself? (During the previous school year in the midwestern city was one in which a new reading series had been arbitrarily adopted by the central administration, without consultation with classroom teachers. Many teachers resented this. Three of the four teachers in the viewing session came from that school district. Once in Philadelphia when I showed footage from Ms. Walsh's classroom to public school teachers I was asked by one of them, "How recently has there been a teacher strike in this district?")

What all this suggests is that, especially for experienced teachers, minimally edited video footage, rather than simply or directly providing an open window on someone else's teaching practice, functions more like a projective test—an inherently ambiguous and incomplete stimulus that invites reaction and speculation ranging far beyond the information that is potentially available in the video clip itself. The experienced teachers' pedagogical commitments seemed to be held, in part, as narrative expectations for what classroom interaction "should" look like. Their strategic and tactical sense of timing in pedagogy—the yearliness, unit-ness, and daily-ness of trajectories of instruction (literally we say "a course of instruction") was another aspect of their narrative understanding of teaching. And their sense of the influence of what lay outside the classroom—at the school building level, the school district level, community and family lives of their students and parents—also influenced what they heard and saw in the video clips. These dimensions or aspects of narrative understanding provided powerful frames of interpretation for video viewing. (Such narrative understanding, and the focus of particular pedagogical commitments, also influenced what the teachers paid attention to in the midst of their own teaching—this was a major finding of the overall study.) The novices to teaching, undergraduate teacher education students, had not yet acquired these kinds of narrative understanding as practitioners and so, while they could track some behavior fairly accurately on the videotapes, they could not make such observation cohere as storylike, interpretive understanding of teaching practice.

CONCLUSION

In a recent article on video data analysis (Erickson, 2006) I argued that videotapes are better regarded as sources for data than as data in themselves. Just as other primary documentary records in qualitative research are not data but are information sources out of which data can be constructed—field notes, interview transcripts, site documents—so audiovisual records of social interaction are information sources. From such records, data can be defined, analytically. But it seems to me that it is naive realism to think of them as data themselves.

This same insight seems to apply to video viewers who are not seeing themselves as scientists trying to do formal data analysis, but as practitioners of teaching who are watching video examples of someone else's teaching with the aim of learning something new about their own teaching and about their own students' learning. I believe that exciting possibilities lie ahead for using minimally edited video to support experienced teachers in continuing professional education programs or novices in teacher education programs in making "virtual visits" to other teachers' classrooms and in the process, developing richer understandings of their own teaching practice and that of others. But we will not be able to take full advantage of these new possibilities, if as developers of digital multimedia archives or "libraries" of teaching practice, we do not take account of the phenomenology of viewing minimally edited footage. Without the scaffolding for interpretive sense making that is provided for the viewer in the genres of fictional cinema and documentary film, without being dragged along hermeneutically by the nose through the visual conventions that are built by editing into narrative film, the viewer doesn't simply forego a narrative understanding of the minimally edited video footage. Rather, the viewer constructs his or her own narrative understanding of the footage on the basis of prior experience.

We have a few reports of cross-cultural viewing of video clips of teaching practice, for example, Jacobs and Morita 2002; Spindler and Spindler 1993, 1994; Stigler and Gallimore, 2003; Tobin Wu, and Davidson, 1989. In showing German teachers and American teachers clips from silent 8mm film of each others' classrooms and inviting them to discuss what they saw, Spindler discovered the importance of what I've called here *pedagogical commitments* that entail deeply held, partially implicit cultural assumptions about the nature of children, learning, and teaching. Tobin has had similar experience in showing footage from Japanese preschool classrooms to American teachers, and vice versa, and something like differences in pedagogical commitments (scriptlike expectations combined with justificatory reasons) is also reported by Jacobs and Morita for Japanese and American mathematics teachers. I first began to show minimally edited research footage to informants in my early research on gate keeping interviews—job interviews and academic advising interviews (Erickson & Shultz, 1982). I would videotape an interview and then show it to each of the two participants in separate viewing sessions, asking them to stop the tape and comment on it "whenever something new or important happened." Sometimes the interviewer and interviewee would stop the tape at different points in the original interview. Sometimes they would stop the tape at virtually the same instant, but in separate viewing sessions. When that happened, sometimes their comments would be similar—they would remark on similar things. But occasionally what they said at the same stopping point seemed as if they had been participating in two different events phenomenologically, not the "same" one, as recorded on the tape. They would notice and comment on differing behavioral details, attribute intentions very differently, in ways that continually surprised me. (When Beckman & Frankel, 1984 used a similar procedure with intern physicians and patients in clinical medical interviews, he found an analogous divergence in what the physician and patient saw and heard in the same videotape. See also West & Frankel, 1991.) Over the years, I have been showing classroom footage to teachers. I have been impressed that their reactions to these complex scenes of social life are sometimes even more divergent than the responses by participants in dyadic interaction—and the teachers' ways of viewing are always influenced by their pedagogical commitments.

It follows that if we want to develop video libraries of teaching practice (see the discussion in Stigler & Gallimore 2003), we must take seriously the "eye of the beholder" issues I've been discussing. In multimedia arrays, we may be able to invent new kinds of framing for looking—using freeze frames with audio commentary over them, prospective and retrospective headline still frames—reinventing those used previously in silent cinema films, adding dynamic links to transcript and interpretive commentary to accompany the video. I'd rather go that way in design—invent new means of scaffolding for viewer attention and interpretation—rather than to simplify the visual image of the classroom by using close-ups, or by shortening clips, or by using clips in which everything seems to be happening smoothly and ideally. That is, to make teaching and learning look like sets of discrete "moves" that are directly imitable— "replicable" from one teacher to the next—and that I see as an engineering notion that is dangerous as a subtext in materials that purport to help one learn to teach better. (I consider teaching practice as something that must be grown anew by each practitioner through reflection and repeated trials; a locally situated tactical accomplishment, not something that can be imitated with "fidelity" to an external model.)

However, as we proceed in these design research attempts, further research on the phenomenology of viewing seems warranted, as done by viewers with differing life experiences and differing pedagogical commitments. In my social interaction documentation and analysis laboratory at Harvard in the early 1970s, we put up a sign on the wall that said "In Video Veritas." But we meant it ironically then. In the ensuing 30 years, I've been exploring some of the implications of that irony. At this writing, I am confident that many more implications of it remain to be discovered.

REFERENCES

Arnheim, R. (1969). *Visual thinking*. Berkeley, CA: University of California Press.

Bartlett, F. (1932). *Remembering.* Cambridge, England: Cambridge University Press.

Beckman, H., & Frankel, R. (1984). The effect of physician behavior on the collection of data. *Annals of Internal Medicine, 101,* 692–696.

Bruner, J. (1991). *Acts of meaning.* Cambridge, MA: Harvard University Press.

Bruner, J. (2002). *Making stories: Law, literature, life.* New York: Farrar, Strauss & Giroux.

Erickson, F. (1986). Listening and speaking. In D. Tannen & J. Alatis (Eds.), *Language and linguistics: The interdependence of theory, data, and application.* Washington, DC: Georgetown University Press.

Erickson, F. (1996). Going for the zone: The social and cognitive ecology of Student–teacher interaction in classroom conversations. In D. Hicks (Ed.), *Discourse, learning, and schooling* (pp. 29–63). New York: Cambridge University Press.

Erickson, F. (2004). *Talk and social theory: Ecologies of speaking in everyday life.* Cambridge, England: Polity Press.

Erickson, F. (2006). Definition and analysis of data from videotape: Some research procedures and their rationales. In J. Green, G. Camilli, & P. Ellmore (Eds.), *Handbook of complimentary methods in education research* (pp. 177–191). Mahwah, NJ: Lawrence Erlbaum Associates.

Erickson, F., & Mohatt, G. (1982). Cultural organization of participation structures in two classrooms of Indian students. In G. Spindler (Ed.), *Doing the ethnography of schooling* (pp. 132–174). New York: Holt, Rinehart, & Winston.

Erickson, F., & Shultz, J. (1982). *The counselor as gatekeeper: Social interaction in interviews.* New York: Academic Press.

Fish, S. (1980). *Is there a text in this class?: The authority of interpretive communities.* Cambridge, MA: Harvard University Press.

Florio, S., & Walsh, M. (1981). The teacher as colleague in classroom research. In H. Trueba, G. Guthrie, & K. Au (Eds.), *Culture in the bilingual classroom: Studies in classroom ethnography* (pp. 87–101). Rowley, MA: Newbury House.

Garfinkel, H. (1967). *Studies in ethnomethodology.* Englewood Cliffs, NJ: Prentice-Hall.

Gibson, J. (1979). *The ecological approach to visual perception.* Hillsdale, NJ: Lawrence Erlbaum Associates.

Goodwin, C. (1981). *Conversational organization: Interaction between speakers and hearers.* New York: Academic Press.

Hall, E. (1964). Adumbration in intercultural communication. *American Anthropologist, 66*(6), 154–163.

Jacobs, J., & Morita, E. (2002). Japanese and American teachers' evaluations of videotaped mathematics lessons. *Journal for Research in Mathematics Education, 33*(3), 154–175.

Newell, A., & Simon, H. (1972). *Human problem solving.* Englewood Cliffs, NJ: Prentice-Hall.

Philips, S. (1983). *The invisible culture: Communication in classroom and community on the Warm Springs Indian reservation.* New York: Longman.

Schank, R., & Abelson, R. (1977). *Scripts, plans, goals, and understanding.* Hillsdale, NJ: Lawrence Erlbaum Associates.

Simon, H. (1972). Complexity and the representation of patterned sequences of symbols. *Psychological Review, 79,* 369–382.

Spindler, G., & Spindler, L. (1993). The process of culture and person: Cultural therapy with teachers and students. In P. Phelan & A. Davidson (Eds.), *Renegotiating cultural diversity in American schools* (pp. 27–51). New York: Teachers College Press.

Spindler, G., & Spindler, L. (Eds.). (1994). What is cultural therapy? *Pathways to cultural awareness: Cultural therapy with teachers and students* (pp. 1–33). Thousand Oaks, CA: Corwin.

Stigler, J., & Gallimore, R. (2003). Closing the teaching gap. In C. Richardson (Ed.), *Whither assessment?* (pp. 26–36). London: Qualifications & Curriculum Authority.

Tobin, J., Wu, D., & Davidson, D. (1989). *Preschool in three cultures: Japan, China, and the United States.* New Haven, CT: Yale University Press.

Wittgenstein, L. (1953). *Philosophical investigations* (G. E. Anscombe, Trans.). New York: Macmillan.

Part Two

Video Research on Peer, Family, and Informal Learning

Video as a Tool to Advance Understanding of Learning and Development in Peer, Family, and Other Informal Learning Contexts

Brigid Barron
Stanford University

Video records have several properties that fundamentally change the way that inquiry takes place and video is now the standard data collection tool for studies of human interaction. This section of the book focuses on the contribution of video-based research to our understanding of learning and development in peer, family, and informal learning contexts. The authors who made contributions to this section are taking up fundamental questions about the processes and outcomes of learning as they emerge in the context of interactions between people, and between people and their physical and cultural environments. We are fortunate that these researchers were willing to share both their struggles in collecting and analyzing video records and the strategies, insights, and techniques they have developed after years of working with video as a data source. In this prefatory chapter, I begin with a discussion of how video has been an important data source for research investigating learning. I provide a summary of some of the theoretical insights that have emerged from studies that relied on film or video, drawing on the published literature including early efforts to use video as an analytic and rhetorical tool by anthropologists, developmental and social psychologists, and sociologists. In the second section, I summarize some of the challenges that video data presents, again drawing on the chapters and the broader literature. In the third section, I share four main methodological and analytical suggestions that emerged across the seven chapters and connect these to more general insights on qualitative research. These ideas should help both novice and seasoned researchers design and carry out research that includes video records as a data source. No rules are offered here; rather, the goal is to collectively enrich our methodological and analytic creativity and become smarter about some of the challenges that video-derived data presents. Across the group of chapters, we have access to a range of approaches and there are an endless number of strategies that might be developed to fruitfully use video as a data source.

VIDEO AS A TOOL FOR INVESTIGATIONS OF PEER, FAMILY, AND INFORMAL LEARNING

Methods used in the analysis of videotaped records are rooted in practices of disciplined observation, a core feature of the scientific method. Independent of the advent of film, social scientists developed approaches that allowed them to document, analyze, and report human behavior to their colleagues. For example, scientists interested in child development created formal approaches to looking at, recording, and describing the natural world in ways that were convincing to others who followed positivist empirical traditions. Systematic observational approaches relied on pre-established coding schemes and were designed to yield reliable judgments by independent observers of behavior taking place in naturalistic settings. Techniques for narrowing the foci of observation through methods such as time sampling, event sampling, or focal person approaches were articulated and used in many of the early studies of child development. For example, early studies of children's play often relied on what was called repeated short samples (Goodenough, 1928) where a child would be observed for one minute a day and their play coded into one of six mutually exclusive categories (Parten, 1932). After a substantial number of observations were made, proportions could be computed in order to draw conclusions about how a particular child spends their playtime. Statistical approaches for determining interrater reliability were key innovations that allowed researchers to determine whether their coding approaches led to similar observations across human coders. These methods require that the focus of inquiry and the coding systems be well worked out before the collection of data. Coding systems also need to be simple enough for two or more observers to achieve interrater reliability.

Video relieves these constraints. The persistence of the record allows researchers to move away from completely predetermined coding systems and instead, develop categorization approaches after examples are carefully studied. It allows the analyst to speed up, slow down, or stop subtle aspects of interaction that normally occur on such a short time scale that they go unnoticed. Tone, eye gaze, affect, gesture, use of material resources, attention, and physical posture can all be studied together or as separate streams. New phenomena can be named, categories described, and when appropriate, coded and quantified. Video records can also be revisited over time with new research questions and new theoretical frames, or through the eyes of researchers who come from different disciplinary traditions.

Collectively, these properties have generated a great deal of excitement among social scientists. In particular, film studies have been taken up by researchers whose intellectual projects involve understanding the details of how communication proceeds in the context of face-to-face interaction. Given the time consuming nature of interactional analyses, one might ask whether the excitement over what film offers to researchers is justified. So, what have we learned from studies that use film?

Insights Based on Film Studies of Human Interaction

A cursory review of the history of the use of film as a data source reveals that those social scientists drawn to it frequently had in common an interest in understanding issues of interdependency, mutuality, and reciprocity in human interaction. The contributions of observational studies to how we understand learning and development have been multidisciplinary and interdisciplinary. Anthropologists, psychologists, sociologists, linguists have all taken up observational work, sometimes coming together but often with very different theoretical frames. In some research programs, the focus has been restricted to questions about the immediate interactional context (e.g., when do infants first demonstrate joint attention and what are its characteristics). In others, the goal has been to articulate the relations between the details of interaction as revealed by microanalyses and larger cultural patterns. This review is by no means exhaustive but is meant to be illustrative of what has been learned, to provide some historical context for the use of film or video as a data source, and to help readers locate published research that might serve as examples of the variety of ways that video-based studies are reported.

Interactional Patterns Within and Across Cultures.   Early on, anthropologists took up the camera as a research tool for the study of culture (see Collier & Collier, 1986; De Brigard, 1975; and El Guindi, 1998 for reviews of the history of the use of film in anthropology). The first example may have been footage of Wolof pottery making (Lajard & Regnault, as cited in Grimshaw, 1982a). The Cambridge Expedition to the Torres Straight in 1898 brought along 16 mm cameras to capture everyday life, ceremonies, and other cultural practices seen on the island. The anthropologist Franz Boas used film in 1930 to capture the everyday life of the Kwakiutl, and attempted to have several of his colleagues analyze it. One of the first clear examples of a scholar using the unique properties of film to advance understanding of a scientific question through detailed microanalyses is a study carried out by one of Boas' students, David Efron. Efron's dissertation, published in a book (Efron, 1941) took up the question of whether gestural style was the result of cultural influences or was biologically based. This nature versus nurture question was animated by Nazi claims that gestures were genetically determined. Efron's approach involved comparative studies of the gestures of men who differed in the degree to which they had adopted American culture. He filmed the natural interactions of first generation Italian immigrants on street corners in New York City as well as other immigrant groups. His analysis of gesture was made possible by his collaboration with an artist who used the film to create frame-by-frame illustrations that allowed for precise descriptions of the form and functions of gesture. He was able to show loss of traditional gestural styles over time and in relation to the degree to which there were deeply felt ties to tradition, strongly countering the claim that gestural style was biologically based.[1] He also showed the situational specificity of gesture use, for example, showing that in some settings traditional gestures were more likely to be used.

Around the same time, between 1936–1939, Margaret Mead and Gregory Bateson used film and photographs in combination with field notes to document parent–child interaction, ceremonial dance and practices, and gesture in Bali (Bateson & Mead, 1942). These documentary approaches were interwoven on a daily basis and Mead produced field notes to annotate the film and still photographs. These records were used to analyze gesture and body movement but they also produced several films describing patterns of child development for use in classes and as a means to share their work with the general public. The records were used in meetings with an interdisciplinary group of scholars to elicit their unique perspectives on the phenomena captured in the film.[2] In these films, Mead narrates and points out phenomena to the viewer and some text is provided that gives additional context. Occasional slow motion is used to emphasize a particular aspect of movement (e.g., Mead & Bateson, 1952).

Margaret Mead reportedly influenced younger anthropologists with these films including Ray L. Birdwhistell whose primary research focused on kinesics (Birdwhistell, 1952, 1970; Davis, 2002). As part of a larger network of researchers interested in the role of context in human activity, he studied gesture, emotional expression during family interactions, and was particularly focused on cross-cultural studies as a way to show the particularity and context specificity of expression in relation to meaning. Along with others, he used film analysis to contribute to an understanding that body movements and verbal communication are linked in multiple ways and can be contradictory, complementary, or reinforcing. In an interdisciplinary project that began in 1959, he collaborated with Bateson and a varied group of psychiatrists, linguists, and anthropologists to study family interactions during a therapy session. This work was reported in a still unpublished book, available only on microfiche, called the *Natural History of the Interview* (Bateson et al., 1971). Birdwhistell and Bateson collaborated on additional studies of parent–child communication, in particular analyzing the interactions of schizophrenic children and their parents. It was in this work that phenomenon of the "double bind"—defined as the delivery of contradictory messages—was articulated. Finally, Birdwhistell also created teaching films to share his approach and major insights, for example in "Microcultural Incidents in Ten Zoos," he compared the interactions of families visiting zoos from a variety of countries including the United States, Italy, and France (Birdwhistell, 1971). The film focused on the approaches and responses by families to the elephants kept in each zoo and this method helped to highlight cultural differences. In a review of the film (C. Bateson, 1972), it was noted that this production made clear that anthropological films had potential to go way beyond the entertainment function that they typically held.

---

[1]The National Anthropological Archives recently acquired several hundred original captioned illustrations created by Stuyvesant Van Veen, the artist who collaborated with David Efron.

[2]The entire collection of their film is archived by the Library of Congress as part of the Margaret Mead Collection along with maps, photographs, field notes, and art.

Interactional Spaces as Sites That Maintain or Challenge Inequity
Other social scientists used film and video to take up the question of how inequity emerges and persists. McDermott and Roth (1978) provide a review of work on the social organization of behavior and use it to argue for the reformulation of the micro and macro distinction so prevalent in the social sciences. They highlight how studies of people in interaction allow analysts to see the mechanisms by which people organize one another's behaviors to produce larger patterns of interaction that traditional macro studies treat more generally with terms like status, gender, ethnicity, or role. In their words, "Whatever form of inequality people are doing to each other, they do it in facing formations and with talk" (p. 338). Collectively, the studies they review provide evidence for how what ultimately happens between people is emergent, and depends on what happens from moment to moment. A key insight they emphasize, quoting Bateson, is that contexts are made up of actions, and each action can be thought of "as part of the ecological subsystem called context and not as the product or effect of what remains of the context after the piece which we want to explain has been cut out from it" (Bateson, 1972, p. 338).

Many studies of face-to-face interaction in educational settings were motivated by controversial claims that achievement gaps between ethnic groups were biologically based (e.g., Jenson, 1969) and that the speech of lower class people was linguistically inferior and inadequate to the demands of modern schooling. Anthropologists, linguists, and sociologists drew on the early accomplishments of interaction analysts to produce finely detailed studies using video and audiotape as data sources to counter these generalizations. As Mehan (1998) highlights, these methods raised very different questions than those articulated by traditional studies of inequality that focused on relationships between distal variables (e.g., what demographic characteristics of families are correlated with achievement). In contrast, interactional studies framed questions in ways that asked for detailed behavioral descriptions of people in face-to-face interaction (e.g., how is inequality in turn taking arranged for in small groups). For example, McDermott (1976) showed through detailed analysis of the interactions of a low-achieving reading group how both teachers and students contributed to interactions that resulted in the absence of help for those who needed it most and overall differential access to learning opportunities relative to those in top reading groups. Another classic demonstration of how interactions arrange for the reproduction of inequality was a study of college counselors interacting with students (Erickson, 1975). This work represents a new paradigm for understanding inequality, shifting the view of its grounds as states or traits to dynamic and mutually constitutive relationships between people and their environments that can change from moment to moment and that are subject to repair (Mehan, 1998).

Language Socialization and Socialization Through Language In the late 1970s, the field of language socialization emerged from the broader area of anthropology (Schieffelin & Ochs, 1996). It is an interdisciplinary field to which scholars from communication, education, and psychology have contributed. Researchers in this field are committed to linking microanalytic accounts of everyday, mundane conversation to broader ethnographic accounts of the activities, beliefs, and practices as revealed by studies of families and communities. This field assumes that language is a primary means through which children are socialized to the values, practices, and worldviews of the communities within which they develop and that children are socialized to use language in particular ways that reflect these deeper underpinnings of a community. Language socialization then is implicated in broader spheres of human development including cognitive development, identity, and gender roles. The goals of this program of research include understanding how communicative practice is organized at the level of routine events, how language practices change or remain stable across situations, and how culture can be understood to be reflected in these patterns. This commitment to link levels of analysis is what distinguishes language socialization from more established fields that also focus on language such as language acquisition and developmental pragmatics (Schieffelin & Ochs, 1996). For example, analyses of dinner time conversation in families offers evidence that epistemologies are socialized even as the immediate activity at hand is organized for other purposes (Ochs, Taylor, Rudolph, & Smith, 1992). The patterns that were observed in this study included the frequent construction and evaluation of theories as families engaged in problem solving narratives. The authors argue that these episodes of collaborative problem solving were particularly rich for intellectual development as they required cognitive decentering that was facilitated by the intimacy and trust that family relationships afford. These dinner time conversation studies have also taken up questions about how dinner time narratives reflect and contribute to gender socialization (Ochs & Taylor, 1996) and to socialization of self and identity (Forrester, 2001).

Ecologies of Peer Interaction   It is probably no accident that one of the first psychologists to use film to advance our understanding of peer interaction and child development was Kurt Lewin (Luck, 1997). Lewin was an experimental and applied psychologist who took up studies of group dynamics, leadership, and the role of the environment in child development. As early as 1926, Lewin began making films of natural child interaction using a 16mm Kinamo, a hand-held camera designed for the nonprofessional filmmaker (Van Elteren, 1992). He followed his own children through their daily activities as well as those of friends and relatives. His goal was to illustrate aspects of his developing theory of the life space, and film was a representational medium that could do this well. Lewin argued that the psychological life space resulted from the interdependencies between the environment and the person and that to understand behavior, both of these must be taken into account (Lewin, 1936, 1951). Prior experience and current tensions and needs were all part of what the person brought to the situation and "field of the child" is created by both the environment and the child's current psychological state.

To illustrate these concepts during a presentation in 1930 at Yale University, Lewin showed a short film about a pair of toddlers, each trying to sit on a stone for the first time. In this film, he shows the persistent strategies of the toddlers who both try to find a way to sit down while not taking their eyes off of the stone. Apparently the film made a significant impact on the audience and the personality psychologist Gordon Allport claimed that it forced some of the American psychologists in the audience to "revise their own theories of the nature of intelligent behavior and learning" (Marrow, 1969, p. 50). In collaboration with professional filmmakers, Lewin went on to create a feature length documentary capturing the life and landscapes of a child in an urban setting titled "Das Kind und die Welt," translated as "The Child and the World" (Lewin, 1931). This project illustrates his use of the medium of film to communicate his ideas about interdependencies between the child and his or her life spaces. Lewin also used film to document the interactions that resulted from experimental work on group dynamics. In an article published in 1939 in the *Journal of Social Psychology* entitled "Patterns of aggressive behavior in experimentally created social climates," Lewin, Lippitt, and White (1939) report on their experiments carried out to compare the effect of three different leadership styles—democratic, autocratic, or laissez-faire, on children's social interactions and industry during collaborative craft activities. Although most of the data for analysis was collected through naturalistic observations, the film was used to confirm the patterns of data as we see in the following report from their article:

> There are the judgments of observers who found themselves using terms such as "dull," "lifeless," submissive, repressed, and apathetic in describing the non-aggressive reaction to autocracy. There was little smiling, joking, freedom of movement, freedom of initiating new projects, etc.; talk was largely confined to the immediate activity in progress, and bodily tension was often manifested. *Moving pictures tell the same story*. (italics added, p. 283)

The methodology section of this article reports a heroic effort to record the boy's interactions. Continuous stenographic records of conversation were collected, a predetermined coding scheme was used to capture directives and responses, a descriptive running record of activity in each subgroup was recorded, and a minute-by-minute structure analysis of the groups was made. Four observers were present, sitting behind a low burlap wall, and each had their own observational task. Film was used to "make movie records of several segments of club life" (p. 274). Like many anthropological films, the purposes of filming were to confirm patterns that were observed and coded, and to illustrate types of interactions rather than to serve as a data source per se. A 31 minute narrated film was created based on these experiments and is still available for purchase (Lewin, 1938). The film shows prototypical group interactions in the three experimental conditions. It also includes graphs summarizing the coded data collected during the experiment and appears to have been produced to share with both research colleagues and nonspecialist audiences.

More recently, video has been used by sociolinguists, psychologists, and linguistic anthropologists to obtain records of children's talk and interaction with minimal adult presence (Ervin-Tripp, 2000; Topper & Boultan, 2002). Video-based studies of play episodes among peers have helped to understand how conflicts can be regulated through explanation of resistance to the proposals of partners (Gottman, 1983); how cooperative play sessions are sequentially related to parallel play between peers (Bakeman & Brownlee, 1982) and how competence in peer interaction and parent– child interaction quality are related to adapting to a new sibling (Kramer & Gottman, 1992). Studies of the emergence of a negative reputation and subsequent rejection by aggressive boys has been studied by analyzing the patterns of interaction in play groups of previously unacquainted peers over time as they relate to the development of reputation and status in the group (Coie et al., 1999). Some research is beginning to make links between institutional level practices and between peer interaction. For example, Ochs, Kremer-Sadlik, Solomon, and Sirota (2001) videotaped interactions between autistic children and their peers at several schools. One of their findings was the role of student atmosphere and classroom practices on positive inclusion in activities. Similarly, Matusov, Bell, and Rogoff (2002) found between school differences in the quality of peer collaboration and were able to link it to classroom practices.

In a recent review, Kyratzis (2004) summarizes research that has examined how children elaborate games, how conflict talk contributes to peer culture, how identities are talked into being through peer conversation, and how adult culture is resisted in peer activity. These studies demonstrate the ways that children socialize one another and oppose established norms rather than simply adopting those that are conveyed by the adult world. Collectively, the studies she reviews suggest that constructs such as social competence, need to be elaborated to include linguistic practices that allow children to position themselves and others and to alter participant frameworks. For example, in video-based studies of elementary school children playing hopscotch, the analyses suggested that sophisticated multimodal strategies were used to challenge fouls and establish social order (Goodwin, Goodwin, & Yaeger-Dror, 2002). Intonation, gesture, body position, and pitch were all critical to communicating stance and achieving joint recognition of when rules were violated. Grounds for rejecting turns were established through explanation or replaying the move. Other discourse-based studies of peer groups have focused on the ways that status is established through talk that forges alliances, often by excluding others. These ethnographic accounts paint a very different picture of leadership than do studies of social competence that rely on rating scales or responses of children to vignettes of socially ambiguous situations and highlight the importance of interdisciplinary work that can capture peer interaction as it occurs in everyday settings.

The Social Infant. Video studies have made important contributions to our understanding of early emergence of sophisticated social awareness in infancy and the bidirectional influences between caregiver and child (Lewis & Rosenblum, 1974). Trevarthen and Aiken (2001) review the important role that film-based studies had in documenting the coordinated interactions of mothers and their infants during naturalistic interactions. Methods of conversational analyses were adapted to provide accurate measurements of the timing of contributions of mothers and infants. Infants as young as 2 months were found to mutually regulate interests and feelings with their mothers. Video studies have helped document early social competencies of infants such as the capacity for joint attention (Adamson & Bakeman, 1991) social referencing (Walden & Baxter, 1989; Walden & Ogan, 1988), affective engagement with partners (Striano & Berlin, 2005), and imitation (Meltzoff & Moore, 1983).

Intimate Partners and Family Interactions

The work on early infancy also led to new analysis approaches such as sequential and times series methods for describing patterns of interaction that were taken up in studies of adult interaction. In particular, studies of dyad-level differences in interactional quality among married partners were pioneered by Gottman and his colleagues. Gottman and Notarius (2002) review the progression of research on marital relationships that began in the 1950s. They mark the publication of Bateson, Jackson, Haley, and Weakland (1956) on the double bind as a turning point in this line of research. It led to a shift from personality-based explanations to studies that observed couples in interaction and focused on processes of communication. In the 1970s, Gottman developed methods that involved both videotaping interactions of couples at a specially constructed "talk table" while the couples also rated aspects of their own communication. Out of this work came the general finding that unhappy couples' interactions were marked by negative affect and a greater likelihood of reciprocating negativity than those of happier couples. This research was aided by other applications of video, namely the study of facial expressions of emotion that led to a coding system called the Facial Action Coding System (Ekman & Friesen, 1978; Ekman, Friesen, & Ellsworth, 1978). At the end of the review, Gottman & Notarious (2002) call for moving the research on couples' interactions out of the lab and into the home. They argued that this move would advance our understanding of the role of emotional regulation in family well being and would provide more ecologically valid accounts of the ways that conflicts emerge and resolve than do lab studies where conflicts are generally stimulated (e.g., couples are asked to discuss a matter that is likely to evoke conflict). This call for more ecological research echoes trends in the social sciences more generally where there is a focus on activity systems as they emerge in real-life contexts. An example of one such effort is the Center on Everyday Lives of Families (see http:// www.celf.ucla.edu/). In this interdisciplinary research effort, one of six centers funded by the Sloan foundation, a video archive is being collected and analyzed in order to understand the ways that families manage the challenges faced by dual-career couples and their children.

Learning Through Activity With Peers, Parents, and Community

The last area of research I'll revview is most closely linked to the contributions in this section. Research programs organized around developing *cultural-historical-activity theory* have used film to carry out microanalyses that focus on the ways that divisions of labor emerge in collective activities and how artifact and cultural tools mediate thinking and learning (Engestrom, 1999; Lave & Wenger, 1991). Analyses of video has contributed to our understanding of learning through game play (e.g., Guberman & Saxe, 1981; Nasir, 2005); learning through apprenticeships; learning through work (e.g., Beach, 1993); learning through collaborative problem solving (Barron, 2000, 2003; Cornelius, & Herrenkohl, 2004; Herrenkohl & Guerra, 1998; Hogan, Nastasi, & Pressley, 1999; Stevens, 2000) and learning from siblings (e.g., Maynard, 2002). Several issues have animated this area of research including the cognitive consequences of schooling leading to comparative studies of learning processes and practices in school and out of school (see Bransford et al., 2006); how cognitive and relational aspects of interaction are intertwined in ways that are consequential for learning (Barron, 2003; Cornelius & Herronkohl, 2004); and how individual cognitive change is coupled to change at the community or societal level (Saxe & Esmonde, 2005). In a recent overview of a related theoretical approach, the situative perspective, Greeno (2006) argues that a productive goal for research on learning would involve the integration of methods and constructs from the traditional cognitive science perspective and from interactional studies of learning. Cognitive science has historically focused on individual cognition and worked to understand how people create mental representations, use them in problem solving, and remember. Interactional studies of learning focus on coordination between people and between the material and informational tools they access. Greeno describes an approach that combines aspects of both interaction analyses and an analysis of information structures that are generated and shared in joint activity. Through analyses of videotaped records, the goal is to produce coordinated accounts of learning across multiple levels and foci of analyses.

In summary, the use of video for microethnography of face to face interaction has contributed to advances in our understanding of the context specificity of behavior, the social nature of learning and development, and relationships between institutional practices and face to face interactions. Early landmark contributions were made by a diverse group of scholars who held very different perspectives. These have been followed by a similarly diverse set of investigations designed to continue to draw out the implications of earlier work. Film-based studies have also contributed to the generation of more interdisciplinary approaches to the study of human learning and development as the medium of film serves as an important boundary object for social scientists from different disciplines (Star & Griesemer, 1989). The authors in this section are contributing to this broad program of research.

CURRENT RESEARCH AGENDAS DRIVING THE USE OF VIDEO AS A DATA SOURCE

In the most general sense, the questions that were pursued by early video analysts are still being pursued today: how do persons and environments interact in the genesis of activity, behavior, and new ideas? Of course the questions have taken different forms and there are new theoretical concepts at play. Ecological perspectives have been developed by theorists who came after Lewin (Brofenbrenner, 1979) and activity theory has emerged that articulates the importance of understanding learning systems (Engestrom, 1999; Greeno, 2006).

As the chapters in this section attest, there is a great deal of interest in merging social and cognitive accounts of learning and in understanding how learning takes place within and across the life spaces of homes, neighborhoods, communities, and through distributed resources such as books and computers (Barron, 2004, 2006). For example, Palmquist and Crowley (this volume) investigate how the "islands of expertise" that children develop through everyday interactions with their families influences conversation. They build on ecological approaches to the study of child development and document the importance of understanding the role of prior family interactions and children's prior knowledge in shaping a visit to a museum. They compare the kinds of interactions and conversations that occur when families are accompanied by a child who has developed more or less expertise in the domain reflected in a museum visit.

Doris Ash also focuses on learning in museums. In her work, the unit of analysis is the family and she is particularly interested in linguistically and culturally diverse families. She asks, "How does learning occur over time in families as shown by their increasing appropriation of canonical scientific discourse?" Drawing on Linde's (1993) insights that conversations can accumulate over time as ideas, and themes are revisited and recombined in new ways, "ideas emerge, submerge, and reappear in morphed forms, traceable over time but only in hindsight." To study these processes, she invites families for repeated visits to the museum setting. She wants to understand how these conversations and the everyday ideas that they contain support the development of academic concepts. Vom Lehn and Heath (this volume) ask questions about the quality of experience at museum exhibits and specifically "what kinds of interaction and communication occur with partners who arrived together and with those who one happens to meet and how does explanation occur, when does it arise, how does it emerge and develop?" Callanan, Valle, and Azmitia also ask about the role of conversation between children and parents in the emergence of understanding of scientific concepts. They too follow families in museums but they also set up laboratory contexts that are designed to stimulate conversation about scientific phenomena. In particular, they are contributing to an exciting line of work on how gesture mediates and reflects learning (e.g., Goldin-Meadow, 2000) and describe how gestures are an important aspect of both child and parent communication about scientific concepts.

Taking a sociocultural perspective, Angelillo, Rogoff, and Chavajay describe their use of video across studies in a research program that seeks to understand how cultural histories and practices shape the way that people engage with one another. They want to go beyond analyses that isolate individual actions to ones that capture how participants in interaction mutually contribute to social events. They want to understand the nature of intersubjective engagement and how it is culturally linked to experiences. To investigate this, they carry out comparative studies of interactions between families who have different culture histories including the extent to which adults have experienced Western schooling. For example, in some of the work they describe in this volume, parents and children are asked to work on puzzles together. The way that family members share or divide the work is investigated.

Building on research in small group learning, Hmelo-Silver, Katic, Nagarajan, and Chernobilsky have used video to develop a case study of a single group of students who are working together on a problem-based learning unit that focuses on learning and cognition. They want to understand what happens in effective groups, how cognition is distributed, and what role different leadership styles and artifacts play in supporting a group's learning interaction. Finally, the chapter by Engle, Contant, and Greeno describes a research project that was organized to develop deeper accounts of the process of conceptual growth in classrooms. They want to articulate how participation in activities accounts for changes in students' abilities to explain concept of adaptation. They too use a case study approach where student's conversations were tracked over several weeks. They use this case to illustrate the methodology they used to explain one group's progressively deeper engagement in scientific content over several weeks.

As the previous summary suggests, scholars who use video-based data in their research draw on a diverse collection of theoretical and methodological traditions. For example, the authors of these chapters include references to ethnomethodology, sociology, conversation analysis, developmental psychology, situated learning perspectives, cognitive psychology, sociocultural theory, and ecological perspectives. Similarly, a broad range of methodologies and approaches to inquiry can be found. For example, video might be productively collected in an experimental context where conditions are controlled in within-subject designs or participants are randomly assigned to conditions (e.g., Karrass & Walden, 2005; Meltzoff & Moore, 1983; Walden & Ogan, 1988) in order to derive behaviorally dependent measures that are predicted to differentiate conditions. Within the set of chapters in this section, we don't have any examples of experimental work but instead, see single case-study approaches (Engle, Greeno, & Contant; Hmelo-Silver; vom Lehn; Christian) or nonexperimental comparative designs (Palmquist & Crowely; Ash; Angelillo, Rogoff, & Chavajay; Callahan). There are also examples of stimulated naturalistic situations. Callahan and her colleagues bring families into a lab where they all look at the same materials. In museum studies, the participants are recruited after they have chosen to be in a particular place such as a museum or classroom. Despite these differences, all researchers who use video face substantial challenges. There is no single community of practice that has organized around video data and there are few guidebooks or conventions. The following is a sampling of some of the common challenges faced by researchers who choose to use video technology in their work.

CHALLENGES OF VIDEO AS A RESEARCH TOOL

Video comes with a dual set of challenges—challenges of capturing good records and challenges of analysis. Early on, there were calls for the creation of standards for capture and analysis that might help address some of the basic challenges (e.g., Grimshaw, 1982a) and this request goes on but remains controversial in intent and scope (e.g., see Derry, this volume). Although standards may be too limiting, it is helpful to at least have a sense of decision points that are made by researchers as often as they are made without thinking through the implications of choices.

Challenges of Collecting High-Quality Video Records

What, Where, When, and How to Capture

Despite the ease of obtaining high-quality equipment, the capture of high-quality video can be a challenge (Roschelle, 2000). Debates about how to capture phenomena of interest that were present from the beginning are ongoing. Even Mead and Bateson disagreed about the ideal way to capture records of interaction. Bateson preferred a hand-held approach that allowed the camera operator to zoom in on interesting events. Mead wanted a stationary camera placed on a tripod that would yield long sequences and records that could be analyzed by scientists who were not there during the filming (see http://www.oikos.org/forgod.htm for a record of a conversation where they discuss their differing views). Several researchers have pointed out that the first theoretical decisions that are made come way before the analysis phase and include issues such as where to point the camera, what to include in the frame, and when to begin and stop recording (Hall, 2000; Lomax & Casey, 1998). These decisions result in a data source that is already theoretically burdened. This fact became obvious to early adopters of film who had a clear sense of what they were looking for. For example, Birdwhistell argued that full body images were needed for the study of kinesics and he noted that camera people were often tempted to zoom and change frame leading to less than ideal footage. In one famous example, the psycholinguist George Miller notes how hiring an educational video producer to collect nursery school video data for studies of linguistic interaction unfortunately yielded video records that were collections of clips zooming in and panning around the nursery, so that audio recordings became a more useful data source (Miller, 1977).

Hall (2000) lays out five examples of how technical arrangements influence the data record that will be available for analysis. One example described how a technology that helped track eye gaze changed the ecology of normal sense making so much that it raised questions about the validity of findings. In the other examples, decisions about what is important in a scene resulted in shots being taken at different distances that included different aspects of a scene of people in interaction. In his work, he has developed the approach of always trying to get multiple perspectives by using both a stationary camera that can take a wide angle shot and hand-held cameras that can be used to zoom in when participants are using resources or creating representations that might be important for the analysis. Hall's examples highlight that the perspective that is captured by the camera is always influenced by the researcher and ironically, is often a perspective that is not one that any one of the participants being filmed would have had. In most cases, the camera will miss some subtle information that would be available to a close observer. For example, emotional expression offered by the eyes. Compared to a live observer, a camera is much more restricted. Whereas an observer can track movement across the room or quickly focus on an object near at hand or across the room, a camera has limited depth of field (Rochelle, 2000). For these reasons, Erickson (1982) argues that engaging in fieldwork is necessary to understand the setting and should ideally occur before videotaping so that there is a better idea of what to capture. He argues that the video records should be collected with the goals of systematic sampling. He suggests identifying the full range of variation in types of events and then establishing the typicality of these events in terms of frequency.

Desire to Capture Over Time and Context

Although in some ways, a camera positioned in one spot is ideal for later analysis, it can also limit what you see. In addition, researchers may be confronted with the problem of their participant moving. Wireless microphones represent one solution but it might be the case that the actions of that participant could be of interest. In this situation, both stationary and hand-held camera approaches are needed or new innovations such as head mounted cameras might be used. Panoramic video capture is being explored as one means for capturing a fuller contextual view of interaction (see Pea & Hoffert, this volume).

The Presence of the Camera

There is also the classic problem of observer or camera effects. That is, does the presence of a camera fundamentally change the behavior of interest? And, do the participants shape what actually gets videotaped in subtle or not so subtle ways (see Lomax & Casey, 1998 for examples). Although most researchers feel that eventually participants forget about the camera, it is always an issue to consider. vom Lehn and Heath (this volume) always use a stationary camera with no camera person as they feel strongly that the presence of a person operating the camera can constrain interaction. They also echo what Birdwhistell and others have noted—that there is a tendency to zoom and pan that can inadvertently diminish the quality of the record.

Issues of Audio, Particularly in Classrooms, Homes, or Other Very Busy Settings

Capturing high-quality audio can also be a challenge, particularly in classroom contexts. The natural acoustics of most classrooms are horrendous. Couple this with the scraping of chairs, the rustling of paper, and multiple people speaking in a narrow space and often at the same time, the researcher is confronted with an extremely difficult situation. Even professional TV film crews used to shooting in all kinds of difficult situations, and now shooting classroom interactions for a professional development company producing video cases, have commented that classrooms present greater challenges than war zones (M. Atkinson, personal communication, December, 2004). The wise researcher will develop methods to check for sound quality early and often. The research team and future transcriptionists will be quite grateful for that effort.

Informed Consent

Finally, there are a number of issues that arise when considering how best to get the informed consent from research participants (Grimshaw, 1982c). Hall (2000) outlines the different communities that might be interested in the video that was collected for a research project including the research group, students in undergraduate or graduate classes, professional colleagues, teachers, and the general public. University Institutional Review boards currently address the issue of multiple uses in very different ways. At some universities, the IRB provides standard forms that ask parents or participants to indicate what specific uses of videotapes they agree to. Categories might include permission for use as a transcribed event where the video is discarded after transcription, use to analyze within the research team or with professional colleagues, or use to illustrate learning phenomena to a general audience or to show at professional conferences. However, many of us now want to share our video records with colleagues via distributed archives for joint analyses. How do we ensure that the data will be safe and how do we communicate the level of safety (e.g., encryption standards) that our team will employ in ways that can be understood? Given how unpredictable future uses might be, how do we best communicate this to our study participants?

There are other complications that can arise. What happens when not everyone in a school, classroom, or family unit gives their consent (Pepler & Craig, 1995)? And what if someone changes their mind about being involved at a later time? Museum researchers may not be bound to the same regulations as researchers who work in a university setting. Still, as described by vom Lehn and Heath (this volume), a great deal of attention is paid to making sure that signs are posted so that visitors know they might be studied and every attempt is made to open communication channels so that unwilling participants have a way to have their records discarded. Palmquist and Crowley (this volume) and Callahan and Azmitia (this volume) also report on the methods they use to invite museum goers to participate in ways that won't disrupt their visits or seem objectionable. For example, they note the importance of understanding deeply the goals of the institution, establishing good relationships with staff, and family desires and preferences. This latter perspective is essential in order to design procedures that will yield a high proportion of families agreeing to participate in the research.

Challenges for Analysts

Volume of Data

The downside of the ease of collection, and the general belief that capturing video will provide a rich data source, is the volume of potential data that can result. There is a tendency to collect the records first and plan for the analysis later. The desire to capture everything can result in office bookshelves filled with tapes, unanalyzed and often without any index as to what is included on each tape other than a label that indicates place and date. Classroom research is particularly likely to generate a huge amount of data as researchers try to capture the implementation of a multiweek curriculum unit and perhaps in more than one classroom. How does one decide which events to look at and of what length? The wise researcher develops at least some basic systematic approach for cataloging the events that occur on different days so that when it comes time for selecting a place to start, one is not relying on memory for what happened on what days.

Complexity and Richness—The Dual Problems of Selecting a Focus and Learning to See

The volume of data leads to the next challenge—how to reduce the data set in some logical way. Engle, Faith, and Greeno nicely summed this challenge up when they noted that the problem generally is not finding something to talk about but choosing among them and fashioning a coherent account (chap. 15, this volume). If the data set is quite large, strategies for coming to an understanding of what the whole corpus represents need to be developed. Units of analysis must be chosen (events that occur over minutes, group work that occurs over days, a curriculum unit that occurs over weeks) and sampling methods need to be designed so that more fine grained analyses can take place. Decisions about how to create an index need to be made.

Erickson (1982) provided a set of recommendations for those interested in studying face-to-face interaction. He described four stages of analysis and suggested that analysts move from reviewing whole events to increasingly shorter exchanges. He described criteria for recognizing boundaries between events and segments within events. Although his suggestions won't apply to everyone, they provide a nice example of how to develop a systematic approach to selecting what to analyze.

The power of video to capture layers of communicative exchange including dialogue, prosody, and posture is wonderful but can also present an enormous challenge to the research team. Erickson (1982) suggests the strategy of using the technological affordances of video to shift one's perceptual stance by choosing a focus of attention (e.g., gesture or talk) for each replay of a segment. He also suggests watching without sound or listening without viewing as ways to obtain more information about an event.

Re-Representation of Interactions With Transcription

Most often for analyses to proceed, the information in the film or video will need to be re-represented in transcript form. Transcription is costly even when one is transcribing a single person speaking. When there are two or more persons who may speak in overlapping turns the challenges are magnified. Time estimates for transcribing the dialogue of two or more speakers suggest a ratio of 4:1 to 10:1, depending on the detail of the transcription needed, degree of speech overlap, and quality of the audio. Thus, for every hour of speech the team should expect 4 to 10 hours of transcription work. And, this is an estimate for transcribing only the speech. Nonverbal behavior such as gesture, posture, emotional expression, and actions might also be described (e.g., Erickson & Shultz, 1982; Goodwin, 2000; Kendon, 1977). Clearly tough choices have to be made about when to transcribe and what to transcribe. The authors contributing to this section have generated a number of unique approaches to this issue (e.g., some have made the decision to code directly from tape, bypassing the need for transcription). There is no single way to address this—the important point is to come up with some rationale for choices made.

Reporting

Although there have been some attempts to create multimedia journals that could include some video as part of the publication (e.g., Sfard & McClain, 2002), there is general agreement that video records must be accompanied by clear written analyses. In most cases, the video records will be left behind in the reporting phase of the project and what was observed must be re-represented. Coding and subsequent quantification is a common approach to reporting results. However, although our ability to code behaviors can rest on the well worked out techniques and methods described earlier, there is still the limitation of losing the whole feeling of an interaction. Narrative description is another method of representation; however, narrative accounts are not credible to many experimentally minded social scientists. One solution to this is to use multiple methods of representation in any data set. For example, in my own studies of group interaction during collaborative problem solving experiences, I was interested in both coding and statistically analyzing aspects of interactive behavior, such as how ideas were responded to in more and less successful groups. I found patterns that reliably differentiated more and less successful groups. However, the ways that these sequences unfolded for individual groups differed in some important ways that were masked by the quantification. In my case, I chose to combine what Bruner (1986) described as a paradigmatic approach (coding and statistical analysis) with narrative approach (that preserved the sequence of interactions). Within the narrative approach, I used three types of representation to convey the complexity of the interaction. First, I used transcripts to illustrate key aspects of dialogue; second, I provided behavioral descriptions that conveyed aspects of the interaction such as facial expression, tone, gesture as they occurred across short periods of time; and third, I used still frames to further illustrate the body positioning of the interacting students at key points. The theoretical explanation for how groups managed to utilize the collective knowledge and cognitive capacities in their group relied on both these narrative accounts where I was able to describe more fully the feeling of the interactions and the codes that were measures of types of reactions and of aspects of joint attention. The problems of re-representing the complexity in video are not trivial and we are in the beginning stages of figuring out field creative ways to do this. We can learn a great deal from one another's attempts to do this well within and across disciplines.

Vividness of Examples and Generalizability

The richness of the video record can lead research teams to be drawn to particular examples that might be especially vivid or compelling. Once chosen for analysis, a huge amount of time can be invested leading to an unwillingness to give up the example or perhaps even to look for counterexamples. This is not necessarily a bad thing, and in fact, what Erickson has referred to as "cherry picking" can be theoretically productive. However, it raises questions about how one can develop methods that decrease the probability of overemphasizing certain examples. I think here, it is helpful to look at the arguments made by ethnographers and other qualitative researchers who invest in case based data. For example, in contrast to yhe idea of statistical generalization is the idea of logical generalization introduced by David Hamilton and discussed by Erickson (2002). Logical generalization is demonstrated not by statistics but by repeated empirical case studies that confirm similarity in processes across cases.

Another issue that can arise is the magnification of events that may really not be significant to the participants. As Lemke (this volume) suggests, some particular interaction that flies by participants in a matter of seconds may not be noticeable to them. The fact that we as analysts can slow it down and study the nuance does not necessarily mean that the phenomenon has huge import for human interaction.

In summary, video capture and analyses can be extremely time consuming and expensive. Before stepping out into the world to capture records of interest, researchers should ask themselves questions about what phenomena they want to capture, how they might want to use the data to communicate or collaborate with new communities, and they should understand the nature of the place in which they plan to film including aspects of the acoustical and physical environment. Plans for indexing tapes as they are collected can save the research team time and frustration. Fortunately, the researcher new to this kind of work can build on the wisdom of more experienced colleagues. The authors in this section agreed to make visible some of their approaches to working with video that helps to tame the complexity (chap. 13, this volume). Below I summarize insights around four general themes that emerged from my reading of the accounts of their work.

Insights on Productive Inquiry Practices

Importance of Theory Inquiry Cycles Before, During, and After Data Is Collected

Designing a plan for research with a set of questions and ideas about the phenomena that one wants to capture and record is a sensible approach whether or not one will collect video. However, it is particularly important when video recording will be the primary tool for data capture. As it was noted earlier, the ease of videotaping interactions makes it tempting to collect the data first and worry about what to do with it later. Our most experienced researchers in this section suggest that substantial time and attention should be devoted to conceptualizing the research questions that might be addressed with the video records in advance. Why is this so important? First, reflection on the kinds of questions that might be pursued may fundamentally change strategies for data collection. For example, if one decided beforehand to organize questions about how material artifacts might facilitate coordination in groups, the decision about where to aim the camera might differ or explicit collection and reproducing of the artifacts might take place. Second, different decisions about sampling interactions might be made. If change in groups over time is of central interest, then adequate samples might be collected from predefined points in a group's work. Third, doing some work beforehand, crafting good questions can be critical for the analysis phase of the research project. Having good orienting questions to begin with helps to maintain a perspective that prevents one from getting lost in the details that video records include.

The concern that was described earlier about generalizability of findings can be countered by explicit attention to the logic of one's inquiry and the processes used to create explanations and generate claims. Issues of reliability and validity of all kinds (internal, convergent, external, descriptive) apply to video based data as they do to any other kind of quantitative or qualitative data analysis. At the same time, one wants to remain open to discovering new phenomena. The chapters in this section offer us examples of a variety of approaches. Some of the authors describe processes that share a family resemblance with an approach to qualitative research more generally called analytic induction, developed by Znaniecki (1934). In analytic induction, a few cases are explored in depth and explanations are developed. New cases are examined for their consistency with the explanations and when they are not consistent, the explanation is revised. For example, Engle, Greeno, and Conant (this volume) suggest an approach they call "progressive refinement of hypotheses." In this approach, a general question is framed and records are collected in an appropriate setting. Once records are collected, more specific hypotheses are formed after some viewing of the records. These hypotheses are then examined in relation to other aspects of the data set leading to more complete explanatory hypotheses. They argue that multiple iterations through hypothesis generation and evaluation lead to greater robustness and increased likelihood that the findings might be replicated in other contexts. As was noted previously, Engle, Greeno, and Conant are interested in conceptual change and designed a data collection plan that would allow them to have pre and post-assessment information that would reflect changes in students' conceptual understanding, and they would have video data of the conversations that were likely to have been generative for that conceptual growth. They articulated a plan for how they would use the records and they represented the problem they were trying to solve in theoretical terms.

Angellio, Rogoff, and Chavaray (this volume) also point out that in their work, the initial questions are first framed at a general level, for example, "what cultural variations and similarities occur in the ways that mothers aid toddlers in problem solving." This general question was then refined as the records were analyzed. Ash describes how, in her program of research, she moved from a general question about how scientific sense making occurs to a series of more focused inquiries about conversations—such as the role of questions or analogies and metaphors in exploring content.

At the same time, one should expect new research questions to emerge from viewing the tapes. Both the Angellio, Rogoff, and Chavaray chapter and the Engle, Conant, and Greeno chapter point to specific findings that emerged that were totally unanticipated. The good questions that they started with were also addressed and the answers informative for their projects but the novel phenomena, they believe, were more theoretically fruitful. vom Lehn and Heath actually begin their analysis directly after they begin collecting records. In what they call a preliminary analysis, they refine their questions and determine whether unanticipated phenomena have emerged for which they might want to develop data capture approaches. They reserve their more intense analysis time for the subsequent records that they collect. This approach is similar to what is considered piloting in experimental approaches to inquiry.

Palmquist and Crowley had also developed specific research questions and a detailed research plan prior to the collection of records. In contrast to Engle, Greeno, and Contant, their research team did make some predictions before collecting data. They predicted that there would be differences between more and less expert children in the roles they took on during their visit to an exhibit called "Dinosaur Hall." To examine this general hypothesis, they decided to collect data on family interaction in the context of this one exhibit hall and designed interviews and assessments to get information about the dinosaur knowledge, interest, and experience of the children and their parents directly after the visit.

Hmelo-Silver, Katic, Nagarajan, and Chernobilsky (this volume) took a single case-study approach. They purposively selected one group to analyze using performance criteria and impressions of the high engagement of group members. They wanted to understand the kinds of interactions that occur in effective groups. Fourteen hours of video were reduced to nine clips that ranged in length from 40 secs to 2 ½ mins. Individual team members generated specific hypotheses after viewing the clips and these hypotheses were refined in whole group discussion. This was a first investigation of the interactions of well-functioning groups and the team plans to design additional studies in order to more fully test the validity of the observations they made about the role of artifacts and leadership as supportive conditions for effective problem-based learning groups.

It is also of note that the authors in this section do not for the most part rely only on the data offered by the video record. As was discussed earlier, the point of view of a camera is always limited. Field notes, photographs of the surrounding field of action, interviews, copies of posted documents, might also be relevant and useful for enriching the video data that will be analyzed at some later point in time and for offering opportunities for triangulation across sources of evidence. For example, vom Lehn and Heath (this volume) collect exhibit specifications, copies of gallery guides, instructions, and carry out interviews not only with visitors but with exhibit designers, curators, museum managers, and educators.

In summary, an experienced researcher may have intuitions about what is going to be interesting and film in a way that is more open ended than the previous advice would suggest and come out with fantastic data. Their intuitions are based on knowledge of the theoretical questions animating the field, familiarity with other empirical findings, and experience in particular settings. Their questions may be more implicit than explicit. For someone who is newer to this kind of analysis, planning up front about what one is after can increase the probability of having interesting contrasts and in collecting the kinds of data that will allow for a systematic analysis that addresses the questions.

For those starting to plan a project that will use video records, it would be wise to focus first on theory-driven questions and develop concrete plans for a first pass at using the video records. Having good questions will help maintain perspective and prevent one from getting lost in detail. At the same time, one should anticipate new discoveries and be ready to articulate questions that can be followed and refined and tested through multiple passes of the video records. Multiple cycles are to be expected and an explicit approach to this can strengthen the likelihood of generating strong findings that are both reliable and valid.

Intermediate Representations Are Critical for Data Selection and Pattern Finding

A second suggestion that can be culled from these chapters is the importance of intermediate representations of the data for identifying which segments to analyze and for understanding patterns within and across segments. Once again, there is a wide range of approaches represented. Some of the authors in this section rely heavily on content logs for identifying segments for analysis. Content logs can be created while video is being collected, for example, in the form of rough field notes. Or, content logs can be created later. They can be extremely detailed, taking a brief standard unit of time (say 3 mins) and describing the major events that took place or they can consist of a several sentence description of the content of a whole hour of instruction. Content logs allow the research team to develop a sense of the corpus of data and facilitate the selection of episodes for further detailed analysis. This kind of indexing should be distinguished from systematic coding. Systematic coding, as we will be discussing, is best done after extensive work has been completed to establish the meaning of codes and the central units that should be coded.

Ash (this volume) has developed a three-part analytical scheme that allows her to analyze scaffolding, everyday knowledge, and biological content and to connect these across time. She begins with a representation she calls the flow chart, which catalogues a family's museum visit from start to finish, including any pre/posttest interviews that were done. The goal is to mark major events and the occurrence of conversations about biological themes. Topics and themes can be coded from this representation to compare families across visits or visits across families. The coding system for identification and categorizing of biological events is itself a complex endeavor that has gone through many iterations. The flow chart representation is also key for selecting the data for her second level of analysis—the significant event. Significant events are selected based on four criteria: 1) They have recognizable beginnings and endings (usually they take place in one exhibit); 2) they have sustained conversational segments; 3) they integrate different sources of knowledge; and 4) they involve inquiry strategies such as questioning, inferring, and predicting. The third level of analysis involves more microlevel analyses of the interactions that occur within significant events. For example, Ash and her team use discourse analytic frameworks to study how ideas are developed over time through the kinds of responses that any utterance affords such as justifying, exemplifying, or reformulating.

Angellio, Rogoff, and Chavaj (this volume) illustrate their approach to representing and re-representing video data in the context of a study that compared mother–child interactions in four distinct cultural communities. The researchers used a protocol that introduced the children to a set of novel objects. The first step that the team took was to generate descriptive accounts of the 1½ hour home visits. These were not event logs but actual descriptions of how mothers helped their toddlers learn about the novel objects in the context of the visit. These accounts were lengthy and often resulted in as much as 30 pages of descriptive writing. These descriptive accounts were not transcripts but were written to help the rest of the research team visualize the sequence of interactions and to capture the purposes and functions of action and dialogue.

As was noted earlier, transcripts of talk and gesture are often needed. Just as it has been argued that the decision about where to point the camera is a theoretical move (Hall, 2000), decisions about what and how to transcribe are argued to be theoretical decisions as well (Ochs, 1979). Even when there is only an audio record, transcripts can vary dramatically in their detail and the kinds of information that is recorded. Like maps of the physical world, the features that are encoded in the representation depend on the purposes of the user of the representation. For example, pauses, overlaps in turns, laughter, intonation, volume, and degree of enunciation are types of information that may or may not be included, in addition to the actual words that people say. Video data adds new possibilities including gesture, posture, visual images, and the like. Multimedia transcription raises new challenges and possibilities. vom Lehn and Heath provide an example of how they use different kinds of transcripts and layouts to capture not only dialogue but gaze direction, posture, and hand movements (also see the examples offered by Goodwin, 2000; Kendon, 1977).

Once transcripts are created, the spatial layout of turns can be designed to make phenomena easier to see. For example, some researchers create conversation maps of various kinds. When completing my studies of peer interaction and problem solving, I was inspired by the representations of conversation used in an article by Resnick, Salmon, Zeitz, Wathen, and Holowchak (1993). They were shared in a two case comparison of groups of college students to show how conversations about the same content could take very different turns depending on the tone and tenor of participants. In my study, transcripts of the turns of each speaker were entered into a unique column and turns were linked by arrows labeled according to an emerging and, at that point, dynamic coding scheme. These maps covered my office walls for a period of months and, although they were not used to communicate findings, they were critical in helping me see patterns of differential responding to problem-solving proposals that were then key to my later quantitative coding and qualitative analysis described earlier (Barron, 2003). Still images from the video were also used as a data source once I had developed the insight that joint attention was a key feature that differentiated more and less successful triads. The images held still aspects of interaction around a particular turn that included a proposal for a solution and allowed me to get a clearer look at the body postures, gestures, and degree of mutual orientation that co-occurred with the dialogue and helped to compare more and less successful turns within and across groups.

Coding and Re-Representation of Video Data Are Critical Processes

It is by no means universal that video records are coded in a way that can yield quantitative data. Many researchers prefer to focus on examples and do not care for counting types of events within or across cases. However, others find coding and quantification a useful aspect of their project. Erickson (1977, 1982, 1986) has written extensively about possible roles of quantification in qualitative research and has a useful discussion of the synergies between approaches. He argues that determining what to count (the qualias or kinds of entities) is more challenging than doing the actual counting. Other excellent discussions of the development and use of observational coding schemes and associated statistical techniques include a primer on the topic of sequential analysis by Bakeman and Gottman (1997) and a paper by Chi (1997).

The authors in this section who discuss their approach to coding describe the development of codes as an iterative process. Like the processes of generating questions or creating representations, the development of a coding approach benefits from iterative cycles, distributed expertise, and moving across levels. For example, Angelillo, Rogoff, and Chavajay begin their chapter by offering a critique of approaches to the study of social interaction that code individual acts by participants and then relate them statistically. They suggest that this approach misses the core of phenomena of social interaction, which is mutual constitution of events by participants, a perspective they share with ethnomethodologists. In their chapter, they describe one approach to investigating patterns of shared engagement that combines qualitative and quantitative approaches. The core of the process involves close ethnographic analysis of a few cases in order to build up a coding scheme based on the observed phenomena that can then be applied to multiple cases. They illustrate this approach using two studies—one that focused on cultural variation in mother's and toddler's contributions to understanding novel objects across four culturally distinct communities, and the other that investigated patterns of joint activity between Guatemalan Mayan mothers and children completing puzzles. The research team went into their cross-cultural analyses with some ideas of the kinds of interactions that might differ across the four cultural groups; for example, the relative reliance on words versus nonverbal demonstration. However, as is the case with many video studies, the video-based data of interactions led to the discovery of new phenomena such as differences in the ways that the mothers from different cultures motivated engagement. Once these phenomena were identified, the team worked to refine the definitions of the categories so that they could be reliably coded. Angelillo, Rogoff, and Chavajay also make the important point that it is important not to be blind to the histories and intentions of research participants. In their work, they seek to gain an insider perspective as they develop their codes. In addition, they continually compared the definitions with individual cases to ensure that they were not distorting the researchers' understanding of the interactions to fit the codes. Once coded, the team used graphs to display codes for individual mother–child dyads in addition to carrying out statistical analysis to confirm that differences were statistically significant. Another representational innovation that turned out to be important for the team was the creation of a diagramming method that allowed the researchers to characterize types of coordination around shared tasks that involved multiple people. In the end, the diagramming resulted in a four-level scheme of types of mutual engagement and these "birds eye" top view diagrams were used to help code videotapes at 1-minute intervals.

Systems of analysis clearly develop over the course of multiple research projects. Ash articulates the changes that have occurred in her coding system and the evolution that resulted in a system they call *tools for observing biological talk over time* (TOBTOT). Through the careful analysis of the talk of families, consultation with biologists, psychologists, and educators, and the work of her research team, they believe they have come up with a system that can be used across projects and not only by their team. She notes that more than a dozen iterations have occurred to get to what they consider now to be a stable and generative system.

Interpretive and Question Generation Activities Benefit From Explicit Social Processes

One of the first papers reflecting on group processes of video analysis described an approach they called *interaction analysis* that included group viewing of video where analysts would be free to stop the tape at any point for discussion and where insights were later harvested by organizers of the analyses from audio recordings of group sessions (Jordon & Henderson, 1995).

Many of the authors in this section explicitly highlight the importance of a variety of social processes as core to their research practice. Rogoff uses the distributed ethnographic analysis of the team members in research meetings to generate codes. In addition, they have developed explicit group activities to support the development of precise questions that will be addressed through their more refined coding. One of these is what they call a "focusing exercise" that involves creation of research questions and coding definitions with the purpose of identifying which could be most productively examined by the data at hand and which were more peripheral. In addition, this exercise is used to help the team think ahead to how they will be summarizing the data. This research team also uses coders who are unaware of the hypotheses of the studies to help articulate the coding scheme and they work to imagine the responses of particular reviewers to their coding systems as a way to push their articulation of clear codes. Although reliability of the coding schemes is critical, these authors make the point that their goal is not to be blind to information about the participants but rather to know their cultural history in as much detail as possible. Because the work involves comparisons of interactions across cultures, insider perspectives are particularly important to bring to the group so that the team can better understand the meanings behind particular actions, gestures, and terminology. Recursive cycling between ethnographic analysis and coding help ensure validity and generalizability for the Rogoff team.

Engle and colleagues also used group viewing sessions relying on the internal research group but also invited outside experts to join in. Ash solicits the expertise of biologists and science educators to elaborate and check the validity of her coding schemes. vom Lehn and Christian occasionally have large viewing meetings where students and museum practitioners join in. Palmquist and Crowley are now collaborating with a sociolinguist who is directing the research team's attention to a number of dimensions that they would not have focused on and they note their ever-changing perception of their data as they sit down with colleagues who come with different interests or disciplinary perspectives.

The Hmelo-Silver, Katic, Nagarajan, and Chernobilsky study provides an example of inviting research participants in to share their reflections on the videotaped records of the interactions to which they contributed. The team carried out 2-hour interviews that involved showing two of the key participants the nine clips that were the focus of analysis. The reflections on these clips were used to enrich the analysis and check on the meaning of certain actions observed in the tape. This approach has been used in studies of teacher decision making and can be considered as a relation to methods developed for use with other kinds of media, such as photographs (see Harper, 2000 for a discussion of photo elicitation techniques). It is an intriguing way to add to the data available for analysis and I expect that its use might become more refined in years to come, as a way to bring both emic and etic perspectives to an inquiry (Pike, 1954).

SUMMARY AND FUTURE DIRECTIONS

Video analysis can be extraordinarily productive as a way to deepen our understanding of learning and human interaction. From early in the history of its development, film was used as a tool to help overcome the limits of real-time human information-processing capacities. Like the invention of the microscope or the telescope, film radically increased our perceptual power, making the invisible visible and subject to analyses (Asch, Marshall, & Spier, 1973; Davis, 2002). Familiar processes can be made strange by slowing them down or speeding them up. It provides social scientists with new ways to test theories and to challenge simplistic explanations of how the world works and perhaps more importantly, it is a vehicle for discovery and encourages interdisciplinary collaboration.

As we look to the future of video research, it seems likely that much will be gained from the development of video collaboratories, (see chapter by Pea & Hoffert, this volume). When researchers come together with common interests and unique data sets, there are rich opportunities for increasing the generalizability and validity of our findings. TalkBank, led by Brian MacWhinney (see his chapter, this volume), is one initiative that may increase our collective capacity for analysis and learning by sharing data. By looking across data sets, we can capitalize on the distributed efforts of researchers across disciplines and advance our understanding of peer, family, and informal learning while setting the stage for the development of more comprehensive and valid theories of learning and development.

REFERENCES

Adamson, L. B., & Bakeman, R. (1991). The development of shared attention during infancy. *Annals of Child Development, 8,* 1–41.

Asch, T., Marshall, J., & Spier, P. (1973). Ethnographic film: Structure and function. *Annual Review of Anthropology, 2,* 179–187.

Barron, B. (2000). Achieving coordination in collaborative problem-solving groups *Journal of the Learning Sciences, 8,* 403–436.

Barron, B. (2003). When smart groups fail. *The Journal of the Learning Sciences, 12,* 307–359.

Barron, B. (2004). Learning ecologies for technological fluency: Gender and experience differences. *Journal of Educational Computing Research, 31,* 1–36

Barron, B. (2006). Interest and self sustained learning as catalysts for development. *Human Development, 49,* 193–224.

Bakeman, R., & Brownlee, J. R. (1982). The strategic use of parallel play: A sequential analysis. *Child Development, 51,* 873–878.

Bakeman, R., & Gottman, J. M. (1997). *Observing interaction: An introduction to sequential analysis.* New York: Cambridge University Press.

Bateson, C. (1972). Film review of microcultural incidents in ten zoos. *American Anthropologist, 74,* 191–192.

Bateson, G. (1972). *Steps to an ecology of mind.* New York: Ballantine.

Bateson, G., Birdwhistell, R. L., Bronsin, H., Hockett, N. A., McQuown, N., & Fromm-Reichmann, F. (1971). The natural history of an interview. *Collection of Manuscripts in Cultural Anthropology* (Series 15, Nos. 95–98). Chicago: University of Chicago Library Microfilm.

Bateson, G., Jackson, D. D., Haley, J., & Weakland, J. H. (1962). A note on the double bind. *Family Process, 2,* 154–161.

Bateson, G., & Mead, M. (1942). *Balinese character: A photographic analysis.* New York: New York Academy of Sciences.

Beach, K. (1993). Becoming a bartender: The role of external memory cues in a work-directed educational activity. *Applied Cognitive Psychology, 7*(3), 191–204.

Birdwhistell, R. L. (1952). *An introduction to kinesics.* Louisville, KY: University of Louisville.

Birdwhistell, R. L. (1970). *Kinesics and context.* Philadelphia: University of Pennsylvania Press.

Birdwhistell, R. L. (1971). *Microcultural incidents in ten zoos* [Motion picture]. (Available from Penn State Media Sales, 237 Outreach Building, University Park, PA 16802-3899.

Bronfenbrenner, U. (1979). *The ecology of human development: Experiments by nature and design.* Cambridge, MA: Harvard University Press.

Bransford, J. D., Barron, B., Pea, R., Meltzoff, A., Kuhl, P., Bell, P., Stevens, R., Schwartz, D., Vye, N., Reeves, B., Roschelle, J., & Sabelli, N. (2006). Foundations and opportunities for an interdisciplinary science of learning. In K. Sawyer (Ed.), *Cambridge Handbook of the learning sciences* (pp. 19–34). New York: Cambridge University Press.

Bruner, J. (1986). *Actual minds, possible worlds.* Cambridge, MA: Harvard University Press.

Chi, M. T. H. (1997). Quantifying qualitative analyses of verbal data: A practical guide. *Journal of the Learning Sciences, 6,* 271–315.

Coie, J. D., Cillessen, A. H. N., Dodge, K. A., Hubbard, J. A., Schwartz, D., Lemerise, E. A., & Bateman, H. (1999). It takes two to fight: A test of relational factors and a method for assessing aggressive dyads. *Developmental Psychology, 35,* 1179–1188.

Collier, J., & Collier, M. (1986). *Visual anthropology: Photography as a research method.* Albuquerque: University of New Mexico Press.

Cornelius, L. L., & Herrenkohl, L. R. (2004). Power in the classroom: How the classroom environment shapes students' relationships with each other and with concepts. *Cognition and Instruction, 22,* 467–498.

Davis, M. (2002). Film projectors as microscopes: Ray L. Birdwhistell and microanalysis of interaction (1955–1975). *Visual Anthropology Review, 17,* 39–49.

De Brigard, E. (1975). The history of ethnographic film. In P. Hockings (Ed.), *Principles of visual anthropology* (pp. 13–44). The Hague, Netherlands: Mouton.

Efron, D. (1941). *Gesture and environment.* New York: King's Crown Press.

Ekman, P., & Friesen, W. V. (1978). *Facial action coding system.* Palo Alto, CA: Consulting Psychologists Press.

Ekman, P., Friesen, W. V., & Ellsworth, P. (1978). *Emotion in the human face: Guidelines for research and an integration of findings.* New York: Pergamon Press.

El Guindi, F. (1998). From pictorializing to visual anthropology. In H. R. Bernard (Ed.), *Handbook of methods in cultural anthropology* (pp. 459–511). London: Sage/Altamira Press.

Engestrom, Y. (1999). Activity theory and individual and social transformation. In Y. Engestrom, R. Meittinen, & R.-L. Punamaki (Eds.), *Perspectives in activity theory* (pp. 19–38). New York: Cambridge University Press.

Erickson, F. (1975). Gatekeeping and the melting pot: Interaction in counseling encounters. *Harvard Educational Review, 45,* 44–70.

Erickson, F. (1977). Some approaches to inquiry in school-community ethnography. *Educational Anthropology Quarterly, 8*(2), 58–69.

Erickson, F. (1982). Audiovisual records as a primary data source. *Sociological Methods & Research, 11,* 213–232.

Erickson, F. (1986). Qualitative methods in research on teaching. In M. C. Wittrock (Ed.), *Handbook of research on teaching* (3rd ed., pp. 119–160). New York: MacMillan.

Erickson, F. (2002). Qualitative research and combined methods. Retrieved January 20, 2006 from http://www.nifl.gov/nifl/webcasts/20020315/transcriptb3–15.htm

Erickson, F., & Schultz, J. (1982). *The counselor as gatekeeper: Social interaction in interviews.* New York: Academic Press.

Ervin-Tripp, S. M. (2000). Studying conversation: How to get natural peer interaction. In L. Menn & N. B. Ratner (Eds.), *Methods for studying language production* (pp. 195–214). Mahwah, NJ: Lawrence Erlbaum Associates.

Forrester, M. A. (2001). The embedding of the self in early interaction. *Infant and Child Development, 10,* 189–202.

Goldin-Meadow, S. (2000). Beyond words: The importance of gesture to researchers and learners. *Child Development, 71,* 231–239.

Goodenough, F. (1928). Measuring behavior traits by means of repeated short samples. *Journal of Juvenile Research, 12,* 230–235.

Goodwin, C. (2000). Action and embodiment within situated human interaction. *Journal of Pragmatics, 32,* 1489–1522.

Goodwin, M., Goodwin, C., & Yaeger-Dror, M. (2002). Multi-modality in girls' game disputes. *Journal of Pragmatics, 34,* 1621–1649.

Gottman, J. M. (1983). How children become friends. *Monographs of the Society for Research in Child Development, 48*(3, Serial No. 201).

Gottman, J. M., & Notarius, C. (2002). Marital research in the 20th century and a research agenda for the 21st century. *Family Process, 41,* 159–197.

Greeno, J. G. (2006). Learning in activity. In K. Sawyer (Ed.), *Cambridge handbook of the learning sciences* (pp. 79–96). New York: Cambridge University Press.

Grimshaw, A. D. (1982a). Forward to special issue on sound-image records in social interaction research. *Sociological Methods & Research, 11,* 115–119.

Grimshaw, A. D. (1982b). Sound-image data records for research on social interaction. *Sociological Methods & Research, 11,* 121–144.

Grimshaw, A. D. (1982c). Whose privacy? What harm? *Sociological Methods & Research, 11,* 233–247.

Guberman, S. R., & Saxe, G. B. (2000). Mathematical problems and goals in children's play of an educational game. *Mind, Culture and Activity, 7,* 201–216.

Hall, R. (2000). Video recording as theory. In D. Lesh & A. Kelley (Eds.), *Handbook of research design in mathematics and science education* (pp. 647–664). Mahwah, NJ: Lawrence Erlbaum Associates.

Harper, D. (2000). Reimagining visual methods. In N. Denzin & Y. Lincoln (Eds.), *Handbook of qualitative research* (pp. 717–730). Thousand Oaks, CA: Sage.

Herrenkohl, L. R., & Guerra, M. R. (1998). Participant structures, scientific discourse, and student engagement in fourth grade. *Cognition and Instruction, 16,* 433–475.

Hogan, K., Nastasi, B. K., & Pressley, M. (1999). Discourse patterns and collaborative scientific reasoning in peer and teacher-guided discussions. *Cognition and Instruction, 17,* 379–432.

Jenson, A. R. (1969). How much can we boost IQ and scholastic achievement? *Harvard Educational Review, 39,* 1–123.

Jordan, B., & Henderson, A. (1995). Interaction analysis: Foundations and practice. *Journal of the Learning Sciences, 4,* 39–103.

Karrass, J., & Walden, T. A. (2005). Effects of nurturing and non-nurturing caregiving on child social initiatives: An experimental investigation of emotion as a mediator of social behavior. *Social Development, 14,* 685–700.

Kendon, A. (1977). *Studies in the behavior of social interaction.* Bloomington: Indiana University Press.

Kramer, L., & Gottman, J. (1992). Becoming a sibling: "With a little help from my friends." *Developmental Psychology, 28,* 685–699.

Kyratzis, A. (2004). Talk and interaction among children and the co-construction of peer groups and peer culture. *Annual Review of Anthropology, 33,* 625–649.

Lave, J., & Wenger, E. (1991). *Situated learning: Legitimate peripheral participation.* Cambridge, England: Cambridge University Press.

Lewin, K. (1931). *Das Kind und die Welt* [The child and the world] [Film].

Lewin, K. (1936). *Principles of topological psychology.* New York: McGraw-Hill.

Lewin, K. (1951). *Field theory in social science: Selected theoretical papers.* New York: Harper & Row.

Lewin, K. (1938). *Experimental studies in the social climates of groups.* Center for Media Production, University of Iowa, 105 Seashore Hall Center, Iowa City, IA, 52242.

Lewin, K., Lippett, R., & White, R. (1939). Patterns of aggressive behaviour in experimentally created "social climates." *Journal of Social Psychology, 10,* 271–299.

Lewis, M., & Rosenblum, L. A. (1974). *The effect of the infant on its caregiver.* New York: Wiley.

Linde, C. (1993). *Life stories: The creation of coherence.* New York: Oxford University Press.

Lomax, H., & Casey, N. (1998). Recording social life: Reflexivity and video methodology. *Sociological Research Online, 3.* Retrieved December 27, 2006 <http://www.socresonline/socresonline.org.uk/3/2/1.html>

Luck, H. E. (1997). Kurt Lewin—filmaker. In W. G. Bringmann, H. E. Luck, R. Miller, & C. E. Early (Eds.), *A pictorial history of psychology* (pp. 282–287). Chicago: Quintessence.

Marrow, A. J. (1969). *The practical theorist: The life and work of Kurt Lewin*. New York: Basic Books.

Matusov, E., Bell, N., & Rogoff, B. (2002). Schooling as a cultural process: Shared thinking and guidance by children from schools differing in collaborative practices. In R. Kail & H. Reese (Eds.), *Advances in child development and behavior* (Vol. 29, pp. 129–160). San Diego, CA: Academic Press.

Maynard, A. E. (2002). Cultural teaching: The development of teaching skills in Maya sibling interactions. *Child Development, 73,* 969–982.

McDermott, R. P., & Roth, D. (1978). Social organization of behavior: Interactional approaches. *Annual Review of Anthropology, 7,* 321–345.

McDermott, R. P. (1976). *Kids make sense*. Unpublished doctoral dissertation, Stanford University, Stanford, CA.

Mead, M., & Bateson, G. (1952). *Trance and dance in Bali* [Film]. Available from Penn State Media Sales, 237 Outreach Building, University Park, PA 16802-3899.

Mehan, H. (1998). The study of social interaction in educational settings: Accomplishments and unresolved issues. *Human Development, 41,* 245–269.

Meltzoff, A. N., & Moore, K. M. (1983). Newborn infants imitate adult facial gestures. *Child Development, 54,* 702–709.

Miller, G. (1977). *Spontaneous apprentices: Children and Language*. New York: Seabury Press.

Nasir, N. S. (2005). Individual cognitive structuring and the sociocultural context: Strategy shifts in the game of dominoes. *The Journal of the Learning Sciences, 14,* 5–34.

Ochs, E. (1979). Transcription as theory. In E. Ochs & B. Schieffelin (Eds.), *Developmental pragmatics* (pp. 43–72). New York: Academic Press.

Ochs, E., Kremer-Sadlik, T., Solomon, O., & Sirota, K. G. (2001). Inclusion as social practice: Views of children with autism. *Social Development, 10,* 399–419.

Ochs, E., & Taylor, C. (1996). "The father knows best" dynamic in family dinner narratives. In K. Hall (Ed.), *Gender articulated: Language and the socially constructed self* (pp. 99–122). London: Routledge.

Ochs, E., Taylor, C., Rudolph, D., & Smith, R. (1992). Storytelling as a theory-building activity. *Discourse Processes, 15*(1), 37–72.

Parten, M. B. (1932). Social participation among preschool children. *Journal of Abnormal and Social Psychology, 27,* 243–269.

Pepler, D. J., & Craig, W. M. (1995). A peek behind the fence: Naturalistic observations of aggressive children with remote audiovisual recording. *Developmental Psychology, 31,* 548–553.

Pike, K. L. (1954). *Language in relation to a unified theory of the structure of human behavior.* The Hague, Netherlands: Mouton.

Resnick, L. B., Salmon, M., Zeitz, C. M., Wathen, S. H., & Holowchak, M. (1993). Reasoning in conversation. *Cognition and Instruction, 11,* 347–364.

Roschelle, J. (2000). Choosing and using video equipment for data collection. In A. Kelly & R. Lesh (Eds.), *Research design in mathematics and science education* (pp. 709–732). Mahwah, NJ: Lawrence Erlbaum Associates.

Saxe, G. B., & Esmonde, I. (2005). Studying cognition in flux: An historical treatment of fu in the shifting structure of Oksapmin mathematics. *Mind, Culture, & Activity, 12,* 171–225.

Schieffelin, B., & Ochs, E. (1996). The microgenesis of competence: Methodology in language socialization. In D. Slobin, J. Gerhardt, A. Kyratzis, & G. Jiansheng (Eds.), *Social interaction, social context and language: Essays in honor of Susan Ervin-Tripp* (pp. 251–264). Hillsdale, NJ: Lawrence Erlbaum Associates.

Sfard, A., & McClain, K. (2002). Analyzing tools: Perspectives on the role of designed artifacts in mathematics learning [Special Issue]. *The Journal of the Learning Sciences, 11*(2&3), 153–161.

Star, S. L., & Griesemer, J. R. (1989). Institutional ecology: Translations and boundary objects: Amateurs and professionals in Berkeley's Museum of Vertebrate Zoology. *Social Studies of Science, 19,* 387–420.

Stevens, R. (2000). Divisions of labor in school and in the workplace: Comparing computer and paper-supported activities across settings. *The Journal of the Learning Sciences, 9*(4), 373–401.

Striano, T., & Berlin, E. (2005). Coordinated affect with mothers and strangers: A longitudinal analysis of joint engagement between 5 and 9 months of age. *Cognition and Emotion, 19,* 781–790.

Tapper, K., & Boultan, M. J. (2002). Studying aggression in school children: The use of a wireless microphone and micro-video camera. *Aggressive Behavior, 28,* 356–365.

Trevarthen, C., & Aitken, K. J. (2001). Infant intersubjectivity: Research, theory and clinical applications. *Journal of Child Psychology and Psychiatry, 42,* 3–48.

Van Elteren, M. (1992). Kurt Lewin as filmmaker and methodologist. *Canadian Psychology, 33,* 599–608.

Walden, T. A., & Baxter, A. (1989). The effect of context and age on social referencing. *Child Development, 60,* 1511–1518.

Walden, T. A., & Ogan, T. A. (1988). The development of social referencing. *Child Development, 59,* 1230–1240.

Znaniecki, F. (1934). *The method of sociology.* New York: Farrar & Rinehart.

Examining Shared Endeavors by Abstracting Video Coding Schemes With Fidelity to Cases

Cathy Angelillo
University of California, Santa Cruz

Barbara Rogoff
University of California, Santa Cruz

Pablo Chavajay
University of New Hampshire

Sociocultural theories of learning treat social interactions as the crucible in which individuals learn and construct the traditions of their cultural communities (Cole, 1996; Lave & Wenger, 1991; Rogoff, 2003; Vygotsky, 1987). Yet, a challenge for research on learning in social contexts is to develop methods to code between-person engagement *explicitly*. Instead, many studies of social interaction have limited coding to a focus on the isolated behaviors of individuals. For example, studies code the number of questions asked or statements made by an adult, which they relate to separate codes of the number of words spoken or errors made by a child with whom the adult was interacting. The relations among such individual behaviors are examined later statistically or simply speculatively as researchers generate interpretations of the data in their discussion section in an attempt to make sense of isolated 'variables' that have been removed from meaningful interpersonal context (Rogoff & Gauvain, 1986). Such coding that breaks down social interaction into a focus on individuals' acts in isolation from other individuals' acts does not directly address the dynamic intersubjective aspects of emerging shared meaning and purposes in group interaction. As a consequence, the coding may not actually address a study's social interactional questions.

Our research team[1] has been developing coding schemes that focus directly on shared engagement among people, examining both the emerging group processes and the contributions of individuals to them. We describe a method of creating coding schemes to examine multiple cases with sufficient precision that different observers can use the schemes with agreement. Our team's method combines the strengths of qualitative and quantitative analyses by abstracting ethnographic accounts of a few cases to examine patterns numerically across cases in a way that maintains fidelity to meanings of individual cases. This approach has yielded several ways of explicitly examining individuals' contributions to shared endeavors as well as the overall form of group engagement.

Moving beyond the limitations of coding isolated behaviors of individuals as if they were independent of each other has been difficult in part because the field has tended to pit qualitative analyses of 'meaningful' social events against quantitative analyses of 'objective' individual behaviors. This unnecessary dichotomy has made it difficult to address the *sharedness* of people's contributions in shared endeavors.

From the sociocultural perspective from which our research team works, the aim is to investigate people's *mutually constituting* contributions to social events as they build on their own and each other's prior contributions. We argue that coding schemes should focus directly on the interactive processes about which researchers want to learn if video studies examine questions of social engagement, intersubjectivity, or mutuality. Coding schemes can directly examine the interrelated processes of group engagements (such as how several people collaborate) or contributions of individuals to emergent social events (such as how one person manages another's attention to an object during a lesson).

This often means that coding schemes require interpretation on the part of coders. However, these interpretations can be cross checked through use of precise definitions and through examination of intercoder reliability. This seems like a more responsible and replicable way of examining social interaction patterns, via explicit checking of the views of several observers, than to leave the interpretation of the relation among coded separate individual behaviors to discussion sections, where authors' interpretations cannot be explicitly checked.

This chapter describes a process for combining the strengths of qualitative and quantitative analyses to examine patterns of shared engagement across cases. We describe our research team's approach to examining mutually constituted social interactions by demonstrating how we develop coding schemes through recursively cycling through various phases of abstraction:

- developing and honing research questions based on the initial focus of the research,
- abstracting ethnographic descriptions of participants' engagements in the activities seen in single cases, and

---

[1]The research team extends beyond the three authors of this chapter to include several 'generations' of graduate students and postdoctoral fellows working with Barbara Rogoff.

- creating and fine tuning coding schemes to apply across multiple cases by abstracting coding categories that retain meanings fitting individual cases.

Using the emerging coding categories on multiple cases then permits graphical and quantitative comparisons across cases, helping researchers to discern differences and similarities in interactive processes across cases and cultural communities (or other comparative questions) in ways that maintain fidelity to the meanings of individual cases.

Unlike individualist approaches, our approach to development of coding schemes addresses both the contributions of individuals to shared endeavors as well as the overall form of shared endeavors themselves. A focus on shared or group processes does not preclude attention to individual contributions—indeed, attending to individual contributions is necessary for the examination of shared or group processes. In this approach, individuals' contributions are viewed as they relate to the contributions of other people, not analyzed as if each individual were acting in isolation.

Using this approach to coding, our research team has investigated similarities and differences across cultural communities in how children's learning is organized. For example, the research team has examined the organization of mothers' and toddlers' interactions as they explored novel objects in four cultural communities (Rogoff, Mistry, Göncü, & Mosier, 1993), patterns of working together among children from U.S. schools that differed in collaborative practices (Matusov, Bell, & Rogoff, 2002), parent volunteers' engagement with children in science activities in a cooperative U.S. school (Matusov & Rogoff, 2002), and group coordination among Guatemalan Mayan mothers varying in schooling experience as they constructed a puzzle with related children (Chavajay & Rogoff, 2002).

In what follows, we describe the process of creating coding schemes to examine mutually constituting social interactions, by abstracting coding categories that permit comparative, numerical analysis across multiple cases from ethnographic descriptions of cases considered one at a time. We begin our description by first emphasizing the importance of focusing and shaping the research question.

WHAT IS THE RESEARCH QUESTION? FOCUSING THE RESEARCH QUESTION AND CODING SCHEME

Key to the process of developing a coding scheme is using the study's working question to guide decisions about what categories should be coded, and how. Without continual reference to and honing of the central research questions, coding schemes run into a number of dangers—especially the risk of trying to capture everything that happens (rather than to focus) and the risk of examining arbitrary variables that do not address the purpose of the research.

A study's research questions may at the outset be rather general, sketching a domain of inquiry that is vague but still useful to guide decisions regarding the coding scheme. For example, Rogoff et al. (1993) began a study in four cultural communities with the broad research question "What cultural variations and similarities occur in the ways that mothers aid toddlers in problem solving?" Such a broad question provides researchers a general focus with latitude that helps them keep an open mind to relevant (and sometimes unexpected) aspects that may appear in their observations of the video data.

In writing and discussing ethnographic accounts of interactions in specific cases, more focused empirical questions may be crafted that address the broader research question, and these more focused empirical questions may serve as drafts for deriving coding categories. For example, one question that Rogoff et al. (1993) derived from the process of analyzing the video cases ethnographically was "How do mothers motivate toddlers' involvement in the activities?" when they noticed that the mothers varied in use of mock excitement or praise with their toddlers. This distinction eventually became a coding category with a precise definition for the final coding scheme.

In several research projects, we have used a "focusing exercise" to bring to the foreground the specific questions that are of most immediate relevance to the broader research question and to articulate the corresponding video evidence that would be required to address them. We show a draft of a research team's engagement in a focusing activity for a study that examined how learning/teaching processes occurred among European American middle-class adult–child dyads varying in experience with computer games (Angelillo, Rogoff, & Tudge, 1997). The focusing exercise in Table 12.1 built on earlier coding schemes developed by Matusov, Bell, and Rogoff (2002) and Matusov and Rogoff (2002).

TABLE 12.1
Focusing Exercise

Main Motivating Research Question	*Specific Empirical Questions*	*Operationalized Categories*
How does the teaching process differ with child versus adult "experts"?	Intersubjectivity (extent of shared thinking)	Collaborative (building on each other's ideas) Division of labor (turn taking; separation of roles) Little to no contact of ideas
	Leadership roles	Mutual (community of learners) Didactic (expert run) Laissez-faire (novice run)
	How does the novice contribute?	Obeys expert Ignores expert Explores on own Draws out expert Reminds expert of need for novice to learn
	Is there guidance from the expert?	Explanation out of context Conversation while collaborating Test questions
	What could learner have learned from guidance?	Goals of game Broad strategies (with purpose beyond immediate decision) Local strategies Just local moves
	How well do they play?	Superficial understanding throughout Some progress Great progress

The process of creating this table helped the research team to cull some aspects of the dyads' interactions that would not address empirical questions related to the main motivating research question (although they might be of interest for some other questions). The process of proposing preliminary definitions for the operationalized categories helped the team focus on the specific aspects of the data to be investigated and to cull categories that overlapped or appeared uninteresting once they were closely examined. This table shows a stage in which the categories that the research team considered were sorted into six empirical questions—which was still overly ambitious for a single study.

In engaging in the recursive cyclical process of honing the research questions and the coding scheme in various studies, our research team has found it useful to ask ourselves periodically whether the draft coding scheme allows us to make conclusions regarding the question under study. We try to think ahead to potential statements summarizing the findings of the study, and without precluding the specifics of the findings, we ask ourselves whether the aspects of the activity that our coding examines will allow us to speak to the topic about which we want to learn enough to make conclusions. In addition to continually checking that the coding categories address the overarching research question, it is also essential to use the coding categories derived from the video data to hone the question itself.

One way to check the appropriateness and clarity of the correspondence between the research questions and coding categories is to pay attention to the confusions of coders-in-training and to anticipate concerns of reviewers. The new perspective offered by coders who are unaware of hypotheses of the study helps not only to articulate the definitions of coding categories but also clarifies the overall coding scheme. The coders' confusions and questions highlight areas of vagueness, overlap, and lack of clarity of purpose in the definitions. Attending closely to these 'glitches' provides crucial information that may help to simplify and focus the study when, for example, it becomes clear that two categories are difficult to distinguish and may need to be combined or new categories need to be added to better capture aspects of interactive processes that are the focus of the study. It is helpful to think of the coders as stand-ins for eventual readers, to whom our coding categories will also need to make sense.

In the next section, we illustrate this approach to analyzing video data by presenting how Rogoff et al. (1993) studied mothers' and children's contributions to shared problem solving in four cultural communities.

CYCLING BETWEEN ETHNOGRAPHIC ANALYSIS OF CASES AND NUMERICAL ANALYSES ACROSS CASES: AN ILLUSTRATION

We illustrate the process of integrating ethnographic analyses of single cases with quantitative analyses of multiple cases with a study that focused on cultural variation in mothers' and toddlers' contributions to distinct forms of guided participation. The study began with extensive ethnographic descriptions of interactions during visits to 14 families in each of four communities (Rogoff et al., 1993). The study eventually utilized dozens of precise coding categories to examine patterns across cases and across communities using case-based graphs and statistical analyses. These quantitative analyses built on in-depth descriptive accounts (averaging about 30 pages each) of the emerging events that occurred across the 1½ hour home visits for each of these 64 cases. The analysis focused especially on how the mothers helped the children figure out how to operate novel objects as well as how to handle problems with some familiar objects (such as getting arms through sleeves and opening a cellophane cookie package).

In beginning to address their loose-starting question, "What cultural variations and similarities occur in the ways that mothers aid toddlers in problem solving?" the researchers (Rogoff et al., 1993) had some idea of what to look for based on the literature and on prior familiarity with the four communities. For example, they planned from the outset to examine the extent of use of words versus nonverbal forms of communication. However, the most interesting features of social interaction that they ended up coding and publishing were ones that came to their notice through writing and discussing their ethnographic descriptions of the videotaped interactions of each family visit.

Abstracting Descriptive Accounts From Raw Video Data

In the descriptive accounts of each case, Rogoff et al. (1993) sought to capture the communicative purposes and functions of the mothers' and toddlers' actions in relation to each other over the course of their interactions. These descriptions were not merely transcriptions of participants' words or accounts of their behavioral moves (although these were included as relevant). Instead, the descriptions reported what the participants said and did to portray how participants coordinated with each other in accomplishing what they did. Here is a sample fragment of the descriptive account from the visit to a 21-month-old middle-class European American child and his family. As they examined a difficult-to-open jar with a peewee doll inside,

> Sandy's mother held the jar up and chirped excitedly, "What is it? What's inside?" and then pointed to the peewee doll inside, "Is that a little person?" When Sandy pulled down on the jar, she suggested, "Can you take the lid off?"
>
> Sandy inspected the round knob on top and said "Da ball."
>
> "Da ball, yeah." his mother confirmed. "Pull the lid," she encouraged, and demonstrated pulling on the knob, "Can you pull?" Sandy put his hand on hers and they pulled the lid off together triumphantly. "What's inside?" asked his mother, and took the peewee out, "Who is that?"

Although the researchers began the process of describing the events with a format that used one column for the mothers' moves and a second column for the child's moves, they found that the meaning of the events was usually better captured by not trying to isolate one partner's moves from the other in such an extreme way. Rather, the participants' moves could be described with respect to others' contributions in a more integrated fashion. The researchers tried to write the descriptions in ways that someone not present could visualize the events that transpired, or even re-enact them (see Rogoff et al., 1993).

Abstracting Coding Categories From the Descriptive Accounts

During the ethnographic analysis, Rogoff et al. (1993) became aware of similarities and differences across all four communities in efforts to bridge between the toddlers' and mothers' understandings of the situation and to structure the toddlers' involvement. For example, they noted striking variations in:

- How mothers and toddlers used language (to give vocabulary lessons or to communicate needed information for shared endeavors);
- How mothers and toddlers attended to ongoing events (simultaneously, in alternation, or even apparently unaware);
- How mothers maintained readiness to assist toddlers;
- How toddlers engaged as members of the group rather than only dyadically or solo;
- How mothers handled toddlers' insistence or refusal to follow the mother's suggestions (by mothers insisting on their own agenda or by allowing the event to follow the toddler's agenda).

Through the abstracted ethnographic analyses and associated discussion of what the researchers discerned in their descriptive accounts of the video cases, Rogoff et al. (1993) began to establish a rough draft of their coding scheme, and then elaborated precise definitions that could be applied across cases and communities. Without the ethnographic analyses, they would have focused only on the categories with which they had begun the study. (The researchers also analyzed and reported those categories, yielding some interesting information, which was not as productive as the categories that emerged from the ethnographic analysis.)

The research team's discussion of the patterns that seemed to emerge from the descriptions was an essential part of the process (along with exploration of the patterns in a preliminary chapter focusing on the ethnographic data· Rogoff, Mosier, Mistry, & Göncü, 1989). The words that the researchers used with each other to try to describe the patterns yielded preliminary ways of describing the coding categories to use to check their impressions systematically across cases and across the four communities. For most of their impressions of the patterns, the preliminary versions of the coding categories required honing to capture the precise meaning so that several of the researchers could look at the same stretch of tape and agree on whether or not they saw the phenomenon in question.

The researchers continually checked their developing coding categories against specific cases, to see whether the categories were capturing what they wanted to address and to make sure they were not introducing misleading contrasts at the same time. They sought categories that would seem "fair" in describing each case, and not impose a meaning on other cases that twisted what was going on.

Employing Graphical and Other Quantitative Comparisons Across Multiple Cases

Quantification of the video data through use of a precise coding scheme permitted the use of graphical as well as descriptive and inferential statistical analyses that revealed striking distinctions in cultural patterns of engagements across the communities. Graphs that maintained information about individual cases were important tools in the comparative, numerical analyses. Utilizing graphs that portray medians, ranges, and middle, upper, and lower quartiles, as well as confidence intervals for comparisons (such as box-and-whisker plots) allowed for examination of the trends across individual cases.

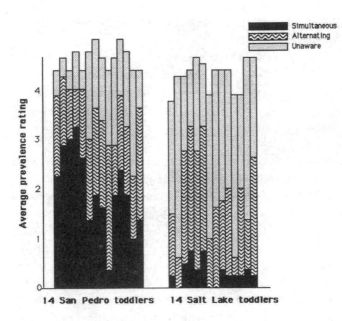

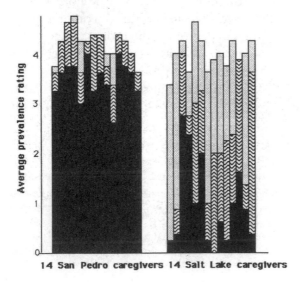

In addition to using box-and-whisker plots to stay close to case-based information, in a number of studies we have developed graphs that display the data for each case in a concise way that encourages examination of patterns across cases. For example, the Rogoff et al. (1993) study presented the case graph shown in Figure 12.1 of 14 individual cases from each of two communities to examine patterns of attentional management during mothers' and toddlers' problem solving. Each bar shows the extent of use of simultaneous and alternating attention, and appearing unaware of events of interest for each case— with the mothers in the bottom two graphs and each toddler displayed directly above their mother.

Figure 12.1.   Case graph of patterns of attention. From "Guided participation in cultural activity by toddlers and caregivers." *Monographs of the Society for Research in Child Development, 58,* Serial No. 236. Reprinted with permission from the Society for Research in Child Development.

These graphs make clear that although there is variability within each community, the ranges barely overlap—for example, the San Pedro mother who used the least simultaneous attention was at the same level as the Salt Lake mother who used the most simultaneous attention.

Rogoff et al. (1993) employed statistical comparisons primarily to check the strength of the patterns discovered in the graphical analyses. The statistics provided confirming support for the strongest patterns in the graphs and provided a cut-off for eliminating discussion of marginal patterns. They were also useful for communicating the extent of the patterns to the readers of the resulting monograph.

USING DIAGRAMS TO CODE SOCIAL ORGANIZATION
OF GROUP PROBLEM SOLVING

Building on the Rogoff et al. (1993) study, our research team sought ways of examining the extent to which people solving a problem together actually think together. The coding categories of Rogoff et al. (1993) focused on the contributions of individuals to shared endeavors, but we wanted also to examine the group process itself, as people engage in shared problem solving. We drew on literature that provided some guidance for explicitly examining intersubjective engagement among participants (such as Bos, 1937; Gauvain & Rogoff, 1989; Glachan & Light, 1982) as well as ideas from the focusing exercise (described earlier) and coding schemes regarding consensus and collaboration versus unilateral processes of problem solving, developed by Matusov, Bell, and Rogoff (2002) and Matusov and Rogoff (2002).

In the process of designing a study that we never carried out, a very useful discussion took place among Eugene Matusov, Chikako Toma, and the three of us as we contrasted coding approaches used in several of our team's previous studies. At some point, we began sketching little diagrams to illustrate the overall form of engagement of a group—how the group coordinated problem solving. These diagrams indicated the roles of each participant vis-à-vis the group's activity, not separately. At the time, the sketches were just for the sake of communication among ourselves, but a couple of years later, they turned into a reorganization of how we code the mutually constituting contributions of participants in a shared activity.

The first full-blown use of diagrammatic coding was in Chavajay and Rogoff's (2002) study examining the social organization of problem solving among Guatemalan Mayan mothers and children. Mothers varying in extent of schooling and three related school-age children were video recorded as they constructed a three-dimensional totem pole jigsaw puzzle. To depict their most prevalent form of coordination during each one-minute coding interval, Chavajay and Rogoff (2002) used "bird's eye view" diagrams that portrayed the extent and type of mutual engagement in problem solving. Their system of diagramming—that served as an ethnographic shorthand—included a number of conventions for visually conveying who was involved with whom and how they were involved, such as acting together, observing, directing others, and playing a supportive role. We summarize the simplified diagrams that were the center of the published findings. [A=mother; 1=oldest child; 2=middle child; 3=youngest child]

In *shared multiparty engagement*, all participants worked together in a coordinated and fluid way, mutually engaged in the same aspect of construction (e.g., same row of the puzzle). Some group members may have been in supporting or observing roles, but all four needed to be involved in the same cohesive focus.

For example, the family group in Figure 12.2 is working together in "shared multiparty engagement." The mother is placing a piece of the puzzle on the totem pole while her three sons are poised ready to help place the next pieces with her. One of the sons helps hold the totem pole together in the process. This segment of shared multiparty engagement would be diagrammed as shown in Figure 12.3.

Figure 12.2.   Example of shared multiparty engagement. (Photograph copyright 1997 by Pablo Chavajay.)

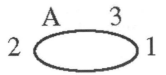

Figure 12.3.   Diagram of shared multiparty engagement show in Figure 12.2.

In *division of labor*, participants worked on different aspects of the puzzle, occasionally checking in with each other. A few of them may have worked together.

For example, the family group in Figure 12.4 is using "division of labor." Mother and niece are constructing the front side of the puzzle together, while the son is constructing the puzzle from the ground up by himself, searching for his next piece on the table. Adjacent to the son, the daughter is constructing the back side of the puzzle on her own. This segment of division of labor would be diagrammed as shown in Figure 12.5.

Figure 12.4.   Example of division of labor. (Photograph copyright 1997 by Pablo Chavajay.)

Figure 12.5.   Diagram of division of labor shown in Figure 12.4.

In *mother directs the children*, the mother unilaterally and explicitly directed the children, as a unit (like a teacher directing a class), to carry out specific aspects of puzzle construction, without conferring with them.

For example, a mother describes the structure of the totem pole puzzle to the children and tells them what pieces to find (see Fig. 12.6).

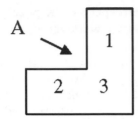

Figure 12.6.   Diagram of mother directing children.

In *noncoordinated engagement*, not all four group members were jointly coordinated with each other, although all members were engaged in puzzle construction. Some individuals or dyads worked without checking in with the others.

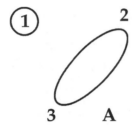

Figure 12.7.   Diagram of noncoordinated engagement.

For example, 1 constructs the second wing of the totem pole by himself, while 3, 2, and A construct the fifth row of the totem pole; neither subgroup connects about their agenda.[2]

By abstracting diagrams from the video data depicting the predominant form of engagement within each one-minute segment, Chavajay and Rogoff (2002) ended up with a couple of pages of small detailed diagrams for each family's group session of approximately 25 minutes. They laid these pages out on a long counter to conduct a visual analysis of the frequency and patterns of diagrams. The clarity of the diagrams allowed them to discern complex patterns in the data in a matter of only a few hours of examination of the data sheets.

In this "eyeball analysis," Chavajay and Rogoff (2002) first examined the diagrams to detect distinctions that could be dropped on the basis of occurring very rarely or could be combined due to being similar conceptually and showing the same patterns across all three levels of maternal schooling (0–2 grades, 6–9 grades, or 12+ grades). (This process yielded the four simplified categories diagrammed earlier.) For example, through visual analysis, Chavajay and Rogoff could see that the division of labor category did not need to maintain the subcategories distinguishing different sizes of work teams, because the subcategories did not show marked differences in pattern and they conceptually made sense to combine into the overarching category of division of labor. (Reliability calculations also helped us determine that some distinctions were too difficult to make, indicating that a distinction should be removed.)

---

[2]From "Schooling and Traditional Collaborative Social Organization of Problem Solving by Mayan Mothers and Children," by P. Chavajay and B. Rogoff, 2002, *Developmental Psychology, 38,* 55–66. Copyright © by the American Psychological Association. Reprinted and adapted with permission of the authors.

Once the categories were simplified into the four mutually exclusive diagrammatic categories described earlier, Chavajay and Rogoff (2002) then visually examined whether any distinct patterns were evident related to mothers' schooling experience. They then tallied the occurrence of each of the diagrammed types of engagement for family groups varying in maternal schooling and constructed graphs of the tallies, taking into account the varying lengths of time each family group took to complete the puzzle. Chavajay and Rogoff (2002) used these graphs to examine the predominant patterns of engagement across the schooling backgrounds and then conducted statistical trend analyses on the data.

The analysis revealed that in Mayan family groups with a mother who had extensive experience in Western schooling (at least 12 grades), group members were more likely to divide the task up among themselves, with the mother at times directing the children, compared with Mayan family groups with a mother who had little experience in Western schooling (0 to 2 grades). These latter groups more often worked collaboratively, with members sharing focus on the same aspects of puzzle construction, in a way that resembles speculations about traditional indigenous ways of organizing community involvement.[3]

As the Chavajay and Rogoff (2002) study illustrates, such diagrams provide a visual format for abstracting the approaches to problem solving used by participants in each family's case, and facilitate distinguishing patterns of engagement. The diagrams are thus a form of ethnographic shorthand, making it possible to visually (and later, statistically) examine participants' coordination of problem solving. Abstracting the cases in this shorthand helps to simplify variations in the diagrams to a few central forms of organization, yielding coding distinctions based on key contrasts among the cases.

Our research team has found this diagrammatic method of coding social organization of participation to be very useful in subsequent studies, to analyze the predominant ways that participants organize their interactions. One study found that during an adult demonstration, triads of Mexican-heritage children whose mothers had little experience in Western schooling (averaging 7 grades) more often worked as a coordinated team in folding Origami figures, whereas Mexican- and European-heritage triads of children whose mothers had extensive schooling more often worked in dyads or solo (Mejía Arauz, Rogoff, Dexter, & Najafi, in press). In another study, the diagrams revealed that triads of European American middle-class siblings more often used turn taking to manage their collaboration in exploring a science exhibit, compared with triads of Mexican-descent siblings who more often collaborated without using turn taking (Angelillo, Rogoff, & Chavajay, 2006).

We have found that diagrammatic coding categories provide a valuable summary view of differences in social organization across cases, while maintaining fidelity to individual cases by representing them directly. Coding of the social organization as a whole may also be accompanied by coding of individuals' contributions to the emerging shared endeavor (such as those employed by Rogoff et al., 1993). Studying both group social organization and individuals' contributions can thus be done in a way that does not simply code individual behaviors as if they occur in isolation.

---

[3]The patterns of social organization among groups involving mothers with 6 to 9 years of schooling fell between those of the groups involving mothers with 0–2 years and 12+ years of schooling.

FINAL REFLECTIONS ON ENGAGING IN RECURSIVE CYCLICAL ANALYSIS OF VIDEO DATA

We have argued for the value of creating coding schemes that address interpersonal processes directly, and described a recursive process cycling between ethnographic analysis of individual cases and abstracted coding schemes capable of portraying patterns across cases while retaining fidelity to the level of meaning of individual cases (see also Rogoff & Gauvain, 1986). By referring to the process as recursive, we emphasize that ethnographic analyses of video data are not simply preliminary. They are also important for elaborating the coding scheme (e.g., providing key examples to explicate categories) and often for disambiguating uncertainties in graphical or statistical patterns found using the coding scheme. The process is a recursive cycling between close-in analysis of a few cases and abstracted analysis and comparisons of numerical patterns across cases. To conclude our chapter, we discuss the importance of larger scale ethnographic understanding and the building of methods across a line of research.

Larger Scale Ethnographic Understanding

Far from trying to be "blind" to phenomena, researchers can use their understanding of how participants' cultural histories shape and are shaped by what they and the researchers do as well as how the participants are used to doing and making sense of related activities. Researchers' and coders' cultural expertise regarding participants' ordinary activities facilitates skilled "reading" of participants' actions, gestures, pauses, posture, and words. In addition, extensive piloting of procedures in researcher-structured situations and decisions involved in making naturalistic observations are essential to informing researchers of the meaning that participants make of the activities under study.

Thus, we argue for the importance of ethnographic study not just in the sense of focusing on individual videotaped cases, but also in the sense of understanding the broader organization and meaning of community events. For example, the development of the questions and coding schemes of several of our studies built on available ethnographic knowledge of the distinct cultural ways that people may use nonverbal and verbal means of communication to manage one another's attention and to coordinate efforts. Some of our research group members had observed everyday activities among Guatemalan Mayan families that tend to rely on subtle and skillful use of nonverbal communication (e.g., gaze, slight changes in posture, touching to call attention) to accomplish everyday tasks such as conveying a message or coordinating a religious ceremony. These observations coupled with other ethnographic studies suggesting extensive use of nonverbal communication among North and MesoAmerican indigenous groups seemed to contrast with observations of more exclusive reliance on verbal communication among some European American groups (e.g., Hall, 1959; Pelletier, 1970; Philips, 1983). This broader ethnographic understanding informed our research on such topics as how mothers and toddlers manage their attention and how groups coordinate their contributions to shared endeavors, in a Guatemalan Mayan and a

European American middle-class community (Chavajay & Rogoff, 1999, 2002; Rogoff et al., 1993).

Ethnographic analysis of video data relies on a deep understanding of the everyday lives of the research participants involved in the study to be able to interpret the video records. For example, in Rogoff et al.'s (1993) study, the development of the research question and the associated coding categories involved the researchers bringing to the discussion their knowledge of local meanings of particular ways of acting. An example of this is the researchers' development of a way to code whether, in the face of toddlers' insistence or refusal to follow the mother's suggestions, the mothers insisted on their own agenda or followed the toddler's agenda. This topic arose in viewing videotapes of the middle-class Turkish community, with which one of the researchers (Göncü) was very familiar, but the other three researchers were not. The researchers all agreed that there was something important distinguishing the communities in this regard, but the idea was vague and initial suggestions for how to code it were off base. Göncü was so familiar with mothers insisting on their own agenda that he did not have words for it, but the other three researchers' initial attempts to suggest a way to code this did not fit with the local meaning. Extensive discussion was required to find a way of describing the phenomenon that got beyond initial suggestions like "intrusiveness," which did not fit the meaning system in the middle-class Turkish community. Knowledge of local meanings in open-minded discussion with colleagues lacking this knowledge was very productive for coming up with a coding category that captured the phenomenon of interest in a way that "fairly" represented its local meaning. (See Rogoff et al., 1993, and Rogoff, Topping, Baker-Sennett, & Lacasa, 2002, for further discussion of this process.)

Building Methods Across a Line of Research

In a program of research, early studies examining a new question are likely to require more cycling between creating ethnographic descriptions, abstracting coding categories from them, and using ethnographic analyses to clarify numerical information. In new lines of investigation, an extensive ethnographic analysis of video data may be necessary to elaborate the study's question and coding scheme in ways that capture the phenomenon, especially if new communities or new aspects of interactive processes are under study. Subsequent related studies can jump into the later phases of this work, tailoring the question and coding scheme to the particular study with savings based on prior studies.

With growing understanding of a phenomenon as well as findings of related prior studies to inform the development of later research questions and coding schemes, a more focused ethnographic analysis may be sufficient. Some of our later studies that have started with a more refined research question have employed limited ethnographic descriptions using a random or targeted sample of cases as a tool for adapting a coding scheme, without the need to create extensive descriptions of all cases. Often the selection of cases has aimed to contrast key aspects of the data (for example, focusing on creating descriptions of three cases from each of the study's conditions or comparison groups). The ethnographic descriptions may also focus on creating summary descriptions or brief notes on specific aspects of social interaction (e.g., how turn taking occurs). Even if less extensive ethnographic analysis is needed in the adaptation of the coding scheme, ethnographic analysis is needed to help interpret the findings.

It is important to note that each new study in a line of research requires some adaptation of the coding scheme built on prior work. Our research team has never found that we can just apply a coding scheme unmodified from one study to the next, because we learn a great deal from each study and need to tailor coding schemes to take into account the next study's specific question and the differing participants, cultural communities, and activities of each study.

A feature of coding that requires rethinking for each study is how the data will be segmented. Some of our team's studies have segmented video data by events, when the flow of an activity provides naturally occurring events. For example, Mejía Arauz, Rogoff, and Paradise (2005) divided the video data into 16 segments according to the 16 folds required to make the Origami figures during an adult's demonstration of Origami paper folding.

For activities that do not involve easily designated segments or when we want to use smaller units, our research team has employed time segments of varying sizes. To determine the length of segments for a particular study, we consider the pace of the activities, the nature of the coding categories, and the "mental load" of the coders. A stretch of a video segment should not be so long that it contains too many of the target events or creates a burden for remembering the details necessary for coding; it should not be so short that it is difficult to achieve an understanding of the meaning of participants' actions.[4] Segmentation of coding for a study needs to be adjusted to the particular activities of the research participants so that the events of interest are caught, without too many of them happening within a particular interval nor too much effort spent in catching them. This requires familiarity with the data (as can be achieved in ethnographic analyses) to design a coding scheme that fits with the specifics of a particular study.

In conclusion, the process of recursive cycling to create ethnographic descriptions and to abstract coding categories of relevance to a research question is a fruitful way to compare interactional patterns across cases for their similarities and differences, while maintaining fidelity to the meaning of events within cases. Checking categories against cases and cases against categories provides a means of making sure that inferences are well founded across cases and well grounded within cases. The result is an approach to analysis that makes the most of both qualitative and quantitative methods to examine the ways that people coordinate their shared endeavors.

These methods can aid in focusing on emerging intersubjective processes that are central to learning through participation in sociocultural activities. We offer these reflections in the hope that they encourage the development of coding schemes that directly address the ways that social groups organize their interactions as well as different ways that individuals contribute to such social processes.

---

[4]In any case, coders consider the whole activity in interpreting an aspect of it. They use surrounding intervals to inform their coding decisions within an interval, so that their coding of an interval takes into account the meaning available in the prior and subsequent moment.

ACKNOWLEDGMENTS

We would like to acknowledge the involvement of other members of our research team in the discussion of these ideas over the years, especially Eugene Matusov, Chikako Toma, Ruth Paradise, Rebeca Mejía Arauz, Jayanthi Mistry, Artin Göncü, Christine Mosier, Pilar Lacasa, Jackie Baker-Sennett, Karen Topping, Maricela Correa-Chávez, Behnosh Najafi, Amy Dexter, Charlotte Nolan, and Katie Silva. Funding to support the preparation of this article came from an endowed chair from the University of California Santa Cruz Foundation (to BR), a grant from the Mayan Educational Foundation (to PC), and a traineeship from the National Institute of Health (MH20025).

REFERENCES

Angelillo, C., Rogoff, B., & Tudge, J. (1997, April). *Teaching/learning processes in computer game activities in which either the adult or the child partner assumes a teaching role: Development of a coding scheme*. Paper presented at the Meetings of the Society for Research in Child Development, Washington, DC.

Angelillo, C., Rogoff, B., & Chavajay, P. (2006). *Coordination among European American middle-class and Mexican-descent siblings engaging with a science exhibit*. Paper presented at the Meeting of the International Society for the Study of Behavioural Development, Melbourne, Australia.

Barron, B. (2003). When smart groups fail. *The Journal of the Learning Sciences, 12*, 307–359.

Bos, M. C. (1937). Experimental study of productive collaboration. *Acta Psychologica, 3*, 315–426.

Chavajay, P., & Rogoff, B. (1999). Cultural variation in management of attention by children and their caregivers. *Developmental Psychology, 35*, 1079–1090.

Chavajay, P., & Rogoff, B. (2002). Schooling and traditional collaborative social organization of problem solving by Mayan mothers and children. *Developmental Psychology, 38*, 55–66.

Cole, M. (1996). *Cultural psychology: A once and future discipline*. Cambridge, MA: Harvard University Press.

Gauvain, M., & Rogoff, B. (1989). Collaborative problem solving and children's planning skills. *Developmental Psychology, 25*, 139–151.

Glachan, N. M., & Light, P. H. (1982). Peer interaction and learning. In G. E. Butterworth & P. H. Light (Eds.), *Social cognition: Studies of the development of understanding* (pp. 238–262). Brighton: Harvester Press.

Hall, E. T. (1959). *The silent language*. Garden City, NY: Doubleday.

Lave, J., & Wenger, E. (1991). *Situated learning: Legitimate peripheral participation*. New York: Cambridge University Press.

Matusov, E., Bell, N., & Rogoff, B. (2002). Schooling as cultural process: Working together and guidance by children from schools differing in collaborative practices. In R. V. Kail & H. W. Reese (Eds.), *Advances in child development and behavior* (Vol. 29, pp. 129–160). New York: Academic Press.

Matusov, E., & Rogoff, B. (2002). Newcomers and old-timers: Educational philosophies-in-action of parent volunteers in a community of learners school. *Anthropology & Education Quarterly, 33*, 415–440.

Mejía Arauz, R., Rogoff, B., Dexter, A., & Najafi, B. (in press). Cultural variation in children's social organization. *Child Development*.

Mejía Arauz, R., Rogoff, B., & Paradise, R. (2005). Cultural variation in children's observation during a demonstration. *International Journal of Behavioral Development, 29*, 282–291.

Pelletier, W. (1970). Childhood in an Indian Village. In S. Repo (Ed.), *This book is about schools* (pp. 18–31). NY: Pantheon Books.

Philips, S. W. (1983). *The invisible culture: Communication in classroom and community on the Warm Springs Indian Reservation*. IL: Waveland Press.

Rogoff, B. (2003). *The cultural nature of human development*. New York: Oxford University Press.

Rogoff, B., & Gauvain, M. (1986). A method for the analysis of patterns, illustrated with data on mother-child instructional interaction. In J. Valsiner (Ed.), *The individual subject and scientific psychology* (pp. 261–290). New York: Plenum.

Rogoff, B., Mistry, J., Göncü, A., & Mosier, C. (1993). Guided participation in cultural activity by toddlers and caregivers. *Monographs of the Society for Research in Child Development, 58*, Serial No. 236.

Rogoff, B., Mosier, C., Mistry, J., & Göncü, A. (1989). Toddlers' guided participation in cultural activity. *Cultural Dynamics, 2*, 209–237.

Rogoff, B., Mosier, C., Mistry, J., & Göncü, A. (1993, revised version). Toddlers' guided participation with their caregivers in cultural activity. In E. Forman, N. Minick, & A. Stone (Eds.), *Context for learning: Sociocultural dynamics in children's development* (pp. 230–253). New York: Oxford University Press.

Rogoff, B., Topping, K., Baker-Sennett, J., & Lacasa, P. (2002). Mutual contributions of individuals, partners, and institutions: Planning to remember in Girl Scout cookie sales. *Social Development, 11*, 266–289.

Vygotsky, L. S. (1987). *Mind in society*. Cambridge, MA: Harvard University Press.

Using Video Data to Capture Discontinuous Science Meaning Making in Nonschool Settings

Doris Ash
University of California, Santa Cruz

I first began to direct my attention to collaborative science dialogue in nonschool settings on a visit to the San Diego Zoo some years ago. As I recall, I sat at the hippopotamus exhibit for quite some time. The exhibit contained several very large tanks, which held a family of hippos; their activity was visible both above and below water. Human families stayed there for a very long time, clearly entranced by being able to watch the animals' behaviors so clearly. They speculated on which hippo was the mother and which the father; they asked each other what hippos eat, how many babies they have, and how much they weigh. I longed for a video recorder in my hand; after all, most of the families had one. I had, of course, noticed similar questioning about animal behavior in classrooms (Ash, 1995; Brown, et al., 1993), but generally those questions were stimulated by secondary sources, such as video, text, pictures, or experts. My current research has grown out of my desire to accurately represent and analyze the actions and meaning-making dialogues created by social groups, such as families, during their visits to informal science learning settings.

I am a biologist, so it is natural for me to capitalize on my fascination with how groups make sense of biology, by using what is set before them, as well as their own native intelligence and their accumulated funds of knowledge. I first became familiar with science learning research in the classroom, and now, I conduct research on science learning in museums. What has changed is the critical mass we have now achieved in both theory and technology; this allows us to confidently trace collaborative talking, reading, questioning, and gesturing through science exhibits. We now actually have in hand the tools to capture meaning making as it occurs over time.

Since that visit to the San Diego Zoo, I have conducted research on collaborative biological sense making in many nonschool settings; at the Exploratorium Frogs exhibit (Ash, 2002), the Splash Zone at the Monterey Bay Aquarium (Ash, 2003b), the African and North American animal dioramas at the Los Angeles County Natural History Museum (Ash, 2004a), and at the Seymour Marine Discovery Center at Santa Cruz.

What do we really know about learning in such science settings, and why do we need to know more? Currently, our country spends considerable funds supporting programs in informal learning (museums, after-school programs, community technology centers, etc.), yet there has been little detailed documentation of the learning that actually occurs in these settings (Ash & Barron, 2005). We do not know the long-term effects of such activities. Moreover, the information we have in hand has focused primarily on the most frequent visitors to museums, that is, middle class European Americans. Our sampling focuses on "in the minute" events, with less understanding of what preceded them or eventuates from them. We do not know how, or if, classroom learning crosses over to museums or vice versa. For all these reasons, it seems imperative to conduct in-depth studies that begin to address these issues. With the advent of new digital collection, management, and analysis tools, we now have the opportunity to shed light on these pressing concerns.

To address these questions, I have been involved in a decade-long research program centered on describing, analyzing, and understanding collaborative groups, in this case families, as they make sense dialogically of core thematic content during life science conversations. This research is consonant with other research efforts, both in and out of the classroom, within which researchers focus on understanding what resources social groups from different cultural and linguistic backgrounds bring to learning settings (Moje, Collazo, & Carillo, 2001; Moschkovich, 2002; Gutiérrez & Rogoff, 2003). Such resources include the everyday understandings each member brings to the nonschool learning settings, and the dialogic inquiry skill of questioning, all of which are used to advance scientific understanding in a variety of learning contexts.

I have argued elsewhere that everyday (Warren, Ballenger, Ogonowski, & Hudicourt-Barnes, 2000) science understanding contains the necessary antecedents for moving toward the canonical science presented in museums and schools (Ash, 2004b). Even though the biological content may not seem "scientific" in the canonical sense, learners do use everyday language to discuss biology and to convey scientific meaning. Such views are influenced by Vygotsky's (1987) notion of spontaneous science concepts (Guberman, 1993; Wells, 1999). Everyday thinking and talking derive from personal and collective culture, language, and history; they are in the moment; and they are observed through the senses. Such ideas are not always correct, but they can provide a foundation for academic concepts, which Vygotsky characterized as coherent, defined, and systematically organized (Ash, 2004b; Guberman, 2003).

Ann Brown said that people need something interesting to talk about. In all my research, I have relied on the underpinnings of deep content principles to guide meaning making. My research, both in and out of the classroom, has relied on the principle of biological adaptation—the modification of living things that fit them for their environment (Eldredge, 1998). I believe that relying on biological principles allows dialogue to become generative, and to endure over a long period of time. Deeply principled content shapes, constrains, and guides science dialogue "about matters that are of interest and concern to the participants" (Wells, 1999, p. xi).

The most daunting challenge in this particular work is capturing and analyzing scientific meaning making as it occurs, interruptedly and discontinuously, over time and across different learning contexts. Such research entails designing theoretically informed, yet practical, systems for data collection and management, as well as for its analysis. Over the past decade, I have come to appreciate the utility, creativity, and power of video data, as well as its limitations as a research tool for meeting these challenges. Like other researchers, I have found many advantages to capturing the detailed verbal and nonverbal actions of people in learning settings, and my analytic methodologies have changed, as new digital media techniques allow new freedom in framing, collecting, archiving, retrieving, and analyzing such data.

Yet, the very richness of video data also presents researchers with new problems. The single greatest problem is complexity; with videotaping, researchers are now faced with overwhelming quantities of data to archive and analyze. Thus arises the countervailing problem of reduction. Data reduction is the current term that refers to the problems of determining how to organize the data, to gain a fruitful overview, and to choose the units of analysis of this organized data that are powerful, manageable, and defensible in a given study. The term data reduction acknowledges the overarching need in research to balance deeply informative microgenetic studies with the larger contextual perspectives within which they are situated. I have come to deeply respect the complexity of nonclassroom learning, which includes variables such as the social interactions (talking, doing, gesturing, listening) between people, multiple exhibits, the social and historical contexts of the families and the institutions, the kinds of information provided by exhibit and family, and the interactions among this diverse information.

The onus is on researchers to establish theoretically based criteria for their selection of units of analysis. This is especially important in light of claims of overgeneralization by research critiques. Researchers must be explicit about their rationale for generalizing from very specific, microgenetically detailed, episode-based analyses, which, typically, are taken from relatively small data sets; in my case, from several families. Often, reviewers and researchers ask how we can justify making generalizations from such small data sets. Small in this case means a small number of subjects (N). One response is to suggest is that the small N is offset by the detailed level of analysis. This is only partially satisfying. A more powerful generalization would include using several levels of analysis in a simultaneous and interlinked way. I call this moving from macro to microanalyses. The challenge, then, is to maintain a general overview of events, while simultaneously isolating detailed and hopefully representative events. Finally, over time, having created detailed analyses of a larger cross section of collaborative sense-making activities, across a variety of settings, we can lay claim to certain generalizations, for example, people across learning settings use the same biological themes, regardless of specific content (Ash et al., in press).

My approach is informed by sociocultural theory, with emphasis on the negotiations of meaning in the zone of proximal development (Vygotsky, 1978) as learners engage in joint productive activity (Tharp & Gallimore, 1988). Complexity, in this context, is a strength rather than a limitation, if it can be tamed. Sociocultural theory offers clear theoretical guidelines, when data reduction needs to occur, by maintaining a clear contextual perspective for the data. Brown's (1992) pivotal paper on design experiments was intended to guide researchers in making just these kinds of decisions. I have adapted Brown's framework in my own research; I use an iterative design cycle that allows for constant refinement of both the video data collection process and the analysis of the ensuing data. In this chapter, I describe how I have begun to balance complexity and reductionism, in order to achieve more accurate and productive data analysis.

In my interpretation of sociocultural theory, I assume that:

1. Collaborative groups are free to come and go, start and stop activities, and walk away whenever they desire, all of which are inherently discontinuous (Linde, 1993) social activities, which some have called this free-choice learning (Falk & Dierking, 2000);
2. Social groups make sense collaboratively (Ash, 2002; Rogoff, 1995; Wells, 1999), relying on a variety of tools and resources (Ash, 2003a; Ash & Wells, 2006);
3. Social groups, such as families, progressively make meaning of science through dialogic inquiry, such as the oral, written, and gestural activities that help advance understanding (Ash, in press; Ash & Wells, 2006), and through scaffolding (Bruner, 1984), as provided by people, signs, and dialogue;
4. Parents are their children's first and life-long teachers, and they can teach us a great deal about teaching and learning; and finally,
5. People of all ages bring their everyday understandings to social learning settings (Warren et al. 2000), and negotiate them in interactions with the exhibit, individual family members and the family's intellectual, social, and cultural background; these everyday understandings ultimately grow into deeper scientific insight (Ash, 2002, 2003a; 2004b).

Scaffolding is an especially important concept for the analysis of dialogue, consistent with Vygotsky's (1978) emphasis on the inherently social nature of learning through his construct of the "zone of proximal development" (zpd). The zpd can be described as "the zone in which an individual is able to achieve more with assistance than he or she can manage alone" (Wells, 1999, p. 4). I am especially interested in successful and unsuccessful scaffolding strategies for fostering scientific meaning making by collaborative social groups. The metaphor of scaffold is described as a " process that enables a child or novice to solve a problem, carry out a task, or achieve a goal that would be beyond his unassisted efforts" (Wood, Bruner, & Ross, 1976, p. 34).

Just as a physical scaffold allows a painter to climb higher along a building wall, the metaphor of interactive scaffolding (Bruner, 1984; Cazden, 2001) allows us to envision how dialogic and gestural scaffolds allow learners to move higher in their zones of understanding. Cazden (2001) provides insights into the origins of scaffolding and cautions us to remember that, unlike physical scaffolding, scaffolds for learning change as the learner changes.

When families move through museums, they can take advantage of a number of resources as scaffolds, discontinuously over time. For example, one family member may learn the name of a particular species by reading a sign and/or looking at a representation designed to foster such familiarity. Typically, naming is the first thing that occurs at an exhibit involving living or stuffed animals (Ash, Tellez, & Crain, in press). The family member tells others that name. Once an animal, such as a leopard shark, has achieved collective identity, the family members are free to use that name as a springboard for asking how the shark eats, has babies, and fights its enemies. Family members are free to compare the shark to the previously named moray eel in the adjoining tank, and so on. Such events are among the most common in our analyses of thematic family talk (Ash & Crain, 2005). The family's path and choice of scaffolds reflect their ideas about what is important and what is interesting, and is typically a negotiation among the desires of different family members. Family talk is easily interrupted, but is taken up again and again, as members interact with exhibits and with each other. The challenge for my research has been to develop methodologies that identify both the biological content and the discourse strategies within dialogic interactions.

CHALLENGES OF DISCONTINUOUS DIALOGUE

Linde (1993) has suggested that it is natural for discourse to progress discontinuously in social settings. A discontinuous timeframe is incomplete in any one of its component parts, but is cumulative over time, as related ideas, actions, and themes come to be combined into a cohesive whole. She has argued that in certain narratives "we confront a unit that is *necessarily* (Linde's italics) discontinuous" (p. 27), and that, in nontask-driven conversations, such as those in informal learning settings, "participants do tend to perform the work required to achieve a resumption of the narrative after an interruption" (p. 25). Discontinuous dialogic events typically begin with the family's everyday scientific understandings and gradually incorporate the museum's more canonical science. This is useful for understanding how, in museum settings, collaborative dialogue becomes more scientific over time. Research is beginning to offer clues as to which conceptual antecedents help move family knowledge toward more canonical science.

Thus, in informal learning environments, such as homes and aquariums, ideas emerge, submerge, and reappear in morphed forms, traceable over time but often only in hindsight. Because meaning is derived from dialogue, activity, representations, and other forms of communication, which do not appear in linear, logical, seamless forms, simple collection and analysis methodologies do not suffice. Only recently have we developed analysis tools (such as NVivo) powerful enough to analyze large data sets discontinuously and with multiple perspectives. These tools, however, are meaningless without theoretically based designs for data analysis that are both broad enough to incorporate scaffolding, everyday knowledge, and biological content, and deep enough to capture discontinuous dialogic meaning-making over time.

Methodologies

To understand scientific sense making over time, we need tools that can track meanings over disconnected dialogic events. It has taken me years to design methods to frame units of analysis that can capture these discontinuous events. I have discovered that I need to focus on three different levels of analysis, and to move back and forth among them, in order to fully follow meaning making in action. In the following sections, I describe the criteria and protocols for three levels of video data analysis; at the end, I give one brief example of discontinuous sense making.

The first level is large grained and holistic; I call this the flow chart. This flow chart provides an overview of one entire visit (typically 40–60 mins), as well as the pre and postinterviews (15–20 mins each). The second level is intermediate; I call it the significant event (SE). The significant event takes one segment of the flow chart and analyzes it in greater detail, emphasizing dialogue, content, and the kinds of tools the groups uses both to make sense of the science and to connect it to their own prior understanding. The SE acts as a knowledge construction event, which inevitably undergoes reconstruction. The third level is microgenetic, and comprises a detailed dialogic analysis. This third level involves analyzing specific SEs in greater detail, ideally using the myriad details available from video data, including dialogue, gestures, gaze, and actions.

These three levels of analysis are interconnected in essential ways; their identification and use rely entirely on the research questions I choose, which in turn are determined by the theoretical framework within which I work. In the paragraphs that follow, I explain the three levels of analysis in more detail, and I explain how interactions between them inform the data analysis. By reflecting on my own analysis of video data, I hope to offer particular tools and criteria that may be useful to others. The particular selection of data I present here is part of a larger research project I have undertaken to study dozens of families in several different learning settings.

The Flow Chart

The purpose of the flow chart is to provide a flexible overview of a single museum visit, from which particular segments can be identified and analyzed in more detail, for example the use of a particular biological theme. In constructing the flow chart, our research team previews the video, audio and transcription data several times; we identify major events within the visit, and select other markers such as use of biological themes or questioning. Figures 13.1 and 13.2 provide examples.

Figure 13.1 is an excerpt from the very beginning of Family J's visit to the Monterey Bay Aquarium's Splash Zone. This particular segment identifies the family members, the interpreter and the researcher, exhibits, and pertinent contextual information. It provides useful background on the daughter's biological interests, even though this dialogue occurs before the actual visit to the Splash Zone.

Family J, Mother, Father, daughter (Eva, 10), son (8), son (Ricardo, 5)
Interpreter (HQ) & Researcher (EB)
June 14, 2001. FC Prep EB

Time	Exhibit	Overview	Content themes
0-30:00	Pre interview with Splash Zone (SZ) cards	Long, rich, interview. Mom talks most. She uses the Splash Zone animal cards to start the conversation. The family has been in USA 10 years.	Frequent museum goers, Museums are fun, better than flea market; all members have favorite animals.
30:00-33:00	Family walks into museum	Eva is the lead, the family follows, all walk to the SZ. Eva is bilingual.	Eva points to the whales on the ceiling, very animated
34:50	Whole family is looking at otters	Some talk about otter behavior Dad/kids lead, MBA is crowded	At the otter tank briefly—Is otter asleep–how would you know?
37:00	Looking at video of whales in the hallways.	Family separates; Mom listens attentively & questions Eva.	Whale is Eva's favorite animal, because they communicate, Blue whale has smaller flippers.
39:02	EB/Eva talk about whales	Eva the lead, walking to the SZ EB listens,	Eva explains her classroom work studying Orca, killer whales
41:00	First coral tank	Looking closely, not in a hurry Dad the lead, asks lots of question, What are those plants in there? Mom leads next with more questions, they are animals?	Conversation about coral —Is it plants or animal; what do they eat? —they eat other animals and that live in layers.

Figure 13.1. Flow chart MBA.

When I constructed my first flow chart 1, early in 2002, I was interested in providing a broad context for each visit by identifying the conversational leader, topic, and location. After we had an overall picture of the visit, certain horizontal rows could be isolated and analyzed in greater detail, for example, the row starting at time marker 41:00, where Family J began to question the nature of coral. With this start, we were encouraged to track dialogue about plant versus. animal, and animal characteristics to its eventual outcome.

Over the years I have kept the basic form of the flow chart, while adapting it to particular needs. For example in Figure 13.2, I have targeted specific areas for later analysis; I closely follow the family's use of biological themes, such as alive versus dead, feeding, breeding, and so forth as well as the types of questions asked. By tracking the vertical columns, I can code particular biological themes and/or questions. I can quantify their use over an entire visit, or compare them to other families, other museums, and so on. Yet, I can still select particular horizontal rows for further analysis, for example, the segment identified with lines 841–902. Using analysis tools, both human and electronic, we can and do select, quantify, and track particular themes, questions, or participation structures.

In my analysis schema, I have generally focused on the kinds of biological themes that parents, children, and museums use to guide understanding. These themes are ubiquitous (Ash, 2002, 2003a, in press) in student and family dialogue, and their change over time provides the researcher with an excellent marker for tracking meaning making. Our research group's coding schemes for biological content have evolved

Family Name: H	SMDC				
Date:	12.17.02				Father, mother, son Eddy(10),daughter Gerry (4)
Lines	Exhibit	Topics	Theme	Type of Question	Context Notes
841-902	Touch Tank	Food Chain & how tank is fed	Feeding	Dad brings in his prior knowledge as a series of questions. The mediator encourages him and he builds his idea based on her confirmation.	Dad uses prior knowledge about food chain, referring to it as something he has "read before" (line 841). He checks his ideas with the mediator. Mom gives examples that help develop the idea that animals don't always eat animals almost their size. Mom also relates back to an ongoing conversation about marine snow and the conversation turns to how animals in the tank are fed vs. animals in the wild.
884-888, 909-929	Touch Tank	Baby animal	Identification & classification		Daughter consistently makes excited references to the "baby"! Both parents respond half-heartedly, then finally engage her. The group makes several references to how the "baby" feels, looks and relates to the daughter.
950-1027	Touch Tank	Anemones/ Solitary Corals	Classification, plant vs. animal		Daughter initiates exchange by noticing that the corals "look like anemone" (line 955). Mediator brings the observation to the family, who join the discussion about the characteristics of corals and anemones. Daughter becomes the expert in this situation.
1136-1155	Touch Tank	Animals take refuge, camouflage	Protection from predators, Aquatic vs. terrestrial	Again, dad presents his ideas as questions, he extends into new ideas.	Dad relates the tank to "the earth" (the land?) to observe that animals must use plants to protect themselves from predators. He makes the analogy to birds' nests and uses the word camouflage (line 1151).

Figure 13.2. Flow chart SMDC.

Alive vs. Dead Breeding Classification/Identification
Feeding Human/ Animal Relations Locomotion Plant vs. Animal
Protection from Predators
Protection from Elements
Sexual Dimorphism Terrestrial vs. Marine

Figure 13.3. Early version of biological theme coding.

over the past decade. Once simple, like the example in Figure 13.3, we have now developed more complex biological coding schemes that are now more reliable. As with all my work, the general biological principle underlying our thinking is adaptation.

It has taken several years of intense work and many iterations to redesign more complex biological thematic coding schemes. Figure 13.4 indicates how biological thematic codes and segmenting are combined within a spreadsheet format. We have spent considerable time designing ways to segment appropriate units of analysis. The segment indicated in the first vertical column has been selected according to topic, much as Gee (1999) uses the term *stanza*; new segments start as new topics are discussed.

Originally, we coded biological content within premarked segments. Over the past year, we have come to realize that, because we are coding for thematic topics, the segment length is determined by that topic's treatment, and cannot effectively be separated. Thus, we have combined biological coding with segment definition. The biological thematic code and segment and other properties are entered into an NVivo platform so that we can more easily quantify, as well as compare and contrast across and within families, and over time.

We assign a code to each segment, and we have detailed rules to guide the process. For example, the code CI (Classification/Identification) reflects attempts to make sense of an organism's name. Other examples of codes are SA (Staying Alive), LC (Life Cycles), R (Reproduction), and (S) Structural or (B) Behavioral adaptations. We have verified the code's content and structure with biologists, psychologists, and science educators over the last dozen or so iterations. Our research team is currently preparing a reflective paper describing our process for selecting and making reliable units of analysis (segmenting), biological coding systems now called Tools for observing biological talk over time (TOBTOT), and the thinking that informed them (Ash et al., in press).

I have written elsewhere (Ash, 1995, 2002, 2004b) about the origins of these biological themes, and about the research that informs our views of the kinds of thematic material that will most attract young students and their families. In past related research, I have focused on adaptations for feeding, breeding (life cycles), protection from predators, and taxonomic relationships, among others. These basic categories of interest have emerged in my analyses of visits in many different settings, after observing many different kinds of families.

Family Name		Mother, father, son (9), daughter (5)		Visit 1
Coder	ABC			
Date	2/28/04			
Segment	Lines	Topic	Code for Theme	Context Notes
1	2 thru 16	shark, sea star	CI OQ B, CI P CN, SA E F B	Dad names shark, daughter names sea star, the mediator introduces the term shark
2	17-35	"	EI HOR PC, SA PW B	Family is curious about the dangerousness of sharks, the mediator talks about how these particular sharks are not dangerous to humans, but are dangerous to other fish
3	36-47	eye	CI OQ S, CI P C, CI P CN	Dad is curious about whether that shark see with its eye

Figure 13.4.   Spreadsheet segment with code and context.

Significant Events

Using the types of flow chart illustrated in Figures 13.1 and 13.2 as a broad out-line of events, the next step is isolating the significant events, or SEs, at the intermediate level of analysis. The SE perspective is neither top down nor bottom up. Using a level intermediate in both size and purpose (White, 1993), I can show how significant events arise from other events and how they subsequently affect future outcomes. An SE marks progress to date, with the understanding that any meaning made at a given moment will subsequently be changed. Each significant event (Ash, 2002) is just large enough to encompass one meaning-making event, and contains:

1. Recognizable beginnings and endings, generally but not always centered on one particular exhibit;
2. Sustained conversational segments that differ from the short, unsustained interactions that can precede and follow SEs;
3. Different sources of knowledge, such as distributed expertise; and
4. Inquiry strategies, such as questioning, inferring, or predicting.

The selection of specific discontinuous SEs necessarily constrains the focus for analysis. SEs provide representative slices in time, illustrating how families negotiate scientific meaning. The operative phrase here is representative. It is essential to have the overview of the flow chart in selecting SEs, which SEs provide glimpses into how families deal with science content by using generative themes that can cut across differ-ent content areas. After examining an entire visit, I seek more detail over small units of time, and then fit these SEs into the larger frame of the entire visit. In short, I work both ends toward the middle. I choose particular criteria for selecting SEs, such as the development of powerful content, or the use of particular resources. I seek to always have a theory-based rationale for the selection of SEs. Figure 13.5 contains a portion of one significant event. It is taken from a latter portion of the flow chart in Figure 13.2. In this SE, the family is looking at the leopard shark tank.

Line	Exhibit	Topic	Questions	Detailed Context
1120-1247	Leopard Shark Tank	Anemones/Solitary Corals	Mediator uses questions to connect this experience with the recent experience at the touch tank. "What is it that you saw of the other side?" (line 1234).	Daughter exclaims excitedly that she has recognized the anemones (from the touch tank). Mediator points out another creature (solitary corals) and dad and son repeat the observations from the touch tank, noticing that they are "little" and "babies" (lines 1125-11126). Group works together to piece together the idea that they look like small anemones (son calls them "little cousins of anemones" line 1235) but are really corals.

Figure 13.5.   SE 1, SMDC.

Here, the 4-year-old daughter has recognized the sea anemones she first saw in the touch tank. The mediator points out that they are really another creature (coral), and the father and son repeat observations they made at the touch tank, noticing the features of "little" and "babies." The relationship between coral and anemone seems obvious to this family; this fits into the biological theme classification. This event is significant because there is biological content; there is distributed expertise; the group uses questioning and inferring; and the conversation is sustained over time. The group cooperates to piece together the idea that, even though these new creatures look like small anemones, "little cousins of anemones," they are really corals.

Dialogic Analyses

Using the SE as the frame, I can then focus on a fine-grained dialogic analysis of carefully selected segments of talk and gestural activities. To do so, I have developed several tools. Typically, I provide a simple framework for identifying the utterances within an SE, and then pair these utterances with the actual function they serve within the SE, as in Figure 13.6. This format is adapted from Roth (2002).

I have also used a discourse analysis frame (Ash, 2002), adapted from Wells (1999), which identifies the range of possible follow-up responses afforded by a particular utterance. The follow-up functions can include expansion on a given response, justification, exemplification, explanation, or reformulation. Such discourse moves can expand the range of possible responses, rather than constraining them, and can serve as strategies for sustaining conversation beyond the simple IRF (initiation, response, follow-up) structure, the so-called triadic dialogue (Lemke, 1990). In the fol-

Utterance	Function in context

Mom: Es animal?
    It's an animal?
 Exp: Se llaman coral. Se llaman coral. Y el coral es un animal que crece arriba de los esqueletos de los antepasados.   Y son muy chiquititos ------
    It's called coral. It's called coral. And coral is an animal that grows on top of the skeletons of its ancestors. And they're very teeny tiny …-----
 	
 Mom: Entonces, ¿Estas también son animales estas plantas, todas los que están aquí, son animales?
    So, are all of these plants animals, all the ones that are here, are they animals?
 Int.: Sí, son animales.
    Yes, they're animals.
 Mom: Pero, ¿No le hacen daño al hombre? ¿Cómo a los buzos?
    But, don't they hurt man? Like the scuba divers?
 Int.: Se llama coral y su nombre es coral.
    It's called coral and its name is coral.
	The mother questions and needs to verify the fact that this is an animal.
	
	 The mediator explains how coral lives. She tells about their growth patterns and is providing more background for a complex story of coral.
	
	 The mother asks again. The interpreter repeats. The mother now asks if coral can sting humans.
	
	 The interpreter repeats that it is coral but does not answer the question.

Figure 13.6.   MBA, Dialogic analysis.

lowing text, I indicate where the speaker follows up to expand the possible outcomes. These responses are valuable in maintaining conversations and in helping to clarify the participant's intent for the next speaker. One rubric I have used to identify the form and function of the follow-up move within dialogue is illustrated in Figure 13.7.

Family of three at the California Academy of Science, Change through Time exhibition
Mother, son (5), daughter (3)
At a diorama of flightless birds being attacked by small predatory mammals

---

Theme	Skill	Move	Function
(Biological theme)	(Inquiry skill)	(Initiate Respond, Follow-up)	(amplify reinterpret extend, etc.)

BOY:
I think that is the daddy [bird], because she [mother bird] is shorter.

Sexual dimorphism	Hypothesizing	R (to Mom's Q)		Intro new theme

MOM:
That is a good point. Most of the animals have smaller females. You are right.

Protecting babies	Interpreting	F	amplifying

What is that, do you think? [small mammal predator in the diorama]

Protecting babies	Questioning	I	re-engaging

Mother invites new follow-up response

They [family of flightless birds] sure don't look very happy, do they?

Personification	Inferring	F	re-interpreting

Figure 13.7.   Follow-up functional analysis.

This type of analysis provides a more detailed account of the shape the dialogue takes and serves to highlight aspects of thematic content, questioning, and other details.

Using All Three Levels of Analysis to Describe Meaning Making Across Discontinuous Events

I now illustrate how I use all three levels together to analyze meaning making over time. The H family visited the Seymour Marine Discovery Center in Santa Cruz twice; the first visit was represented in Figure 13.2. During that first visit, the H family explored the Marine Snow exhibit. Marine snow consists of tiny bits of dead plant, animal, and other material, which constantly shower down toward the deep ocean floor. Dr. Silver, the researcher whose work the exhibit explores, has studied its origins and make up, destination, and role in the marine ecosystem. At the beginning of the family's first visit to Marine Snow, the mediator had explained that "marine snow" is a mixture of dead things and live things; the father then said "They're like natural wastes." The mediator supported this by saying, "Yes, it is a kind of 'natural recycling'." A few moments later, the mother related it to "fertilizer, maybe." The father then asked if any little animals might eat marine snow. Figure 13.8 shows this family at the actual marine snow exhibit. The exhibit includes the large microscope, plexiglass container of marine snow, a picture and text about the researcher, Mary Silver, and a video of Dr. Silver explaining her work.

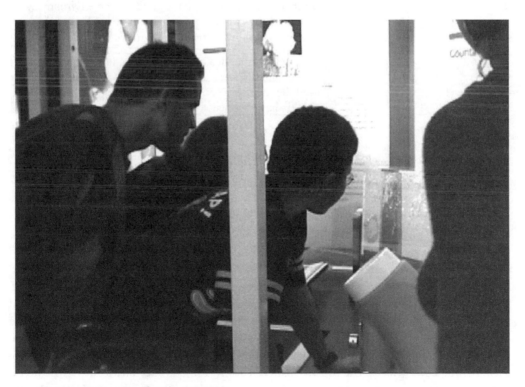

Figure 13.8.   One family at SMDC.

All family members were intrigued with the idea of marine snow; the son and daughter handled the microscope and the plexiglass display, while the mother and father looked, gestured, and asked questions of the bilingual mediator in Spanish. The signs and the video were in English. The father compared marine snow to how humans shed their hair and skin when they bathe, and so forth. The mother explained to her son what the mediator had just been teaching the parents; the parents then encouraged both children to become actively involved in looking at the marine snow and talking about it. The children in turn made analogies; for example, the daughter likened marine snow to "little papers." The son recognized that the objects under the microscope were the same things that were inside the plexiglass box. The subsequent portion of visit 1 is detailed below in Figure 13.9.

Lines	Topics	Theme	Context Notes
1204-1234	Marine snow	Alive/dead	The parents look into the microscope and have many questions about marine snow, including "is it natural?" and "is it alive?" Mediator helps them along with short explanations about what is in marine snow. The children become interested when "dead animals" come up and want to look, and even notice "some little animals."
1235-1260	recycling	Recycling/	When mediator describes the snow as "as mixture of dead things and live things," dad makes an analogy to "natural wastes" and begins to talk about what's good for the environment. Their ideas of "natural" are connected to their ideas about what is good for the environment. Finally Mom wonders if marine snow is like fertilizer.
1261-1264	MS	Feeding	Dad asks if animals eat marine snow.
1265-1276	cycle of life	Recycling	Dad relates the idea of feeding to the idea of a "circle," indicating his ideas about the way life is connected. He and the mediator verify that he often relates what he sees to this larger concept.

Figure 13.9.   Marine snow, Family H, visit 1.

A portion of the actual dialogue is shown in Figure 13.10.

The marine snow was the last exhibit seen in the H family's first visit to the Seymour Marine Discovery Center. The family spent over 10 minutes there, discussing the meaning of marine snow with the mediator and with each other. During the second visit six months later, the family discussed marine snow at the first and subsequent exhibits, even though they were very far away from the actual marine snow exhibit. Throughout the second visit, in fact, and without any prompting, the parents showed themselves to be extremely aware of marine snow. They noticed marine snow at the first exhibit, the guitar fish. Later, at the rocky reef tank, the father again pointed out the marine snow to the mediator, while the mother read about marine snow from the informational cards. When the mother asked if some animals eat marine snow, the mediator asked her which animal she thought might do so, and the mother and father began

Utterance	Theme
...At marine snow	
1. Dad: Pero, ¿Es natural eso?	The mother, and father
2. **But, is this natural?**	begin to question the
3. Med: Eso es natural. Se encuentra en el océano.	mediator about the
4. **This is natural. It can be found in the ocean.**	nature of the particles,
5. Mom: Pero, ¿Es vivo? ¿O es....?	alive, dead, natural or
6. But, is it alive? Or is it.....?	man-made.
7. Med: /Son partículas muertas ----	
8. **They are dead particles -----**	These questions center
9. Med ---- de animales. Más, más que todo,	on properties of life.
10. son esa partículas que se /caen de la superficie.	
11. **----dead animals. More, more than anything,**	It is important to them
12. **these are particles that /fall from the surface.**	that some are alive.
...	
13. Med: Pero también hay... animales vivos.	All three members are
14. **But there are also.... live animals./**	actively engaged in
15. Eddy: /Mira, unos animalitos./	these questions
16. **Look, some little animals./**	
17. Mom: Vivos.	
Live.	

Figure 13.10.   Dialogic analysis.

to speculate. The father guessed sea stars, and then he wondered about some type of sucking fish. Finally, the mother thought it might be "those things that look like flowers" (anemones). When the mediator asked her why she guessed that, the mother surmised that it could be because they seem to catch things that fall.

In Figure 13.11 is a portion of the marine snow dialogue during Visit 2.

Line	Exhibit	Theme	Detailed Context
2091 - 2097	Marine Snow	Classification	Dad is testing son to see if he remembers marine snow from the last visit. They successfully identify part of what makes up marine snow as "remains of animals"
2122 - 2151	Marine Snow	Human/animal relations	The mother is questioning the role of scientists, as her perceptions of them were shaken when she realized no one had noticed marine snow as something important until 1970.
2152 - 2183	Marine Snow	Feeding, Terrestrial/marine	Using knowledge from the previous visit and the info card, the mom and dad discuss the importance of marine snow. Dad is struck by the perfect analogy, "its like the cycle of the, the leaves that fall from the trees . . ." A very complex strain of reasoning is brought to the surface after a lot of discontinuous discussion of marine snow.

Figure 13.11.   Marine snow, Family H, Visit 2.

Once the family actually arrived at the marine snow exhibit for the second visit, the mediator asked them why marine snow is important. The mother immediately said that animals, such as anemones, eat it, while the father used the powerful analogy that marine snow must be like the "cycle of the leaves that fall from the trees in the jungles" and "It seems like they don't do anything but they help the regeneration of the soil … and to protect from erosion."

The parents have thus come full circle from the first visit, when the father thought of marine snow as "natural wastes," and the mother thought of it as "fertilizer." In the second visit, the family expanded these original inferences toward comparing aquatic and terrestrial environments, by focusing on the role of marine snow in the regeneration of the ocean bottom as well as its function as food. It is clear that the family remembered and built on their prior understandings in important ways. Without a more complex analytic scheme, using several levels of analysis, the important details of the gradual interweaving of information into a larger marine snow narrative, an example of the discontinuous nature of meaning making, might have been missed.

DISCUSSION

Throughout this chapter, I have made two main arguments. First, theory must inform all design decisions. For example, methodological choices at each key decision point, including unit of analysis, target group, learning context, and analytical entry points, have to be guided by theory, rather than arbitrarily selected. In my research, I have relied on sociocultural theory to inform my choices of social group and social activity, including talk and actions.

Second, once the research goals are clearly articulated within the relevant theoretical frame, then the role of video data can logically follow. In short, the methodologies need to match the goals. Referring to Brown's design experiment paper (1992), design choices must be grounded in first principles; they can be tested iteratively, but they must always be informed by clear goals. These goals need to be narrow enough to be accomplished and broad enough to inform the field.

To use a biological metaphor for the relationship of methodology to goal, form must follow function. For example, one of my research goals is to determine the amount and quality of everyday understanding that social groups bring to scientific meaning making in nonclassroom settings. To accomplish this, I need to collect and analyze dialogic events in a variety of settings, including interviews, actual exhibit visits, and questionnaires. The results of data collection from these events then need to be analyzed at a metalevel, to enable more complex insights. Merely observing families, or asking their opinion of their interests, is not very informative by itself. Audiotaped information is also insufficient. However, by using a combination of remote microphones attached to a video recorder, as well as several audio recorders for the times when family members take separate courses through exhibits, I have been able to capture detailed dialogue and nonverbal communication as it occurs naturally, in situ.

Designing analytic tools for discontinuous dialogue has required accounting for three levels of analysis, in order for me to accomplish my research goals. Reasoned decisions are necessary in the selection of levels of analysis. I first chose significant events as an entry point because this intermediate level of analysis allowed me to begin track-ing dialogic and gestural activities over time, without losing detail or perspective. By starting at the middle level, I have been able to shift up or down or both, and still main-tain an overview of the entire research field. When starting at the top or at the bottom, I often found myself getting lost, either in details of dialogue or in the larger social con-text of the events I was analyzing.

Significant events may not be entirely representative of one or more visits, but they can provide valuable insights into understanding collaborative meaning mak-ing. For example, in a related study, we chose one particular key question, "What does it eat?," to track meaning making throughout one family's visit (Ash, Loomis, & Hohenstein, 2005). The question, first seen in analyzing several significant events, arose repeatedly during the 2 hours the family spent in the aquarium. By stringing together related significant event dialogue, with a coding scheme that recorded the location of the dialogue in the transcript, the speaker, the utterances, and an inter-pretation of those utterances, we provided a coherent analysis of one common par-ticipation structure in informal settings. Such analyses provide insights into the issue of generalization or "representativeness." One powerful outcome of tracking a par-ticular question, which in turn is linked to important content (feeding), is the provi-sion of an existence of proof and a methodology for linking individual significant events together in important and discontinuous ways. Thus, it is possible to use one particular SE as an exemplar of a larger pattern.

It would be foolish to overgeneralize data from single significant events for single families. However, after exhaustive analyses at three locations, with more than 20 fami-lies, we are confident that the TOBTOT coding scheme will be useful across settings. We are now able to trace discontinuous talk over time, using biological content (Ash & Crain, 2005), questioning (Ash et al., 2005), and most recently metalinguistic markers as the focus. In each of these research-related efforts, working with the same data set, our three-part analysis scheme and our methods for collecting, copying, and managing digital data serve as the foundation for continual interrogation of the data.

Over time, using the three levels of analysis, we have determined several impor-tant outcomes. First, all families bring a variety of resources, including identifiable (codable) biological themes, to life science exhibits. Everyday science provides a foun-dation for more formal, canonical science (Ash, 2004b). Families, individually and col-lectively, use thematic resources in flexible ways. They also use a variety of inquiry skills, particularly questioning, to scaffold their own understanding. Rather than being linear, meaning making is circuitous and iterative, and it is interwoven into the fabric of social talk and action (Ash & Wells, 2006). Ideas, questions, and themes surface and go underground, only to re-surface in a slightly morphed form (Ash, 2004b). Choosing an intermediate level of analysis allows a mechanism for providing correlation and dis-agreement between detailed analyses (micro), and generalization (macro). It is critical to provide such checks on analysis of complex data with a small N, in order to make larger claims.

In my research, I started with a general question aimed at characterizing scientific sense-making, focused on principled content. Over the years, this larger question has translated into several smaller research questions, including deeper analyses of the structure and function of everyday understanding, the structure and function of questions, and determining how they interact in the discontinuous dialogue that is representative of collaborative and noncollaborative activity in complex social settings. I expect more questions to arise over time. One area we are currently exploring in greater detail is the role analogy and metaphor play in dialogue.

I have argued that the three levels of analysis I have selected allow movement from the macro to the micro, providing multiple entry points, and a means for both quantitative and qualitative analyses. These choices are informed by theory, reflecting the goal of multiple perspectives on meaning making. Using new analytic tools, it is relatively easy to quantify selected items, such as questions, or the use of certain terms, such as alive or dead. Our team codes these items, thus providing cumulative measures. Yet, by using the significant event as focal point, we also track meaning as it shifts and turns over time in unexpected ways.

Yet, there are still methodological challenges that come with this type of research. For example, using such detailed levels of analysis, it is difficult to say, with certainty, that convincingly larger numbers of family dialogues can be analyzed. We know, however, that without a fine level of detail, we have only a superficial understanding of progressive meaning making. Striking the right balance between larger context and detail is an ongoing challenge.

CONCLUSION

In this chapter, I have suggested several methods for collecting, handling, and analyzing complex digital video. These methodologies have served as tools for me to understand complex dialogic data and are based on sociocultural theory. I have discussed the role of video data collection and analysis in interpreting how social groups, in this case the family, make sense of science dialogically, over time. I have illustrated both the fruitful outcomes and the challenges to researchers in interpreting collaborative scientific meaning making, using discontinuous dialogic events as the analytic frame. These methodological issues matter to researchers, because they enable us to analyze complex sets of dialogic data without oversimplifying their meaning, either in detail or contextually. As our digital media technologies increase in power, so, too, must our ability to deepen our analyses grows exponentially.

REFERENCES

Ash, D. (1995). *From functional reasoning to an adaptionist: Children's transition toward deep biology.* Unpublished doctoral dissertation. University of California, Berkeley.

Ash, D. (2002). Negotiation of biological thematic conversations in informal learning settings. In G. Leinhardt, K. Crowley, & K. Knutson (Eds.), *Learning conversations in museums* (pp. 357–400). Mahwah, NJ: Lawrence Erlbaum Associates.

Ash, D. (2003b). Dialogic inquiry and biological themes and principles: Implications for exhibit design. *Journal of Museum Education, 28*(1), 8–13.

Ash, D. (2003a). Dialogic inquiry in life science conversations of family groups in museums. *Journal of Research in Science Teaching, 40*(2), 138–162.

Ash, D. (2004b). Reflective scientific sense-making dialogue in two languages: The science in the dialogue and the dialogue in the science. *Science Education, 88,* 855–884.

Ash, D. (2004a). Knowing the right question to ask: Everyday science, necessary antecedents, and diversity. *Curator, 47*(1), 8–100.

Ash, D. (in press). Making sense of living things: The need for both essentialism and activity theory. Accepted for publication in *Cognition & Instruction*.

Ash, D., & Barron, B. (2005, April). *Taxonomies, timescales, and learning outcomes in the study of informal learning*. Structured poster session presented at the Annual Meeting of the American Education Research Association, Montreal.

Ash, D., Crain, R., Crain, R., Brandt, C., Wheaton, M., & Bennett, C. (in press). Talk, tools, and tensions: Observing biological talk over time. *International Journal for Science Education*.

Ash, D., & Crain, R., (2005, April). *Building biological understanding over time with everyday talk*. Poster presented at the Annual Meeting of the American Education Research Association, Montreal.

Ash, D., Tellez, K., & Crain, R. (in press). The importance of objects in learning and talking science in and out of the classroom. In K. Gomez & K. R. Bruno (Eds.), *Talking science, writing science*. Mahwah, NJ: Lawrence Erlbaum Associates.

Ash, D., Loomis, M., & Hohenstein, J. (2005). "What does it eat?": Questions as resources for bilingual families making sense of science in a Marine Discovery Center. *Sinectica, 26*, 51–64.

Ash, D,. & Wells, G. (2006). Dialogic inquiry in classroom and museum: Actions, tools and talk. In Z. Bekerman, N. C. Burbules, & D. S. Keller (Eds.), *Learning in places: The informal education reader* (pp. 35–54). New York: Peter Lang Press.

Brown, A. L. (1992). Design experiments: Theoretical and methodical challenges in creating complex interventions in classroom settings. *Journal of Learning Sciences, 2*(2), 141–178.

Brown, A. L., Ash, D., Rutherford, M., Nakagawa, K., Gordon, A., & Campione, J. C. (1993). Distributed expertise in the classroom. In G. Salomon (Ed.), *Distributed cognitions: Psychological and educational considerations* (pp. 188–228). New York: Cambridge University Press.

Bruner, J. S. (1990). *Acts of meaning*. Cambridge, MA: Harvard University Press.

Cazden, C. (2001). *Classroom discourse: The language of teaching and learning*. Portsmouth, NH:Heinemann. (Original work published 1988)

Eldredge, N. (1998). *The pattern of evolution*. New York: Freeman.

Falk, J., & Dierking, L. (2000). *Learning from museums: Visitor experiences and the making of meaning*. Walnut Creek, CA: Alta Mira Press.

Gee, J. P. (1999). *An introduction to discourse analysis: Theory and method*. New York: Routledge.

Guberman, S. (1993, April). *How parents help children connect everyday and academic concepts in a museum setting*. Paper presented at the Annual Meeting of the American Educational Research Association, New Orleans.

Gutlérrez, K., & Rogoff, B. (2003). Cultural ways of learning: Individual traits or repertoires of practice. *Educational Researcher, 32*(5), 19–25.

Lemke, J. L. (1990). *Talking science: Language, learning, and values*. Norwood, NJ: Ablex.

Linde, C. (1993). *Life stories: The creation of coherence*. New York: Oxford University Press.

Moje, E. B., Collazo, T., & Carillo, R. (2001). "Maestro, what is 'quality'?": Language, literacy, and discourse in project-based science. *Journal of Research in Science Teaching, 38*(4), 469–98.

Moschkovich, J. N. (2002). A situated and sociocultural perspective on bilingual mathematics learners. *Mathematical Thinking and Learning, 4*(2&3), 159–212.

Rogoff, B. (1995). Observing sociocultural activity on three planes: Participatory appropriation, guided participation, and apprenticeship. In J. V. Wertsch, P. del Rio, & A. Alvarez (Eds.), *Sociocultural studies of mind* (pp. 139–164). Cambridge, England: Cambridge University Press.

Roth, K. (2002). Talking to understand science In J. Brophy (Ed.), *Social constructivist teaching: Affordances and constraints series (Advances in research on teaching)* (pp. 197–261). Oxford: Elsevier Science Ltd.

Tharp, R. G., & Gallimore, R. (1988). *Rousing minds to life: Teaching, learning, and schooling in social context.* Cambridge, England: Cambridge University Press.

Vygotsky, L. S. (1978). *Mind in society: The development of higher psychological processes.* Cambridge, MA: Harvard University Press.

Vygotsky, L. S. (1987). Thinking and speech. In R. W. Rieber & A. S. Carton (Eds.), *The collected works of L. S. Vygotsky. Vol. 1: Problems of general psychology* (pp. 35–218). New York: Plenum. (Original work published 1934)

Warren, B., Ballenger, C., Ogonowski, M., & Hudicourt-Barnes, J. (2000). Re-thinking diversity in learning science: The logic of everyday languages. *Journal of Research in Science Teaching, 38*(5), 529–552.

Wells, G. (1999). *Dialogic inquiry: Towards a sociocultural practice and theory of education.* New York: Cambridge University Press.

White, B. (1993). Intermediate causal models: A missing link for successful science education. In R. Glaser (Ed.), *Advances in instructional psychology* (pp. 177–252). Hillsdale, NJ: Lawrence Erlbaum Associates.

Wood, D., Bruner, J., & Ross, G. (1976). The role of tutoring in problem solving. *Journal of Child Psychology and Psychiatry, 17*, 89–100.

14

Expanding Studies of Family Conversations About Science Through Video Analysis

Maureen Callanan
University of California, Santa Cruz

Araceli Valle
University of California, Santa Cruz

Margarita Azmitia
University of California, Santa Cruz

Our research explores how conversations with parents help children to develop conceptual understanding and ways of thinking in science domains. Over the past few decades, research on science education has increasingly considered the importance of informal science learning at home, in museums, and in other nonschool contexts, as well as science learning in formal classroom settings (see Leinhardt, Crowley, & Knutson, 2002; Paris, 2002). This shift in the education literature meshes well with trends in the field of cognitive developmental psychology, where sociocultural approaches have refocused the study of cognitive development. Rather than focusing on the child's knowledge as a measurable individual entity, these theories have encouraged attention on the essentially social processes by which conceptual change occurs as children engage in everyday cultural practices (see Cole, 2002; Rogoff, 2003).

In our work, we have blended these different approaches, attempting to capture children's early science understanding through analysis of everyday conversations they have with their parents. We argue that these conversations are important in that they can potentially uncover the processes through which children construct and negotiate their science understanding. Our research explores ways in which parents contribute to their children's understanding of specific scientific phenomena. We are especially interested in the kinds of explanations children hear and produce when questions about natural events arise in everyday conversation. We have used video technology to record naturally occurring conversations about science topics in children's museum settings and in naturalistic laboratory settings.

In this chapter, we discuss the ways in which video analysis has facilitated and improved this research. The advantages of video will be discussed mainly through two examples from our research. First, we consider a set of studies we have conducted on family conversations about science exhibits in a children's museum setting. Next, we consider a study of family conversations about earthquakes in a naturalistic laboratory setting. In both cases, we discuss the advantages of using video, both in terms of capturing the phenomena we seek and in terms of uncovering patterns in the data that would not have been available otherwise. In the chapter, we emphasize in turn three key opportunities that video provides for addressing research questions: (1) the goal of capturing spontaneous conversations while keeping the recording of data as unobtrusive as possible, (2) the ability to study patterns in the data that were not anticipated when the study began, and (3) the ability to capture both the nonverbal as well as verbal aspects of conversation. Subsequently, we discuss two challenges that we face in this work and some of the analytic solutions we have developed in response to these challenges. The first challenge has to do with assessing how "true" a representation of everyday action we have captured in our video. The second challenge is that of moving from an ongoing video record to a data set that allows us to address research questions; we present some of the steps along the way from video to analysis.

CAPTURING SPONTANEOUS CONVERSATION

Children's museums are an ideal context for capturing spontaneous family conversations about science. Museum visits are activities that are usually freely chosen rather than required, and yet, museums provide opportunities for learning that families may want to take advantage of. Our initial interest in observing museum visits was to ask about the ways that families negotiate an understanding of science topics in a complex setting that occurs naturally in their everyday lives. For this reason, we wanted to be as unobtrusive as possible. Video recording allowed us to do this.

We have developed a protocol for obtaining consent while reducing our interference in the family visit. This work, which began in collaboration with Kevin Crowley (see Crowley, Callanan, Tenenbaum, & Allen, 2001), and with the staff at the Children's Discovery Museum of San Jose, involves asking parents for informed consent to participate in the research just after they leave the admission desk. Our researchers explain that we are studying children's learning in the museum and that we are also working with the museum to improve the exhibits. For families who agree to participate, we ask children to wear stickers that will identify them as participants in the study. The stickers are also coded in subtle ways to inform us of the child's age. When children with stickers approach an exhibit where we are taping that day, our camera operators know that they have permission to turn on the camera. This protocol has resulted in very high success rates in recruitment—usually above 90% of the families agree to participate.

The video-recording techniques that we use in museum settings allow us to record conversations that would be extremely difficult to hear using any other techniques. Children's museums are noisy and busy places, and human observers would have an overwhelming task if they were attempting to listen in on individual conversations. It is often necessary to rewind and replay a brief interaction several times to transcribe or code all of the relevant conversation and action. Our remote microphones, placed unobtrusively in the exhibit, allow us to record audio right at the source of the conversations, while recording video from a camera that is a few feet away. Thus, we are able to record the family members' conversations with reasonably good clarity while reducing the intrusiveness of the camera.

One other potential advantage of video recording in the museum is that it may increase our chances to study populations who are not easily recruited to engage in laboratory-based research studies. Visitors to the San Jose Children's Discovery Museum are quite diverse, especially in terms of language spoken and cultural background of the family. It is more difficult to determine the variation in economic and educational background, and yet the museum is engaged in a number of outreach activities targeted at increasing the diversity of visitors on these dimensions, and they have a flexible admission policy to help make it affordable for all visitors to attend.

In our research, we have also brought families to the museum who might not have gone there otherwise, including one study of Mexican-descent families from a variety of educational backgrounds (Siegel, Esterly, Callanan, & Wright, in press; Tenenbaum, Callanan, Alba-Speyer, & Sandoval, 2002). Although the museum may be an unusual environment for some of these families, many of our participants have been very interested in going on this "field trip" with us. In one exhibit we have observed, visitors interacted with a globe where they could focus a small camera on sections of the globe and view them on a screen. Not surprisingly, many families sought out their country of origin and used this activity as an opportunity to talk with their children about their relatives who lived far away. In addition to these personally meaningful conversations, however, we also observed conversations about the shape of the earth, about how the earth turns in space, and other scientific topics.

Exploring Unanticipated Research Questions

An especially important advantage of video analysis is that it allows us to examine large numbers of interactions and to look for patterns, even with questions that were not part of the original research question. For example, the original museum project set out to study family conversations about science and in particular, the nature of questions and explanations that arose in families' conversations about the exhibits. It was not a study of gender differences, and yet, because boys and girls participated, the video record allowed us to pursue serendipitous findings or research questions that were not in the initial plan.

As they operated the cameras, several of our student researchers developed a hunch that parents conversed and interacted differently with daughters and sons. At their urging, we went back to the transcripts and videotapes to ask whether the different types of explanations we were coding varied with the gender of the child or the parent. We were surprised to find a large gender difference, such that parents explained scientific phenomena to boys much more often than to girls. This finding became an important new focus of the research (see Crowley et al., 2001). The video record also allowed us to go back and recode interactions in order to ask important clarification questions. For example, we needed to know whether this difference in explanations could be accounted for in terms of the number of questions asked by boys versus girls (as it turned out, there was no difference) or in terms of the likelihood that boys versus girls would initiate engagement with the exhibit (again, there was no difference). Using other data recording techniques, we would have been able to address the gender question, but it would have been a much more laborious process involving new data collection.

Another example involves a study of young children's early symbolic understanding. An important topic in the developmental literature is that of how young children come to develop understanding that photographs, videos, and models can be seen as representations for their referent objects. Working with videotapes that had been collected as part of the larger study of parent–child explanations, we were able to go back to investigate conversations at particular exhibits that included video as part of the display. A new coding scheme was applied to these extant videotapes, and an entirely new analysis was undertaken—with a focus on how parents talk with very young children about the notion of video as a representation (see Soennichsen & Callanan, 2006). In sum, video allows for rich new analyses on data collected for completely different purposes.

Capturing Both Nonverbal and Verbal Aspects of Conversation

Despite the advantages of the museum context, there are also advantages to conducting studies in which we can observe family conversations in a more controlled setting. In particular, it is important to be able to observe more extended activities where there are not so many other activities competing for the participants' attention. Laboratory settings also allow participants to control how much time they spend in the activity because they are not feeling pressure to share them with other families. In this section, we focus on how use of video has allowed us to obtain a more complete picture than would otherwise be possible, of children's conceptual understanding and of the nature of social interaction contributing to that understanding; in particular, how verbal and nonverbal communication contribute to children's and their parents' co-construction of conceptual understanding.

As Goldin-Meadow and her colleagues have shown (Church & Goldin-Meadow, 1986; Goldin-Meadow & Alibali, 1995), gesture can be an extremely sensitive guide to children's conceptual understanding. In particular, these researchers have found evidence that there is sometimes a mismatch between children's words and their gestures, especially when they are in a transitional stage with regard to a particular problem-solving task. For example, children who are in a transition stage in their understanding of the Piagetian number conservation problem will sometimes use gestures that indicate the right answer even though their verbal answers are incorrect. By allowing us to access both the verbal and nonverbal aspects of explanation, video analysis has given us the opportunity to examine the gestures used by both children and parents as they are negotiating an understanding of scientific concepts and to assess the role of these gestures in this negotiation.

In a study of children's understanding of earthquakes, we invited families to come to our campus laboratory to participate. Forty-eight parents and their 4- or 8-year-old children were asked to look through an illustrated book specially designed to encourage discussion about earthquakes, and to use the book as a way to start talking about earthquakes the way they might at home. The book introduced two children who had heard about recent earthquakes in the news and were asking their parent some questions. This was followed by eight pages each containing a hand-drawn picture and one or two questions. The questions related to scientific causes of earthquakes, earthquake preparedness, the parent's personal experiences with earthquakes, and the effects of earthquakes on buildings of different types.

Again, we were especially interested in explanations that occurred in these conversations. From the pilot study, we anticipated that gestures would be an important part of some of these explanations. For example, explanations of plate movements under the earth were often accompanied by gestures. The following explanation occurred during a discussion of fault lines between a mother and an 8-year-old girl:

> Parent: So there's a lot of pressure and sometimes pressures moving
> against the line like this [places her hands together and pushes
> them toward each other] and sometimes it's moving in one direc-
> tion or another. And sometimes there's so, so much pressure like if
> you went like this, [takes child's hand and pushes hers against it]
> you know and we pushed as hard as we could, as hard as we could,
> somebody is going to have to give up eventually. And kind of jolt
> off, so push really hard. [Their hands push against each other until
> they slip apart and move]. So that might happen, look what hap-
> pens. There's energy that leaves.
>
> Parent: (after repeating the demonstration at the child's request) See what
> happens, kind of like that happens, there's so much pressure
> [places fists together and pushes them toward each other] that
> eventually something has to give and that's what happens.

Use of gesture was very common in these family conversations about earthquakes. Overall, parents in 75% of the dyads and children in 54% of the dyads used gestures to support at least one verbal explanation. In some cases, such as the previous example, gesture was an integral part of the explanation, whereas in other cases, gestures played a more supportive role.

Gestures were particularly common in the explanations for certain phenomena. One page of the book posed the following question: "Will buildings made of brick fall more during an earthquake than buildings made of wood?" We found that families often used gestures to explain the relative flexibility of brick versus wood. This science principle was used to explain why wood buildings are less likely to fall than brick buildings by 36 (75%) of the dyads, and 72% of discussions of this principle included a supporting gesture by the parent. In the following example, a mother explains the importance of flexibility to an 8-year-old girl:

Parent: Wood stretches when there is an earthquake, the wood will bend [sways hands back and forth] and the bricks won't bend. And so the fact that they won't bend means that they will break instead. You know how like if you take, like a stick, a hard stick, and you start bending it this way, this way, this way, this way, what happens? [simulates bending a stick]

Child: Break.

Parent: Right, if you take a piece of grass and you start bending it this way, this way, this way, what happens? [demonstrates bending a blade of grass.]

Child: It goes up again.

Parent: Right, yeah. So that's why, so wood will do that. It will bend. If you bend it far enough it will break but with brick you can't bend it at all. So that's why it breaks and falls apart.

Dyads also discussed other properties of the materials, for example, the brittleness of brick versus the flexibility of wood. These discussions often were augmented by gestures, as in the following explanation by a mother to a 4-year-old girl (dyad 9):

Parent: Well because (bricks are) put together with what's called mortar which is kind of like a cement and they are kind of like glued together to make a box. And when they shake they don't give like wood, you know how wood sometimes can bend a little bit, well brick is so hard that it can't bend. So in an earthquake there is no sway to it and so it has more tendency to even break and fall down. [uses fingers to show how bricks are glued together and how they shake. She sways her body from side to side to demonstrate wood shaking.]

Another question posed in the book that was likely to generate explanations accompanied by gestures was: "Will taller or shorter buildings fall more often during an earthquake?" A quarter of the dyads mentioned relative stability of shorter versus taller buildings as a reason the taller buildings would be more likely to fall in an earthquake. In the following example, an 8-year-old boy incorrectly replied that shorter buildings were more likely to fall, however, the reason and the gestures supported the opposite response: "Shorter because taller they're more, they're not stable [gestures wide shaking at top of a tall building]." The mother proceeded to clarify the child's response and later provided a more elaborate explanation that incorporates both words and gestures:

Parent: Because you said the taller one's less, the taller one is less stable … It shakes, it shakes the top part of it shakes. In fact, what, I guess there's nothing in here that we can play with. But maybe later when you build something, if you take something that's really tall and you shake it a little bit at the bottom. You take a little thing that's squatty but not very tall, and shake it a little bit at the bottom. If you look at the top of those two things I think that you'll see that the top of the tall building could actually move quite a bit. [uses hands to show how the tall and short buildings would shake during an earthquake.]

As this example shows, our video recordings also allowed us to identify places where there is a mismatch between gesture and verbal explanation. Goldin-Meadow and Alibali's (1995) work suggests that these mismatches are often indicative that children are in the process of figuring out a new concept. Mismatches can also occur in another setting—gesture may sometimes give children a way to express understanding of a concept even if they can't or don't explain it verbally. For example, when asked by her father why she thought taller buildings were more likely to fall than shorter ones, one 8-year-old girl said "Because when the earth shakes, there isn't, it won't be swaying as much if it's shorter." This verbal explanation does not do justice to the level of understanding of the concept of stability that is evident in her use of gestures that accompany this simple statement. Parents are perhaps likely to pick up on these mismatches between gesture and talk and use them to interpret what the child is trying to convey (see Goldin-Meadow & Momeni Sandhofer, 1999). In the cases where the child's gesture indicates misunderstanding, parents may even explain the concept in more detail. These are intriguing questions for future research.

In addition to these particular uses of gesture, children sometimes used the form of an explanation, but without access to the video record, it would have been impossible to figure out what the child was trying to say. For example, sometimes children simply said "because it goes like this," then demonstrated the difference in the pattern of shaking in a tall versus short building. Our use of video allowed us to code these explanations.

Gestures were also used to construct explanations that considered multiple factors, such as the following given by a mother to her 8-year-old daughter:

Parent: And you know what, it doesn't matter if it's tall or short, umm, well because of the construction nowadays, but if you have this house and this house and they were exactly the same, same construction, same everything, if you were up here you would be shaking a whole lot more then you would be shaking down here.

Child: Aha.

Parent: Do you know, do you know a little about that? Have you ever thought about that?

Child: I would think you would not be shaking as much up here [point to top of tall building], and more down here [points to short building].

> Parent: Yeah, it's the way the movement travels like, like on my arm [raises her arm and shakes it, while holding her elbow still]. If you start shaking down here, look I am just shaking a little bit of my elbow but look what my hand is doing up here. It's kind of going and what happened in our house is our house kind of goes like this [moves her hand very quickly back and forth].

Parents sometimes responded to children's gestures by providing additional explanation or appropriate terminology as in this example of an 8-year-old girl and her father discussing what an earthquake is:

> Parent: But when the earth, when the two pieces of the earth are moving against each other, they shake. They kind of rumble.
>
> Child: And sometimes one of those pieces of land, they will collide and then one will go like that [uses hands to demonstrate].
>
> Parent: Yeah that's right. That's called subduction, when one goes under the other one. It starts to slide under. Those two pieces are shoving against each other and then slips. [demonstrates with his hands]. That's when the earthquake happens.

As this example suggests, parents' sensitivity and response to this nonverbal communication may be an important way in which they help children learn culturally appropriate ways of talking about natural phenomena. Using video, it is possible to systematically consider not only the ways that gestures are used in the context of explanations about scientific phenomena, but also the ways that explanations are dynamically negotiated in parent–child discourse. These results also point to the importance of considering nonverbal factors in studies of what children know and how they come to understand concepts about the natural world. Gestures appear to remain important even in adolescence. For example, Roth (1999) found that high-school students used gesture and bodily movement to understand physics concepts. As in Goldin-Meadow and Alibali's work mentioned earlier, knowledge about the physical world may be expressed and even understood in both nonverbal and verbal ways, and insistence on detailed verbal explanation may be uncommon outside a formal school setting. More central to the goals of this volume is the point that these issues would be very difficult to analyze in detail without the use of video recording.

How "Representative" Is Videotaped Action of Everyday Action?

Anyone who observes "natural" interaction must ask themselves about the impact of the observer on the activities that are observed. In the case of videotaped data, the role of the camera must be considered, and yet it can never be fully examined.

Perhaps somewhat surprisingly, our experience suggests to us that cameras seem to make visitors less self-conscious than do human observers. Admittedly, families know that they are being recorded in our studies, and there is individual variation in how nervous or avoidant of the camera people are. Anecdotally, however, we find that families in the museum setting seem more inhibited when a researcher is looking into the camera's viewfinder than when the camera is standing and recording on its own. Typically, families get engaged in the activity and although they notice the camera, they seem to quickly forget about its presence. Thus, while we can not be sure that our participants behave on camera as they would off camera, our experience suggests that there are not major changes in behavior when people move in front of the camera.

Moving From Video to Answering Research Questions

Perhaps the biggest challenge in using video data is that of determining what to look at and analyze from the hours of continuous activity, and how to use these data to answer specific research questions. We have developed a number of intermediate steps in moving from video to interpretation, and the methods we use vary depending on the research question. After we gather our videotapes, we often transcribe them and then code the conversations while referring simultaneously to the videotape and the transcript. Typically, although we begin a study with some preliminary codes in mind, we refine and augment the codes as we examine a few of the tapes together. In this beginning phase of the coding, the research team meets routinely not only to discuss and refine the coding scheme, but also to share their observations of data patterns that are pertinent to the original research questions as well as emergent themes that can take the research in a new direction. When, after several iterations between independent coding and group discussion, a consensus is reached on the coding scheme, we establish intercoder reliability using Cohen's kappa (on approximately 25% of the data set) and the finalized, reliable code is then applied to the remaining tapes

For most of the research questions we address, we move from our coding schemes to quantitative analyses. We take issue with a common dichotomy that suggests that one either uses numbers or uses meaningful data. For example, Schegloff (1993) argues that quantification raises several potential problems in the analysis of conversation, and states that "quantification is no substitute for analysis" (p. 114). We agree with the concerns Schegloff raises related to the importance of not overlooking the rich context of each conversational turn at talk. We would argue, however, that these are potential problems for any coding of talk, whether or not it is followed by quantification. In particular, Schegloff advises those who talk about percentages of events in conversation that they need an "analytically defensible notion of the denominator, ... the numerator, ... and ... of the domain or universe being characterized" (p. 103). In other words, he argues that one shouldn't blindly count events without knowing how they relate to the larger context. We would argue that our coding schemes are very carefully designed and refined to capture intentions and actions that are very meaningful in our participants' conversations with one another. We are counting, but we are counting something interesting, subtle, and meaningfully connected to our specific research question, that takes into account the larger context in which this research question and the participants' social interactions are embedded (see also Azmitia & Crowley, 2001).

CONCLUSION

We have provided examples of two types of studies from our research on family conversations about science where video recording and analysis have played a crucial role. We have discussed three major ways that video offers opportunities to address certain types of research questions. First, we argue that video recording gives us a better chance of obtaining access to spontaneous family conversations. Second, we argue that video data facilitates the investigation of unanticipated research questions. And finally, we document the analysis of nonverbal aspects of conversation that would be difficult or impossible without a video record. Because these three issues are substantive, we suggest that the use of video is more than just a methodological issue. Analysis of conversation has become more and more sophisticated as technological innovations have emerged, and new technology continues to be designed in order to meet the needs of researchers. Perhaps this volume will contribute to this mutual, positive influence between the design of tools and the design of research studies.

We also discussed two crucial challenges that must be faced in video-based research; the problem of how well video data represents everyday life, and the question of how best to transform video into data that can productively address questions about human communication and development. While no immediate solutions to the first challenge is provided here, some examples of how to transform video data to answer questions were presented. The ideas discussed will hopefully provide food for further thought and discussion. Again, this volume—with its focus on multiple analytical perspectives—is likely to contribute to positive progress on these issues.

REFERENCES

Azmitia, M., & Crowley, K. (2001). The rhythms of scientific thinking: A study of collaboration in an earthquake microworld. In K. Crowley, C. Schunn, & T. Okada (Eds.), *Designing for science: Implications from everyday, classroom, and professional settings* (pp. 51–81). Mahwah, NJ: Lawrence Erlbaum Associates.

Church, R., & Goldin-Meadow, S. (1986). The mismatch between gesture and speech as an index of transitional knowledge. *Cognition, 23*, 43–71.

Crowley, K., Callanan, M. A., Tenenbaum, H. R., & Allen, E. (2001). Parents explain more often to boys than to girls during shared scientific thinking. *Psychological Science, 12*, 258–261.

Cole, M. (2002). Culture and development. In H. Keller & Y. H. Poortinga (Eds.), *Between culture and biology: Perspectives on ontogenetic development* (pp. 303–319). New York: Cambridge University Press.

Goldin-Meadow, S., & Alibali, M. W. (1995). Mechanisms of transition: Learning with a helping hand. In D. L. Medin (Ed.), *The psychology of learning and motivation: Advances in research and theory* (Vol. 33, pp. 115–157). San Diego, CA: Academic Press.

Goldin-Meadow, S., & Momeni Sandhofer, C. (1999). Gestures convey substantive information about a child's thoughts to ordinary listeners. *Developmental Science, 2*, 67–74.

Leinhardt, G., Crowley, K., & Knutson, K. (2002). *Learning conversations in museums*. Mahwah, NJ: Lawrence Erlbaum Associates.

Paris, S. (2002). *Perspectives on object-centered learning in museums*. Mahwah, NJ: Lawrence Erlbaum Associates.

Rogoff, B. (2003). *The cultural nature of human development*. New York: Oxford University Press.

Roth, W.-M. (1999). Discourse and agency in school science laboratories. *Discourse Processes, 28*(1), 27–60.

Schegloff, E. A. (1993). Reflections on quantification in the study of conversation. *Research on Language and Social Interaction, 26*, 99–128.

Siegel, D., Esterly, J., Callanan, M., & Wright, R. (in press). Conversations about science across activities in Mexican-descent families. *International Journal of Science Education.*

Soennichsen, M., & Callanan, M. (2004). *Children's emerging understanding of video as a symbolic medium: The role of parent–child conversation.* Unpublished manuscript.

Tenenbaum, H., Callanan, M., Alba-Speyer, C., & Sandoval, L. (2002). The role of educational background, activity, and past experiences in Mexican-descent families' science conversations. *Hispanic Journal of the Behavioral Sciences, 24*, 225–248.

Progressive Refinement of Hypotheses in Video-Supported Research

Randi A. Engle
University of California, Berkeley

Faith R. Conant
Five Colleges, Inc.

James G. Greeno
University of Pittsburgh

Scientific inquiry in the learning sciences often includes hypothesis generation and evaluation together, in a kind of dialectic process. We believe that understanding such inquiry can be enhanced by focusing on the interaction between hypothesis generation and evaluation, rather than by separating them as some accounts of science have in the past (e.g., Popper's, 1959 distinction between the "context of discovery" and the "context of justification"). Therefore, much research in the learning sciences involves not just classic hypothesis testing in which scientists specify hypotheses in advance of collecting data and design experiments to provide definitive refutation or tentative confirmation of those hypotheses. At the same time, what many of us are doing is not characterized by a naïve view of ethnographic inquiry in which investigators attempt to begin with a theoretical clean slate, developing hypotheses only in response to patterns in the data. Instead, the kind of research many of us are doing is characterized by the progressive refinement of hypotheses, with theory and data interacting throughout the process.

Specifically, in the process of *progressive refinement of hypotheses,* an investigator begins with a general question and then decides to collect empirical records in a relevant setting with an initial plan for how to use them to learn more. Initial analysis of these records informs more specific hypotheses that may then be addressed in other aspects of the data. Those results then provide even more specific questions that are addressed in some other part of the data, leading to explanatory hypotheses that can then be evaluated in still other aspects of the data, and modified or made more specific in relation to those data, and so on. The development and evaluation of hypotheses involves a kind of constraint satisfaction, in which the coherence between hypotheses and the consistency of hypotheses with data are both taken into account.

We believe that much progress that is made in scientific research actually follows some version of this pattern. Commonly, the pattern plays out over multiple studies, with data and theoretical proposals provided by multiple investigators. However, with the advent of records like video that both are relatively comprehensive and support multiple re-viewings, it is now often possible to take several steps in the progression using the same set of records. Thus, video records can allow a single study to progress through multiple iterations of hypothesis generation and evaluation, making the resulting findings more robust than they might have been otherwise. By more robust, we mean that the findings have been evaluated and modified in relation to more aspects of the data, and therefore, are more likely to stand up to further investigation.

In this chapter, we discuss a research project that illustrates the use of video records to support the progressive refinement of hypotheses about student engagement and learning. We first discuss how our overall project was structured to support this process, with a focus on how video provided useful affordances for it. In the core of the chapter, we then discuss a specific study we did (Engle & Conant, 2002) to show how the process of progressive refinement can work over the course of a video-based study. As part of this, we show how video-based case studies like ours can be used both alone and in concert with later studies to support hypotheses of wider generality. We believe that this study sheds some light on ways that development and evaluation of hypotheses can interact productively, especially in research that includes extensive video records as a source of data.

SUPPORT FOR PROGRESSIVE REFINEMENT OF HYPOTHESES IN THE DESIGN OF A RESEARCH PROJECT

When we (Greeno & Engle) proposed a research project about the role of discourse in conceptual learning, we had not formulated any hypotheses that could be falsified by the data we proposed to obtain. However, we did much more than simply identify a potentially interesting setting for study (Fostering Communities of Learners classrooms) in the hope that worthwhile findings would somehow emerge. We had formulated the issue in theoretical terms, and we had also characterized how we planned to use video records to develop more specific hypotheses and then to evaluate them.

Specifically, the goal of the research that we initiated was to advance understanding of the processes of conceptual growth in project-based classrooms. Our theoretical goal was to develop and evaluate hypotheses about how students' participation in classroom activities could account for the advances in their conceptual understanding. We proposed to combine methods of cognitive developmental psychology to assess students' conceptual understanding before and after a unit of study with methods of discourse analysis and cognitive representations of information to analyze students' and teachers' discussions during the unit.

Video records of classroom interactions were a crucial component of our research plan. We made video records of groups of students working on assessment tasks before and after the unit, and these were the materials we analyzed to characterize the conceptual growth that we planned to document and explain. We also needed to observe classroom activity across several school weeks in order to be able to examine significant changes in students' conceptual understanding. Thus, we also recorded video during the approximately 12 weeks during which two classes worked on the endangered species unit of the Fostering Communities of Learners curriculum (Brown & Campione, 1994; Kohl & Wingate, 1995). In particular, we followed two target groups of students in each class, videotaping both their interactions and whole-class activities.

It was through the examination and analysis of these videos that we hoped to develop and evaluate hypotheses that we could propose as explanations of conceptual growth. Our plan, which we followed, was to use the pre/post-assessments to identify topics on which students' conceptual understanding had advanced, then to examine episodes during the unit in which those topics were discussed, and analyze ways in which discourse in those episodes provided occasions in which students could have advanced their understanding in ways consistent with the changes we had observed. Our expectation was that different groups (and students within these groups) would demonstrate contrasting patterns of change on the assessments that would provide us with contrast cases to constrain the explanations we could generate for how the students' conceptual growth could have occurred.

Thus, although we did not have empirically testable hypotheses at the start, we were committed to developing and evaluating more definite hypotheses in the course of our study. We anticipated findings about the nature of conceptual growth from the assessment part of our study, and we proposed to obtain evidence in the classroom discourse part of the study that we would use to explain the findings of the assessment part. Our plan was plausible, we believed, because it was continuous with methods of successful programs of research in the general domain of our study (e.g., Rosebery, Warren, & Conant, 1992; Yackel, Cobb, & Wood, 1991), and we had participated in projects that used many of the methods that we needed to carry out in our proposed research (e.g., Engle & Greeno, 1994; Greeno & Engle, 1995; Greeno, Engle, Kerr, & Moore, 1993; Linde et al., 1994).

Our plan was consistent with a research strategy of progressive refinement of hypotheses. We needed to conduct and record the pre/post-assessments in order to identify aspects of the conceptual domain on which students advanced in their understanding. Having found some evidence for conceptual growth, we needed to have records of teaching and learning to examine in order to develop and evaluate hypotheses about how that conceptual learning occurred.

Consider what we could have done without records of this kind. If we had not made recordings—either audio or video—we would have had to depend on field

notes or observational coding sheets that we would have been forced to develop before observing the units or identifying the topics on which students changed their understanding. As it turned out, that would have vitiated our study almost entirely.
Although we correctly anticipated the general topics on which the students were
likely to make progress (e.g., a better understanding of habitats and adaptation), we
had no basis for anticipating the specific contents of their conceptual advances. For
example, one of the features of many of the students' conceptual growth was their
consideration of birthrates as quantitative variables that are influenced by multiple
causes and have multiple effects (Engle, 2006b). This is not surprising in retrospect,
but we had no basis beforehand for giving this topic sufficient priority to specifically
code for it in a classroom observation. Therefore, instead, we decided to collect an
extensive record of classroom activities on video, making it likely that, once we did
identify exactly where particular students' conceptual growth occurred, we would
have sufficient material about what happened during the unit to develop justifiable
accounts of its development.

More importantly, an observational coding sheet would have presupposed that
we could anticipate the specific kinds of interactions that would figure in our explanatory hypotheses. Similarly, for field notes to have been successful, we would have had
to recognize potentially significant interactions as they were occurring and make sure
to record the most relevant details. We simply were not in a position to anticipate the
kinds of events that we have decided, in retrospect, help account for the kinds of
conceptual understanding and learning that appear to have occurred. For example, in
retrospect, we have concluded that teachers attributing authorship of ideas and information to students was probably a critical factor in their learning (Engle, 2005b; Engle
& Conant, 2002). Such attribution of authorship also might have been expected, but it
is unlikely that we would have given that characteristic, above many other possibilities,
a high priority in any coding scheme we could have developed in advance. Even if we
had, we cannot imagine how we could have constructed a sufficiently accurate record
of the events we now identify as positioning students as authors without having had a
fairly complete record of the interactions to refer to. With video, however, all that was
possible.

In addition, the video records turned out to enable an analysis of an issue that we
could not have anticipated, as it involved analysis of a series of events that were not a
part of the planned curriculum. Furthermore, our attention was drawn to the issue and
the events by unanticipated opportunities for collaboration.

Following the collection of the video records, as planned Greeno, Engle, and
their colleagues began to analyze records from the assessments and some of the re-
cords of interaction from the unit in order to develop accounts of conceptual growth
(Benke, 1999; Engle, 2006b; Greeno, Benke, Engle, Lachapelle, & Wiebe, 1998;
Greeno, Brown, & Campione, 1999; Lachapelle, 1997; Wiebe, 1999). Faith Conant and
Frederick Erickson were each visiting for the year and interested in working with us.
Erickson said he was particularly interested in exploring the question of "what makes a
conversation take off?" During our collection of the FCL videos, we had noticed that
the target group who was studying whales had become unexpectedly and passionately
engaged in an argument about whether orcas ("killer whales") are whales or dolphins

(what we came to call the "orca controversy"). This occurred despite the fact that the issue of how species should be classified was not part of these students' assigned task, which was simply to explain why whales were endangered.

We had not planned to focus on this series of events as its topic did not correspond with any items in the pre/post-assessments, but Erickson's interest made us realize its potential value for our understanding of conceptual learning. These incidents underscored that conceptual understanding is fundamentally affective and historical as well as informational. For example, when reporting about their work to a student teacher just joining the class, one student's brief mention of the orca issue caused the argument to re-erupt with great emotional intensity, with this student remarking that the group had had a "big ol' argument" about the topic earlier. This implied to us that our account of the students' conceptual understanding must include such historical and affective aspects along with informational content. We had captured the students' activity on our video records, but we were, in turn, captured by those video records, especially by the intellectual and affective intensity of the students' discourse. Thus, this series of arguments became a major focus of analysis.

In the rest of this chapter, we will highlight some of the opportunities that our video records afforded us as we worked with these data to explore Erickson's question of what make a conversation take off, drawing primarily on Engle and Conant (2002), the first article our orca controversy analysis group has written from our work together.[1] First, we explain how we interacted with the videotapes to capture the phenomenon of a conversation taking off. In particular, in this section, we discuss how having this focusing question shaped our initial selection of video records to analyze, how we used transcripts to represent our interpretations of these records, and how we helped check and enrich our initial hypotheses by both sharing the videos with other investigators and performing formal codings of them. Following this, we then discuss how we worked with our video records to develop an explanation for why the conversation took off. In this section, we consider how we approached this process with an eye toward generalization; identified possible explanatory factors; restructured our explanation into a simplified general model; drew on the full set of videotapes to refine and specify it; and finally, investigated whether the theoretical concepts in the model could be used to help account for other cases in the literature.

PROGRESSIVE REFINEMENT WHILE CAPTURING AND COMMUNICATING A PHENOMENON

Selecting Video to Analyze

As is typical with many classroom-based projects that use video, we had collected a whole bookcase worth of videotapes that we could have analyzed. How did we know what to actually analyze? Our research question drove our selections. We remembered the conversation had first "taken off" the day after the students had returned from a trip to Marine World as they were working on their bulletin board, and that the oral re-

---

[1]The core group included Conant, Engle, Erickson, Greeno, and Muffie Wiebe Waterman.

port in which the debate had re-erupted had occurred 7 weeks later. We began our analyses with these two incidents and then searched before and after them to find other occasions in which the students discussed the orca issue.

Searching for episodes of discussion of this topic was feasible because we had made content logs of the videotapes in our collection (Jordan & Henderson, 1995). A content log is written by someone watching a tape with only minimal rewinding in order to provide a time-indexed list of topics being discussed. These content logs were used to select segments of video to watch and transcribe.[2] In addition, we paid careful attention to references that the students themselves made to potentially relevant past events or anticipated future ones. Using these resources, we found out that the orca issue had been discussed by the students on at least eight occasions over the last 8 weeks of the unit. We found that although there were differences in the degree to which particular students continued to be engaged in the orca issue over time, it was clear that it was not a one-time issue for this group, but instead, the focus of sustained examination by them.

Representing Evidence for Interpretations

Compared with field notes and observational coding sheets, video records are a relatively under-interpreted source of information.[3] Because of this, working with video invariably involves a second phase in which researchers grapple with how they are going to interpret the events that have been recorded. In addition, because video materials contain so much information, part of this process involves deciding which of the many things that can be seen and heard are relevant to making interpretations about the phenomenon of interest.

In our case, we were interested in understanding how one can tell that an academic conversation has taken off, which we eventually described by developing the concept of "productive disciplinary engagement." Briefly, productive disciplinary engagement is occurring when students are deeply engaged in an issue ("engagement"),

---

[2]We could also make use of the fact that some rough transcripts had already been made as part of another analysis of this group's learning (Engle, 2006b).

[3]However, although specific interpretations might not be attached to particular video records, the types of interpretations one can most easily make from them can be profoundly affected by choices about where to focus the camera and what kinds of audio information to record (Hall, 2000). Nevertheless, these constraints and affordances on interpretation are dwarfed by those characteristic of observational coding sheets. For them to work, it is necessary to prespecify one's interpretive categories. This means that a key part of the interpretation of events is done before they even occur, with other aspects done in-the-moment, as one fills out one's observation sheet. The degree and types of interpretative moves made in a given set of field notes depend on the skills and methodological commitments of the person making them, but it is neither possible nor desirable to eliminate all interpretation from them. As it is only humanly possible to record a small subset of what one might be able to observe when making such notes; at the very least, one's selection of what to record ends up reflecting one's emerging interpretations about what is going on. Given this selectivity, a relatively "raw" record might not be understandable. Such issues are well recognized by experienced ethnographers and observational researchers so we do not wish to further belabor them here in a book focused on the use of video.

their engagement with it makes significant contact with disciplinary ideas and practices ("disciplinary"), and through this engagement, they make progress on the issue in question ("productive;" see Engle & Conant, 2002, pp. 402–403 for more complete definitions). It seemed evident to us from the tapes that the students were, at the very least, intensely engaged in the orca issue, but we needed to become more explicit about the basis for this and other interpretations to be able to examine them more systematically and share them with others. We did this in a multistep process that involved progressive refinement of our representations of the events and our claims about the nature of the students' engagement in them.

From viewing the videos, we had noticed that the students often fought for the floor, emphasized their points in a dramatic fashion, built on each others' ideas, and celebrated when they had successfully argued a point. However, many of the specific actions we were considering as evidence of the students' engagement—for example, overlapping speech, emotional displays, and emphatic stress—had not been systematically included in the rough transcripts we had. If we were to describe the intensity of engagement that we were seeing to others, we would have to record its relevant features. More detailed transcripts were constructed to include these and other features. For example, a section of transcript from the "Big Ol' Argument" that originally had been transcribed as:

Liana:	AND she probably didn't even know, because she was just there for a little while, and then she just made it up.
Brian:	She was only there for two years!
Liana:	yeah, so she didn't know that much, so she just thought of it then, at that time 'cause she probably was thinking about DOLPHINS!
Brian:	Exactly! (*Brian and Toscan high-five*)

Was revised to:

409 Liana:	(*waving right hand wildly as Brian nods head and shoulders emphatically*)
410	OH! . .oh . .oh . .
411	A:::ND . she probably didn't even <u>know</u> .
412	because she was probably just there for a little whi::le.
413	and then sh just made it [up
414 Brian:	[SHE WAS ONLY THE;RE FOR TWO
415	YEA::[;RS
416 Liana:	[YEAH! . . so she didn't know that much she/she just thought of
417	it that . time cuz she probably was thinking about . DO::LPHINS
418 Brian:	(*big grin*) . . exa::ctly (*Brian "high five's Toscan*) . . . .
...	
419 Liana:	yup
420 Racquel:	(*pats Liana on the back and they exchange nodding looks*)

to represent the degree of emphasis, overlapping speech, emotional displays, and non-speaker involvement that were characteristic of this discussion.[4] In effect, our decisions about what categories of things to record in our transcripts helped us specify some characteristic features that might provide evidence of student engagement. Thus, our transcripts became a record of our emerging hypotheses as embodied in the specifics of a particular interaction (see also Ochs, 1979).

Refining Hypotheses by Making Comparisons, Enlisting Other Analysts, and Coding

Based on sufficiently detailed transcripts, we were able to be much more explicit about what we were noticing in the orca discussions that led us to the conclusion that the students were strongly engaged in them. In order to provide evidence that these features were significant, however, we needed some other discussions to which we could compare the orca discussions. To provide this comparative data, we re-examined the video of the students' report to the new student teacher, which also contained presentations about the rest of the students' work during the unit. With this video, we could compare the students' engagement in these other topics with their engagement in the orca issue in order to access our impression that they were more engaged in the orca issue than in their other research topics. As mentioned earlier, there appeared to be a dramatic shift in the students' engagement the moment that one of the students (Brian) brought up the controversy as part of his report on whale features.[5] We slowed down the tape, sometimes watching it with the sound off, and noticed both how Brian signaled that the topic was coming, and the rapidity with which the other students responded to his signals with evidence of their engagement. For example, as soon as Brian mentioned the orca issue, the whole group squeezed closer together, other members of the group began shaking their heads in agreement or disagreement, and various students began to bid for the floor. (These phenomena, by the way, would have been impossible to observe had we only audiotaped this session.)

Our impression of a shift was then corroborated by a group of colleagues whom we asked to watch the tape and note where they observed a change (if any) in the students' level of engagement. Their proposed locations were close to where we had identified the shift as occurring, giving us additional confidence that it was there. On this basis, Engle and Conant (2002, pp. 417–419) wrote an analytic description of the shift, supported by a revised transcript of Brian's full report (Engle & Conant, pp. 465–473).

To provide evidence that could be abstracted from the video record for readers without access to it, and to illuminate features of interaction that might provide evi-

---

[4]In this transcript, colons represent elongation of vowels; periods between spaces indicate the lengths of pauses; underlining and capitalization each represent additional vocal emphasis; and paired left brackets indicate the beginning of overlapping speech. We are grateful to Fred Erickson for making this version of this transcript as well as for leading the effort to create transcripts that would more systematically record theoretically relevant features of the discussions.

[5]The orca controversy was relevant to whale features as, during their earlier debate, the students had tried to resolve the issue by comparing various features of an orca (especially its dorsal fin) to those characteristic of different species of whales and dolphins.

dence for changes in student engagement, we then coded and statistically compared the 70 secs of Brian's report before he raised the orca issue and the next 70 secs afterward. In particular, we showed that when discussing the orca issue as compared to the rest of Brian's report: (a) speakers were much more equally distributed between the students; (b) students were more likely to address each other with their contributions rather than only the new student teacher; (c) students engaged in activities unrelated to the discussion much less often; (d) there were many more emotional displays by the students; and (e) there were also many more task-relevant spoken overlaps as students fought for the floor or collaboratively completed each others' ideas (see Engle & Conant, 2002, p. 419 for results, and pp. 473–476 for methodological details). This comparative coding provided measurable, behavioral indicators of the students' greater engagement in the orca question than in the rest of their work during the unit. At the same time, it suggested some features that might be useful for assessing engagement in other contexts.

We then participated in similar processes of progressive refinement in order to establish how the students' engagement in the orca issue was also both "disciplinary" and "productive" (see Engle & Conant, 2002, pp. 420–424 and 424–429, respectively).

PROGRESSIVE REFINEMENT WHILE DEVELOPING A GENERALIZABLE EXPLANATION FOR THE PHENOMENON

While we were using the videotapes to clarify how the orca discussions were a case of productive disciplinary engagement, we were simultaneously exploring an additional set of related research questions at another level of analysis. First, we wanted to know why this group of fifth graders had become so engaged in this particular, fairly technical question about the classification of orcas. Given it was not part of the students' assigned task, how did it become relevant to them and why did they continue debating it over the course of the unit? At the same time, we wished to see whether there was anything about the design of the FCL learning environment itself that might account for the students' engagement. If we were successful at doing that, this might help us better understand how FCL classrooms work. Last but not least, we were interested in theorizing much more generally about what needs to be in place in a learning environment if it is likely to support such productive disciplinary engagement. If we could specify aspects of the FCL learning environment that could help account for the students' engagement at a sufficiently broad level of generality, then our research would provide hypotheses about how productive disciplinary engagement could be supported more generally. So we began trying to account for the whale students' engagement in the orca question with an eye to both explaining the specific events that were involved in their discussions while also creating hypotheses that could be adapted for use in other contexts.

Identifying Possible Explanatory Factors Through Watching and Re-Watching

The initial phase of trying to explain the students' engagement involved a lot of collective watching of the videotapes by our group of five while we brainstormed about possible explanations for the students' productive disciplinary engagement.

In much of the video-analysis work that we have done, a phase like this that involves some kind of collective exploration of the videos has proven to be essential. Everyone gets to know the data very well. From watching the same videos on different occasions, multiple interpretations and hypotheses can be specified and then evaluated for agreement both with the data and with each other. If one short circuits this phase, one can easily make premature, unwarranted conclusions. In our case, we added ideas to a whiteboard as we thought of them, eventually identifying as many as a dozen or so possible factors that might have contributed to the students' productive disciplinary engagement in the orca question.

Enhancing Communication and Generality Through a Simplified Model

There were many possible factors that might have contributed to what had happened during the orca controversy. How could we possibly describe—let alone explain—all of them in a talk or article? This problem is endemic to video analysis. Because of the richness of video data, the problem is not usually of having something to say, but of choosing among the many things that one could say and fashioning them into a coherent account.

To address this problem, we took the dozen or so contributing factors we had identified and organized them into four categories:

- *Problematizing*: encouraging the emergence of open intellectual questions that are considered open and of importance to learners while incorporating issues that learners are supposed to be to learning about;

- *Authority*: "authorizing" learners to address the questions in their own wasy (*agency*), publicly giving them "authorship" over their responses (as *stakeholders*), and encouraging them to grow into local "authorities" (*local experts*) whose ideas might begin to shape those of others (*contributors*);

- *Accountability*: encouraging students to "account for" how their contributions are responsive to shared norms for quality contributions as well as to the relevant contributions of others in the local learning environment and beyond,[6] and

- *Resources*: providing or helping learners to find the various resources they need to directly support their productive disciplinary engagement (e.g., sufficient time or instruction in key technical skills to be able to engage productively) as well as to realize the other three principles (e.g., training in how to construct evidence-based arguments in a particular discipline in order to foster accountability to the norm of giving evidence).

---

[6]Notice the contrast between this notion of accountability and the more externally oriented one discussed more commonly in education in which outsiders evaluate the degree to which students have met standards. Here we are discussing a more internally oriented notion of accountability in which students account for how their contributions make sense within the context of their learning environment.

Originally, the purpose of these categories was to provide an uncontroversial package that could embed the specific factors we were using to explain the orca case.

However, as we reviewed relevant literature, we noticed that these four categories that had emerged in our analysis were often discussed separately and in somewhat different terms by different researchers. At the same time, we came to recognize that these four aspects were often realized together in the designs of many innovative and effective learning environments. Thus, we realized that together these four categories were, in effect, a model for what might be involved in supporting productive disciplinary engagement more generally (see especially Engle & Conant, 2002, pp. 408–410). They had the potential to characterize one possible consensus set of design principles implicitly shared by current designers of many successful learning environments. Therefore, by proposing the four categories as guiding principles that might account for productive disciplinary engagement more generally, we also were taking advantage of an opportunity to identify some common threads underlying many—in some cases quite diverse—efforts.

Refining and Specifying the Model With the Larger Data Set

Now that we had a model consisting of principles whose embodiment might foster productive disciplinary engagement, how could we use our videotapes to specify how this model might work? Here, because we had collected a comprehensive record of the unit, we could draw on videotapes not just of the orca controversy discussions themselves, but also of most everything else this group of students had participated in during the unit, including discussions that the teacher held with the class as a whole that helped reinforce the norms of the classroom. With our content logs and transcripts, we could do a systematic search before the controversy erupted to explore how each principle was embodied in the classroom learning environment more generally. We could then combine this with analyses of how the principles were embodied when the controversy erupted, during the critical Big Ol' Argument discussion, and throughout the rest of the unit (see summary tables in Engle & Conant, 2002, pp. 449 and 452 as well as full discussion on pp. 430–447).

We found our explanation compelling to the extent that we could identify linkages between specific aspects of the students' productive disciplinary engagement and the specific ways that each principle had been embodied in this particular learning environment with these particular students around this particular topic (see especially Engle & Conant, pp. 448–451). Our theorizing was significantly constrained by the fact that our observations during data collection suggested that this particular issue was the one that had prompted the strongest level of productive disciplinary engagement during the unit among any of the four target groups. So we could not fashion an explanation that only referred to global factors true about the learning environment in general—instead, we needed to combine analysis of such global factors with those that applied specifically to this group's discussions about the orca issue. By coding how the students positioned themselves and others, we were able to demonstrate that it was not just the teacher and learning environment that had positioned the students as accountable stakeholders in the debate, but that the students had also positioned themselves in this manner (see Engle & Conant, pp. 447–448 for results and pp. 480–483 for methodology).

Generalization by Accounting for Additional Cases

As mentioned earlier, we sought to identify explanatory factors that could account for productive disciplinary engagement in general. Some critics might maintain that successfully achieving such generality is threatened in a research project such as ours, either because of the specificity of our videotaped materials, or because we had not specified falsifiable hypotheses in advance of data collection and analysis. The alternative that we consider more realistic and productive involves using such records to inform the development of concepts and principles capable of providing coherent explanatory accounts of findings in a wide range of activities and settings. From the start, we viewed the orca discussions as a "case" of something more general (see Shulman, 1992), productive disciplinary engagement. We then sought to explain this case of productive disciplinary engagement by proposing four general principles for fostering such engagement. We combined a theoretical argument for why these principles might help account for productive disciplinary engagement in general with detailed empirical analyses of how the principles were embodied in this particular case to help explain specific aspects of it.

We then evaluated the generality of the principles by seeing whether they could be used to help account for several other published cases of productive disciplinary engagement (see Engle & Conant, pp. 451–459). In the paper, we chose to report two of these contrasting cases from the math and science reform literature: (a) the Japanese Hypothesis-Experiment-Instruction approach in which students become engaged in productive discussions around math and science concepts by debating what the correct answer is to a carefully designed multiple choice question (e.g., Hatano & Inagaki, 1991; Wertsch & Toma, 1995), and (b) the Water Taste Test investigation in which a class of Haitian seventh and eighth graders became deeply engaged in scientific investigations of the taste and safety of their school's water supply (Rosebery, Warren, & Conant, 1992).[7]

Since then, we have successfully applied the principles to understand cases that contrast even further from the original three in their institutional settings, topics of instruction, ages of the students, and degree of success in achieving productive disciplinary engagement. For example, Conant and her colleague Marilyn Webster found the principles useful for understanding preschool students' productive engagement in literacy practices in a classroom influenced by the Reggio Emilio approach (Webster & Conant, 2003). Engle and her colleagues have also found the principles helpful for explaining why a set of practices for orchestrating discussions around high-level mathematical tasks are likely to be effective at promoting productive engagement by students in that discipline as well (Stein, Engle, Hughes, & Smith, 2006).

Our model has been deepened still further by applying the principles to understand cases of both strong productive disciplinary engagement and weak productive

---

[7]Because Conant had participated in videotaping and analyzing the Water Taste Test discussions, we were required to account for a wider range of data than is usually available in published papers.

disciplinary engagement by beginning teachers in mathematics education and educational psychology courses (Engle, 2004; Engle & Faux, 2006). In these studies, we have found evidence of productive disciplinary engagement falling short when one or more of the principles are not fully embodied. From this, we have developed hypotheses about what it might take to realize the principles in new learning environments. In particular, we now hypothesize that when learners are already enculturated into traditional schooling practices, it is wise to first embody the principle of student authority before attempting to hold students accountable to the discipline (Engle, 2004b; Engle & Faux, 2006). We observed that when attempts were made to try to embody accountability to the discipline before student authority had been fully established, students seemed to interpret the instructor's efforts to hold them accountable to relevant disciplines as a signal that their own authority did not matter (Engle & Faux, 2006; cf. Hamm & Perry, 2002).

Another use of these ideas was in a study of how students' competence as mathematics learners was constructed in several episodes observed in middle-school mathematics classes (Gresalfi, Martin, Hand, & Greeno, 2005). These students were positioned as having different kinds of capabilities, ranging from being able to simply execute problem-solving methods they had been shown, to being able to generate examples of a mathematical idea, to being able to invent symbols to represent properties of an event. Gresalfi et al. (2005) found that one important factor in the construction of such student competence was the nature of students' interactionally constituted accountabilities: to whom and for what were particular students positioned as being accountable?

The model was used and extended in another analysis of episodes recorded in a seventh-grade biology class (Greeno, 2003). The class was part of a large project by Richard Lehrer and Leona Schauble in which students studied the growth of fast plants.[8] In the episode analyzed by Greeno (and other participants in a workshop on comparative methodologies in interaction research), students constructed representations of a collection of data from measurements of the heights of plants on a single day. The analysis showed a need to differentiate our concepts of authority and accountability. Students were positioned with authority and accountability for constructing a variety of representations, which led them to generate alternative representations that helped problematize issues of representational practice. In the episodes that Greeno analyzed, the students were not positioned with the authority and accountability to reconcile these issues, however. Instead, issues were reconciled through the direction of adults. Greeno concluded that it may be productive to consider reconciling, along with problematizing, as aspects of practice to be included in the model, and to consider the authority and accountability with which students are positioned separately regarding these two aspects of practice.

In addition, we are currently extending these ideas to consider how problematizing, authority, accountability, and resources might also be useful for explaining long-term productive collaborations that are not necessarily focused on discipline-spe-

---

[8]Fast plants complete a life cycle in about 40 days, which helps support inquiry-based investigations of plant biology and related topics. For more information, see http://www.fastplants.org/

cific issues. In particular, Engle and colleagues are seeing ways that these principles might explain the multi-year and generally very productive collaborations that researchers and curriculum developers had with teachers and math-using professionals in designing innovative mathematics curricula as part of the middle-school mathematics through the applications project (Engle, 2006; Engle & Goldman, 2006; Greeno et al., 1999). Using both interviews with participants and videotapes of project meetings, Engle and her colleagues are finding that the project systematically supported all four principles. Here we briefly outline results about how the project supported teachers' *authority*. First, the project gave teachers the *agency* to focus their efforts on project activities most valuable for them, many of which were designing curricula or other materials for other teachers. Second, teachers perceived themselves as true *contributors* to the project, with almost all noting that teachers' feedback was incorporated into later versions of the unit. The projects' support for teachers to be active agents and contributors to the joint enterprise were viewed as key markers of professional respect that encouraged these teachers to continue actively participating in a highly substantive and productive manner.

Last but not least, these ideas are beginning to be taken up by other investigators who have generatively used the principles in their own research. Some of these studies have provided additional cases of productive disciplinary engagement that can be understood using the principles (e.g., Cornelius & Herrenkohl, 2004; Leinhardt & Steele, 2005). Others have developed the principles in new directions, like investigating how scaffolding can support problematizing (Reiser, 2004) and how effective interdisciplinary collaborators negotiate their disciplinary accountabilities (Nikitina, 2005).

Thus, the test for generality that we applied and that will continue to be applied to our principles and similar ones deriving from this kind of video-based research is whether they support coherent accounts in a wide range of relevant activities and settings, what is sometimes referred to as the criterion of "fruitfulness" or "fertility" in philosophy of science circles (e.g., Kuhn, 1977; McMullin, 1976). If they do, they will become established in the field's explanatory discourse. If they do not, they will cease to be attended to. Granted, this is not as clear cut as the method of falsifying hypotheses. However, that apparent clarity is probably illusory, as analyses of actual scientific practice have shown (e.g., Lakatos, 1970).

REFERENCES

Benke, G. (1999, April). The relevance of discourse: A relevance-theoretic approach to learning as discourse practice. In J. G. Greeno, A. L. Brown, & J. C. Campione (Chairs), *Cognitive and social-interactional aspects of conceptual growth.* Symposium presented at the annual meeting of the American Educational Research Association, Montreal, Quebec.

Cornelius, L. L., & Herrenkohl, L. P. (2004). Power in the classroom: How the classroom environment shapes students' relationships with each other and with concepts. *Cognition and Instruction, 22*(4), 467–498.

Engle, R. A. (2004, April). Revisiting previous discussions to deepen teachers' engagement with mathematics and pedagogy. In M. S. Smith (Chair), *Developing a knowledge base for teaching: Learning content and pedagogy in a course on patterns and functions.* Symposium at the annual meeting of the American Educational Research Association, San Diego, CA.

Engle, R. A. (2006a). *Engaging a diverse group of stakeholders in innovative curriculum design: The case of the Middle-school Mathematics through Applications Project (1990–2002).* Pittsburgh, PA: Learning Research and Development Center. Available at http://www.lrds.pitt.edu/metastudy/pub.htm

Engle, R. A. (2006b). Framing interactions to foster generative learning: A situative account of transfer in a community of learners classroom. *The Journal of the Learning Sciences, 15*(4), 451–498.

Engle, R. A., & Conant, F. C. (2002). Guiding principles for fostering productive disciplinary engagement: Explaining an emergent argument in a community of learners classroom. *Cognition and Instruction, 20*(4), 399–483.

Engle, R. A., & Faux, R. B. (2006). Towards productive disciplinary engagement of beginning teachers in educational psychology: Comparing two methods of case-based instruction. *Teaching Educational Psychology, 1*(2), 1–22. Available at http://www.coc.uga.edu/tep/pdf_2006/index.html

Engle, R. A., & Goldman, S. R. (2006, May). *Collaborating to support innovationism theory and practice: The Middle-school Mathematics through Applications Project.* Invited talk, 18th annual convention of the Association for Psychological Science, New York, NY.

Engle, R. A., & Greeno, J. G. (1994). Managing disagreement in intellectual conversations: Coordinating social and conceptual concerns in the collaborative construction of mathematical explanations. In A. Ram & K. Eiselt (Eds.), *Proceedings of the Sixteenth Annual Conference of the Cognitive Science Society* (pp. 266–271). Hillsdale, NJ: Lawrence Erlbaum Associates.

Greeno, J. G. (2003). *A situative perspective on cognition and learning in interaction.* Paper presented at the Theorizing Learning Practices conference, University of Illinois at Urbana-Champaign, IL.

Greeno, J. G., Benke, G., Engle, R. A., Lachapelle, C., & Wiebe, M. (1998). Considering conceptual growth as change in discourse practices. In M. A. Gernsbacher & S. J. Derry (Eds.), *Proceedings of the Twentieth Annual Conference of the Cognitive Science Society* (pp. 442–447). Mahwah, NJ: Lawrence Erlbaum Associates.

Greeno, J. G., Brown, A. L., & Campione, J. C. (1999, April). *Cognitive and social-interactional aspects of conceptual growth.* Symposium presented at the annual meeting of the American Educational Research Association, Montreal, Quebec.

Greeno, J. G., & Engle, R. A. (1995). Combining analyses of cognitive processes, meanings, and social participation: Understanding symbolic representations. In J. D. Moore & J. F. Lehman (Eds.), *Proceedings of the Seventeenth Annual Conference of the Cognitive Science Society* (pp. 591–596). Hillsdale, NJ: Lawrence Erlbaum Associates.

Greeno, J. G., Engle, R. A., Kerr, L. K., & Moore, J. L. (1993). Understanding symbols: A situativity-theory analysis of constructing mathematical meaning. *Proceedings of the Fifteenth Annual Conference of the Cognitive Science Society* (pp. 504–509). Hillsdale, NJ: Lawrence Erlbaum Associates.

Gresalfi, M. S., Martin, T., Hand, V., & Greeno, J. (2005). *Constructing competence: An analysis of student participation in the activity systems of mathematics classrooms.* Unpublished manuscript.

Hall, R. (2000). Video recording as theory. In D. Lesh & E. Kelley (Eds.), *Handbook of research design in mathematics and science education* (pp. 647–664). Mahwah, NJ: Lawrence Erlbaum Associates.

Hamm, J. V., & Perry, M. (2002). Learning mathematics in first-grade classrooms: On whose authority? *Journal of Educational Psychology, 94*(1), 126–137.

Hatano, G., & Inagaki, K. (1991). Sharing cognition through collective comprehension activity. In L. B. Resnick, J. M. Levine, & S. D. Teasley (Eds.), *Perspectives on socially shared cognition* (pp. 331–348). Washington, DC: American Psychological Association.

Jordan, B., & Henderson, A. (1995). Interaction analysis: Foundations and practice. *The Journal of the Learning Sciences, 4*(1), 39–103.

Kuhn, T. S. (1977). Objectivity, value judgment, and theory choice. In T. S. Kuhn (Ed.), *The essential tension* (pp. 320–339). Chicago: University of Chicago Press.

Lachapelle, C. (1997). Looking at changes in student understanding using a situation model analysis of discourse. In M. G. Shafto & P. Langley (Eds.), *Proceedings of the*

Nineteenth Annual Conference of the Cognitive Science Society (p. 978). Mahwah, NJ: Lawrence Erlbaum Associates.

Lakatos, I. (1970). Falsification and the methodology of scientific research programmes. In I. Lakatos & A. Musgrave (Eds.), *Criticism and the growth of knowledge* (pp. 91–196). Cambridge, England: Cambridge University Press.

Leinhardt, G., & Steele, M. D. (2005). Seeing the complexity of standing to the side: Instructional dialogues. *Cognition and Instruction, 23*(1), 87–163.

Linde, C., Greeno, J. G., Roschelle, J., Brereton, M., Lewis, J., & Stevens, R. (1994, April). *Explanations in discourse: Considering explanation as a social process.* Paper presented at the annual meeting of the American Educational Research Association, New Orleans, LA.

McMullin, E. (1976). The fertility of theory and the unit for appraisal in science. In R. S. Cohen, P. K. Feyerabend, & M. W. Wartofsky (Eds.), *Essays in memory of Imre Lakatos* (pp. 395–432). Dordrecht, Netherlands: Reidel.

Nikitina, S. (2005). Pathways of interdisciplinary cognition. *Cognition and Instruction, 23*(3), 389–425.

Ochs, E. (1979). Transcription as theory. In E. Ochs & B. Schieffelin (Eds.), *Developmental pragmatics* (pp. 43–72). New York: Academic Press.

Popper, K. R. (1959). *The logic of scientific discovery.* New York: Harper & Row.

Reiser, B. (2004). Scaffolding complex learning: The mechanisms of structuring and problematizing student work. *The Journal of the Learning Sciences, 13*(3), 273–304.

Rosebery, A., Warren, B., & Conant, F. (1992). Appropriating scientific discourse: Findings from language minority classrooms. *The Journal of the Learning Sciences, 2*, 61–94.

Shulman, L. S. (1992). Toward a pedagogy of cases. In J. H. Shulman (Ed.), *Case methods in teacher education* (pp. 1–30). New York: Teachers College Press.

Stein, M. K., Engle, R. A., Hughes, E., & Smith, M. S. (2006). *Orchestrating productive mathematical discussions: Helping teachers learn to better incorporate student thinking.* Unpublished manuscript.

Webster, M., & Conant, F. R. (2003). *"It says 'soap' now": Understanding young children's approaches to text in signs* (Tech. Rep.). Pioneer Valley Teacher Research Collaborative, Amherst, MA.

Wertsch, J. V., & Toma, C. (1985). Discourse and learning in the classroom: A sociocultural approach. In L. P. Steffe & J. Gale (Eds.), *Constructivism in education* (pp. 159–174). Hillsdale, NJ: Lawrence Erlbaum Associates.

Wiebe, M. (1999). Learning as transformation of participation: One student group's changing practice of doing research. In J. G. Greeno, A. L. Brown, & J. C. Campione, (Chairs). *Cognitive and social-interactional aspects of conceptual growth.* Symposium presented at the annual meeting of the American Educational Research Association, Montreal, Quebec.

Yackel, E., Cobb, P., & Wood, T. (1991). Small-group interactions as a source of learning opportunities in second-grade mathematics. *Journal for Research in Mathematics Education, 22*(5), 390–408.

Soft Leaders, Hard Artifacts, and the Groups We Rarely See: Using Video to Understand Peer Learning Processes

Cindy E. Hmelo-Silver
Rutgers University

Elvira Katić
Ramapo College

Anandi Nagarajan
Rutgers University

Ellina Chernobilsky
Rutgers University

Facilitating student-centered peer learning is difficult and the limited instructional resources often mandate allocating this resource where it is most needed—to groups that are having some difficulty. In a class with six or seven groups, well-functioning groups are rarely the focus of instructor facilitation. This chapter will look at what happens in effective groups—the ones we rarely see. Video data provides the opportunity for in-depth analysis of and reflection on learning in such groups. Although some uses of video data have focused on the development of ideas (e.g., Hmelo-Silver & Barrows, 2005; Powell, Francisco, & Maher, 2003), in this work, we have purposely selected video to develop theory about factors involved in effective social knowledge construction (Hall, 2000). Interaction analysis (Jordan & Henderson, 1995) has helped us to identify two key features of such groups; 1) "soft leaders," individuals who help facilitate the group, and 2) "hard artifacts," tools that serve as a focus for negotiations. This approach to data analysis is consistent with the view that cognition is situated in social interactions. Understanding how cognition is distributed in effective groups has important implications for scaffolding collaborative learning, and video analysis is a key tool in understanding how cognition is distributed.

RESEARCH CONTEXT

The context for this group interaction was a problem-based educational psychology course for preservice teachers. The students completed five problem-based learning (PBL) activities during the semester. Some of the problems were solved in face-to-face class meetings and some in a combination of class and online meetings using the eSTEP system (Derry, 2006). The eSTEP system is an integrated learning environment that provides video cases of classroom practice, an online learning sciences hypertext, and a collaborative PBL environment. In the PBL environment, students could interact in a group whiteboard and threaded discussion. Our analysis focuses on the second and fourth problems the students worked on, and only on face-to-face interactions.

In the second problem, the students first engaged in solving a mathematical proof problem and then watched a 20-min video of a child engaged in the same task (Maher, 1998). Using unifix cubes, the students needed to prove that they could make a certain number of towers, four cubes high, with two colors. They were then asked to demonstrate how many different pizzas they could make with four different toppings. After working on this problem, students watched a video of a fourth-grade child, Brandon, working on the pizza problem using a self-generated matrix representation and then making an analogy to the towers problem. Their overall goals were to engage in a cognitive analysis of the tasks and to identify Brandon's learning and reasoning strategies.

The fourth problem required students to redesign a physics lesson. The students viewed a video of a physics teacher teaching a traditional unit on static electricity. They were told that he wished to redesign his instruction based on a workshop that he attended. Students then viewed the teaching methodology of the workshop leader, who used a more constructivist approach to teaching. This second video served as a contrasting case to help students differentiate among different instructional approaches (Derry, 2006; Schwartz & Bransford, 1998). The students worked on each problem for approximately 3 weeks. The first author (CHS) was the course facilitator and the third author (AN) was the TA facilitator.

APPROACH TO DATA ANALYSIS

Our approach to analyzing this data is driven by sociocultural conceptions of learning that view knowledge as collaboratively constructed and mediated by various tools and artifacts (Cole, 1996; Hmelo-Silver, Chernobilsky, & DaCosta, 2004; Pea, 1993). Video is an ideal medium for studying collaborative and mediated knowledge construction because it allows analysts to take differing perspectives, depending on their goals (Goldman-Segall & Maxwell, 2003). In some sense, our analysis is both reflexive and recursive (in terms of connecting the lines between theory and practice) because we analyze video of students working with video of classrooms.

Our data corpus consisted of approximately 14 hr of video of one student group in a PBL course in educational psychology. Two cameras were used; one fixed on the group from a distance and one handheld with the focus on speaking students or constructed/referenced artifacts. Because our goals were to understand the leadership roles of particular students and how artifacts mediated collaboration, we wanted to sample the videotape for more intensive analysis. The corpus was reduced to nine short video clips that ranged between 40 sec and 2:35 min each for the interaction analysis (Jordan & Henderson, 1995). A brief synopsis of each of the video clips is summarized in Table 16.1.

We chose this analysis method partly for pragmatic reasons. Our research group had a great deal of video data and our usual approach was to transcribe and engage in traditional formal coding processes (e.g., Chi, 1997; Hmelo-Silver, 2002). In this study, we wanted to "test" a methodology that might provide a more manageable way to work with our data. Also, our research questions were related to social and material aspects of group collaboration, which are readily discernible in macrolevel interactions. The second author viewed the entire video data corpus several times in order to identify potentially significant moments. The clips that were chosen indicated themes that significantly and concisely related to our research goals. These clips represented between 2% and 5% of the corpus. Although we could have randomly selected video clips, clips chosen in that manner might not have shown the aspects of interaction in which we were most interested for this analysis.

The goal of interaction analysis (IA) was to "identify regularities in the ways in which participants utilize the resources of the complex social and material world of actors and objects within which they operate" (Jordan & Henderson, 1995, p. 41). Observations and hypotheses were generated individually by all members of the IA group during and immediately following each clip. Our driving question was how collaboration and artifacts mediated the groups' activity and subsequent learning. Transcripts of all the clips were available to the IA group and clips were replayed if requested. The group then discussed their observations and ideas. These were later integrated into a single document by the first author and circulated to the IA group for revision. Hypotheses were constructed both inductively through viewing the data as well as deductively by exploring "hunches" that had developed from experience with teaching the course. Some hypotheses were grounded close to the data, such as Caitlin mapping between the numbers and the physical artifacts as she explained her thinking to the group. Others were higher level, such as noting that the blocks seemed to drive the activities as the students alternated between playing with the blocks and using them to test their ideas with other students. There were also hypotheses related to the kinds of roles that different group members adopted. As part of this process, we (the IA group) noticed many similarities but also brought different

TABLE 16.1
Summary of Clips

Clip	Synopsis	Clip length
Problem 2		
1	Caitlin suggests exponential calculation (2^n, etc.) as a way to describe and predict the numbers of blocks and towers in the tower problem. The group quickly does a sequence of calculations and their results match the correct number of towers for the first four combinations. They conclude that they have broken down the block pattern to a mathematical formula.	1:32
2	The group's big question in this clip is "WHY?" They question their solution to the tower problem that Brandon faced and their methods of obtaining it. Matt explains his reasoning, while other group members question him extensively.	2:00
3	The group tries to explain their solution to the tower problem in mathematical terms and search for an equation to describe their solution.	2:11
4	The group talks to the facilitator. Bob states that the group was tipped off by the previous lecture on transfer unlike Brandon, who solved the problem without any such help. When the facilitator asks what transfer is, another group member explains.	:40
5	The instructor visits with the group and the group questions her about the details of the video, in particular, the students want to know if Brandon requested the blocks while solving the pizza problem or if they were provided.	2:09
6	The group uses their artifacts to report their solution. During the presentation, Bob, who presents the solution, mentions that both performance and pedagogical tools were used and discusses the distinctions between them.	1:36
Problem 4		
7	The group discusses various formative and summative assessments that they could offer the physics teacher presented in the video clip.	1:23
8	Caitlin and Bob use the group artifact to explain the final solution to problem four and provide the group's rationale for the proposed changes to the lesson. They stress that prediction and evolving understanding are important aspects of learning.	1:32
9	The group is on the spot during the class discussion of their solution. The rest of the class is not happy with the group's statement about promising but never administering a quiz. The group stands together in backing up their proposal, during the discussion, alters their position about giving the quiz but not assigning a grade.	2:35

viewpoints to our observations based on our backgrounds. The propinquity of watching the video and working together as an IA group lent itself to very fluid and diverse hypothesis generation. The general tendency was for each team member to identify local hypotheses but as a group, we tried to select coherent themes that cut across the local observations. For example, several group members noticed specific aspects of the interaction, such as specific discourse moves that Bob made, and as a group, we tried to characterize these at a higher level, such as the actions of soft leaders in facilitating interaction. These hypotheses were often refined and elaborated as we watched subsequent video clips. The synthesis of the IA sessions included all viewpoints and accompanying elaborations. We resolved any differences through discussion.

In addition to these data, two key members (Bob and Caitlin) of the student group were interviewed as they reviewed the selected video clips. We hoped to interview all group members and get as many perspectives as possible on the group interaction, but other group members did not respond to requests to set up interview appointments. The second author, who was not involved in the course, interviewed Bob and Caitlin.[1] The semistructured, stimulated-recall interview protocol was developed to gain a better understanding of the significance of the events for each of the two key participants. Open-ended questions elicited responses from the key participants as to the "sense of each others' actions as meaningful, orderly, and projectable" (Jordan & Henderson, 1995, p. 41) within the setting of the group. Jordan and Henderson (1995) state that comprehensive interaction is a collaborative achievement. In asking the participants to clarify the nature and significance of their actions and collaboration, we looked for ways in which the participants identified and used social and material resources in order to make this happen.

Bob and Caitlin were interviewed separately, with each interview averaging about 2 hours. Prior to the interview, each participant reviewed a copy of the interview protocol with the second author. The nine significant clips were played singly. Bob and Caitlin were encouraged to stop, pause, replay, and otherwise manipulate the digital clips as they wished while discussing the clips and the protocol questions. After discussion about a clip was exhausted, the next clip was played. The interviews were transcribed and segmented based on the clip being discussed. We then highlighted talk related to the major themes of collaboration and the role of artifacts. These data were used as a supplement to the IA analysis to better understand what happened on the clips from the participants' perspectives.

RESULTS

One question that the video allowed us to address was why this group worked so well. In past courses, we had randomly chosen groups to videotape but this did not always answer our questions about effective collaboration. We chose this group based on their performance on the first problem and our perceptions that all the group mem-

---

[1]All group members are identified by pseudonym.

bers were engaged in the PBL activities. This perception was supported by the group members themselves as these quotes from Bob and Caitlin show:

> Bob: Yeah, so I think we all got along and it was nice that you know, all of us, you know, like I said it, it, it all, we all gave each other a chance to speak and we all respected each other's opinions from the beginning. Like we had this mutual respect for each other beforehand, before we all even worked together and I think that's what really helped out in the whole group work (Interview: Bob)
>
> Caitlin: Everybody asked each other questions ... everybody listened to ... I mean for the most part, everybody listened to everybody's suggestions ... like you know, when Matt was explaining, you can see that ... most people were paying attention to what he was saying, like Bob was paying attention, Liz was really, really paying attention (Interview: Caitlin).

Although as course instructors we could observe the group's final problem solutions, the video allowed us to examine group process in depth. In the selection of clips that we chose, we could see that the majority of students were involved in social knowledge construction. Although there are many indicators of positive group functioning, we focus on; (a) norms of interaction that facilitate joint attention and social knowledge construction, (b) the role of artifacts, and (c) the role of soft leaders. Our main findings demonstrate that groups develop norms of interaction that help students coordinate their joint attention and allow a respectful give and take of ideas among the group members. Artifacts played an important role in several ways; as communication tools, as instruments for problem solving, and as a focus for joint attention. Finally, students who take on the role of gentle leadership, what we call *soft leaders*, play a key role in facilitating the group's interaction by helping the group manage the agenda and pushing their peers to think deeply about the content.

Norms That Facilitate Joint Attention and Social Knowledge Construction

For example, in the first clip, most students either contributed to the discussion or wrote on the whiteboards (student-generated artifacts that recorded the group's ongoing ideas). The video allowed us to examine the gaze of students who were quiet and to observe the direction of their gaze. This suggests that students were coordinating their joint attention, which, as Barron (2000) demonstrated, is a key aspect of collaborative problem solving. The students built on each other's thinking as they tried to figure out how many different combinations of four block towers they could make using two different colors of blocks:

1 Liz: Yeah ... I guess, I guess if you have five, yeah it will probably be ...
2 Bob: Be more.
3 Helen: You would add another one ... so it will be—

4 Liz: Wait a minute … maybe it's 2 to the third … 2 to the third is, is 8 … like you have three to the four—

5 Carla: Don't start talking math!

6 Liz: No, but listen, listen, listen … I might be right though, if you have two to the third, right, three blocks, that's eight, two to the fourth, is sixteen.

7 Bob: s sixteen.

8 Carla: Ahhh! (drops something and reaches down to pick it up)

9 Liz: Maybe two to the fifth—

10 Helen: 25.

11 Liz: 25.

12 Matt: Let's find out (clip 1).

In this excerpt, Liz wondered what would happen if the tower were five blocks high (Turn 1). Bob completed her first sentence in the next turn. Carla had been involved with manipulating the blocks and she was comfortable voicing her reservations about the group's new line of inquiry (Turn 5). After Bob, Helen, and Liz made suggestions in Turns 6–11, Matt suggested that the group try investigating their proposals (12). The group remained engaged as Liz and Helen made some computations. Carla continued to be involved in the discussion and proposed, "Well … let's try it to see if it works." The group then decided to see how their ideas about exponents held up if they only had two blocks per tower. Finally, they jointly realized that they had solved the towers problem in a way that would hold for a number of different cases.

Throughout this episode from Clip 1, the students were respectful and allowed all group members to voice their ideas. There appeared to be genuine give and take with regards to group collaboration among members. Students were comfortable admitting their lack of understanding and worked to achieve consensus.

The group developed several norms that contributed to their effective collaboration. First, there was mutual respect among the group members. Bob noted, "Everybody had a say and sometimes, you know, something you, you uh researched didn't make it and I think all of us gave you know, all of us had manners and were able to, you know, give everyone a chance to speak" (Interview: Bob).

Additionally, they developed norms for questioning and explaining: "We were always worried about trying to find … We always had to back up everything so we were like, well let's make sure we know the reasons why" (Interview: Bob). The group was comfortable even early in the semester asking many questions as Carla asked Bob how he came up with his answer of 16 and then Caitlin asked whether he had used the blocks.

13 Bob: No, I mentally thought it up, seriously.

14 Caitlin: Well, no you couldn't have … just said … 'Um, let me pick a number, I'm gonna pick 16.' There had to be a reason why you picked 16.

15 Liz: Is there any … is there any mathematical—?

16 Bob: No, no, no, … I, I looked here and I started adding up different possibilities—

17 Carla: What is the method to your madness?

18 Liz: Is there any mathematical … ?

19 Caitlin: So is it, is it the same thing as this? (Points to block towers).

20 Matt: Yeah … it's … it's basically the same thing.

21 Carla: No, it can't be because there is four different combinations … this is only two different combinations.

22 Caitlin: No, it can't be.

23 Matt: I know but, but like … each level represents a different topping, instead of … like this (points to bottom block of block tower) would represent, like let's say, sausage and this (points to third block of same tower) would represent pepperoni … and like all the ones with blue on the bottom will have sausage on it, and all the ones without it (clip 2).

This excerpt shows a somewhat playful give and take during social knowledge construction discourse. The students tried to understand how the block problem was related to the pizza problem in Turns 19–23. The group members questioned Bob's solution to the block problem (e.g., Turns 17–18). The members readily voiced their ideas to each other and their collective voice ended up in the artifacts they created.

The Role of Artifacts

The students used a variety of material artifacts. We focus here on two examples; blocks and paper tools. Blocks were the manipulatives used primarily in Clips 1, 2, and 3 while this group worked on the block tower and pizza topping combinations problems. Various members of the group used the blocks while formulating ideas that they thought might solve the problem. In clip 3, Caitlin tried to formulate an algebraic explanation that would calculate the total number of towers (based on the number and type of blocks in each tower, shown in Turn 24):

24 Caitlin: Oh! Four <u>squared</u>. So, you're saying 4 squared …?

25 Bob: So it's, it's squared, this is—

26 Matt: You go over this … and I'm like, there is a reason.

27 Liz: Yeah, because it's 2, because you're using 2. Alright, so when you square it, it's not so much the—

28 Caitlin: But then when you had three cubes, it wasn't … three … squared …

29 Matt   It was 2 to the third
and Caitlin: (simultaneously)

30 Helen: Yeah, when you have three toppings … It's also—

31 Matt: So the number of cubes is the exponent.

32 Caitlin: Yeah … and then for this one the same … the 2 is the base … and n is the number of toppings. But we just don't know why it's 2. Yeah, like, I mean it works out both ways but why is it 2? I feel … you know what I mean, does 2 represent a color? Like, I don't know (clip 3).

Different members of the group introduced different bits of mathematical reasoning (Turns 25–27) and Caitlin seemed to direct the discourse toward the mathematics of their suggestions in Turn 28. She tried to prove an exponential theory that had been suggested by finding a reason as to why a base of "2" (with an exponent that refers to the number of blocks in a tower) happened to work when calculating combinations. She indicated that the base might have some connection to the two different colors of the blocks in the towers (Turn 32). During the collaborative discourse in this example from Clip 3, Caitlin gestured and touched the blocks stacked on her desk while making nondeclamatory statements to the group that ended with rising intonation (indicating a tentative conclusion and inviting others to contribute) in the last turn in this example.

In this episode, the group tried to explain why the suggestions they had made thus far applied to the combination solution that they had formulated. The blocks served both communicative and instrumental functions. In her stimulated-recall interview, Caitlin recalled that the group used the blocks as "processing manipulatives," but that she had not used the blocks much in her personal problem-solving process:

> I think if I was by myself, maybe I don't think I would need the blocks, but I think that if you are trying to explain to someone it is easier to have the visual to explain to more than one person … 'cause you, I think we have them all rearranged on the … on the table … it's a combination … I don't know.

Although Caitlin initially stated that she did not refer to the blocks as much as other group members, when prompted to clarify her verbal and gestural contributions to the group collaboration taking place during Clip 3, she replied, "I think it was the only thing that we could connect … it connected with the visual that we were looking at … That is the only place that we could pull at '2' from … we had two different colors." When reminded of her initial statement about her minimal use of the blocks, she seemed somewhat surprised at her words and actions during a repeat viewing of clip 3:

> Yeah, no that was probably definitely … 'cause I mean, I can see from … from looking at this (watching clip and pointing to computer monitor), I was definitely using the tools, maybe I just didn't even realize that I was using it at all. (Interview: Caitlin)

During Clip 3, it appeared that representations mediated the group collaboration process by serving as accessories in real-time, instrumental interaction. For example, the group began by building block "opposites" as a way to move into a discussion of block combinations. In addition, the group directly manipulated or referenced the blocks while trying to identify the components of a proposed mathematical calculation to support the block towers combinations. At a social knowledge-building juncture such as this one, both verbal and physical contributions to group collaboration take turns in the discourse. Manipulation of the blocks mediated conversation to a point where the concrete arrangements of the blocks remained the focus of many consecutive verbal turns and thus enhanced the group's problem-solving processes. Artifacts served to either develop or disregard problem-solving ideas. Their presence during foundational collaborative talk about solutions to the problems served as formative influences on the direction of discussions to come.

Based on the data, representations most frequently functioned as a means to visually explain some idea or solution. Less frequently, they functioned as reference or justification of the group's final conclusions. During the gallery walk (through the poster presentations), one or two members from each group presented their solution for the "walking" students—the remaining students from all groups in class. This group's posters were, as a rule, quite detailed, neatly executed, and usually embellished with hand-drawn illustrations that corresponded to the conclusions they had reached.

In Clip 6, Bob used a poster created by his group in order to mediate his presentation during a gallery walk. He used it as an explanatory tool to visually demonstrate the formulated ideas and solutions the group had reached through their collaborative problem-solving discourse. As the group had made intensive use of tools and artifacts during their collaborative talk, the poster (drawn by Carla) prominently featured several illustrations. One illustration showed several block tower combinations in red and blue and another showed a pizza with the words "We Deliver!" written along the side of it. As Bob explained the collective understandings of his group, he read from the poster, gestured at different areas/aspects of the poster, touched the poster illustrations and turned his head to address the audience, while keeping his body turned and right hand outstretched toward the poster.

The poster mediated Bob's presentation as he looked to the poster for (1) a comprehensive account of the group's collaborative work and (2) the order and delivery of this information to his audience. The poster also served as a supportive prop and visually reinforced appropriate points from Bob's explanation:

> Bob: And likewise, when he went back to the um ... block problem (gestures at stacking blocks illustration), he was able to actually understand it further, because he used his information from the pizza problem (moves hand down to pizza illustration) and said, 'Hey wait ... this graph is just the same as these blocks (moves hand back up to stacking blocks illustration, shown in Fig. 16.1) ... like the whole ah ... like ... one, yeah know, zero zero zero, ... So ... ah, (reads verbatim from poster) number four, what activities did Brandon use to contribute to his learning ... strategies? ... His activities ... included the use of tools. Um, (looks at poster) pedagogical

tools which are, ya know, he created a chart to organize his
thoughts ... So you can say that the blocks (points to blocks picture) were actually ah ... performance tools ... and then the pedagogical tools which focus on ... changing the user's competence ah,
example, a simulation designed to change the literate understanding of math, mathematical concepts ... which was the pizza problem (moves hand down to point at pizza illustration, shown in Fig.
16.2; Clip 6).

The group's poster was central to this episode; it mediated not only Bob's behavior, but also directed the student audience's behavior and attention. While watching
this clip during his interview, Bob reflected, "I thought like, you know, wow, this is a
great way of explaining the tools we used to help explain solving the problem." Having
a comprehensive visual representation indicates that the group had reached some
agreement regarding the results of their collaborative problem solving. The representation served as a benchmark, an indicator of what had been discussed and what solutions had been agreed upon. At the same time, it served as a stepping stone to the next
level of problem-solving discourse. The group did not have to revisit the material they
had covered, except as referential information pertinent to new discussion. The poster
mediated the framework of future collaboration in a chronological sense, punctuated
by completed discussion topics.

Bob used the poster to focus on the results of the group deliberations, not on
their process. When Bob read from the poster (or from a nearby textbook), he indicated that the information on the poster (as in the textbook) was of a finished, definitive nature; something that was complete to such a degree that it served as a reference,
rather than a work in progress. In this way, Bob indicated that the group's problem-solving processes had moved from processing to processed. Another key mediating factor of social knowledge construction, other than the artifacts, was the roles that
Bob and Caitlin adopted in their group.

The Role of Soft Leaders

Bob and Caitlin played critical roles in this group. They engaged in many of the
behaviors and adopted many of the goals observed of experienced PBL facilitators
(Hmelo-Silver, 2002; Hmelo-Silver & Barrows, 2005). They pushed the group for explanations and monitored the group progress and dynamics. Unlike the PBL facilitator,
they needed to be involved participants in the group's knowledge construction efforts.
There is a gentle interplay between these two kinds of roles that Bob and Caitlin
adopted—hence the term soft leaders. Very early in the process, Bob and Caitlin were
comfortable pushing their group members for explanations as when Caitlin challenged Bob's solution to Problem 2: "Why, why did you say it was 16 at first?"

Another way in which they helped group dynamics was by inviting other students
to contribute to the conversation. When Caitlin tried to check the generality of her solution in Clip 3, she said, "Yeah ... so I don't know, because, usually, because 4 squared
does work, because it's 16. But then ... it doesn't work otherwise, like if, ok, if you have

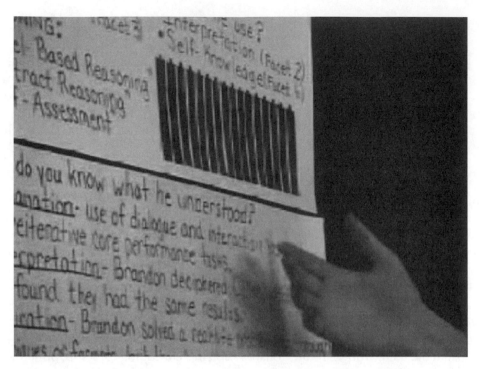

Figure 16.1   Using the poster to mediate explanation of block problem.

Figure 16.2.   Using poster to mediate explanation of pizza problem.

four rows squared it's 16 but if you have three rows squared it's 9?" The rising intonation suggests her uncertainty and provided an opportunity for other members to add to the discussion. Similarly, Bob asked for permission to interrupt in Clip 7 as they were working on Problem 4: "I'd like to say something, but ya know I usually get jumped on." This provided an opening for Liz to make a contribution based on an earlier discussion with Bob:

> We like the portfolio idea, like last time, especially because they are going to be coming up with their own experimental things like that, we felt that that could be really useful umm, the ability to interpret their own results when doing their experiments ummm, explain the results in reason to the teacher and as well as to other students, apply their knowledge by being able to predict new but related experiments and an end to the unit test … like to have some kind of summative test (Clip 7).

Caitlin marked the significance of Liz' contribution as she said, "Whoa, that's a lot, ok." Up until this exchange, the group had been focused on tests for assessment. Bob shifted the direction of the conversation by creating an opportunity for Liz. While watching this clip, Bob noted that, "Because … she wrote down more eloquently than me. I wrote down—I'm like you know, more, you know, [unintelligible] than she'd be, so … I thought she had a better approach" (Interview: Bob). Both Bob and Caitlin had an understanding of the distributed expertise in the group, which may have contributed to knowing when to create an opening:

Caitlin: And … umm … I think a lot of times, whatever, whatever our specialties was that we tried to connect it … and sometimes we tried to make it fit, you know, like trying to shove a puzzle piece in that just doesn't fit, and we would try to make everything connect …

Elvira: Why was there, why do you think you wanted to try and make that piece fit?

Caitlin: … Everybody had their little … areas of specialty that we researched … I can't remember, but I know mine was formative and summative assessment … and every topic that we did had something to do with formative and summative assessment for me … I could have connected for every single topic … but sometimes it just didn't fit … and you would want it to because that is what you know … so you could give something into the group … (Interview: Caitlin).

In addition to making sure that other group members were involved, Bob and Caitlin actively monitored the group's progress on tasks. At one point during Clip 5, the group was becoming bogged down in details. They developed a pattern of working through the details of single ideas before considering a range of ideas. As the group explained what they were doing to the instructor, Bob evaluated the group's progress saying, "I think we got too in details, over details … so you know … we got to the point where we were missing like the whole interest of the ... like the picture ... because of all these, you know ... ideas. We weren't covering the plot itself, we found a side story" (Clip 5). He pointed out that they were so focused on details that they lost the big picture of what was going on in the video.

Another soft leadership characteristic is managing task requirements. Caitlin frequently did this as she helped the group clarify that what they were calling "assessments" were the same as the "evidence of understanding" that they needed to include in their lesson design: "Ok, evidence of understanding, for understanding static electricity ... What do you, like ... Is that the same thing as assessing understanding?" (Clip 7). Bob and Caitlin's behaviors helped support deep thinking, asking for explanations, managing tasks, and functioning well as a group. When asked why the group functioned well, Bob noted that:

> Bob: I, I think it's dynamics ... and who uh, steps up to the plate ... : I haven't seen all the clips yet, but I know it seemed like Caitlin was the leader to a point, she was the main person. And I would help her out a lot. And then it seemed to like fall down the chain, like Liz is probably the next person ... And it wasn't like anybody exerted their authority, saying like, 'I'm the group leader and you do that' (Interview: Bob).

Part of adopting the roles of soft leaders was a matter of confidence. Caitlin noted that the other group members did not want to do the presentations in the gallery walk:

> Nobody felt comfortable to do it. And I remember ... during our third meeting, because we had to meet outside the class to do the posters ... and I remember saying, 'Ok, well Bob did it and I did it so who wants to do it next?' And I remember ... there was a good minute pause or so, and you know, everyone kind of came up with excuse like you know, for this and that, and ... it just ended up being us again (Interview: Caitlin).

As soft leaders, these two students helped mark task boundaries, model explanations, and provide opportunities to ascertain common ground. Bob told us that Caitlin would type up summaries of their out-of-class meetings that the group would later use as starting points for their in-class discussions. The facilitation function was distributed between the two of them. Often Bob would focus on redirecting the group while Caitlin focused on making sure tasks were being addressed. They engaged in these behaviors without dominating the group or standing on the sidelines. They were both engaged in the process of social knowledge construction, and softly, helped the entire group share in that engagement.

CONCLUSIONS

In our many PBL classes, we have observed groups that are more or less effective. The demands of classroom research often require the researcher to be in the dual role of teacher, and to attend to those groups that are having difficulty, making it hard to understand what makes some groups function particularly well. In addition, the interactions in a group are complex and cannot be understood at a glance. The research team needed the opportunity to carefully consider how the chosen group worked together and what mediated their learning. Although it was the researchers' intuition that leadership was a factor in this peer learning situation, it was hard to say what that meant without analyzing the video (and using the video as a stimulus for reflection).

Interaction analysis allowed us to explore the video in productive ways to examine factors related to social interactions, and using the video as a tool for stimulating recall, allowed us to gather additional information about the group's interaction. A limitation of this research was that we viewed purposely selected pieces of video. We believe that for the research questions that we were addressing, we selected representative examples from the video corpus but other analysts might differ and the choice of video clips was very much influenced by the specific phenomena of interest. Despite these limitations, the IA approach illuminated the specific aspects of social interaction that were our focus.

The group that we videotaped and analyzed in this chapter was a successful group. The students in the group were enthusiastic and comfortably distributed their expertise. They were respectful of each other's ideas and worked to construct joint understanding. There were two key factors that mediated this group's success. First, the artifacts played an important role in mediating their learning, explanations, and communication. The blocks and whiteboard served as concrete referents for student discourse. They reified the students' ideas and served as a focus for negotiation. The blocks allowed the students to physically test their ideas and refine their thinking. The final poster provided support for explanation as a record of the negotiated understanding. Second, two of the students took on the role of soft leaders. These students distributed some of the facilitation functions among themselves as they pushed their group to think deeply, ensured that all group members were involved, and helped manage the agenda while at the same time being part and parcel of the knowledge construction process.

This case provides in-depth information about how one successful group functioned and what roles artifacts had in mediating their learning. One of the challenges that video analysts face relates to what one can learn from a single case. Generalizing from the results of one case is always limited. Future research needs to examine video of other more and less successful groups to determine whether these results converge on more generalizable conclusions. For example, it may be that less successful groups use artifacts in similar ways to this well-functioning group and perhaps it is the leadership issue or development of norms that are the key factors. This analysis allows us to generate hypotheses that can be tested with other groups. Moreover, it would have been useful to have the perspectives of other group members to check on Bob and Caitlin's interpretation of the video as well as the interpretation of the research team. Video is critical in providing a record of group interactions that can be replayed and viewed from multiple perspectives. Another way to test these results is to try and scaffold students in "soft leadership" roles and appropriate ways of using artifacts to support their learning. As we demonstrated in this study, video can be a powerful lens with which to train the analytical gaze and to further clarify social and material aspects of collaborative learning. Video proved to be a flexible medium that gave us the chance to scrutinize the data from a variety of angles.

ACKNOWLEDGMENTS

This research was funded by NSF ROLE grant #0107032. Any opinions, findings, and conclusions or recommendations expressed in this material are those of the authors and do not necessarily reflect the views of NSF. We are indebted to the members of this group who allowed us to videotape them, and especially, those participants who returned for follow-up interviews.

REFERENCES

Barron, B. J. S. (2000). Achieving coordination in collaborative problem-solving groups. *Journal of the Learning Sciences, 8,* 403–436.

Chi, M. T. H. (1997). Quantifying qualitative analyses of verbal data: A practical guide. *Journal of the Learning Sciences, 6,* 271–315.

Cole, M. (1996). *Cultural psychology: A once and future discipline.* Cambridge MA: Harvard University Press.

Derry, S. (2006). eSTEP as a case of theory-based web course design. In A. M. O'Donnell, C. E. Hmelo-Silver, & G. Erkens (Eds.), *Collaborative learning, reasoning, and technology.* Mahwah NJ: Lawrence Erlbaum Associates.

Goldman-Segall, R., & Maxwell, J. (2003). Computers, the Internet, and new media for learning. In W. M. Reynolds & G. E. Miller (Eds.), *Handbook of psychology: Educational psychology* (Vol. 7, pp. 393–427). New York: Wiley.

Hall, R. (2000). Video recording as theory. In D. Lesh & A. Kelley (Eds.), *Handbook of research design in mathematics and science education* (pp. 647–664). Mahwah, NJ: Lawrence Erlbaum Associates.

Hmelo-Silver, C. E. (2002). Collaborative ways of knowing: issues in facilitation. In G. Stahl (Ed.), *Proceedings of Computer-Supported Collaborative Learning 2002* (pp. 199–208). Mahwah, NJ: Lawrence Erlbaum Associates.

Hmelo-Silver, C. E., & Barrows, H. S. (2005). *Facilitating collaborative ways of knowing.* Manuscript submitted for publication.

Hmelo-Silver, C. E., Chernobilsky, E., & DaCosta, M. C. (2004). Psychological tools in problem-based learning. In O. Tan (Ed.), *Problem-based learning and enhancing thinking: Understanding key cognitive processes in action* (pp. 17–37). Thomson Learning: Singapore.

Jordan, B., & Henderson, A. (1995). Interaction analysis: Foundations and practice. *Journal of the Learning Sciences, 4,* 39–103.

Maher, C. A. (1998). *Can teachers help children make convincing arguments? A glimpse into the process.* Rio de Janeiro, Brazil: Universidade Santa Ursula.

Pea, R. D. (1993). Practices of distributed intelligence and designs for education. In G. Salomon & D. Perkins (Eds.), *Distributed cognitions: Psychological and educational considerations* (pp. 47–87). New York: Cambridge University Press.

Powell, A. B., Francisco, J., & Maher, C. A. (2003). An analytical model for studying the development of learners' mathematical ideas and reasoning using videotape data. *Journal of Mathematical Behavior, 22,* 405–435.

Schwartz, D. L., & Bransford, J. D. (1998). A time for telling. *Cognition and Instruction, 16,* 475–522.

Studying Dinosaur Learning on an Island of Expertise

Sasha D. Palmquist
University of Pittsburgh

Kevin Crowley
University of Pittsburgh

Videotaping family interactions provides researchers with unique access to the kinds of experiences that collectively shape the everyday cognitive ecology of childhood. In this chapter, we will discuss some selected findings about the ways that islands of expertise change how families talk in museums. We will briefly outline some of the key theoretical motivations for this work as well as some of the specific challenges associated with conducting research in museum settings. Through a discussion of our methods and a pair of example transcripts, we will illustrate how video can reveal the rich character and texture of parent–child communication and provide essential data that helps us understand how families learn about science in informal settings.

Children's experience of the world around them is often filtered through a lens of curiosity and wonder. Whenever a new idea, object, or experience captures a child's attention, there is an opportunity for learning. In the context of ordinary activities, children begin to practice observation and inquiry skills that develop into the fundamental components of scientific thinking. Our work is focused on the idea that children's attitudes and beliefs about science and scientific thinking can be influenced by early experiences in informal settings with parents and peers. Long before encountering their first formal science lessons, children are actively engaged in gathering information about topics, artifacts, and processes that spark both their interest and imagination. Embedded in these everyday moments may be the seeds of intuitive scientific inquiry.

What develops from this inquiry? In the long run, we believe that everyday activity can serve as the basis for things like scientific reasoning and conceptual development, however the immediate outcome of children's exploration is usually something much more concrete. Children don't often learn about hypothesis testing or experimental design directly before entering school. They more often begin to learn science by getting to know all about planets, robots, insects, horses, marine mammals, dinosaurs, or another topic that captures their interest. Crowley and Jacobs (2002) talk about the enrichment of these focused interests as children developing an island of expertise. These islands are built from topics that are easy to access in the cognitive ecology of family life. They are rooted in everyday family activities like reading books, visiting museums, playing with toys, watching DVDs, and using the web. When children initially express that a topic has captured their attention, parents often provide additional opportunities that help to develop that interest. Over time, children and their parents can accumulate an unusually rich body of shared knowledge and experience around these interests. We refer to that knowledge and experience as an island of expertise, because it is often a unique and relatively complex set of information possessed by children who otherwise are universal novices. As an island matures, it may create an opportunity for the first time for a child to know enough technical vocabulary and access enough disciplinary examples to support some relatively powerful conversations about big ideas in science. For children who develop an island of expertise focused on dinosaurs, big ideas in science that might become accessible include understanding functional morphology, theories of extinction, and evolutionary relationships. The recognition of a child's expertise through conversations with parents, caregivers, and peers can provide early positive experiences with science concepts that may endure long after a particular island of content knowledge recedes.

Families begin to structure learning experiences around children's developing interests when children are toddlers and continue throughout grade school. Classic research by Chi and Koeske (1983) demonstrated that a 5-year-old child is capable of developing an extensive knowledge base of the declarative domain of dinosaurs before entering Kindergarten. Incorporating and extending this finding, Chi, Hutchinson, and Robin (1989) discussed a series of studies illustrating how children structure knowledge in a declarative domain like dinosaurs and how that organization impacts their subsequent application of knowledge. In declarative or conceptual forms of expertise, the ability to correctly identify salient features of objects and to use them to categorize instances at basic and subordinate levels is a critical skill for knowledge acquisition. In this set of studies, Chi and colleagues found that childhood dinosaur experts were able to accurately categorize a novel dinosaur based on highly integrated, hierarchical family relations. In contrast, dinosaur novices seemed to rely on surface attributes, greatly limiting the amount of information available to inform categorical placement of a novel dinosaur.

Discussions of expertise in conceptual/object-centered domains suggest that the distinctions between novices and experts are dependent on the development of perceptual sensitivity for object attributes, familiarity with subordinate-level category names (species), and more generalized verbal and intelligence characteristics (Johnson & Mervis, 1994; Mervis, Johnson, & Mervis, 1994). Continuing with this approach to investigating childhood expertise, Johnson and Eilers (1998) conducted a study to determine the impact of domain-specific knowledge and level of development on subordinate-level category formation and transfer of expert categorization skills to another domain. Through an investigation of both children and adults with high, moderate, and low levels of dinosaur knowledge, they found that high dinosaur knowledge adults and children were able to accurately differentiate categories at the subordinate level, were less likely to overextend category assignment, and more likely to underextend categories. In comparison, child and adult novices performed more poorly on category differentiation, were more likely to overextend category assignment, and less likely to make errors of underextension than were more knowledgeable groups. Differences between knowledge groups' performance on outcome measures were a result of prior knowledge, perceptual acuity, and recognition of salient physical attributes between category examples.

More recently, Johnson, Scott, and Mervis (2004) have begun to move beyond studies of expert performance on categorization tasks and have conducted investigations of how adult and child dinosaur experts use their associated conceptual knowledge both within and between biological domains. Johnson and colleagues found that children primarily rely on surface attributes to make selections and to justify their reasoning on both familiar and unfamiliar domains, regardless of level of knowledge. In contrast, high knowledge adults consistently use more domain-general, metacognitive strategies to successfully complete inference tasks. Johnson and colleagues conclude that for most children, expert levels of knowledge are necessary for more sophisticated reasoning about conceptual relationships, however, knowledge alone is not sufficient to extend and transfer this knowledge to other domains.

Although this research supports the idea that children who are on an island of expertise have access to sophisticated representations of detailed knowledge, the existing literature has not addressed the question of how these islands function in everyday activity. How do families talk when they are on an island? What role do children and parents have in accessing the shared information resources of the island? How does new learning occur on the island? Is the island used as a platform for more advanced discipline-specific learning?

In this chapter, we will focus on these questions using data from a study of families visiting a dinosaur hall in a natural history museum. Each family's visit was videotaped and then parents and children completed separate assessments focusing on knowledge, interest, and experience with dinosaurs. During an average 10 minute visit to Dinosaur Hall, parents and children engage in constant negotiations of the exhibit space as they verbally and nonverbally communicate and pursue their interests. The combination of the questionnaire and the interview allowed us to examine these rich interactions through the lens of domain expertise. The use of video recording in these settings allows us to capture and preserve the kinds of detailed parent–child interactions that contribute to the development of islands of expertise with in the everyday cognitive ecology of childhood.

PREDICTING THE INFLUENCE OF PRIOR KNOWLEDGE ON FAMILY INTERACTIONS IN DINOSAUR HALL

When families visit the Carnegie Museum of Natural History in Pittsburgh, Pennsylvania, one of the main attractions is Dinosaur Hall. Parents and children come to Dinosaur Hall to share the experience of walking among the fossils of creatures that lived millions of years ago. As families move through the exhibit space, they choose how deeply and in what ways they want to engage with the implicit and explicit information in Dinosaur Hall. Before beginning data collection, we predicted that families who were working on a dinosaur island of expertise might demonstrate different patterns of talk than families who were not on a dinosaur island. For novice children and their parents, we expected that a visit to Dinosaur Hall would support conversations about the obvious surface features of the fossils on display (e.g., "Look how long that one is, look at the big one, look at those teeth!"). Most of their visit would likely be devoted to figuring out how to talk about dinosaurs while looking at fossilized remains. Because that might be a relatively novel topic for the family, we expected that the parents with novice children might assume primary responsibility for structuring the experience in the hall by controlling the pace of movement through the exhibit, choosing when to stop and discuss a dinosaur mount, and deciding what kinds of things to talk about at each stop. In this setting, we anticipated that parents might adopt a guiding, coaching, or teaching role in a domain that their children did not know much about.

For expert children and their parents, we expected that a visit to Dinosaur Hall would likely involve talk about the surface features of the fossils on display, but that the conversations would go beyond simple description and begin to connect with more advanced disciplinary ideas such as extinction, form and function relationships, and behavioral adaptations. As the family already brings a lot of dinosaur knowledge to the museum, the visit might support conversations where children explicitly rehearse existing knowledge about the fossil specimens on display. We expected to see parents and children engage in reciprocal questions and answers as well as find opportunities to integrate the information presented in the exhibit with prior examples of co-constructed dinosaur knowledge (e.g., "There's *T. rex*. He's my favorite because he is called the king of dinosaurs and he was a great hunter. He even hunted dinosaurs with strong defenses like *Triceratops*. I read about that in my dinosaur book."). We also expected that the interests and knowledge of expert children would more directly influence families' movement through the hall, either with parents navigating the visit around their children's favorite dinosaurs or with the child leading the parent through Dinosaur Hall.

Data Collection in the Carnegie Museum of Natural History

Conducting successful research in a museum setting like the Carnegie Museum of Natural History requires that the research team develop a professional relationship with the museum prior to beginning data collection. This is critical when designing and conducting in-vivo museum research that requires data collection to take place on the museum floor. In many cases, the presence of a research team in the museum will the style devised by Josef von Sternberg for Marlene Dietrich in her Paramount films in the early 1930s (see, particularly, *The Devil Is a Woman* [1935]).

Todd, who began in the British cinema in 1932, enjoyed great success as the masochistic, emotionally fragile pianist Francesca in *The Seventh Veil* (1945), and her performance as Madeleine Smith exploits, and subverts, the submissive basis of her earlier films. Early scenes in the film seemingly emphasize this trait, as she appears vulnerable before the two males in her life—her lover, Emile L'Anglier, and her father, James Smith. However, Madeleine Smith is not a submissive Scottish lady torn between the forbidden pleasures offered by Emile and the social demands of Victorian society. Madeleine subtly resists the pressure from her father to marry the wealthy, but dull, William Minnoch and tries to seemingly live within the constrictions imposed by Victorian society while indulging her sexual desire for Emile. After Emile rejects Madeleine's offer to run away and get married, she realizes that her French lover is more interested in her family's social position and wealth than in marriage to her.

Emile's rejection follows the sexual consummation of their relationship, and Lean leaves little doubt as to their sexual behavior through his judicious insertion of a passionate dance sequence at a highland festival as Madeleine and Emile engage in foreplay. When Madeleine falls over in a dark forest and looks up with anticipation at Emile towering over her, Lean cuts to the frenzy of the dancers performing in the village below. Following the purchase of arsenic by Madeleine from a local chemist, Emile suffers two bouts of poisoning, with the last one being fatal. Madeleine is tried for murder, and the jury finds the charge "not proven." The film concludes with a voice-over pointing out that she is deemed neither innocent nor guilty, but Lean's final close-up of Madeleine's sly smile as she leaves the court leaves little doubt as to her guilt.

Lean, together with script writers Stanley Haynes and Nicholas Phipps, transformed a real court case that took place in Glasgow in 1857 into a study of the destructive power of Victorian patriarchy. The film makes it quite clear that the real motivation for Madeleine's actions emanate from James Smith's authoritarian control over his family, in particular, the (lack of) rights accorded to women. The repression of female desire leads to betrayal and murder. The scenes between Todd and Banks are both revealing and powerful as the patriarch tries to impose his will on Madeleine. She, in turn, subverts his authority.

Geoff Mayer

THE MAN FROM COLORADO (Columbia, 1948). *Director:* Henry Levin. *Producer:* Jules Schermer. *Script:* Robert Hardy Andrews and Ben Maddow, based on an original story by Borden Chase. *Cinematography:* William Snyder. *Music:* George Duning. *Cast:* Glenn Ford (Owen Devereaux), William Holden (Del Stewart), Ellen Drew (Caroline Emmett), Ray Collins (Big Ed Carter), Jerome Courtland (Johnny Howard), James Millican (Jericho Howard), William "Bill"

Phillips (York), Denver Pyle (Easy Jarrett), James Bush (Dickson), Ian MacDonald (Jack Rawson), Myron Healey (Powers).

In the late 1940s a cycle of westerns presented a radically different view of the West by shifting from its traditionally optimistic view of human behavior to a more problematic presentation of psychologically disturbed characters fighting their inner demons. This cycle of westerns combined elements of the traditional western with an emphasis on entrapment that was more generally associated with urban noir crime films. A key film in this cycle was the 1947 Warner Bros. production *Pursued*, starring Robert Mitchum as a young man haunted by a childhood trauma. This somber film, influenced by Hollywood's fascination with psychoanalysis, transformed the familiar western setting of Monument Valley in Utah, the location for many of John Ford's westerns, into a nightmarish landscape by cinematographer James Wong Howe. *Pursued* was accompanied in 1947 with another noir western, *Ramrod*, directed by Andre De Toth and starring Joel McCrea as the cowboy caught up in the destructive behavior of an obsessed female landowner, played by Veronica Lake.

Blood on the Moon, also starring Robert Mitchum, was released the following year by RKO, and director Robert Wise, assisted by cinematographer Nicholas Musuraca (who photographed many of RKO's noir films, including *Out of the Past* [1947]), presented a claustrophobic environment that reinforced the sense of personal entrapment. Columbia's contribution to this cycle in 1948 was *The Man from Colorado*, starring two popular contract players at the studio, Glenn Ford and William Holden. In 1941, at the start of their careers, both men had costarred in *Texas*, an upbeat western that presented Ford and Holden as two cowboys who end up on different sides of the law. Seven years later, *The Man from Colorado* presented a more troubling view of human behavior, and although it does not have the superior direction of and expressive cinematography of *Pursued* or *Blood on the Moon*, thematically, it is the most extreme noir western of the late 1940s.

If *The Man from Colorado* did not have a western setting and mid-nineteenth-century background, it would have been a typical film noir in the late 1940s as it dramatizes the fears and anxieties, especially involving a strong sense of disillusionment and dislocation, found in many urban noir films released between 1946 and 1949 (including *The Blue Dahlia* [1946], *Crossfire* [1947], *Ride the Pink Horse* [1947], *Dead Reckoning* [1947], and *Act of Violence* [1949]). *The Man from Colorado* adapts, and transforms, these aspects from 1945 to 1865 and the post–Civil War period, and the film suggests that whatever the war, its destructive effects are the same as some men never recover from its aberrant condition, where killing is presented as morally and socially acceptable.

The film begins on the final day of the Civil War as a strong force of Union soldiers, led by Colonel Owen Devereaux, corner a small band of Confederate soldiers. When the Confederates try to surrender by flying a white flag, Devereaux pretends that he does not see their surrender and opens fire with his artillery. A close-up of Devereaux, with Glenn Ford uncharacteristically presented with curly hair and a grim visage, shows his evident pleasure in the suffering inflicted by his artillery on the helpless soldiers.

After the war, Devereaux, accompanied by his close friend Captain Del Stewart, returns to his hometown in Colorado, where he is welcomed as a war hero. Devereaux accepts an appointment as a federal judge, while Stewart becomes the town marshal. As a federal judge, Devereaux is able to operate without scrutiny whereby he can inflict pain without the fear of prosecution or legal retribution. However, his biased support for the major capitalist in the region, Big Ed Carter, who is exploiting and persecuting the local miners who have lost their rights while fighting in the Civil War, forces Del Stewart to abandon the law and wage guerrilla warfare against Carter and Devereaux.

The conventional western plot, which sides with the miners against the corporate greed of Carter, is pushed into the background as the film focuses on Devereaux's fragile mental condition. Devereaux, unlike most other western characters, is a divided figure, fighting his inner turmoil, which surfaces in a need to inflict pain. This link between psychosis and war makes *The Man from Colorado* a topical film in 1948, when many ex-army personnel were required to adjust to peacetime following an extended experience with the abnormal and atrocities of war.

The film, however, extends this link by also including a thinly veiled attack on large-scale capitalism. Hence the insane federal judge (Devereaux) uses his position to support the unscrupulous entrepreneur (Big Ed Carter). This link is strengthened by the nominal hero's (Stewart) vigilante in rejecting the law as impotent in protecting the helpless miners from Devereaux and Carter.

In some ways Glen Ford's behavior in *The Man from Colorado* represented a continuation of the unstable character he portrayed in *Gilda* (1946). His psychotic judge in this western contained many of the qualities of the neurotic, tormented lover he played in *Gilda*. Both men are unable to find comfort and solace with their lovers—in *The Man from Colorado* Devereaux's willful actions destroy his marriage. Whether this illness is a direct result of the war or whether the war merely provides the opportunity for such tendencies to erupt is not clear in either *Gilda* or *The Man from Colorado*. However, while audiences in the 1940s were prepared to enjoy such qualities in a contemporary melodrama such as *Gilda*, which was Columbia's most commercially successful film in 1946, they were more hesitant to see them displayed in a western such as *The Man from Colorado*.

Geoff Mayer

MANN, ANTHONY (1906–1967). Mann was one of the best and most significant Hollywood directors of the middle decades of the twentieth century. A superb storyteller and a fine stylist, he made compelling films in several genres. Perhaps best remembered today for the dark westerns he made with James Stewart in the 1950s, Mann began his directing career with a number of memorable contributions to the classical film noir cycle. Sometime shocking in their bursts of brutality, these films were always well paced and replete with striking visuals, thought-provoking subject matter, and strong acting performances. Mann was born Emil Bundmann (some sources say Bundsmann) in San Diego, California. From his late teens he worked on the theater stage in New York and became a successful Broadway director. Mann went to Hollywood in 1940 and was soon directing low-budget films for Republic and RKO. At this time Mann worked as an assistant director on Preston Sturges's classic, self-reflexive comedy of Hollywood life *Sullivan's Travels* (1941). He began his own directing career with films that have since been defined as noirs, then moved on to make a number of westerns, which also had a noticeably dark tone to them. Mann, early on, exhibited a very expressive visual style. His films featured lots of tracking shots as early as *Dr. Broadway* (1942) and *The Great Flamarion* (1945). Already, too, he was showing a thematic preoccupation with characters who were trapped in a bleak and hostile world and who could find no escape other than death or madness. What some now consider as being his first full-fledged film noir was *Desperate* (1947). This small-budget feature displayed Mann's excellent use of light, which was to reach its culmination in *T-Men* (1947) and *Raw Deal* (1948), where he worked with John Alton, perhaps the greatest of the many superb cinematographers who worked in noir. *Desperate* had a memorable suspenseful set piece as a group waits in a darkened building for the midnight execution of a young thug. *T-Men* combined a semidocumentary story structure with bold chiaroscuro lighting and chillingly subjective depictions of despair and impotence in the face of violence. *Raw Deal* included a rare female voice-over in a tale that featured outbreaks of sadistic viciousness. These are among the finest of all classical noirs: their dark world of pessimism (even of nihilism) is crammed with danger, tension, violence, and drama. With *Border Incident* (1949), Mann continued his skilful exploitation of the potential of noir story lines in unusual settings through his location shooting on the U.S.–Mexican frontier. A continuing and sustained theme of revenge can be found in his work, beginning with *Raw Deal* and going through to the dark James Stewart westerns such as *The Naked Spur* (1953) and *The Man from Laramie* (1955). The bridging films between his noir period and the thematically similar westerns were *Border Incident* (1949) and *The Devil's Doorway* (1950). The former showed the western landscape in a dark way (including a hideous encounter with quicksand), and the latter's somber story of an American Indian's tragic fate has undeniably noirish elements. Mann died comparatively young while filming the drama *A Dandy in Aspic* in 1967.

Selected Noir Films: *Dr. Broadway* (1942), *The Great Flamarion* (1945), *Two O'Clock Courage* (1945), *Strange Impersonation* (1946), *Desperate* (1947, and story), *Railroaded!* (1947), *T-Men* (1947), *Raw Deal* (1948), *He Walked by Night* (1948, uncredited), *Follow Me Quietly* (1949, uncredited and story), *Reign of Terror* (a.k.a. *The Black Book*, 1949), *Border Incident* (1949), *Side Street* (1950), *The Furies* (1950), *Devil's Doorway* (1950), *The Tall Target* (1951).

Brian McDonnell

THE MARK OF CAIN (Two Cities/Rank, 1947). *Director:* Brian Desmond Hurst. *Producer:* W. P. Lipscomb. *Script:* W. P. Lipscomb (adaptation), Francis Crowdy, and Christianna Brand, based on the novel *Airing in a Closed Carriage* by Joseph Shearing. *Cinematography:* Erwin Hillier. *Music:* Bernard Stevens. *Cast:* Eric Portman (Richard Howard), Sally Gray (Sarah Bonheur), Patrick Holt (John Howard), Dermot Walsh (Jerome Thorn), Denis O'Dea (Sir William Godfrey), Edward Lexy (Lord Rochford), Therese Giehse (Sister Seraphine), Maureen Delaney (Daisy Cobb), Helen Cherry (Mary), Vida Hope (Jennie), James Hayter (Dr. White).

Four novels written by Joseph Shearing, a pseudonym for Gabrielle Margaret Vere Long, were produced as films in 1947 and 1948. Whilst *Moss Rose* was produced in Hollywood, *Blanche Fury* (1948), *So Evil My love* (1948), and *The Mark of Cain* were all produced in England. All four films involve female protagonists in situations involving betrayal, murder, and unhappy relationships. *The Mark of Cain* is consistent with this pattern, although its female protagonist is not as complex, assertive, or interesting as her counterparts in *So Evil My Love* and *Blanche Fury*. Instead, the virtues of *The Mark of Cain* reside more in Alex Vetchinsky's striking Victorian sets and Erwin Hillier's low-key photography.

Two brothers, Richard and John Howard from the north of England, fall in love with French-raised, British-born Sarah Bonheur when they visit France to purchase cotton for their mill. Sarah is initially attracted to Richard's high-culture leanings, although she ultimately favors the masculine prowess of John. Although the marriage produces a child, Sarah is unhappy with John's dictatorial, boorish behavior and, encouraged by Richard, investigates the possibilities of a divorce. When Sarah learns that Victorian law favors the husband and that she will have to leave her daughter with John if she abandons the marriage, she decides to stay. Ironically, as John's health and overt masculine power deteriorate, Sarah is increasingly attracted to her husband. Richard, thwarted by Sarah's reconciliation with John, poisons his brother and frames Sarah for the crime, planning to save her at the trial. Richard's bizarre performance in the courtroom fails to save Sarah, and she is found guilty. Only the intervention of Jerome Thorn, who is attracted to Sarah, saves her from the gallows when he plays on Richard's mental disintegration.

The Mark of Cain is less interesting than either *Blanche Fury* or *So Evil My Love* due to the fact that Sally Gray's Sarah Bonheur is little more than a victim, or pawn, of the sibling rivalry involving Richard and John. Nevertheless, director Brian Desmond Hurst, supported by Erwin Hillier's superb cinematography, provides a strong dramatic visual basis for this elemental melodrama and its biblical subtext. Primarily, it is a story of virtue (Bonheur) struggling to survive, and overcome, the demands of the Howard family in particular, and the iniquities of Victorian patriarchy in general.

Geoff Mayer

MASON, JAMES (1909–1984). James Mason's film career was built on the enthusiastic reaction, mainly from women, to his sadistic treatment of a number of young women in a series of films in the early and mid-1940s. His "dangerous" persona at that time was a revelation to British audiences more accustomed to the cultured, sexually restrained style of theatrically trained actors such as Laurence Olivier, Ralph Richardson, and John Gielgud. Mason, on the other hand, was tall, dark, and brooding, and he became, for a period, every woman's favorite brute, which made him a huge box office draw.

After studying architecture at Cambridge, Mason made his professional stage debut with a repertory company in Croydon. In the 1930s, after stage work with the Old Vic and Dublin's Gate Company, Mason appeared in low-budget "quota quickies," such as *Late Extra* (1935), with supporting roles in better films such as *Fire over England* (1936) and *The Return of the Scarlet Pimpernel* (1937). In 1941 he had a key role as the sensitive hero in Lance Comfort's gothic noir film *Hatter's Castle* (1941). More important to the development of his "dangerous" persona was his starring role as the brooding war-affected composer in *The Night Has Eyes* (1942) who arouses strong passions in the vulnerable, repressed heroine (Joyce Howard). The film's pervasive sense of danger was reinforced by the fact that the characters were trapped in an isolated mansion on the Yorkshire moors.

Mason's ability to convincingly project a feeling of repressed violence and sexual sadism, coupled with suggestions of emotional disturbance, was further developed in his breakthrough role as the sadistic Lord Rohan in *The Man in Grey* (1943). While Rohan was an unmitigated villain, Mason's character as Nicholas, Ann Todd's perverse mentor in *The Seventh Veil* (1946), was also disturbing, especially in his overt hostility to women, as he explains to Todd in their first encounter. However, Todd, like Howard in *The Night Has Eyes*, welcomes Nicholas's firm control, and the immense popularity of this film reinforced the desire of producers to continue casting Mason in similar roles. In both *The Seventh Veil* and *The Man in Grey* Mason inflicts physical damage on his women—beating Margaret Lockwood to death in *The Man in Grey* and thrashing the delicate fingers of pianist Todd in *The Seventh Veil*. Mason's highwayman, Captain Jackson, opposite Lockwood in *The Wicked Lady* (1945), was more of the same for Mason; the only difference is that this time, Lockwood kills him. There was also his persecution of Phyllis Calvert in *Fanny by Gaslight* (1944).

A welcome change for Mason was his performance as the vulnerable Northern Irish gunman Johnny McQueen in Carol Reed's *Odd Man Out* (1947). This critically celebrated film depicts the last hours in McQueen's life after he has been fatally wounded in a raid on a linen mill to obtain money for the IRA. Before he left for Hollywood, following the success of *Odd Man Out*, Mason gave a less showy performance as a doctor avenging the death of his lover in Lawrence Huntington's film noir *The Upturned Glass* (1947), where Mason costarred with his wife, Pamela Kellino, whom he murders in the film.

Mason was voted number one male star in Britain in the mid-1940s. It took time to establish himself in Hollywood—an early film role was as a sympathetic doctor with a working-class clientele in Max Ophuls's noir film *Caught* (1949), with Mason, now the hero, opposite the archetypal screen villain Robert Ryan. A more significant noir performance by Mason was in his next film for Ophuls, *The Reckless Moment* (1949), as Martin Donnelly, a small-time criminal who tries to blackmail Joan Bennett but gradually falls in love with her. His regeneration is complete when he protects Bennett from his boss and dies taking the blame for an earlier murder he did not commit to prevent Bennett and her daughter from facing an investigation from the police. This was Mason's most important role in a film noir produced in Hollywood.

After essaying the famed German General Rommel in two films for Twentieth Century Fox, *The Desert Fox* (1951) and *The Desert Rats* (1953), and the notorious espionage agent Ulysses Diello ("Cicero") in *Five Fingers* (1952), Mason played Norman Maine, the fallen movie star, in the Technicolor remake of *A Star Is Born* (1954). This role earned him his first Academy Award nomination. As Mason gradually moved into character parts in Hollywood in the 1950s, he produced a number of his films, including Nicholas Ray's bold melodrama *Bigger Than Life* (1956), a story of a teacher who terrorizes all around him after becoming affected by prescription drugs. Later, Mason was cast as Professor Humbert, who becomes infatuated with underage Sue Lyon, in Stanley Kubrick's *Lolita* (1962). In 1969 Mason traveled to Australia to portray a painter, loosely based on the activities of artist Norman Lindsay, who gains inspiration from young Helen Mirren in *Age of Consent* (Mason also produced this film, directed by Michael Powell).

The 1970s and early 1980s were less rewarding as he walked through numerous European coproductions, although he gave a mannered performance as the cruel plantation owner in *Mandingo* (1975), and he was a fine Dr. Watson in *Murder by Decree* (1979). The standout performance by Mason in his final decade was as the amoral lawyer Edward Concannon in Sidney Lumet's courtroom drama *The Verdict* (1982), demonstrating again that when the script and direction were right, James Mason was a fine actor. He died in Switzerland in 1984. Aside from his nomination for *A Star Is Born*, Mason also received Oscar nominations for *Georgy Girl* (1966) and *The Verdict* (1982). Mason was nominated for best British actor at the 1963 BAFTA Film Awards for *Lolita* (1962) and at the 1968 Awards for *The Deadly Affair* (1967).

Selected Noir Films: *Late Extra* (1935), *Troubled Waters* (1936), *Twice Branded* (1936), *Prison Breaker* (1936), *Catch As Catch Can* (1937), *I Met a Murderer* (1939), *Hatter's Castle* (1941), *The Night Has Eyes* (1942), *Alibi* (1942), *The Man in Grey* (1943), *They Met in the Dark* (1943), *Fanny by Gaslight* (1944), *A Place of One's Own* (1945), *The Seventh Veil* (1945), *Odd Man Out* (1947), *The Upturned Glass* (1947), *Caught* (1949), *The Reckless Moment* (1949), *One Way Street* (1950), *Lady Possessed* (1952), *The Man Between* (1953), *Bigger Than Life* (1956), *Cry Terror!* (1956), *The Deadly Affair* (1966),

11 Harrowhouse (1974), *The Marseille Contract* (1974), *Murder by Decree* (1978), *A Dangerous Summer* (1982), *The Verdict* (1982).

Geoff Mayer

MEMENTO (Newmarket/Summit Entertainment, 2000). *Director:* Christopher Nolan. *Producers:* Suzanne Todd and Jennifer Todd. *Script:* Christopher Nolan, from a short story by Jonathon Nolan. *Cinematography:* Wally Pfister. *Music:* David Julyan. *Cast:* Guy Pearce (Leonard), Carrie-Anne Moss (Natalie), Joe Pantoliano (Teddy), Mark Boone Junior (Burt), Jorja Fox (Leonard's Wife), Stephen Tobolowsky (Sammy), Harriet Sansom Harris (Mrs. Jankis).

Insurance investigator Leonard Shelby, protagonist of Christopher Nolan's *Memento*, suffers from a special type of short-term amnesia that prevents him creating any new memories. This disability struck him at the time of the rape and murder of his wife. Despite the obstacles his amnesia creates, he tries to solve the mystery of her death by tattooing his torso with important instructional statements and by recording precautionary instructions and details using annotated Polaroid photographs. As if to mimic his situation, the film's plot unfolds in reverse chronology. Early in the film, but late in Leonard's story, Leonard shoots dead a man named Teddy. He had been guided to Teddy by a mysterious woman named Natalie, whose motives in doing so are highly suspect. The core of the film involves Leonard's bewildering interactions with Teddy, Natalie, and a drug-dealing associate named Dodds, all this mixed up with another narrative line depicting the plight of a second short-term amnesia victim, Sammy Jankis, whose situation is analogous to Leonard's. It is possible that Leonard is being set up as a convenient fall guy/assassin by Teddy, who exploits his condition to persuade him to kill off Teddy's rivals and then to promptly forget what he has done. Can Leonard possibly extricate himself from this entrapped existence?

Memento has been widely praised for its ingenuity, originality, and complexity and its innovative, experimental story structure. As a murder mystery whose narrative has been dismembered and reassembled so that its plot is presented in reverse chronological order, *Memento* recalls the time-reversal devices of Harold Pinter's *Betrayal* (1983) and the fractured storytelling of John Boorman's 1967 classic revenge movie *Point Blank*. In making amnesia its central subject matter, *Memento* also looks back to the classical film noir tradition of stories about amnesiac World War II veterans and about men and women accused of crimes they had possibly committed prior to temporary memory loss. The ambitions of *Memento*, however, go far beyond those earlier films. In this film the narrative building blocks are laid out in reverse order, and such a reverse assembly requires the viewer to deal with what is, in effect, the opposite of the usual cause-and-effect narrative logic. In order to make sense of this back-to-front storytelling, the viewer has to take each successive scene and mentally place it before the preceding scene. *Memento* has a very complex story line that is further complicated by its unique narrative structure. It is a film that in its overall story content conforms to the thriller tradition, but it is a thriller that is not about what happens *next*, but about what happened *before*, and why what we see happening now has come about. Director Nolan says it is about the "existential conditions of identity," and its amnesiac hero, Leonard, has a virtually unmatched opportunity to freely choose his own destiny, if only he can escape other people's traps.

In thematic terms, the film is thus clearly an example of the neo-noir genre. The plot of *Memento*, like those of many film noirs both old and new, has the framework of a (distorted) detective story. A man is trying to track down the killer of his wife to exact revenge. In experiencing the film, we in the audience are forced into the same position as Leonard as protagonist, a process helped by Guy Pearce's compelling performance in the leading role. Seen generically in this way, the film fits well the noir pattern of extreme subjectivity. Typically enough, *Memento* starts with the killing of Teddy, and by the end we know how that death has come about and why. Seen from this viewpoint, the plot is a process of the accumulation of evidence until we have sufficient knowledge to explain the enigmas of the story. In setting, too, we get an appropriate (even classical) film noir landscape: the unspecific, unmemorable, interchangeable, and architecturally anonymous world of Los Angeles motels and shopping strip streets in the northern suburbs of the San Fernando Valley. Los Angeles has always been the *locus classicus* of urban film noir, exemplified either by its rain-slicked nighttime streets or its burned-out, semiurban no-man's-land.

Memento is also very similar thematically to a cycle of wartime and (more commonly) postwar film noirs on the subject of amnesia such as *Street of Chance* (1942), *Spellbound* (1945), *Black Angel* (1946), *Crack-Up* (1946), *Somewhere in the Night* (1946), *Fear in the Night* (1947), *High Wall* (1947), *The Clay Pigeon* (1949), *The Crooked Way* (1949), and *The Blue Gardenia* (1953). Several of these films concern returning World War II veterans rendered amnesiac by wartime traumas. In those examples of the sub-genre where women are the central characters, the memory loss may be short-term through blacking out due to alcohol or being drugged. Men in some such films are likewise framed with drugs by unscrupulous villains or have committed crimes while drunk, only to forget them after blacking out. Their lead characters frequently have to become quasi-detectives in order to discover the truth about how they had actually come to be in the pickles in which they currently find themselves. Leonard's story in *Memento* takes this a step further because his amnesia is of a different order. So, too, is his predicament, which could be framed as a question: how do you manage to pursue a quest when you keep forgetting "what" and "why"? Leonard's life has come to resemble a nightmare where he cannot progress. Many detectives in such classical and neo-noir films are seeker-heroes who have quests, mysteries to solve. Leonard, as the main character of *Memento*, is one such, but he is also a noir-style fall guy, the unwitting tool of both Natalie's and Teddy's duplicity. His peculiar situation therefore makes him as much a victim-hero as a seeker-hero. But in this instance the tables can be turned, on Teddy at least: Leonard gets revenge on him for his exploitative deceit. In effect, Teddy has, by setting Leonard up as a hunter of criminals, created a Frankenstein monster who then kills him.

Brian McDonnell

MILDRED PIERCE (Warner Bros., 1945). *Director:* Michael Curtiz. *Producer:* Jerry Wald. *Script:* Ranald MacDougall, based on the novel by James M. Cain. *Cinematography:* Ernest Haller. *Music:* Max Steiner. *Cast:* Joan Crawford (Mildred Pierce), Jack Carson (Wally Fay), Zachary Scott (Monte Beragon), Eve Arden (Ida Corwin), Ann Blyth (Veda Piece), Bruce Bennett (Bert Pierce), Lee Patrick (Maggie Biederhof), Moroni Olsen (Inspector Peterson), Veda Ann Borg (Miriam Ellis), Butterfly McQueen (Lottie), John Compton (Ted Forrester).

In the middle and late 1940s, there was a cycle of noir films that focused on middle-class dissatisfaction with the so-called American dream. In films such as *The Woman in the Window* (1945), *Nora Prentiss* (1947), and, especially, *Pitfall* (1948), normal, reliable men were presented as dissatisfied with their steady jobs, loving partners, and traditional families. These protagonists were male, and their problem was usually instigated by an infatuation with *la belle dame sans merci*, the "fatal woman." However, on a deeper level, these films represented a change in the balance of power between men and women. As the men became less heroic and more vulnerable, the women were seen as more independent and assertive. *Mildred Pierce* represented an assimilation of this trend with the twist that the protagonist was a woman, and the basis of the film represented a critique of the very basis of society, the normal love of a parent for a child.

This theme is established in the film's opening narration as Mildred expresses her dissatisfaction with the monotony and routine of domesticity:

> We lived on Corvallis Street, where all the houses looked alike. I was always in the kitchen. I felt as though I had been born in a kitchen and lived there all my life, except for the few hours it took to get married.
> I married Bert when I was 17. I never knew any other kind of life. Just cooking, ironing, and washing.

When Bert loses his job as a real estate broker, and Mildred suspects he is having an affair with a wealthy woman, she forces him out of the house. Mildred, determined that her daughters will avoid the domestic drudgery that dominates her life, works hard baking pies to raise money for music and ballet lessons for her children Veda and Kay. Mildred takes a job as a waitress, and after learning the restaurant business, she joins with Bert's former real estate partner Wally Fay to open her own restaurant. This is achieved when playboy Monte Beragon provides the land.

The restaurant is successful, but the price that Mildred has to pay for deviating from social norms is high. While Mildred is making love to Monte Beragon in his beach house, her youngest daughter, Kay, dies of pneumonia. Although both events are geographically separate, the editing establishes a thematic link suggesting that maternal neglect was motivated by illicit sex as Mildred and Bert are still married. Although Mildred works long hours, and eventually establishes a chain of restaurants, she cannot acquire what she wants most—Veda's love and respect. Instead, her daughter openly despises her mother and her work ethic, preferring the company of dilettante Monte Beragon, Mildred's erstwhile lover. When Veda seduces a wealthy young man, Ted Forrester, into marriage and then forces his family into a financial settlement after a false claim of pregnancy, Mildred confronts her daughter in the film's most powerful scene. Here Veda's thinly veiled contempt for Mildred explodes in her display of overt hatred. Mildred tells Veda to leave and pays off Beragon, who has been romancing both women.

Mildred's obsession with Veda, however, is so warped that she humiliates herself by marrying Beragon in an attempt to attract Veda back into her home. The price of the marriage, a third share in Mildred's restaurant chain, eventually leads to her financial ruin when Beragon sells his share and Mildred goes into debt providing for her daughter's materialistic needs. Mildred's final humiliation occurs when she discovers Veda and Beragon making love in the same beach house where she had sexual intercourse on the night that Kay died. When Beragon refuses to marry Veda, she shoots him, and although Mildred attempts to take the blame, the police arrest Veda as she attempts to flee the country.

By the close of the film Mildred has lost her once-thriving business and her family has been destroyed—Veda has been arrested for murder, and Kay is dead. Her only alternative is to return to Bert, the ineffectual husband she left at the beginning of the film. As they leave the police station, director Michael Curtiz provides one of Hollywood's most devastating gender images, showing two women on their knees scrubbing the floor in the right-hand corner of the frame. This sad, prophetic symbol of Mildred's futile attempt to break away from her domestic prison suggests not only her fate, but also points to the power of patriarchal norms, which can turn the normal love for a child into an unhealthy obsession.

Joan Crawford deservedly won an Academy Award for her performance as Mildred Pierce, and the role was the high point of a long career. Born into poverty as Lucille Fay LeSueur in San Antonio, Texas, she was recreated by MGM in the 1920s as Joan Crawford. However, in 1943, it appeared that her days of a star were over when, approaching 40 years of age, MGM did not renew her contract. Crawford signed with Warner Bros., and after a two-year wait for a suitable role, *Mildred Pierce* represented one of the best comebacks in the history of Hollywood. It extended her acting career until 1970 and the low-budget British horror film *Trog*.

Geoff Mayer

MINE OWN EXECUTIONER (London Film Productions, 1947). *Director:* Anthony Kimmins. *Producers:* Anthony Kimmins and Jack Kitchin. *Script:* Nigel Balchin, based on his novel. *Cinematography:* Wilkie Cooper. *Music:* Benjamin

Frankel. *Cast:* Burgess Meredith (Felix Milne), Dulcie Gray (Patricia Milne), Kieron Moore (Adam Lucian), Christine Norden (Barbara Edge), Barbara White (Molly Lucian), John Laurie (Dr. James Garsten), Michael Shepley (Peter Edge), Walter Fitzgerald (Dr. Norris Pile), Edgar Norfolk (Sir George Freethorne).

Following the end of World War II, a cycle of films was produced in Hollywood and London that examined the psychological and physical effects of the war on the combatants. *Mine Own Executioner,* produced by Alexander Korda's London Films, was a strong example of this cycle, and despite overt signs of studio (and possibly censorship interference) affecting the character motivation and narrative coherence, the film retained its bitter, downbeat climax. After Adam Lucian tries to murder his wife, Molly, she approaches London psychoanalyst Felix Milne to treat her husband. Milne reluctantly accepts the case and quickly realizes that Adam Lucian is schizophrenic as a result of the torture and mental anguish inflicted on him by the Japanese during the war. While Milne seemingly makes rapid progress with Lucia, the psychoanalyst realizes that his early success with his patient is misleading as Lucian's problems are deep seated and will not respond easily to treatment.

There are a number of fascinating, if underdeveloped, subplots in *Mine Own Executioner.* These include the personal and professional tensions emanating from that fact that Milne is not a qualified doctor and only practices due to the support of eminent Harley Street specialist Dr. James Garsten. Milne's lack of qualifications in turn threatens the funding of his institute. Milne also suffers from his own form of schizophrenic behavior with regard to his marriage. While he loves his long-suffering wife, Patricia, under pressure, he is drawn to the sensual charms and forbidden excitement offered by a family friend, Barbara Edge. Patricia Milne is aware of this relationship and, in an unlikely plot device, accepts Felix's relationship with Barbara Edge, despite her anguish and misgivings.

These narrative strands come together when Milne, feeling poorly because of the flu, fails to follow his intuition when he allows Lucian to postpone treatment until the next day. Milne, realizing that Lucian's mental condition represents an immediate threat to his wife, Molly, is distracted because of his own problems, including the news that the institute will be denied funding because of his lack of formal qualifications. As the pressure on the psychoanalyst intensifies, Milne resumes his affair with Barbara Edge, and just as they are about to have sex, he receives news (from his wife, Patricia, who knows that he is spending the night with Edge) that Lucian has fired four bullets into Molly. Milne tries to redeem himself by climbing a long fire ladder to speak to Lucian, who is poised on the roof of a building. However, this fails, and Lucian shoots himself in the head. Only the support of Dr. Garsten at the subsequent inquest saves Milne's position at the institute.

After the death of Lucian and Milne's humiliation at the inquest, the ending of the film is less than satisfactory. Although Milne has failed to heal Lucian, which resolves the main narrative thread, other issues, notably the funding of the institute and, more important, Milne's relationship with wife, Patricia, and mistress, Barbara, are basically ignored. This, however, did not weaken support for the film, which was both a critical and commercial success.

Geoff Mayer

MITCHUM, ROBERT (1917–1997). The night after Robert Mitchum died on July 1, 1997, James Stewart died, and Stewart's death overshadowed the media coverage of Mitchum's death. And Robert Mitchum would not have had it any other way. He, unlike Stewart or Wayne, was perceived as unpredictable, edgy, even dangerous, qualities that made him such a fine noir actor. While his range was extensive, from the passive schoolteacher in David Lean's *Ryan's Daughter* (1970) to the psychotic religious fanatic in Charles Laughton's *The Night of the Hunter* (1955), from the Irish-Australian itinerant worker in Fred Zinnemann's *The Sundowners* (1960) to Raymond Chandler's detective Philip Marlowe in *Farewell, My Lovely* (1975) and *The Big Sleep* (1978), he was most known in the 1940s and early 1950s as the preeminent film noir actor with morally problematic characters.

Throughout his career Mitchum expressed contempt for Hollywood and found little pleasure in acting. Yet he was the consummate professional actor—well prepared and reliable—and he made acting deceptively easy. These qualities, however, resulted in critics underestimating his performances. He was born Robert Charles Durman Mitchum in Bridgeport, Connecticut, and his father was one-quarter Scottish, one-quarter Irish, and half Blackfoot Indian, while his mother was a Norwegian immigrant. His father, a railroad worker, died in an accident at work in 1917. Aged 11, Mitchum ran away from home but was soon caught. In 1928 he was expelled from Felton High School for fouling a girl's shower cap, and although he was eventually reinstated, on the eve of his graduation, after being named Felton High's valedictorian, he left for good after failing to collect his diploma. In the next few years he worked as a deckhand; rode the boxcars; was arrested for vagrancy in Savannah, where he escaped the chain gang; and worked as a dishwasher, truck driver, forest laborer, coal miner, bouncer, and part-time prize fighter while drifting throughout the United States.

In 1937, following pressure from his sister, he joined the Long Beach Player's Guild and began appearing on stage—including the role of gangster Duke Mantee in *The Petrified Forest*—and in 1939 he wrote and directed two children's plays. In 1940 he married Dorothy Spence and took a job at the Lockheed aircraft factory in Burbank while continuing to appear on stage. In 1942, while working as a shoe salesman, he debuted on the screen as a model in the short film *The Magic of Make-Up*. The same year, the producer of the Hopalong Cassidy series, Harry Sherman, hired Mitchum for supporting roles—mainly villains. He also appeared in a number of nonwesterns, such as *Cry Havoc* and *Gung-Ho*, both released in 1943. In 1944 RKO selected Mitchum to replace Tim Holt, who was in the military, to star in their western series based on novels by Zane Grey. Mitchum appeared in two of these films, *Nevada* (1944) and *West of the Pecos* (1945). He also starred in two low-budget films for Monogram, the film noir *When Strangers Marry* (1944), as the murderer, and the topical comedy *Johnny Doesn't Live Here Anymore* (1944).

Mitchum's big break came in 1945 when he was cast as Lieutenant Walker in William Wellman's *The Story of GI Joe*. This prestigious film was based on war correspondent Ernie Pyle's factual account of his experiences with Company C of the 18th U.S. Infantry during their North African campaign. In Italy, Walker is promoted to captain and dies when the unit attacks a mountaintop monastery—a battle that was also featured in John Huston's documentary *Battle of San Pietro* (1944). In the film Mitchum gives a moving performance as a leader who has seen too many of his men die. He was rewarded by an Oscar nomination for best supporting actor, although, characteristically, he did not attend the ceremony. This was his only nomination in a long career. Before the film was released, Mitchum was drafted into the military and served for eight months.

With the release of *The Story of GI Joe*, Mitchum's status in Hollywood was strong, and he was offered major roles in big-budget films, such as *Undercurrent* (1946), as well as lead roles in medium-budget films at RKO such as *Out of the Past* (1947) and *Blood on the Moon* (1948). It was during this period that he developed a screen image of a problematic figure who was physically strong but emotionally and morally vulnerable. In *The Locket* (1947) he commits suicide when he is tormented by a psychotic woman; in *Out of the Past* he jeopardizes his vocation and life when he falls under Jane Greer's spell; in *Pursued* (1947) he is tormented by a childhood trauma that emanates from the repressed memory of his father's adulterous affair with Judith Anderson; in *Where Danger Lives* (1950) he suffers from concussion and the manipulative behavior of Faith Domergue; and in *Angel Face* (1952) he cannot disentangle himself from the psychotic behavior of a young woman determined to kill her stepmother.

This list does not include noir-inflected westerns, such as *Blood on the Moon* and *Track of the Cat* (1954), as well as Nicholas Ray's elegiac rodeo film *The Lusty Men* (1952), where Mitchum plays the veteran rider whose death causes an arrogant man, played by Arthur Kennedy, to reassess his life on the rodeo circuit. When this cycle ended in 1954, Mitchum began appearing in more straightforward action-romantic films in less neurotic roles such as the *River of No Return* (1954) with Marilyn Monroe and *Heaven Knows, Mr. Allison* (1957) with Deborah Kerr.

While Mitchum continued acting until his death in 1997, with a number of roles in television films and miniseries in the 1980s, his career as a film star lasted only until the late 1970s. During the 1960s and 1970s he returned to film noir, beginning with his sadistic Max Cady in *Cape Fear* (1962), the middle-aged small-time criminal in *The Friends of Eddie Coyle* (1973), and the retired private eye in Sidney Pollack's underrated film *The Yakuza* (1975), which is set in Japan and parallels *Chinatown* (1974) in that its detective (Mitchum) inadvertently causes pain to the innocent. Mitchum also gave a world-weary interpretation of Raymond Chandler's detective Philip Marlowe in *Farewell, My Lovely* (1975) and *The Big Sleep* (1978).

Selected Noir Films: *When Strangers Marry* (1944), *Undercurrent* (1946), *The Locket* (1946), *Pursued* (1947), *Crossfire* (1947), *Out of the Past* (1947), *Blood on the Moon* (1948), *The Big Steal* (1949), *Where Danger Lives* (1950), *His Kind of Woman* (1951), *The Racket* (1951), *Macao* (1952), *The Lusty Men* (1952), *Angel Face* (1952), *Second Chance* (1953), *Track of the Cat* (1954), *The Night of the Hunter* (1955), *Cape Fear* (1962), *The Friends of Eddie Coyle* (1973), *The Yakuza* (1975), *Farewell, My Lovely* (1975), *The Big Sleep* (1978), *Cape Fear* (1991).

Geoff Mayer

THE MONEY TRAP (MGM, 1965). *Director:* Burt Kennedy. *Producers:* Max E. Youngstein and David Karr. *Script:* Walter Bernstein, based on a novel by Lionel White. *Cinematography:* Paul C. Vogel. *Music:* Hal Schaefer. *Cast:* Glenn Ford (Detective Joe Baron), Elke Sommer (Lisa Baron), Rita Hayworth (Rosalie Kelly), Joseph Cotten (Horace van Tilden), Ricardo Montalban (Detective Pete Delanos), James Mitchum (Detective Wolski), Ted de Corsia (Police Captain), Eugene Iglesias (Father), Teri Lynn Sandoval (Daughter), Than Wyenn (Phil Kenny).

Released at a time (1965) when there were very few noir films, *The Money Trap* has been ignored by most studies of film noir. Aside from the fact that its plot was a relatively common one in the late 1940s and 1950s, the film was also notable for being one of the last studio films shot in both CinemaScope and black and white. *The Money Trap*, based on a novel by Lionel White, who also wrote the source novel for Stanley Kubrick's *The Killing* (1956), was directed by Burt Kennedy, a writer more famous for his scripts for Budd Boetticher, who directed Randolph Scott in a series of westerns such as *Seven Men from Now* (1956) and *The Tall T* (1957). Kennedy also directed a number of successful comedy westerns such as *The Rounders* (1964) and *Support Your Local Sheriff* (1968).

The Money Trap was a rare excursion for Kennedy into film noir, and the film reworks the familiar noir plot of an essentially good man (Joe Baron) lured into crime by his need to satisfy the materialistic needs of an attractive young woman, in this case, his wife, Lisa. The film's moral basis, which emphasizes the folly of obsessing about material possessions, is conveyed in the final moment, which ends in Joe's opulent house, complete with a spectacular swimming pool, large garden, and spacious interior, which is extravagant for a man whose annual income of $9,200 barely supports the servants. Initially, Joe is financially dependent on his wife's inheritance, and when her dividends fail, he is faced with the prospect of returning to his working-class neighborhood and, possibly, losing his young wife. Instead, he decides to rob the safe of a crooked doctor, Horace van Tilden, who has been trafficking in heroin. However, when his partner in the police force, Pete Delanos, discovers the plan, he insists on joining in.

The film begins with the failed attempt of a drug addict, Phil Kenny, to rob van Tilden. The investigation takes Joe back to his working-class neighborhood and the wife of the dead man, Rosalie Kenny, Joe's childhood sweetheart, who now works in a bar and takes home men to supplement her meager income. Actress

Rita Hayworth, in one of her final roles as Rosalie, is reunited with Glenn Ford, her costar in the memorable 1946 film noir *Gilda*. Twenty years later, both actors are allowed one night of tired passion as a sad reminder of the sparkling energy, and torment, they inflicted on each other in *Gilda*. In *The Money Trap*, they are disillusioned and doomed.

Van Tilden kills Rosalie and traps Joe and Pete when they rob his safe. Pete, obsessed by the prospect of money after years of hard work and little personal or material reward, dies from a gunshot wound, and Joe, also wounded, kills van Tilden before limping home to his wife. While awaiting death, or imprisonment, he tells Lisa that it is "never the money" that corrupts, "it's people and the things they want."

The Money Trap is one of the final noir films in a cycle that began in the 1940s to trace the corruption of a basically good man who, through sexual desire, greed, or even boredom, falls from grace. Other films in the cycle include *Night Editor* (1946), *Pitfall* (1948), and *The File on Thelma Jordan* (1949). There is a subplot within *The Money Trap* that reinforces this theme and shows a loving husband who murders his wife when she tries to supplement their income by working as a prostitute. Joe traps the man at an amusement park when he takes his young daughter on a trip for her birthday. Happiness, in this filmic universe, is never lasting.

Geoff Mayer

MR. PERRIN AND MR. TRAILL (Two Cities, 1948). *Director:* Lawrence Huntington. *Producer:* Alexander Galperson. *Script:* T. J. Morrison and L.A.G. Strong, based on the novel by Hugh Walpole. *Cinematography:* Erwin Hillier. *Music:* Allan Gray. *Cast:* David Farrar (David Traill), Marius Goring (Vincent Perrin), Greta Gynt (Isobel Lester), Raymond Huntley (Moy-Thompson), Mary Jerrold (Mrs. Perrin), Edward Chapman (Birkland), Finlay Currie (Sir Joshua Varley), Ralph Truman (Comber), Viola Lyel (Mrs. Comber), Don Barclay (Rogers).

Although *Mr. Perrin and Mr. Traill* is largely forgotten today, it exemplifies many of the virtues of the British film industry in the 1940s and early 1950s. The story concerns a new teacher, David Traill, fresh from the army and successes on the rugby field, who fails to pay appropriate respect to the obsessive rituals of Vincent Perrin, a master in an elite private boarding school. The conflict between the two teachers escalates from small issues, such as Perrin's perceived right to the first reading of the daily newspaper and the sole occupancy of the communal bathroom, to larger issues, such as romancing local nurse Isobel Lester. The news of Traill and Lester's engagement results in the total breakdown of Perrin's fragile grasp on reality.

The film refuses to replicate the sentimental stereotypes in films such as *Goodbye, Mr. Chips* (1939), and it also does not present the conflict between Perrin and Traill as one of virtue versus evil. In fact, Traill, the film's so-called normal character, gains most of his sympathy from the craven behavior of the other masters who, faced with a tyrannical headmaster, lack his spirit. On the other hand,

Perrin's virtues, such as his dedication to the students and their education, are overwhelmed by his obsession with the trivialities of daily life—until the climax of the film, when circumstances transform him into a hero who saves Traill's life after he falls down the side of a cliff. After Perrin loses his life while rescuing his nemesis, Traill exposes the real villain, the despotic headmaster Moy-Thompson, by accusing him of systematically breaking the spirit of Perrin and the other masters.

Geoff Mayer

MURDER, MY SWEET (RKO, 1944). *Director:* Edward Dmytryk. *Producer:* Adrian Scott. *Script:* John Paxton, from the novel *Farewell, My Lovely* by Raymond Chandler. *Cinematography:* Harry J. Wild. *Music:* Roy Webb. *Cast:* Dick Powell (Philip Marlowe), Claire Trevor (Velma/Mrs. Grayle), Anne Shirley (Ann), Otto Kruger (Amthor), Mike Mazurki (Moose Molloy), Miles Mandor (Mr. Grayle), Don Douglas (Lieutenant Randall), Ralf Harolde (Dr. Sonderborg).

Along with *Double Indemnity, Laura, The Woman in the Window,* and *Phantom Lady, Murder, My Sweet* marks the release, in 1944, of the first substantial group of Hollywood films that helped establish what we now call the classical noir cycle. It was the first film adaptation of one of Raymond Chandler's Philip Marlowe novels, and it built on the start made with *The Maltese Falcon* in setting up the tradition of noir private eye stories. In addition, through Paxton's script and Dmytryk's directorial style, it went far beyond the 1941 film in developing the narrative form, the lighting policy, and the general look that has since come to be associated with film noir. In regard to the contrast between these trailblazing films, in his book *Film Noir,* Andrew Spicer writes that the team of producer Adrian Scott, script writer John Paxton, and director Edward Dmytryk was "determined to make its mark through a radically different approach that would approximate much more closely to the novel's cynical, dispassionate take on American society" (p. 54). *Murder, My Sweet* also successfully transformed crooner Dick Powell into a convincing tough guy who could walk the mean streets of the dark city along with the best of them.

In plot the film employs the standard private eye formula whereby the detective is approached by a bizarre client, is given the quest of finding a missing person, and then becomes personally involved in all the adventures and risks such a task can entail. The giant figure of Moose Molloy appears in Marlowe's Los Angeles office wanting to find long-lost love Velma. But the path of the subsequent investigation is far from straight. Marlowe finds himself with multiple clients, each seeking contradictory outcomes and all prepared to lie to him to further their own ends. In the course of his investigation he is often in personal jeopardy and is beaten, drugged, slugged, and kissed by a variety of devious men and women. Frequently he must use his wits, his smart talk, and his presence of mind to keep ahead of both the bad guys and the abrasively unsupportive police. There are shocks, twists, and turns, seemingly unconnected cases finally blend, and Marlowe's realization that he has been used comes only just in time to allow him to put all the tangled

Murder My Sweet (1944). Directed by Edward Dmytryk. Shown from left: Dick Powell (blindfolded as Philip Marlowe), Don Douglas (far right as Lieutenant Randall). RKO/ Photofest.

pieces together and emerge relatively unscathed. From time to time he is forced to reexamine his own code of conduct by which he tries to function, despite the mix of danger and temptation he confronts. The chronological unfolding of the quest is contained (in flashback form) within a frame where a blindfolded Marlowe is being interrogated by police detectives, and thus the audience is kept intrigued by never knowing more than the hero at any stage.

Apart from Marlowe, the film has a rich complement of well-drawn supporting characters, from Moose himself; through the super-rich Mr. Grayle, who is a far-from-virile old man with a sexually voracious and desirable younger wife, the "paltry foppish," sexually ambiguous Marriott; the spunky daughter Ann; the urbane and ruthless blackmailer Anthor; and, naturally, one of the finest of noir's femmes fatales, Velma herself (or Helen, as she calls herself for much of the film), who is first seen with much of her legs dangling bare for Marlowe's delectation. Expressionistic lighting is used right from the credits, and subjective camera work is foregrounded in the scenes where Marlowe receives blows to the head or is pumped full of drugs at a sinister private clinic. Nightmare sequences feature visual and aural distortions, and throughout, Marlowe's voice-over narration consciously conveys the hard-boiled tone of Chandler's original prose style. In the film's climactic scene at an expensive beach house, all the plot complexities are explained in a rush of bewildering dialogue, but of much more impact are the strong visuals as gunfire flashes in the darkened room, temporarily blinding Marlowe. In a gush of violent death, all at once, Moose, Grayle, and Velma are dead. The cynical Marlowe is collaterally maimed, and the one pure soul, Ann, is left to inherit the family fortune and to bestow a final, tender kiss on knight errant Marlowe.

Brian McDonnell

MY BROTHER'S KEEPER (Gainsborough, 1948). *Director:* Alfred Roome. *Producers:* Anthony Darnborough and Sidney Box (executive). *Script:* Frank Harvey, based on the story by Maurice Wiltshire. *Cinematography:* Gordon Lang. *Music:* Clifton Parker. *Cast:* Jack Warner (George Martin), Jane Hylton (Nora Lawrence), George Cole (Willie Stannard), David Tomlinson (Ronnie Waring), Yvonne Owen (Meg Waring), Raymond Lovell (Wainwright), Bill Owen (Syd Evans), Beatrice Varley (Mrs. Martin), Garry Marsh (Brewster), Wilfrid Hyde-White (Harding), Brenda Bruce (Winnie Forman), Susan Shaw (Beryl).

Alfred Roome, one of Britain's top editors for more than 40 years, directed only two films: *My Brother's Keeper* and *It's Not Cricket* (1948). Roome, uncomfortable directing actors, was assisted by Roy Rich on both films. Although *My Brother's Keeper* is marred by a distracting subplot concerning a young journalist (Ronnie Waring) on his honeymoon, the film is tough and unsentimental. *My Brother's Keeper* also provided Jack Warner with his most complex and, arguably, best screen performance as the hardened criminal George Martin.

My Brother's Keeper is, despite the light-hearted scenes showing the marital and vocational problems of Ronnie Waring, a bleak film. It presents a postwar society devoid of compassion and lacking the communal spirit of the war years. The film is sometimes associated with the racial drama *The Defiant Ones* (1958) because both films share the basic premise of two contrasting convicts handcuffed together on the run. However, *My Brother's Keeper* has little in common with Stanley Kramer's 1958 Hollywood production, and it is a much tougher film, marked by an absence of sentimentality. Only with the offer of Martin's mistress, Nora Lawrence, to testify to the police on behalf of the hapless Willie Stannard, and thereby incriminate herself, does the film display any hint of human generosity. In all other respects the basic institutions of society, including the media and the police, lack compassion, as does George Martin.

The dramatic focus in *My Brother's Keeper* is the complex character of George Martin, a man who can use his literary knowledge and street cunning to charm people, including Willie Stannard, who idolizes Martin; his mistress, Nora Lawrence, who assists the convict; and Martin's wife (Beatrice Varley), who tells Nora at the end of the film that in her 22-year marriage to Martin, she never really knew her husband as he spent 14 years in jail, 5 years in the army, and only 3 years with her (during which time she, presumably, shared Martin with Nora Lawrence). Both Stannard and Lawrence eventually realize that Martin cares for no one except himself. As Martin tells Stannard, "There's no such things as friends. People are either useful or useless. You're useless!"

Most of the film is concerned with Martin's and Stannard's attempts to elude capture. After Martin is able to free himself from Stannard, who is presented as basically an innocent young man, the film focuses on Martin's flight from the police. In this situation, audience sympathy normally gravitates toward the hunted, but director Alfred Roome steadfastly refuses to present Martin as a victim—the closest we come to understanding his lack of compassion for other people comes from his wife's suggestion that the army, and his years in jail, played a part in molding his character. Essentially, the film is content to reinforce her conclusion that "there's something in his mind that's not quite right. He's just a misfit."

Geoff Mayer

MY NAME IS JULIA ROSS (Columbia, 1945). *Director:* Joseph H. Lewis. *Producer:* Wallace MacDonald. *Screenplay:* Muriel Roy Bolton, based on the novel *The Woman in Red* by Anthony Gilbert, the pseudonym of Lucy Beatrice Malleson. *Cinematography:* Burnett Guffey. *Music:* Mischa Bakaleinikoff. *Cast:* Nina Foch (Julia Ross), Dame May Whitty (Mrs. Hughes), George Macready (Ralph Hughes), Roland Varno (Dennis Bruce), Anita Bolster (Sparkes), Doris Lloyd (Mrs. Mackie), Leonard Mudie (Peters).

Director Joseph H. Lewis made more than 20 films prior to *My Name Is Julia Ross*. Most critical studies of Lewis, however, begin with this noir/gothic hybrid. The gothic influence on film noir is evident in the 1940s, particularly in the early years of the decade with films such as *Among the Living* (1941), *Experiment Perilous* (1944), and *Gaslight* (both the 1940 British version directed by Thorold Dickinson and the more well-known but inferior MGM version directed by George Cukor, and released in 1944 starring Ingrid Berman and Charles Boyer). *My Name Is Julia Ross*, with only a fraction of the budget lavished on the MGM film, is superior and confirmed the ability of Lewis to inject vitality and a visual sophistication into films that, in the hands of a lesser director, would be utterly conventional and banal.

With a small budget, less than $150,000, and a short shooting schedule of 10 days, this 64-minute film was intended as merely another disposable B film from Columbia Studios. Lewis, however, had other ideas and was able marginally to extend the shooting period and increase the budget after Harry Cohn, the head of production at Columbia, was impressed by footage from the first day of shooting. Also, Lewis insisted that the film receive a preview, which was rare for a B film

My Name Is Julia Ross (1945). Directed by Joseph H. Lewis. Shown: George Macready (as Ralph Hughes), Nina Foch (as Julia Ross). Columbia Pictures/Photofest.

from Columbia. The result was that *My Name Is Julia Ross* was a financial and critical success and gave Lewis the career breakthrough he had been seeking after more than a decade working on low-budget films.

Julia Ross is unemployed in England when she receives an offer to act as secretary to Mrs. Hughes and her son Ralph. Julia is in a vulnerable state due to the fact that the man she loves, Dennis Bruce, has rejected her so that he can marry another woman. After Mrs. Hughes administers a drug, Julia wakes up in bed in a large house situated on a lonely stretch of the Cornwall coast. Being a young woman alone and threatened by sinister forces in a large, isolated house is conventional material to those familiar with gothic stories. When Hughes and Ralph attempt to eliminate Julia's identity by calling her "Marian," the name of Ralph's dead wife, the film's gothic basis is reinforced.

Julia learns that Mrs. Hughes and Ralph plan to kill her as part of the cover-up for Ralph's murder of his wife, and the body of the film is concerned with Julia's attempts to escape the Hughes household. Her inability to convince the locals that she is not Ralph's dead wife generates a powerful sense of frustration in the viewer, leading to an equally powerful sense of catharsis when her attempt to contact Dennis in London succeeds and he arrives just as the psychotic Ralph is about to smash in Julia's head with a rock.

This is almost the perfect low-budget film. Taut, suspenseful, with strong performances from Nina Foch as the resolute heroine; George Macready, who, by 1945, had perfected the mannerisms of the gleefully sadistic, psychologically unstable villain, traits that he fully displayed the following year in *Gilda* (1946); and Dame May Whitty, cast against type as the ruthless mother determined to protect her insane son from a murder charge.

Geoff Mayer

THE NAKED CITY (Universal International, 1948). *Director:* Jules Dassin. *Producer:* Mark Hellinger. *Script:* Albert Maltz and Marvin Wald, based on a story by Marvin Wald. *Cinematography:* William Daniels. *Music:* Miklos Rozsa and Frank Skinner. *Cast:* Barry Fitzgerald (Lt. Dan Muldoon), Howard Duff (Frank Niles), Dorothy Hart (Ruth Morrison), Don Taylor (Jimmy Halloran), Ted de Corsia (Garzah), House Jameson (Dr. Stoneman), Anne Sargent (Mrs. Halloran), Mark Hellinger (Narrator).

The Naked City is perhaps the best known of a group of films made in the immediate postwar period that are commonly referred to as semidocumentary film noirs or as semidocumentary crime films. Some historians exclude them from the canon of classical film noir for ideological reasons, claiming that their political values are conservative rather than radical, while others are more inclusive and consider them an important subgroup within the overall genre. It is widely accepted that during World War II, there was a softening of the control that the Production Code Administration exerted over the content of Hollywood films, especially of violence. As part of the war effort, newsreel footage of battles, along with somber documentaries showing such enemy atrocities as concentration camps, included images of brutality that would not previously have been allowed on American screens. These nonfiction films, with their authoritative (even strident) voice-over narrations, helped foster a desire in the public for a greater realism in fictional movies. The classical film noir cycle helped satisfy that appetite.

Louis De Rochemont had been a producer on the weekly newsreel series *The March of Time*. At the end of the war he moved to the Twentieth Century Fox

The Naked City (1948). Directed by Jules Dassin. Shown: Ted de Corsia (as Garzah). Universal Pictures/Photofest.

studio, and with the encouragement of production head Darryl F. Zanuck, he began to make fictional films that employed techniques taken from documentaries and newsreels. The first of these was a spy film titled *The House on 92nd Street* (1945), directed by Henry Hathaway. This film combined actual FBI surveillance footage with dramatized scenes involving actors shot on actual locations and narrated by the stentorian voice of a newsreel announcer. Later examples, which were also influenced by such Italian neo-realist films as *Rome: Open City* (1945), included *Boomerang!* (1947) and *The Street with No Name* (1948). The popularity and success of these rather cheap films spurred other studios to follow suit. Theatrical journalist and film producer Mark Hellinger, who had worked as a war correspondent, produced *The Naked City* at Universal and also narrated it (he died of a heart attack before its general release). Its visual style was influenced by realist photographer Weegee (Arthur Fellig), who had published *Naked City*, a book of photographs of crime scenes and other stark urban images, in 1945, although his contribution to the film is not credited.

Essentially a police procedural, *The Naked City* traces the investigation into the murder of a young woman in New York, leading up to the identification and pursuit of her killer. Many of the scenes are filmed in the standard dramatic style of

Hollywood filmmaking, but the story is framed by documentary-style shots of the city that place the woman's individual fate in the context of the organic life of the whole metropolis. The narration generalizes about work and crime and psychology, personalizes the city itself as an emotional entity, and even verges on self-congratulation in its description of the way in which the production has used the actual city locations of the story's events. The opening aerial shots of Manhattan give way to a montage of vignettes of ordinary New Yorkers going about their lives. This begins in the predawn hours of a hot summer's night, and among its shots of night workers and revelers, it shows the murder of a young clothing model by two men. Later, one of the men ruthlessly kills his accomplice, who has shown weakness and remorse. A veteran police lieutenant Dan Muldoon (played with an air of wisdom and world weariness by Barry Fitzgerald) leads the investigation. The film emphasizes the unglamorous routines of Muldoon and his team, their doggedness and the way in which progress is made by legwork and thoroughness rather than through spectacular insightfulness or violent action. His associates are mostly young family men, and the film conveys a sense of America reconstituting itself as the baby boom generation settles energetically into postwar domesticity.

The first suspect grilled about the killing is an interesting character and represents at least one way in which *The Naked City* could be said to be authentically a film noir. Frank Niles (Howard Duff) is a close friend of the dead woman and appears at first to be an all-American nice guy from a respectable family and with a sound war record. It soon emerges, though, that he is actually a pathological liar, a con man, an unfaithful Lothario engaged to another woman, and at least indirectly connected with the murder. Always ready with a new story or a plausible excuse, he has a superficial charm and, as Duff plays him, is a complex mixture of strength and weakness. In a film with many stark contrasts between good and bad, he has the morally ambivalent nature of many characters in film noir. It turns out that Niles (in league with the dead woman) had been running an operation involving high-society jewel thefts. As a source of intelligence on desirable gem collections, he had used a socially prominent doctor who unwittingly provided helpful information on the dead woman, who was his mistress. Two thugs in Niles's gang had committed the murder for money, and *The Naked City* culminates in a fairly conventional chase sequence where the surviving killer (Garzah, a harmonica-playing wrestler) is pursued up into the superstructure of the Williamsburg Bridge until he plummets to his death.

These climactic scenes make excellent use of the real bridge and its environs, and their verisimilitude must have been a novelty on the movie's release in 1948. The milling multiethnic crowds filmed in the teeming streets of the Lower East Side have the feel of an Italian neo-realist film by Roberto Rossellini or Vittorio De Sica. The whole enterprise is also greatly aided by the underplaying and whimsicality of Fitzgerald in the main role. Certainly, like other semidocumentaries, *The Naked City* is plainly on the side of authority and shows scant sympathy for the predicaments of its lawbreakers, but it does have a place within a broad definition of film noir. Over the closing credits, Hellinger's narration famously states that "there are eight million stories in the naked city—this has been one of them." The popularity of the film and of this catchphrase led to the production of a long-running (1958–1963) and high-quality television crime-drama series of the same name.

Brian McDonnell

NEAL, TOM (1914–1972). Tom Neal is not much more than a bizarre footnote to the history of film noir, although his real life parallels aspects of his most famous film, *Detour* (1945). Neal was born into a middle-class family; his father was a banker in Evanston, Illinois. While the family favored a career in the legal profession for Tom, he, a keen boxer, preferred acting and entered summer stock at West Falmouth.

Neal made his Broadway debut in 1935, and he soon left Broadway for Hollywood, where he was offered a contract by MGM, who saw him as another Gable due to his athletic build and rugged good looks. MGM started Neal in small parts in series films, such as *Out West with the Hardys* (1938) and *Another Thin Man* (1939), as well as in more significant roles in B films such as Jacques Tourneur's *They All Come Out* (1939). However, by 1941, Neal had left MGM and was freelancing. His first lead role was opposite Frances Gifford in *Jungle Girl* (1941), Republic's exciting 15-episode serial directed by William Witney and John English. He also had a prominent role as Dave Williams in *The Courageous Dr. Christian* (1940), the second in RKO's series of six Dr. Christian films starring Jean Hersholt.

Neal alternated between small roles in big-budget studio films, such as Warner Bros.'s war film *Air Force* (1943), and lead roles in low-budget films such as *Two Man Submarine* (1944), where he costarred with Ann Savage, his nemesis in *Detour*. Neal's career received a (brief) boost after starring in RKO's low-budget exploitation film *First Yank into Tokyo* (1945), where he played Major Steve Ross, whose face is transformed by plastic surgery so that he can move freely in Japan during World War II. However, after principal photography was finished, the studio shot extra material so as to change the reason for Ross's motivation to enter Japan. In the original screenplay he seeks an American scientist who has invented a new gun. This was changed to the A-bomb, making *First Yank into Tokyo* the first Hollywood feature film to use the atomic bomb as part of its plot. The film was a financial success as it was released one month after atom bombs were dropped on Hiroshima and Nagasaki.

Neal's first starring role in a film noir was as crime reporter Jim Riley in *Crime, Inc.* (1945), who befriends a crime boss fighting the growing power of a crime syndicate composed of businessmen and criminals. When he refuses to reveal his sources, Riley is arrested and threatened with jail by the leader of the syndicate, who poses as a respectable businessman and controls the jury investigating Riley. *Crime, Inc.* enjoyed a larger budget than most films produced at PRC, and Neal followed it with *Detour*, another film produced at PRC on a miniscule budget. Nevertheless, it gave Neal his most famous role as the hapless Al Roberts who, through a combination of personal weakness and fate, falls under the control of Ann Savage in *Detour*. This ends in her death and his arrest. Unfortunately, the film's virtues were not recognized at the time, and Neal did not progress beyond low-budget films for the next few years. He followed *Detour* with *Blonde Alibi* (1946), another low-budget noir in which he played Rick Lavery, a pilot accused of murdering his ex-girlfriend's rich lover. By 1949 he was no further advanced than he was in 1941 as his most significant role that year was the lead in Columbia's 15-episode serial *Bruce Gentry—Daredevil of the Skies*.

The following year, he was reduced to an appearance in a low-budget Lash La Rue western, *King of the Bullwhip* (1950). However, worse was to come, and his off-screen behavior effectively ended his career. In 1951, Neal, who had a reputation for brawling, was involved in a scandal when his girlfriend, actress Barbara Payton, decided to leave him and marry Franchot Tone. Neal reacted violently and smashed Tone's cheekbone, broke his nose, and inflicted concussion. Fifty-three days after the marriage, Payton decided to return to Neal. She started divorce proceedings against Tone, and Neal was in the headlines again as he was named in the suit. Except for a few appearances in television series, such as the *Adventures of Wild Bill Hickok*, he was finished in the entertainment industry, and Neal left Hollywood and moved to Palm Springs, where he married Patricia Fenton and started a landscaping business. Neal tried a comeback in the late 1950s but was unable to find work, except appearances in the television series *Tales of Wells Fargo* and *Mike Hammer*. Fenton died of cancer a year after giving birth to their son in 1958, and in 1960, Neal married Gale Kloke. However, in 1965, Neal killed her by putting a .45-caliber bullet through the back of her head, and while the prosecution sought the death penalty, the jury returned a verdict of involuntary manslaughter. Neal was sentenced to 10 years in jail, and on December 7, 1971, he was released on parole after serving six years. Tom Neal, aged 58 years of age, died less than nine months later after suffering heart failure.

Selected Noir Films: *Sky Murder* (1940), *Under Age* (1941), *The Racket Man* (1944), *Crime, Inc.* (1945), *Detour* (1945), *Blonde Alibi* (1945), *The Hat Box Mystery* (1947), *The Case of the Baby Sitter* (1947), *Fingerprints Don't Lie* (1951), *Danger Zone* (1951).

Geoff Mayer

NIGHT AND THE CITY (Twentieth Century Fox, 1950). *Director:* Jules Dassin. *Producer:* Samuel G. Engel. *Script:* Jo Eisinger, based on the novel by Gerald Kersh. *Cinematography:* Max Greene. *Music:* Franz Waxman. *Cast:* Richard Widmark (Harry Fabian), Gene Tierney (Mary Bristol), Googie Withers (Helen Nosseross), Hugh Marlowe (Adam Dunne), Francis L. Sullivan (Phillip Nosseross), Herbert Lom (Kristo), Stanislaus Zbyszko (Gregorious), Mike Mazurki (the Strangler), Edward Chapman (Hoskins), Maureen Delaney (Anna O'Leary), James Hayter (Figler).

Night and the City (1950). Directed by Jules Dassin. Shown: Richard Widmark (as Harry Fabian). Twentieth Century Fox Film Corporation/Photofest.

This key noir film was filmed in London in 1949. Darryl F. Zanuck, the head of production at Twentieth Century Fox, sent director Jules Dassin to Britain as he was about to be expelled from the studio following orders from New York because of his left-wing political sympathies. Zanuck told Dassin to start filming Jo Eisinger's script for *Night and the City* as soon as he could, and he also told Dassin to film the most expensive scenes first so that it would be costly for the studio to remove him from the film. Zanuck also asked Dassin if he could develop a role for one of the studio's most important female stars, Gene Tierney, as he wanted to get her away from Hollywood following a failed romance.

Dassin did not have time to read Gerald Kersh's book, published in 1938, and his interest in the project was both formal and ideological. He wanted to present London as an urban nightmare with night-for-night shooting at a time when it was still difficult to generate sufficient light for extended night scenes, especially those filmed in long shot. Dassin, however, received the cooperation of many London businesses, who agreed to leave their lights on at night so as to assist the filming. As a result, *Night and the City* is one of the strongest examples of film noir expressionism, and it presents London as an urban hell—a world of dark shadows, desperate individuals, and derelict buildings. Tourist landmarks such as Trafalgar Square and Piccadilly Circus, along with other parts of the city, were transformed into a consistent vision of urban hell, a perfect encapsulation of a dark, threatening world permeated by betrayal, fall guys, and moral corruption.

Dassin was also attracted to the film's overarching theme based on the destructive effect of money and ambition, and *Night and the City* is one of the toughest, bleakest films ever produced by a major Hollywood studio. The film's opening sequence was developed by Zanuck, who jettisoned the more conventional, and softer, opening scenes in Eisinger's script. Zanuck wanted to emphasize Fabian's vulnerability from the start. The film begins with Harry Fabian, a cheap American-born scam artist, running through the desolate streets of London, and the film ends in the same way, with Fabian running for his life through the same wasteland until he is executed by his nemesis, the Strangler, with his body dumped into the Thames at Hammersmith. In between these events, the film traces the downward spiral of Fabian as he tries to live down failed investments and "be somebody." In the past Fabian's activities have caused suffering to his girlfriend, Mary Bristol. Now he is doomed. He overreaches himself when he tries to compete with men such as Kristo when, striving to lift himself out of the world of small-time crime, he manipulates himself into the position of wrestling promoter when Kristo's father, Gregorius, and his wrestling protégé Nikolas become disenchanted by Kristo's demeaning exploitation of the wrestling business. Fabian exploits this rift by promising Gregorius that he will promote classical Greco-Roman wrestling, but short of funds, Fabian gets caught between Helen Nosseross's desire to leave her husband and start up her own nightclub and Phil's jealousy and sexual frustration. Fabian accepts money from both parties, and this eventually leads to his downfall when, in financial desperation, he tries to provoke Gregorius into fighting the Strangler. Fabian loses control of the situation, and when Gregorius dies after subduing the Strangler, Kristo sets the London underworld onto Fabian with the promise of a bounty for his head.

This sets up the film's magnificent final act as Harry seeks refuge among the denizens of London's underworld only to discover that, except for his surrogate mother, Anna, and Mary, nobody will help him. His unsentimental death lacks any sense of glamour. Fabian, as Dassin constantly reminds us with his mise-en-scène, is doomed from the start. He is a tragic figure who, as one character tells him, is "an artist without art," who overreaches himself. Fabian's grasp of an unstable world is shown to be untenable right from the start, and at the film's conclusion he runs through the nightmarish streets lamenting that he "was so close to being on top." The film concludes with his death as his body is dumped into the Thames.

At times, the doomed protagonists of film noir assume some of the dramatic characteristics of tragedy, particularly when they overstretch themselves. Richard Widmark's Harry Fabian at times assumes this tragic persona. At other times he approximates his giggling psychopath persona from his trademark performance as Tommy Uddo in his debut film *Kiss of Death* (1947). Overall, he is a slacker, a con man sent out to The American Bar to persuade gullible American tourists to follow him back to Nosseross's Silver Fox club, where "hostesses," trained and drilled by Helen Nosseross, can fleece their victims. He dies when he tries to move out of this limited sphere. In his attempt to "be somebody" and raise money to promote a legitimate wrestling match, Fabian takes the audience on a tour of London's underbelly as he visits, first, the Fiddler, who runs a scam involving beggars with fake disabilities (the Fiddler, who eventually betrays Harry near the end of the film, offers to set Harry up with his own operation involving "a few good beggars"), then Googin, who forges birth certificates, passports, and medical licenses, and finally, Anna O'Leary, who deals in stolen nylons and cigarettes. This is a world devoid of so-called normal people.

Night and the City was a startling production from a major Hollywood studio due largely to its almost total lack of sentimentality. American director Jules Dassin and actors Richard Widmark, Gene Tierney, and Hugh Malowe joined talented British actors such as Francis L. Sullivan as the love-stricken Phillip Nosseross and Googie Withers as his venal wife, Helen, and German cinematographer Max Greene, who gave Dassin the depth of field and unusual compositions he wanted. Greene and Dassin filmed many scenes just prior to sunrise so as to accentuate the film's sense of fatalism.

When Dassin returned to the United States for postproduction work on *Night and the City*, he was, due to the fact that his left-wing past had become public, prevented from entering the studio and had to convey his ideas with regard to the film's postproduction to editors Nick De Maggio and Sidney Stone and composer Franz Waxman by phone as they were too frightened to meet him in person due to the possibility that any direct association with Dassin might have damaged their careers. The film received mostly negative reviews in the United States and Britain, possibly affected by the political climate, and performed poorly at the box office. Dassin did not direct another film, until the wonderful *Rififi*, for five years. *Night and the City* was remade in 1992 with Robert DeNiro as the doomed protagonist, but the change of setting to New York, and a more sentimental perspective, weakened the film, and it is an inferior version.

Geoff Mayer

THE NIGHT HAS EYES (Associated British Picture Corporation, 1942). *Director:* Leslie Arliss. *Producer:* John Argyle. *Script:* Leslie Arliss, based on the novel by Alan Kennington. *Cinematography:* Gunther Krampf. *Music:* Charles Williams. *Cast:* James Mason (Stephen Deremid), Wilfred Lawson (Jim Sturrock), Mary Clare (Mrs. Ranger), Joyce Howard (Marian Ives), Tucker McGuire (Doris), John Fernald (Barry Randall).

The damaging social and or psychological effects of war were utilized as a plot element in a number of films such as *Mine Own Executioner* (1947) and *They Made Me a Fugitive* (1947). *The Night Has Eyes*, in this respect, is a rarity as the film was produced at a time (1942) when most British and American films were focused not on the destructive physical and emotional effects of war, but on the need for communal solidarity to defeat the Axis powers. James Mason's Stephen Deremid in *The Night Has Eyes* is a tormented, broken man following his return from fighting for the Republican side in the Spanish Civil War. With his career as a composer and pianist in tatters, Deremid is bitter and estranged from normal society. He is nursed back to health by his housekeeper, Mrs. Ranger, although he still fears that he has not conquered his murderous impulses.

He lives an isolated existence on the Yorkshire moors, and he is not happy when two schoolteachers on holiday, Marian Ives and her American friend Doris, seek the safety of his house during a storm. Marian, however, has her own emotional problems due to the loss of a close female friend who disappeared in the area 12 months earlier. Marian, sexually repressed and introverted, is immediately attracted to Deremid, especially after he explains to her what a "queer fascination cruelty has." He begins dressing her in his grandmother's clothes, physically carrying her around, and, on occasion, humiliating her with vicious taunts: "What've you got? No beauty. No brains. Just a lot of half-digested ideas about life picked up in a teachers' common room." This does not, however, deter Marian.

This bizarre premise is let down by the film's perfunctory climax. Deremid's aberrant mental condition, aggravated by the regular appearance of dead animals, is shown to be caused by the actions of his greedy housekeeper, Mrs. Ranger, and the handyman, Jim Sturrock, who are determined to keep Deremid dependent on them. They killed Marian's friend when she threatened to upset their scheme by bringing in a medical specialist. Deremid, after saving Marian from Ranger and Sturrock, drives the crooked pair to their deaths on the moors.

The Night Has Eyes was released twice in the United States under two different titles. In 1943 PRC released the film as *Terror House*. In 1949, after James Mason had left England and was working successfully in Hollywood, Cosmopolitan Pictures released the film as *Moonlight Madness*.

Geoff Mayer

NIGHT MOVES (Warner Bros., 1975). *Director:* Arthur Penn. *Producer:* Robert M. Sherman. *Script:* Alan Sharp. *Cinematography:* Bruce Surtees. *Music:* Michael Small. *Cast:* Gene Hackman (Harry Moseby), Jennifer Warren (Paula), Edward Binns (Joey Ziegler), James Woods (Quentin), Melanie Griffith (Delly Grastner), Susan Clark (Ellen), Janet Ward (Arlene Iverson).

Much has been made of the fact that the neo-noir genre emerged in American film in the early 1970s when a public mood of pessimism was being identified by many social commentators. Much of this mood was ascribed to a kind of psychological hangover from the heady excesses of the 1960s, to ambivalence about the U.S. military failure in Vietnam, and to a feeling of political disenchantment associated with the damaged Nixon presidency after the Watergate scandal. Comparisons have been made to the disillusionment that accompanied the end of World War II and that saw the establishment of the classical film noir cycle. Characteristic neo-noir movies released in the early 1970s, full of references to paranoia and futility, included *The Parallax View* (1974), *Chinatown* (1974), *Three Days of the Condor* (1975), and Arthur Penn's private eye drama *Night Moves*. Penn had tapped into the violent American zeitgeist with *Bonnie and Clyde* in 1967, and he has spoken about the way the tone of *Night Moves* reflected a national malaise. He told *Sight and Sound*'s Tag Gallagher in 1975, when asked about the film's pessimism, "I feel that way about the country . . . I really think we're bankrupt, and that the Watergate experience was just the coup de grâce." At that time the ideological content of many Hollywood genres was also undergoing reexamination, prompted partly by changes in the film industry itself. It is not surprising then, given such social and industrial ferment, that in 1975 Penn should produce a private eye film noir that is as morally murky as *Night Moves*.

Harry Moseby (intelligently played by Gene Hackman with antiheroic traits similar to those he employed in 1974's *The Conversation*), is a Los Angeles private detective feeling alienated from his regular case load and dealing with the shock of his wife, Ellen's, infidelity. In the film's opening sequence he is given the standard sort of commission that noir private eyes have been offered ever since Sam Spade and Philip Marlowe: to find a missing teenage daughter. An aging minor actress named Arlene says that her 16-year-old daughter Delly has run away, and Harry quickly discovers from Delly's friend Quentin that she had been involved in an affair with a movie stuntman on location in New Mexico. Thus the plot of *Night Moves* touches on the edges of Hollywood itself. It appears that Delly may be attempting to punish her mother by having sex with Arlene's ex-lovers, including Delly's own stepfather, Tom Iverson. As Tom lives in the Florida Keys, Harry heads off there, and much of the film's action takes place in and around those subtropical isles. Delly is indeed living there, and the task of restoring her to her mother seems straightforward, but the plot thickens when Harry becomes entangled with Tom's girlfriend Paula and when a body is spotted in a sunken light aircraft during a nighttime boating excursion.

Soon after Delly returns to L.A., and after he has attempted a reconciliation with his wife, Harry is shocked to learn that Delly has been killed in a car crash resulting from a movie set stunt gone wrong. He suspects her mother may be involved since Delly was due to inherit a large sum of money from her dead father when she turned 18. Further information from Quentin that the body in the sunken aircraft was Delly's stuntman lover leads Harry to fly to Florida once again to unravel the mystery of her death. The final quarter of *Night Moves* is crowded with confusion, twists and turns, and abrupt revelations about hidden motivations. It transpires that the stuntmen, in association with Tom, have been smuggling Indo-American artifacts out of the Yucatan into Florida and that Delly has been killed to prevent her exposing the illegal trafficking. Harry fights with Tom and travels out with Paula to the area of the sea where the smuggled objects are hidden. While Paula dives down to the site of the concealed statuary, Harry is attacked in the boat by a seaplane. In a rapid, culminating series of violent acts, Harry is wounded in the thigh by submachine gun fire, Paula is killed by the pontoon of the plane as it touches down on the water, and the aircraft itself crashes and sinks after colliding with a large statue. Harry sees that it is piloted by the stuntman involved in Delly's car crash. The film's final shot shows the boat alone on the empty ocean, circling in a compelling symbol of futility as Harry struggles to reach its controls.

Harry Moseby thus becomes another of those frustrated detectives found in many of the films of the early modernist phase of neo-noir. He is particularly comparable to Jake Gittes in *Chinatown* (1974) in his inability to intervene successfully in the complex events and schemes in which he finds himself embroiled. Harry complains bitterly to Paula near the end that he has actually solved nothing. As well, he had talked earlier in the film of his fondness for chess and used an anecdote about a famous chess maneuver to raise the topic of lost opportunities. His abortive attempts to save Delly, or to catch her killers, seem analogous to those of a stymied chess player. The film's title *Night Moves*, with its pun on the chess piece the knight, hints at this notion of noble intentions going awry and also plays on the metaphor of the detective as knight errant found as far back as the novels of Raymond Chandler. By the mid-1970s, the film seems to claim, such questing figures can only be depicted with irony, and the lasting impression of the investigator in *Night Moves* remains that encapsulated by the image of a tiny craft circling in existential futility.

Brian McDonnell

NIGHTMARE ALLEY (Twentieth Century Fox, 1947). *Director:* Edmund Goulding. *Producer:* George Jessel. *Script:* Jules Furthman, from the novel by William Lindsay Gresham. *Cinematography:* Lee Garmes. *Music:* Cyril Mockridge. *Cast:* Tyrone Power (Stanton "Stan" Carlisle), Joan Blondell (Zeena Krumbein), Colleen Gray (Molly), Helen Walker (Dr. Lilith Ritter), Taylor Holmes (Ezra Grindle), Mike Mazurki (Bruno), Ian Keith (Pete Krumbein), Julia Dean (Addie Peabody), James Flavin (Hoatley), Roy Roberts (McGraw), James Burke (Town Marshall).

Tyrone Power was Twentieth Century Fox's main matinee idol in the late 1930s and early 1940s. He was Jesse James in *Jesse James* (1939) and Zorro in *The Mark of Zorro* (1940) as well as the star of *Blood and Sand* (1941), Fox's remake of the Valentino silent film. Prior to leaving Hollywood for military service, he was a swashbuckler in *The Black Swan* (1942), and he returned briefly from active duty for the submarine melodrama *Crash Dive* (1943). After the war, Power wanted a change from adventure films, and in 1946 he starred in Fox's adaptation of Somerset Maugham's *The Razor's Edge*. He also persuaded the studio to purchase the rights to *Nightmare Alley*, William Lindsay Gresham's powerful exposé of carnival life. However, the head of production at the studio, Darryl Zanuck, was reluctant to risk Power in this role of an amoral con man who seduces two women and tries

Nightmare Alley (1947). Directed by Edmund Goulding. Shown: Joan Blondell (as Zeena Krumbein), Tyrone Power (as Stanton "Stan" Carlisle). 20th Century Fox/ Photofest.

to fleece a wealthy businessman when he forms an alliance with a ruthless psychiatrist, Dr. Lilith Ritter.

The film begins with Stan working in a mind-reading act in a second-rate carnival. Away from the stage, he amuses himself with an affair with Zeena, a middle-aged married woman and star of the act. When Stan learns that Zeena and her alcoholic husband Pete were once headliners in major nightclubs with a mind-reading act, he initiates a scheme to acquire the code from Zeena. When Pete dies after drinking wood alcohol given to him by Stan, Zeena teaches him the code. However, when Stan seduces another carnival woman, Molly, her partner forces him to marry the young woman.

Stan, now Stanton, takes his act to Chicago and, with the assistance of Molly, is a great success in a major nightclub. During one of his performances he meets Dr. Ritter, an unscrupulous psychiatrist, and they devise a plan to swindle a wealthy businessman, Ezra Grindle, by using information given to Ritter in her professional relationship with Grindle. However, their plan disintegrates when Molly, dressed as Grindle's long-lost lover, fails to complete her part of the scam. Stan is forced to flee from the police and sends Molly back to the carnival. Hiding out, and in disgrace, he begins drinking heavily. Finally, in need of alcohol and desperate for money, he accepts a position only available in small, disreputable carnivals—the geek who, in exchange for a bottle of bourbon, bites off the heads of live chickens.

This morality tale is Power's best role. His striking features and easy charm seduces both Zeena into giving him the code and also Molly into running away with him. For most of the film he is the homme fatal, the duplicitous and destructive counterpart of noir's femme fatale. In the latter part of the film, however, he occupies the more traditional role of the noir protagonist who falls victim to his overwhelming ambition.

The film is faithful, except for the last scene between Stan and Molly, to the tone of Gresham's novel. Gresham was fascinated by the dark underbelly of the entertainment industry in general, and carnival life in particular, and he uses his story to question the basis of religious faith and the gullibility of its adherents. Nevertheless, the film places Gresham's story within a moral context, and Stan's scam collapses when Molly, the only virtuous character, realizes that there is a difference between the entertainment of low-level carnival tricks and the fraud that Stan and Ritter try to perpetuate on Grindle.

Two of the film's recurring motifs—alcohol and the geek—open and close the film and, in the process, document Stan's fall from grace. At the beginning he watches the geek run wildly in a drunken frenzy and wonders how a man can be reduced to such a pathetic state. By the end of the film he is the geek, running in a crazed fashion from carnival workers who are trying to calm him down. The geek, the film's metaphor for human weakness, is universal, and when Stan's ambition exceeds accepted moral codes, his "inner geek" appears.

Nightmare Alley emerged from filmmakers not generally associated with film noir. Director Edmund Goulding was responsible for a series of literate, so-called quality films such as *Grand Hotel* (1932), *Dark Victory* (1939), and *The Razor's Edge* (1946). Much of the film's credit can be attributed to cinematographer Lee Garmes's imagery—most notably when Stan confronts Ritter in her office just before they form a partnership to deceive Grindle. Garmes's lighting fractures the frame, creating a distorted world, which is reinforced by Ritter's masculine attire, consisting of a black business suit. Although Stan believes he is in control of the situation, the lighting and costume indicate otherwise. Finally, screenwriter Jules Furthman wrote many of Josef von Sternberg's 1930s films starring Marlene Dietrich such as *Morocco* (1930) and *Shanghai Express* (1932).

Geoff Mayer

99 RIVER STREET (United Artists, 1953). *Director:* Phil Karlson. *Producer:* Edward Small. *Script:* Robert Smith, based on a story by George Zuckerman. *Cinematography:* Franz Planer. *Music:* Emil Newman and Arthur Lange. *Cast:* John Payne (Ernie Driscoll), Evelyn Keyes (Linda James), Brad Dexter (Victor Rawlins), Frank Faylen (Stan Hogan), Peggie Castle (Pauline Driscoll), Jay Adler (Christopher), Jack Lambert (Mickey), Eddy Waller (Pop Dudkee), Glen Langan (Lloyd Morgan),

John Day (Bud), Ian Wolfe (Walde Daggett), Peter Leeds (Nat Finley), William Tannen (Director), Gene Reynolds (Chuck).

Director Phil Karlson was one of the best, if not *the* best, director of low- to medium-budget black-and-white film noirs in the early 1950s. He could take what would be conventional stories in the hands of a less talented director and imbue his films with a pervasive sense that the world was an endlessly violent, arbitrary place, and *99 River Street* is a strong example of his style and worldview. He even presents a story within a story when actress Linda James convinces taxi driver Ernie Driscoll that she has killed a man who tried to molest her. This melodrama within a melodrama not only does not jeopardize the film's main story line, but also serves to complement the film's main premise, which involves murder and betrayal.

A similar deception occurs at the start of the film. It begins with a boxing match between the champ and Ernie Driscoll. Driscoll, who comes within one second of winning the title, eventually loses the fight when a punch opens up his eye. However, after he is defeated, the camera pulls back to reveal Driscoll watching this match on a television set—the fight was three years ago, and now Driscoll is a taxi driver in an unhappy marriage to his wife, Pauline. This deception, however, performs a number of functions in the film: it conveys Pauline's disappointment that Ernie, as a taxi driver and no longer a prizefighter, cannot meet her material aspirations. Second, Driscoll's disappointment at losing the fight functions to spur him on during the film's climax when all appears lost—when Pauline's lover, gangster Victor Rawlins, shoots Driscoll, he draws on the fight to summon sufficient strength to eventually overpower Rawlins.

Near the start of the film, Ernie, in an attempt to rekindle his relationship with Pauline, arrives at her workplace with a box of candy. However, he sees his wife kissing a man (Victor Rawlins) and, in a scene handled effectively by Karlson, arrives with a satisfied smile on his face, and as he moves out of his cab into a close-up, he records Driscoll's reaction to his wife's betrayal. Angry, he allows actress Linda James to take him back to the theater, where she claims that she has killed a casting director. This also proves to be a lie designed to show off James's acting skill for the director of a new play. Doubly humiliated, Driscoll storms off after starting a fight.

Rawlins kills Pauline and frames Driscoll for the murder, and the second half of the film sees Driscoll on the run trying to clear himself. This is pure melodrama with pulp dialogue, pulp characters, and a noir sensibility. Driscoll has only one choice—to fight for his survival—and he has two allies: Linda James, who is remorseful after her attempt to exploit his kindness, and the dispatcher at Driscoll's taxi company, Stan Hogan, who diverts the police away from Driscoll. Driscoll, with their assistance, exposes Rawlins as the killer and finds love with a good woman, Linda James.

When *99 River Street* was released in 1953, it provoked discussion because of its realistic violence. Aside from the boxing match that opens the film, there are two prolonged fight scenes—between Driscoll and a killer, Mickey, in a small apartment, and the climactic fight between Driscoll and Rawlins. Both are, for the early 1950s, savage. Karlson even wanted to film these sequences in slow motion to maximize the violent effect—similar to the way that, 16 years later, Sam Peckinpah used this device to capture the bloodletting in *The Wild Bunch* (1969). However, producer Edward Small would not allow Karlson to use slow motion and told the director that it was only suitable for comedy shorts, not a feature-length drama. Nevertheless, Karlson, with enthusiastic support of lead actor John Payne and veteran cinematographer Franz Planer, transformed Robert Smith's familiar story into a powerful film noir that bypassed the conventional basis of the story line to emphasize the pain of marginalized people such as Ernie Driscoll and Linda James—people who aspire to reach the top but, through circumstances beyond their control, realize that they will never make it and that they have to adjust to this realization.

Geoff Mayer

NO ORCHIDS FOR MISS BLANDISH (Renown Pictures, 1948). *Director:* St. John Leigh Clowes. *Producers:* Oswald Mitchell and A. R. Shipman. *Script:* St. John Leigh Clowes. *Cinematography:* Gerald Gibbs. *Music:* George Melachrino. *Cast:* Jack La Rue (Slim Grisson), Linden Travers (Miss Blandish), Hugh McDermott (Fenner), Walter Crisham (Eddie), Danny Green (Flyn), Lila Molnar (Ma Grisson), Richard Nelson (Riley), Frances Marsden (Anna Borg), Percy Marmont (Mr. Blandish), Leslie Bradley (Bailey), Zoe Gail (Margo), Charles Goldner (Louis).

The hostile critical reception that *No Orchids for Miss Blandish* received in 1948 was only surpassed by the hysterical critical reaction to Michael Powell's *Peeping Tom* 12 years later. The film was little more than a low-budget gangster film that tried to replicate many of the conventions of its Hollywood counterparts. The weaknesses of the film are obvious—the phony American dialogue and accents; the small budget that renders much of the film static by limiting the number of interior sets; the mediocre night club acts and variability of the performances, as some British actors, such as the South African-born Sid James, struggle to duplicate the speech patterns of Hollywood gangsters. Perennial American B actor Jack La Rue, as the love-smitten gangster Slim Grisson, having spent the bulk of his career in such films, flourishes in the role, as does Linden Travers, who reprises her 1942 stage role as the wealthy Miss Blandish, who prefers the masculine pleasures of gangster Slim Grisson to the dull respectability and sexless existence of the Blandish family.

The casual way the film utilizes violence, especially early in the film, when Richard Nelson's "Riley" tries to maim most members of the cast, caused problems for the film. However, it is a gangster film, and most of the violence is contained within its generic context. It is mainly implicit and conveyed by the reaction of the other actors. What may have really upset the British critics was the film's attempts to replicate what was perceived as a lowbrow strand of American popular culture: the hard-boiled novel, which author James Hadley Chase tried to emulate.

Chase—in actuality, Rene Raymond, a librarian and wholesale bookseller—wrote a more lurid story than the film, focusing on the sexual harassment of a heiress who is kidnapped and repeatedly raped by gang members. Eventually, she falls prey to the attentions of Slim Grisson, a sexually disturbed member of the Grisson gang, which is headed by his mother, Ma Grisson. The novel concludes with Blandish's death. She kills herself because she feels degraded by the experience. The 1948 film changes the motivation for Blandish's suicide, a factor that may have inflamed the critics of the film even more. She chooses death in the film because she is separated from her lover. She would rather die than return to her family. Significantly, Miss Blandish, after the initial violation of other gang members, is shown enjoying and soliciting Slim Grisson's attentions, even after he offers her the opportunity to escape.

Geoff Mayer

O

O'BRIEN, EDMOND (1915–1985). O'Brien was a rather ordinary-looking character actor who had a surprising number of leading roles through both the 1940s and 1950s. He was energetic and credible in a wide variety of roles and used his strong, expressive voice well. O'Brien played the lead in a number of the most interesting and significant films of the classical noir cycle. Solidly built, he was not handsome, with his heavy facial features lending themselves to scowling rather than a winning smile. Hence O'Brien had few romantic roles, but specialized in playing intelligent versions of men who show moral ambiguity. Born in New York, he briefly attended Fordham University, then went on to study acting in New York and played a number of roles on Broadway. O'Brien was a member of Orson Welles's Mercury Theater group on radio and stage, then moved out to Hollywood after gaining a contract at RKO. *The Killers,* directed by Robert Siodmak, was his first noir in 1946. In this film O'Brien played the crucial linking narrative role of the insurance investigator who tries to piece together the reasons for the death of the Swede (Burt Lancaster), a man gunned down by hired assassins. O'Brien had a major role in the James Cagney classic *White Heat* (1949) for Warner Bros., where he played an undercover man infiltrating Cagney's gang. From that point on O'Brien freelanced at a number of different studios. He was eminently employable because his screen persona was that of a likeable, Everyman figure. This is clearly evident in the very important noir *D.O.A.* (1950), a film whose relentless narrative momentum he effectively drives by himself. O'Brien plays Frank Bigelow, a man poisoned with a drug that will inevitably kill him but who fights to solve his own murder, despite the pressures of a literal deadline. He was also highly effective as the morally equivocal protagonist of the undervalued film noir *711 Ocean Drive* (1950), in which he evinces a complex set of values interacting with financial and sexual temptations. One of his most intriguing projects was in *The Bigamist* (1953), a marginal noir directed by Ida Lupino, who also directed O'Brien in *The Hitch-Hiker* the same year. *The Bigamist* is a very unusual and interesting film for its time. As its title indicates, it concerns a man (O'Brien) who is bound up in the continuing crime of bigamy, but the film treats his situation with sympathy and a subtle approach seemingly at odds with the Production Code. Around this time, O'Brien played a number of tough-cop roles, as in *Shield for Murder* (1954), where he is an example of the corrupt policeman figure often seen in the 1950s (e.g., *Rogue Cop, Pushover,* and *Private Hell 36,* all from 1954). The minor film noirs *Man in the Dark* and *The 3rd Voice* also show him at his best. O'Brien won the best supporting actor Oscar for *The Barefoot Contessa* in 1954, and despite never becoming a first-rank star, he made memorable character acting appearances through the 1960s in films such as *The Wild Bunch* (1969).

Selected Noir Films: *The Killers* (1946), *The Web* (1947), *Brute Force* (1947, uncredited), *A Double Life* (1947), *An Act of Murder* (1948), *White Heat* (1949), *Backfire* (1950), *D.O.A.* (1950), *711 Ocean Drive* (1950), *Between Midnight and Dawn* (1950), *Two of a Kind* (1951), *The Turning Point* (1952), *Man in the Dark* (1953), *The Hitch-Hiker* (1953), *The Bigamist* (1953), *Shield for Murder* (1954, and codirector), *A Cry in the Night* (1956), *The 3rd Voice* (1960).

Brian McDonnell

OBSESSION (An Independent Sovereign Film, 1949). *Director:* Edward Dmytryk. *Producer:* N. A. Bronsten. *Script:* Alec Coppel, based on his novel *A Man About a Dog. Cinematography:* C. M. Pennington-Richards. *Music:* Nino Rota. *Cast:* Robert Newton (Dr. Clive Riordan), Sally Gray (Storm Riordan), Phil Brown (Bill Kronin), Naunton Wayne (Superintendent Finsbury), James Harcourt (Aitken), Ronald Adam (Clubman), Michael Balfour (American Sailor).

In the late 1940s Edward Dmytryk left Hollywood for London due to pressure from anti-Communist forces within the United States. He was a director familiar with the conventions of film noir, although the term had not yet been used in Hollywood. *Obsession* offered Dmytryk an opportunity to rework this type of film in Britain. The outcome is a fascinating mixture of styles and narrative modes as American film noir mutates into a different mode due to a different cultural context. Yet, despite the film's civilized tone, which emphasizes the superficial display of civility and manners between the three central characters (and a characteristically perceptive, if eccentric, English police detective), *Obsession* preserves one of the most common themes within film noir: the destructive power of sexual repression. Similarly, the characters belong to the hard-boiled world of film noir—the bitter husband (Riordan) who wants to preserve his marriage by killing his latest rival (Kronin), despite the fact that his adulterous wife (Storm) despises him.

The basic dramatic situation in *Obsession* is a familiar one—Harley Street specialist Dr. Clive Riordan resents the succession of lovers attracted to his wife, Storm. Thus Riordan kidnaps Storm's latest lover, American Bill Kronin, and chains him to the wall in a bombed-out, abandoned building. Riordan's plan is to keep Kronin alive so that if things go wrong, he can produce him to the police. In the meantime, Riordan brings Kronin food, drinks (martinis), and the newspaper as he prepares to kill the American in an acid bath—in a characteristic English touch, each time Riordan visits Kronin with supplies, he also carries acid in a hot water bottle.

It is worth noting that the nationality of Storm's lover (American) is not an accident as early scenes involving Riordan at his private club show the elderly men discussing world politics, where they bitterly resent the decline of the British empire and the dependency of Britain on American aid. This aspect is not fully developed as audience sympathy, almost by default, drifts toward the hapless American chained to the wall (symbolic revenge on the part of the British?) as Riordan's superior, calculating manner destroys any semblance of audience support for his marital situation. Similarly, Storm is depicted as shallow and totally self-absorbed. At the end of the film, when Storm visits Kronin in a hospital, she tells him that she plans to recover from her ordeal with a South American holiday. Kronin, who is fortuitously saved by the actions of the Riordan family dog and an alert policeman, indicates to Storm that he prefers the family dog to her company—a satisfying end to a superior film noir.

Geoff Mayer

THE OCTOBER MAN (Two Cities, 1947). *Director:* Roy Baker. *Producer:* Eric Ambler. *Script:* Eric Ambler. *Cinematography:* Erwin Hillier. *Music:* William Alwyn. *Cast:* John Mills (Jim Ackland), Joan Greenwood (Jenny Carden), Edward Chapman (Peachey), Kay Walsh (Molly Newman), Joyce Carey (Mrs. Vinton), Catherine Lacey (Miss Selby), Patrick Holt (Harry Carden), Jack Melford (Wilcox), Felix Aylmer (Dr. Martin), Frederick Piper (Detective Inspector Godby), James Hayter (Garage Man), Juliet Mills (Child).

Jim Ackland, an industrial chemist, suffers both physically and mentally following a bus crash that kills a young child in his care. After a year in the hospital, Ackland leaves his home in Sheffield for a new start in London. He moves into a second-rate establishment, the Brockhurst Common Hotel, on the edge of Clapham Common. Although Ackland still suffers from periodic bouts of depression and thoughts of suicide, his life gradually turns around, particularly after falling in love with Joan, the sister of a fellow worker (Harry Carden). However, when a resident of his boarding house, Molly Newman, is strangled on the common, the police, and a vindictive boarder (Mrs. Vinton), quickly assume that Ackland killed Newman.

The identity of the killer is unclear until late in the film, when another boarder, Peachey, reveals that he killed Newman after she rejected him. Prior to this confession, Ackland suffers regular bouts of self-doubt as no one, except Joan Carden, will believe he did not murder the woman. However, *The October Man* is more than just a crime thriller. It is a story of psychological trauma and regeneration as Ackland fights to control his impulse to commit suicide by jumping from a railway bridge as a train passes below. He visits the bridge three times in the film, and each time, he is tempted to jump. Director Roy Baker utilizes the motif of Ackland's habit of twisting his handkerchief into the symbol of a rabbit as an indication of his torment, and it is used to signify his victory over this impulse in his final visit to the bridge.

The October Man was Baker's first film as director. It was a somewhat troubled project, and producer/writer Ambler lost faith in Baker due to the slow pace of production (the film was scheduled for 12 weeks and took 17). Nevertheless, it is a remarkably assured debut film. Baker uses composition, lighting, and sound to advance the story, and he generates a degree of romantic warmth through editing and composition to show the immediate attraction between Ackland and Joan Carden at a dinner dance. While Joan Carden, the only person who supports Jim throughout the film, is a relatively conventional character, Joan Greenwood imbues the role with her customary sensuality and screen presence.

The October Man represented a significant shift in the British crime film/melodrama as noir attributes found their way into the genre. For example, a plaque on the wall in Ackland's barren room reads "From ghoulies, ghosties and short-legged beasts that go bump in the night, Good Lord deliver us." Unfortunately, *The October Man*, like other postwar British films, reveals a universe where such hopes for personal security are shown to be simplistic and obsolete.

Geoff Mayer

ODDS AGAINST TOMORROW (United Artists, 1959). *Director/Producer:* Robert Wise. *Script:* Abraham Polonsky (as John O. Killens) and Nelson Giddin, from the novel by William P. McGivern. *Cinematography:* Joseph Brun. *Music:* John Lewis. *Cast:* Harry Belafonte (Johnny Ingram), Robert Ryan (Earl Slater), Shelley Winters (Lorry), Ed Begley (Burke), Gloria Grahame (Helen), Will Kuluva (Bacco).

Released in 1959, *Odds Against Tomorrow* is a very late and very accomplished example of the noir genre. Indeed, it is often thought of as the last film of the classical cycle, supplanting in the view of many historians Orson Welles's 1958 *Touch of Evil* in that role. Looking back from the perspective of the early twenty-first century, *Odds Against Tomorrow* is clearly marked by precise historical placing in the transition from the 1950s to the 1960s. The presence of old hands Robert Ryan, Shelley Winters, Gloria Grahame, and Ed Begley links it to earlier films. However, characters also talk of Cape Canaveral and Sputnik. The credit sequence is abstract and, like much of the film, is accompanied by a very 1960s jazz score featuring a vibraphone. There is also considerable evidence of the impact of the Production

Code having weakened from earlier years. Bullets, for instance, are seen to actually hit people and leave bloody wounds. Sex is treated more openly than in previous noirs and is even linked to violence, as in Helen's erotic interest in Earl's killing. Gloria Grahame's torso is seen clad only in a bra, and a minor hood named Coco is flamboyantly homosexual. The film's interest in racial questions too looks forward to the flowering of the civil rights movement in the 1960s. Unsurprisingly, since it was made by HarBel (Harry Belafonte's own independent production company), there is a sense of a complex negro culture rather than just the inclusion of the odd black maid or Pullman attendant. James Naramore, in his book *More Than Night: Film Noir in Its Contexts* (p. 241), claims that script writer Killens was black, but recent evidence indicates that this name was a pseudonym for the blacklisted Abraham Polonsky.

Odds Against Tomorrow is an unsentimental and fatalistic example of the heist (or caper) subgenre of film noir, and this makes it similar to famous earlier versions of that type such as *The Asphalt Jungle* and *The Killing*. All three main characters have a beef against the system. Their lives have varying degrees of isolation: the most extreme is Burke, who lives alone with his dog in a room full of photos from the irretrievable past. He is old, corrupt, and embittered. Each man has goals; they are looking for just "one roll of the dice." Ryan plays Earl, an aging man (Ryan was 50 when the film was made) pressured by time passing him by: "They're not gonna junk me like an old car." He is an Okie (similar to Ryan's early noir character in *Crossfire*), an example of damaged masculinity, as seen in an incident where he strikes a young soldier in a bar or in his picking up of dry-cleaning or his babysitting. Earl's wife earns more than him. He has a criminal record of manslaughter and is a racist: "You didn't tell me the third man was gonna be a nigger." Earl is correct to have misgivings about how he will go with the "colored boy." Johnny is an affluent stylish negro with a sports car and snappy clothes. Johnny is a gambler with debts as well as alimony payments. A musician, he views himself as a "bone player in a four-man cemetery." His wife mixes with whites, whom he calls "ofey." Burke likes Johnny and kills himself at the end to try to save him. The film is virtually allegorical in its thematic intentions, with its three main characters being representative of both clear social placement and common noir situations. The irrelevance of ethnicity is highlighted when at the film's end the burned bodies of Earl and Johnny (incinerated in an explosion of huge gas tanks) are indistinguishable, with an ambulance officer unsure which is which. The social compact of World War II is plainly now over, despite references to it and to wartime commonality across the race divide. The film's view of criminal morality is summed up by the gangster Bacco: "I'm a little bit on the inside, a little bit on the outside" of the law.

The film has a typical heist film narrative structure, being a film of two halves: 60 minutes of the story is set in New York, where the raid is planned, 30 minutes on the robbery itself in the upstate town of Melton. As the team is assembled, their initial reluctance is overcome by the various pressures on their lives. Things go wrong (of course) during the heist, chiefly through the thieves falling out with each other, as has happened in almost every crime story since Chaucer's "The Pardoner's Tale". Director Robert Wise creates a consistent and intriguingly developed visual style through his coldly alienating urban and rural landscapes. There is a motif of murky water featuring images of ominous flotsam and jetsam such as broken dolls, and even Central Park looks inhospitable. The sound track adds to the perilous atmosphere with a shock balloon burst in a phone booth or raucous zoo noises. The location shooting includes some very early use of a zoom lens and of an infrared filter to emphasize the wintry sky. On an overnight reconnoiter of the bank, a hotel window view is used for quick exposition, as in Kurosawa's *Yojimbo*. Symbolism dominates the allegorical epilogue filmed in a wasteland of gas tanks reminiscent of *Touch of Evil*. A post closely resembles a cross, and the words *dead end* on a road sign are portentous. There is almost a pre-echo of James Baldwin's 1963 book *The Fire Next Time* in the cataclysm at the end. Thus, while *Odds Against Tomorrow* neatly closes off the classical noir cycle, it helps usher in the mood of the 1960s both in terms of Hollywood and of American culture in general.

Brian McDonnell

ON DANGEROUS GROUND (RKO, 1952). *Director:* Nicholas Ray. *Producer:* John Houseman. *Script:* A. I. Bezzerides, adapted by A. I. Bezzerides and Nicholas Ray from the novel *Mad with Much Heart* by Gerald Butler. *Cinematography:* George E. Diskant. *Music:* Bernard Herrmann. *Cast:* Ida Lupino (Mary Walden), Robert Ryan (Jim Wilson), Ward Bond (Walter Brent), Charles Kemper (Bill Daly), Anthony Ross (Pete Santos), Ed Begley (Captain Brawley), Ian Wolfe (Sheriff Carrey), Summer Williams (Danny Malden), Gus Schilling (Lucky), Frank Ferguson (Willows), Cleo Moore (Myrna Bowers), Olive Carey (Mrs. Brent).

On Dangerous Ground is a film of two halves, although both are thematically related. The first section of the film focuses on an urban feel and the way it affects one man, New York policeman Jim Wilson. The second half shows Wilson's regeneration when he leaves New York for a case in a rural area upstate. The transformation from the grimy, hothouse streets and apartments to the snow-covered landscape mirrors Wilson's change, and this provides the setting for his redemption. Director Nicholas Ray uses the landscape as not only a physical but also an emotional setting to show how the healing power of nature overcomes the psychic pain that the large city invokes in Wilson's tormented cop.

Although Ray was unhappy with the film, it represents one of his best. The film's first image, a close-up of a black gun on the white linen of a bed, is an effective visual motif that dominates the film and reveals the transformation in the character whereby the purity of the snow eventually subjugates the arbitrary violence of the gun. Three policemen, Jim Wilson, Pete Santos, and "Pop" Day, investigating the killing of a cop, receive information from a prostitute, Myrna Bowers, regarding the whereabouts of her former boyfriend, Bernie Tucker. Tucker, the police

On Dangerous Ground (1952). Directed by Nicholas Ray. Shown from left: Robert Ryan (as Jim Wilson), Cleo Moore (as Myrna Bowers), Anthony Ross (as Pete Santos). RKO Radio Pictures/Photofest.

believe, will be able to help them solve the murder. Wilson sends Santos away and "interrogates" Myrna alone, and the film's depiction of sadomasochism is startling for a studio film in the early 1950s. Wilson's eagerness to exploit Myrna's masochistic tendencies with his own sadism and anger, which, the film suggests, emanate from his disgust with the promiscuous behavior of young women, is very raw.

Earlier, when a girl flirts with him in a bar, he has her removed, and when the topics of marriage or family are raised by his colleagues, he ignores or rejects their comments. So when Myrna reveals that she likes it rough, Wilson accommodates her on the pretext that he is gathering vital information. After showing him the bruises on her arm inflicted by Bernie Tucker, Myrna tells Wilson that "he [Tucker] was real cute" and that Wilson "is cute, too." She then invites more of the same by telling him, "You'll make me talk, you'll squeeze it out of me with those big strong arms . . . won't you?" Wilson, who is standing next to Myrna, replies in a soft voice, "That's right, sister." Wilson the policeman is, at this point in the film, no different from Tucker the hoodlum—both enjoy inflicting pain on Myrna.

As a result of information supplied by Myrna, Wilson corners Bernie in a dingy apartment. With the policeman towering over the gangster, who is lying on a bed, Wilson attacks, after which the hood taunts him with "go on, hit me, hit me." As

Wilson smashes into Tucker's body, the cop cries out, "Why do you make me do it? You know you're gonna talk. I'm gonna make you talk. I always make you talk. Why do you do it? Why? Why?"

Wilson lives alone in a sparsely decorated apartment, and after 11 years in the force, his existence is defined by his obsession to apprehend, and hurt, the criminals. After chasing the "garbage" down an alley and responding with characteristic violence, Pop Daly tells Wilson that he should know "the kind of job it is" and come to terms with it. Wilson, however, asks Daly how he manages to divorce himself from this life when he goes home ("How do you live with yourself?"). Daly tells Wilson that, unlike him, he "lives with other people" and that when he goes home, he does not take "this stuff" with him.

The urban landscape, composed of expressionist lighting, rapid editing, and a prowling camera captures Wilson's torment, and only when he reluctantly goes upstate does he achieve some kind of equilibrium in his life. While the rural destination, Westham, shelters its own perverse killer, Wilson, when he confronts the killer, a young boy, is able to show compassion. This occurs due to his regeneration through a relationship with a blind woman, Mary Walden.

During the last section of the film, the noir elements recede in favor of classical melodrama, as purity and virtue (mainly through the effect Mary has on Wilson) combine to regenerate the tormented cop so that, by the end of the film, he is rational and humane. The price for Mary, however, is the death of Danny, her shy, retarded brother who has killed a young girl. When the boy invites Wilson to hit him, the policeman responds by offering to protect him from the father of the dead girl, Walter Brent. However, after the boy falls to his death, Mary asks Wilson to return to the city. During the trip back Wilson recalls Mary's warning about loneliness as well as Pop Daly's advice about the violence within him. Wilson turns the car around and goes back to Mary. This was not the ending Ray favored. He wanted Wilson to return to the streets of New York, with the suggestion that he may even return to his old ways. This would have been a true noir ending. However, the studio and the two stars favored a more optimistic ending showing Wilson's redemption through his love for Mary.

While this decision ruptures the noir basis of the film, it is a logical ending as it completes Wilson's spiritual regeneration. The film is especially noteworthy for Robert Ryan's performance as the tormented policeman. Following powerful performances in *Crossfire* (1947) and *The Set-Up* (1949), *On Dangerous Ground* confirmed Ryan as one of Hollywood's best, and most underrated, actors, especially in his ability to convey a sense of inner torment and repressed violence. This quality is also reinforced by Bernard Herrmann's superb score, which captures the seething violence of the urban sequences, the pulsating excitement of the chase across the snow, and the lyrical, tender moments between the blind woman and the big-city policeman.

Geoff Mayer

OUT OF THE PAST (RKO, 1947). *Director:* Jacques Tourneur. *Producer:* Warren Duff. *Script:* Geoffrey Homes, from his novel *Build My Gallows High. Cinematography:* Nicholas Musuraca. *Music:* Roy Webb. *Cast:* Robert Mitchum (Jeff Bailey/Jeff Markham), Jane Greer (Kathie Moffett), Kirk Douglas (Whit Sterling), Rhonda Fleming (Meta Carson), Richard Webb (Jim), Steve Brodie (Fisher), Virginia Huston (Ann), Paul Valentine (Joe), Dickie Moore (The Kid), Ken Niles (Eels).

Often described as the quintessential film noir, Jacques Tourneur's 1947 classic *Out of the Past* is certainly one of the finest, most accomplished, and most moving films in the genre, representing the combined talents of many individuals working at their peak with the support and encouragement of the RKO production team system. Tourneur was a master of atmosphere and tone, Nicholas Musuraca created a luminous black-and-white aesthetic both in his location and studio-bound cinematography, and the three leads (Robert Mitchum, Jane Greer, and Kirk Douglas) forge characterizations that have become the standard for particular genre types: the fated man, the duplicitous femme fatale, and the rich and ruthless criminal boss. In addition, *Out of the Past* examines what Alain Silver and Elizabeth Ward call, in their book *Film Noir: An Encyclopedic Reference to the American*

Out of the Past (1947). Directed by Jacques Tourneur. Shown: Robert Mitchum (as Jeff Bailey), Jane Greer (as Kathie Moffett). RKO/Photofest.

Style, a subject central to the noir genre: the "destruction of a basically good man by a corrupt woman he loves" (p. 218).

Mitchum plays this "basically good man" as someone seeking a moral life but unable to escape his fate. As Jeff Bailey, garage owner, he dwells in the present in a small Californian mountain town, living quietly and wooing a wholesome local lass named Ann. But, as the title warns us, into this apparent idyll comes a figure from his past to upset the equilibrium. Joe Stephanos is a messenger from a wealthy gambler, Whit Sterling, who wants Jeff to do one more job for him. Jeff explains the background to this surprise development to Ann as they drive to Whit's compound on Lake Tahoe. Jeff's story, narrated as a 30-minute flashback, tells how he had once been a detective in New York and had been hired by Whit to find errant girlfriend Kathie Moffett, who had shot Whit and decamped with $40,000. Jeff tracks her down in Acapulco and, perhaps predictably, falls in love with her. They evade Whit and return from Mexico to the United States, where they hide out in San Francisco. All is well until Jeff's business partner Fisher finds them and is shot dead by Kathie, who then drives off and out of Jeff's life.

In the story's present day, Ann leaves Jeff at Tahoe, where he finds that Kathie has reunited with Whit and that Whit wants him to retrieve some incriminating tax documents from his accountant Eels in San Francisco. Jeff's suspicions that he is being framed are confirmed when the accountant turns up dead, and Jeff is determined not to become a fall guy. Events become very complicated as double-crossings multiply and several characters are killed. Jeff has to juggle his genuine love for Ann with his increasing mistrust of Kathie's motives. Back in the mountains, he pretends to want to escape justice with Kathie, who has murdered Whit, but both figures perish in a police roadblock arranged by a pessimistic, even fatalistic, Jeff. Ann is left free to go on with her life in the belief that Jeff was planning to abandon her for Kathie.

The chief theme that emerges from this convoluted story line is a mainstay of film noir: the presence of a fate that you cannot escape, a fate embodied by a figure from your past. Jeff Markham is such a fated figure but one marked by a streak of passivity. At one stage he responds to a question of Kathie's with the remark, "Baby, I just don't care." A sense of how chance plays a part in his destiny can be seen in the fact that he finds the Acapulco telegraph office closed when he goes to wire Whit that he has found Kathie. In effect, he shrugs his shoulders, and his life changes course. It would, though, be an oversimplification to consider him merely passive. At times he can take charge in an assertive way, as when he breaks free of the frame Whit and others attempt to bind him in over the murder of Eels. In this case he effortlessly turns the tables on the double-crossers. He is also able to manipulate Kathie at the end so that she is trapped by the authorities. But he more or less accepts that in doing this, he will lose his own life and his chance of happiness with Ann. His philosophy could be summed up in his rueful claim that "if I have to [die], I'm going to die last."

responsibility on those who select and use such cases, to make sure that what they show is in fact representative of the larger set of phenomena from which they are drawn. How one can go about establishing representativeness is discussed later in this chapter, after we consider the broader question of how viewers understand class-room video cases.

MAKING SENSE OF VIDEO CASES

The fact that video cases are compelling might seem to imply that they are iconic, with a clear meaning accessible to all. Such is manifestly not the case. Roschelle (2000, p. 723) vividly describes a problem that many readers of this chapter will have experienced from one or both perspectives; one he reports having witnessed "too many times":

> A researcher attends a prestigious conference armed with a project video to show. After brief introductory remarks, the researcher says, "I am going to let the data speak for themselves." But contrary to his or her expectation, the audience sees events in the video that did not appear in the researcher's analysis. Soon the session is spinning out of control, with the researcher unable to inject his or her point of view into what is becoming a charged and confrontational atmosphere.

Because researchers who present video cases have selected them from a larger pool of materials and have watched them repeatedly, the meaning of those cases is clear to them in a way unlikely to extend to new viewers. In addition to this familiarity gap, there are many other factors that could potentially affect what viewers attend to in watching a video case, including cultural background, expertise, and educational philosophy. In this section, we will describe recent research looking at who notices what when watching classroom video, as well as work on the time course of impression formation and implications for video cases as an educational tool.

Who Notices What?

We recently completed a study comparing what elementary school teachers and college students in China and the United States noticed when they watched a series of short classroom video segments (Miller, Zhou, Sims, Perry, & Fang, 2007). After viewers watched each segment, they were asked to write a description and evaluation of what they found noteworthy in the video case they had just seen. We developed a coding system to categorize the resulting narrative descriptions. These codes describe aspects of both the teacher and the lesson. The categories are teacher personality, interpersonal/affective, lesson presentation, student participation, motivation, physical classroom environment, classroom management, lesson content, lesson tools, lesson structure, student understanding, teacher questions, teacher knowledge, and general description. We found striking differences between viewers of the two cultures, as well as some smaller differences between teachers and students within cultures.

Figure 20.1 shows the set of codes for comments that U.S. viewers were more likely to make than were Chinese viewers. U.S. viewers were significantly more likely to comment on teacher personality (e.g., "the teacher was energetic and warm" or "she was not the nicest person to her students") than were Chinese viewers. U.S. viewers were also more likely to comment on aspects of teaching involving interpersonal relations.

Social psychologists starting with Ross (1977) have used the term "fundamental attribution error" to describe the tendency to overestimate the role of personal or dispositional factors in accounting for behavior (compared with situational factors). Westerners are more likely to emphasize personal attributes as the cause of behavior than are their East Asian counterparts (Morris & Peng, 1994). Thus, the finding that our U.S. viewers were particularly prone to comment on personal dispositions of teachers was not surprising, but it may have important consequences for efforts to use video cases in teacher education. To the extent that viewers a) focus on such personal attributes, and b) view them as stable traits, they may be less likely to notice aspects of the instructional approach that could be applied to improving instruction.

U.S. viewers, and particularly U.S. teachers, also made significantly more comments about general pedagogical issues, such as classroom management, interpersonal relations with students, presentation style, participation, classroom structure,

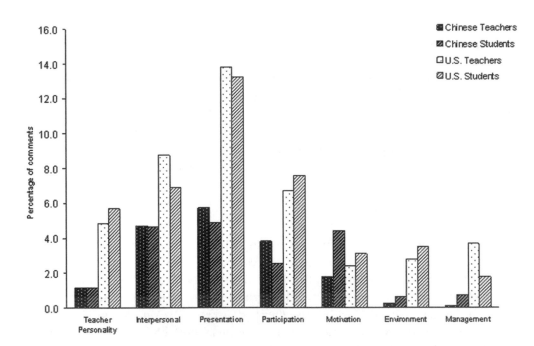

Figure 20.1.   Percentage of comments from free description of classroom viewing that were coded into various categories. U.S. viewers were more likely than Chinese viewers to comment on aspects of the teacher's personality, interpersonal relations with students, presentation style, student participation, aspects of the classroom environment (such as organization of desks), and classroom management (adapted with permission from Miller et al., 2007).

and motivational strategies (e.g., "she was very effective, had control of the class," or "this game is a little boring and children will not long show interest toward it."). These are clearly important aspects of classroom interaction; the differences between U.S. and Chinese viewers watching the same classroom video suggests that the former were more prepared to notice and comment on them.

Figure 20.2 shows the set of codes for comments that Chinese viewers were more likely to make than were U.S. viewers. Chinese viewers commented more on the mathematical content of the classes (e.g., "the content of the knowledge learned in this class is little, and the level is low") and on the kind of knowledge that Shulman (1987) termed "pedagogical content knowledge." Pedagogical content knowledge includes information about the kinds of difficulties students might have and strategies for helping them overcome these obstacles. It differs from general teaching strategies primarily by being domain specific and focusing on student understanding. Examples of pedagogical content knowledge mentioned more by Chinese viewers are the use of lesson tools to facilitate learning (e.g., "there should have been more examples of equivalent bars" and "this method combined the class content with everyday life") and comments on student understanding (e.g., "they may find this easy to understand, and have a deep impression and solid memory of the knowledge").

Student viewers commented more than teachers on student motivation (e.g., "this game is a little boring and children will not show long interest towards it"). The differences between teachers' and college students' comments seem to reflect a differ-

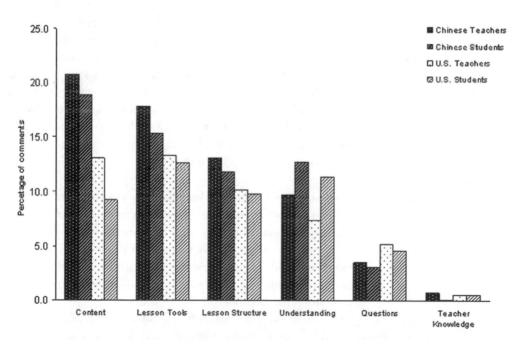

Figure 20.2. Percentage of comments from free description of classroom viewing that were coded into various categories. Chinese viewers were more likely than U.S. viewers to comment on aspects of the lesson content, tools such as manipulative devices that were used, lesson structure, and student understanding (adapted with permission from Miller et al., 2007).

ence in perspective, with students finding it easier to think about what it would be like to be a young student in this classroom, and teachers comparing their methods to what they see the teacher doing.

U.S. and Chinese viewers saw the same set of classroom videos, but in many cases, noticed different aspects of those cases. These differences are consistent with previous reports of differences in the training and beliefs of U.S. and Chinese teachers. Fore example, the difference between Chinese and U.S. viewers (particularly teachers) in their focus on content-specific aspects of teaching versus general pedagogical issues is consistent with differences in their training and the organization of their professional lives (e.g., Ma, 1999). Chinese elementary school teachers are usually specialized in reading and mathematics, and their training typically focuses on content-related instruction from the start of their teacher preparation.

Whatever the source of these differences, their existence serves as a reminder that viewers are likely to see video cases through significantly different lenses depending on the experiences they bring to the task of watching. The next section will look at one key aspect of this larger issue—exploring the time course of impression formation from viewing classroom video and the extent to which task matters in affecting what viewers notice.

The Time Course of Learning From Video

Ambady and Rosenthal (1993) showed untrained judges three 10-sec silent videos of instruction by college instructors and high school teachers. Judges were asked to rate the instructors on a variety of dimensions (active, confident, dominant, enthusiastic, likable, and optimistic, among others). The adjectives listed correlated significantly (with r's > .70) with end-of-semester course evaluations of the college instructors from (different) students, despite the fact that one set of judgments was based on 30 sec exposure and the other on a semester of classroom experience. Similar results were found for videotaped thin slices of instruction by high school teachers, using principal's evaluations as the criterion (although in this case, "attentive" and "empathic" were added to the highly correlated list and "confident" and "dominant" were absent).

The "thin slice" research results are potentially bad news for anyone interested in using classroom video as an instructional technique. They seem to imply that viewers will form quick and persistent impressions of the materials shown that will resist change as a result of further experience. Alternatively, what viewers are asked to judge may affect what they notice and the time course of impression formation. The second possibility would suggest that viewers can be guided to engage with classroom video in a way that goes beyond impressions formed from brief encounters.

Zhou and her colleagues (Zhou, Miller, Sims, & Perry, 2007) used a modification of the thin slice paradigm to look at how instructions affected the time course of impression formation in watching classroom video. College students watched one of two U.S. elementary school math lessons. The video was paused at 10 sec, 30 sec, 1 min, 2 min, 5 min, 10 min, and ended after about 20 min. At each pause, viewers filled out a quick rating scale evaluating aspects of either teacher personality (using eight highly correlated items from Ambady & Rosenthal, 1993) or instructional processes (using seven most commented-on items from Miller et al., 2007). At the conclusion of the entire video, each viewer wrote a narrative description of what they noticed and filled out the rating scale used by the other group.

Viewers' ratings at each time the video stopped were compared to their ratings on the same item after watching the entire video. As with previous work using this technique (Ambday & Rosenthal, 1992, 1993; Ambady, Bernieri, & Richeson, 2000), we found that viewers quickly formed evaluations of the teachers' personality and these evaluations tended to remain stable over time. Figure 20.3 shows correlations between college students' ratings of personality variables at each stop and their ratings on the same scale after watching the whole video. The dashed line shows the critical value for statistical significance. Remarkably, after only 10 sec of viewing, seven of eight ratings on the teacher's personality were already significantly correlated with the final judgment. Ratings of teacher personality features were also quite persistent, showing little change across the viewing episode.

Judgments of instructional processes, on the other hand, were more likely to change over the course of the video case. Figure 20.4 shows the correlation between college students' ratings of instructional variables at each stop and their ratings on the same scale after watching the whole video. The dashed line again shows the critical value for statistical significance. The pattern is quite different from that obtained for

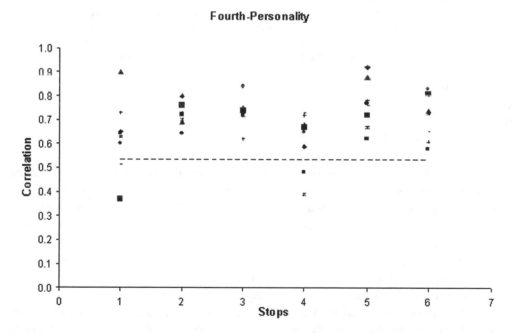

Figure 20.3.   Correlation between ratings on different aspects of teacher personality at each stop (10 sec, 30 sec, 1 min, 2 min, 5 min, and 10 min) with the final rating after 20 min of viewing. Dashed line shows statistical significance. Judgments of personality were formed remarkably quickly and were generally quite stable across the entire viewing experience (adapted with permission from Zhou et al., 2007).

Fourth-Instruction

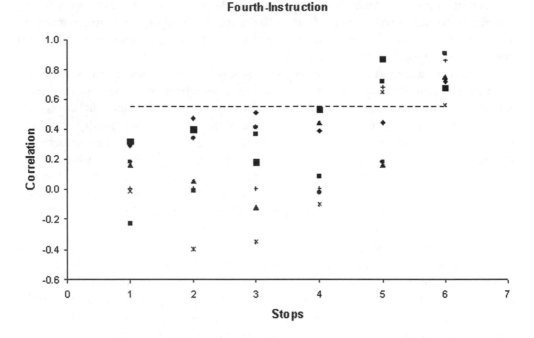

Figure 20.4.   Correlation between ratings on different aspects of classroom instruction at each stop (10 sec, 30 sec, 1 min, 2 min, 5 min, and 10 min) with the final rating after 20 min of viewing. Dashed line shows statistical significance. In contrast with the results obtained for personality judgments, judgments of classroom instruction were more variable, indicating that viewers did not form a quick, stable impression of classroom processes (adapted from Zhou et al., 2007).

personality ratings. No instructional variable was significantly correlated until viewers watched a full 10 min of the lesson.

Viewing task also had significant effects for the kinds of open-ended comments and descriptions viewers wrote after watching the entire case. Viewers who had rated teacher personality were significantly more likely to make comments on the teachers' personal attributes than were those who had focused on instruction. And those who had focused on instruction made more comments on instructional features of the lesson. This study suggests that even very simple changes in the viewing task can have big consequences for what viewers take from a video case.

Our results indicate that Ambady and colleagues are right in identifying an extremely rapid process by which viewers make judgments of teacher personality based on very thin slices of behavior. Recall also that U.S. viewers were more likely than Chinese viewers to comment on these features when they were asked to report what they noticed after watching video cases. This suggests that, left to their own devices, U.S. viewers are likely to take away from viewing video cases the kinds of inferences that could be drawn from a few seconds of viewing.

But there is also a more hopeful message that emerges from these results. Viewers who were explicitly instructed to rate the quality of instruction showed a very different profile of impression formation when watching video cases, and the instructions carried over to affect the general descriptions and comments they wrote after watching the entire video case. Viewers may need to be given a task when watching classroom video if we wish them to go beyond noticing teacher personality, but simple variations in viewing instructions can lead to very different experiences with the same video case.

We are at a relatively early stage in developing a pedagogy of classroom video, in part because of the apparent iconicity of what are in fact deeply complex educational materials. In the remainder of this chapter, we will discuss two issues that affect the educational use of video cases—the importance of establishing their representativeness and some ways that these materials can be Used to develop reflective practitioners.

VIDEO CASES IN EDUCATIONAL PRACTICE: A PEDAGOGICAL PROLEMOMENA

Using educational research or as instructional materials involves "taming the anecdote," that is, collecting cases in a comprehensive and systematic way, presenting them as well-described samples from a larger domain of instructional experience, establishing their representativeness, and presenting them in a manner that enables viewers to comprehend them. The specific requirements for providing context will vary with the audience and intended use. The emerging methodology of video surveys (Stigler & Hiebert, 1999) provides a good context for discussing each of these issues, although they apply as well to other approaches that involve the use of case reports in educational research.

Issues in Collecting Cases

Video methods provide a powerful method for collecting and presenting classroom interactions in their complexity, but not in all of their complexity. Decisions about what to record have important consequences for what viewers could learn from the resulting record. Hall (2000) provides a set of clear examples showing how beliefs about what is important in classroom interaction will constrain the kind of information collected and the kind of inferences that can be drawn. A video record that focuses tightly on the teacher, for example, will limit viewers' ability to understand students' contributions to classroom processes. Multiple camera views can provide additional information about classroom interactions, and techniques for such integrating multiple perspectives are increasingly accessible to researchers (Kumar & Miller, 2005). But the reality is that researchers will be constrained to record a limited sample of the complexity of classroom processes for at least the foreseeable future. Decisions about what to record in a classroom need to be made, and the protocols for recording and selecting cases should be explicitly described.

A good example of such a description is that provided by the TIMSS video study (Stigler, Gonzales, Kawanaka, & Serrano, 1999). The procedure for sampling classrooms from within the larger TIMSS sample is described, as is the procedure for sampling lessons within the school year in the three participating countries, including differences between them (for example, the Japanese sample was collected over a shorter time period than was the case for German and the United States). Because this study was part of the larger TIMSS project, the researchers were able to select classes from a national probability sample in each country, where each class had an equal chance of being selected. The classes were filmed with a single camera, and the videographer was given a series of principles to use in deciding where to point that camera, chief among them the instruction to:

> Assume the perspective of an ideal student in the class, then point the camera toward that which should be the focus of the ideal student at any given time. An ideal student is one who is always attentive to the lesson at hand and always occupied with the learning tasks assigned by the teacher. An ideal student will attend to individual work when assigned to work alone, will attend to the teacher when he or she addresses the class, and will attend to peers when they ask questions or present their work or ideas to the whole class. (Stigler & Hiebert, 1999, p. 35)

This "ideal student perspective" is well suited to the intended purpose of the TIMSS video survey, getting a rich representation of a lesson as delivered by a teacher to an entire class, but it would fail to capture other information that educational researchers might be interested in. For example, it does not provide a good representation of the teacher's viewpoint in the classroom interaction, which may be an important perspective to capture in trying to understand teacher decision making. If one is interested in variation among students' experiences, then a focus on an ideal student is unlikely to capture the range of experiences that different students might have of the same lesson. Finally, to the extent that a classroom involves multiple children and groups of children engaging in different activities, there may not be any single "ideal student" perspective to be found.

Because of the complexity of classroom interactions, any representation of those processes will necessarily provide a limited perspective.

Establishing Representativeness

A second issue arises when researchers select cases for presentation to a larger audience. These cases are nearly always short segments of interaction drawn from larger cases, which in turn are samples from the underlying classroom phenomena. The dangers of selection bias are magnified in cases where rich case material is presented, because whether or not the cases are representative of what the researcher saw, they will be representative of what the viewer sees.

The problem of representativeness was brought home in a project where we have been videotaping mathematics classes in China and the United States. The first pilot video we collected in Beijing was particularly striking. A second-grade teacher taught her students a gestural schema for representing addition and subtraction problems, in which they identified three numbers—the larger number, the smaller number, and the difference between them—all represented by separate gestures. Then they developed addition and subtraction problems, acting out the various operations one could do on these numbers, such as taking the smaller number away from the larger one, or adding the difference between the numbers to the smaller one in order to get the larger one. The approach was engaging, with students actively participating in constructing problems, acting out the solutions, and commenting on each other's ideas.

U.S. viewers found this case interesting and thought provoking but a group of Chinese elementary school teachers had a different reaction. They were very critical, complaining that this was an old-fashioned teaching approach that is rarely used any more. As we collected more cases, we discovered they were right. Thus, presenting it in situations where viewers were unfamiliar with Chinese mathematics teaching would do those viewers a real disservice because they would be likely to assume that what they saw was typical of mathematics teaching approaches used in China. Because of these considerations, we have been forced to stop showing this case in that context. In other contexts, particularly where viewers could see enough examples to get a sense of the range of approaches used in teaching mathematics in China, it would be entirely appropriate to include this case. But the power and vividness of video cases impose a responsibility on those who collect and use them to ensure that what viewers will take as representative in fact is representative.

In order to ensure that cases are representative of underlying phenomena, researchers need to explicitly describe how these cases were selected and to provide evidence supporting the claim that they are representative cases. We will describe two approaches to establishing the representativeness of cases, one very straightforward and the other more formal. Each has advantages, but either should suffice for addressing the concern that cases may provide a misleading picture of the phenomena addressed.

The simplest way of establishing the representativeness of cases is to rely on expert opinion from persons who are closely familiar with the phenomenon being described. This approach was used in the pioneering video study of preschool classrooms in China, Japan, and the United States by Tobin, Wu, and Davidson (1989). Tobin and his colleagues videotaped preschool classrooms in these three countries and produced edited videos depicting a "typical day" in each setting. They then showed them to teachers and principals in each country for confirmation that the events shown were typical and representative of what went on in their country's preschool. Where there were disagreements (for example, viewers in Beijing felt that the Chinese preschool might represent rural preschools but did not represent urban settings), these were described as well.

A similar approach was taken by Stigler and Hiebert (1999) in the TIMSS video study. The teachers who were videotaped were asked to assess the typicality of their videotaped lesson and to identify any atypical aspects of the class.

A second approach to identifying the typicality of cases relies on quantitative coding of the relevant features of cases. Once the cases have been coded in terms of the categories of interest, statistical procedures such as HOMALS (Gifi, 1990) can analyze the similarities among the individual profiles of codes, providing a spatial representation of the typicality of individual cases. This can provide a quantitative way of identifying cases that are typical of a larger sample.

Either approach requires that the purveyors of video cases confront the question of whether or not the materials they are showing are representative of the phenomena of interest. Because viewers are likely to assume implicitly that the cases are representative of the larger universe of educational phenomena, it is critical that the representativeness of those cases be established. Journals have developed standards for describing who research participants are and how they came to be included in a study; similar practices will need to be developed for video cases, so that viewers can understand the degree to which the cases they watch represent a larger reality.

CONCLUSIONS

What makes video cases compelling is their ability, partly real and partly illusory, to communicate to viewers something of the chaos and complexity of classroom interactions. Evidence reviewed in this chapter suggests that such presentations may be far more compelling than are more traditional ways of communicating educational phenomena. The greater persuasiveness of video cases imposes a responsibility on those who would use such cases, however, to evaluate the representativeness of the examples shown and situate them in a larger educational context.

What makes learning from video cases hard is the fact that viewers bring a variety of different kinds of filters to the task of viewing classroom video. Some of these reflect background experiences, aspirations, and the viewers' construal of the task of watching video. Understanding both the perspectives viewers bring and the ways that viewing tasks affect what viewers learn from video cases will be essential to developing effective instructional techniques based on video cases. The fact that one can watch the same interaction repeatedly, in different viewing contexts or with different tasks or questions in mind, is at the heart of the unique power of video representations of classroom processes, but as with other educational experiences, it is the sense that viewers construct from engaging with these materials that determine the value of the experience of working with classroom video.

Some of the problems described in this chapter are likely to diminish as the use of video cases becomes more widespread. The more classroom videos one sees, the more one is able to place new viewing experiences in a larger context. Thus it is very likely that the problem of learning from video is a moving target, in terms of what viewers bring to watching classroom video as well as in educators' understanding of how to present representative and effective video cases.

New technology has made feasible the rich presentation of classroom processes in a way that can capture much of the complexity inherent in education. Yet, because video cases provide an illusion of direct experience of some phenomenon, those of us who attempt to capture classroom processes and use them for instructional purposes take on a burden of responsibility to present them in a way that fairly represents the underlying phenomena. Issues of producing and consuming video cases are ineluctably intertwined. If this technology is to fulfill its promise to provide new windows into classroom processes, the technology will need to be matched by the development of methodological standards to ensure that the picture presented provides a representative depiction of the underlying phenomena and provides contextual support so that viewers will come to a deeper understanding of the phenomena represented.

ACKNOWLEDGMENT

This chapter is based upon work supported by the National Science Foundation under Grant No. 0089293. Address correspondence to: Kevin F. Miller, Combined Program in Education and Psychology, University of Michigan, 601 E. University, Ann Arbor, MI 48109–1259 (email: kevinmil@umich.edu).

REFERENCES

Ambady, N., Bernieri, F. J., & Richeson, J. A. (2000). Toward a histology of social behavior: Judgmental accuracy from thin slices of the behavioral stream. In M. P. Zanna (Ed.), *Advances in experimental social psychology* (Vol. 32, pp. 201–271). San Diego, CA: Academic Press.

Ambady, N., & Rosenthal, R. (1992). Thin slices of expressive behavior as predictors of interpersonal consequences: A meta-analysis. *Psychological Bulletin, 111,* 256–274.

Ambady, N., & Rosenthal, R. (1993). Half a minute: Predicting teacher evaluations from thin slices of nonverbal behavior and physical attractiveness. *Journal of Personality and Social Psychology, 64,* 431–441.

Borgida, E., & Nisbett, R. (1977). The differential impact of abstract vs. concrete information on decisions. *Journal of Applied Social Psychology, 7,* 257–271.

Cacioppo, J. T., Petty, R. E., & Morris, K. J. (1983). Effects of need for cognition on message evaluation, recall, and persuasion. *Journal of Personality and Social Psychology, 45,* 805–818.

Gifi, A. (1990). *Nonlinear multivariate analysis.* Chichester, England: Wiley.

Hall, R. (2000). Videorecording as theory. In A. E. Kelly & R. A. Lesh (Eds.), *Handbook of research design in mathematics and science education* (pp. 647–664). Mahwah, NJ: Lawrence Erlbaum Associates.

Hiebert, J., Gallimore, R., & Stigler, J. W. (2002). A knowledge base for the teaching profession: What would it look like and how can we get one? *Educational Researcher, 31*(5), 3–15.

Hirsch, E. D. (1997, April 10). *Address to California State Board of Education.* Retrieved October 20, 2005 from http://www.coreknowledge.org/CK/about/articles/CAStBrd.htm.

Kumar, S., & Miller, K. F. (2005). Let SMIL be your umbrella: Software tools for transcribing, coding, and presenting digital video in behavioral research. *Behavior Research Methods, Instruments, & Computers, 37,* 359–367.

Ma, L. (1999). *Knowing and teaching elementary mathematics.* Mahwah, NJ: Lawrence Erlbaum Associates.

Miller, K. F., Zhou, X., Perry, M., Sims, L., & Fang, G. (2005). *Do you see what I see? Effects of culture and expertise on attention to classroom video.* Unpublished manuscript, University of Michigan.

Morris, M. W., & Peng, K. (1994). Culture and cause: American and Chinese attributions for social and physical events. *Journal of Personality & Social Psychology, 67*(6), 949–971.

Nisbett, R. E., & Ross, L. D. (1980). *Human inference: Strategies and shortcomings of social judgement.* Englewood Cliffs, NJ: Prentice-Hall.

Petty, R. E., & Cacioppo, J. T. (1984). The effects of involvement on responses to argument quantity and quality: Central and peripheral routes to persuasion. Journal of Personality and Social Psychology, 46, 69–81.

Roschelle, J. (2000). Choosing and using video equipment for data collection. In A. E. Kelly & R. A. Lesh (Eds.), *Handbook of research design in mathematics and science education* (pp. 709–729). Mahwah, NJ: Lawrence Erlbaum Associates.

Ross, L. (1977). The intuitive psychologist and his shortcomings: Distortions in the attribuiton process. In L. Berkowitz (Ed.), *Advances in experimental social psychology* (Vol. 10, pp. 173–220). New York: Academic Press.

Shulman, L. (1987). Knowledge and teaching: Foundations of the new reform. *Harvard Educational Review, 57*, 1–22.

Slater, M. D., & Rouner, D. (1996). Value-affirmative and value-protective processing of alcohol education messages that include statistical evidence or anecdotes. *Communication Research, 23*, 210–235.

Stigler, J. W., Gonzales, P., Kawanaka, T., Knoll, S., & Serrano, A. (1999). *The TIMSS videotape classroom study: Methods and findings from an exploratory research project on eighth-grade mathematics instruction in Germany, Japan, and the United States, NCES 99-074*. Washington, DC: National Center for Education Statistics.

Stigler, J. W., & Hiebert, J. (1999). *The teaching gap*. New York: Free Press.

Tobin, J. J., Wu, D. Y. H., & Davidson, D. (1989). *Preschool in three cultures*. New Haven: Yale.

Zhou, X., Miller, K. F., Sims, L., & Perry, M. (2007). *It does matter how you slice it: Effects of viewing task on attention to classroom video*. Unpublished manuscript, University of Michigan.

It's Not Television Anymore: Designing Digital Video for Learning and Assessment

Daniel L. Schwartz
Stanford University

Kevin Hartman
Stanford University

When used effectively, video is a powerful technology for learning. Researchers can examine videotapes to learn about patterns of classroom interaction. Inservice teachers can review videos of their own teaching to reveal their strengths and weaknesses as instructors. In these instances, the video captures naturally occurring events that often elude the naked eye when seen in person but can become clearer upon review. In this chapter, we consider a different use of video for learning. We describe the use of designed video, where the author of a video decides on its components and features beforehand. For example, take the case of a scripted video of a child incorrectly solving a math problem. A researcher can ask other children to watch the video and comment on the errors they notice. When used this way, the video is designed as an assessment that helps researchers learn what the children know. Designed video can also help students learn. For example, a professor might use the same video clip to help explain common mathematical errors to an entire class. Designed video can support learning in many ways. In this chapter, we provide a simple framework for mapping uses of video into desired and observable learning outcomes. We then show how this framework can be applied when designing video embedded in multimedia environments.

In the not too distant past, videos for learning were often underbudgeted, highly didactic efforts with laughable production values. How could they compete with prime time? Advances in technology, however, have made it so that designed video is no longer in the sole province of broadcast television or dependent on a full-fledged production studio. At a public high school that we frequent, students produce daily newscasts with digital camcorders. What the segments lack in high-end production, they make up in immediacy and prove that effective applications of video are within reach.

As instructors of courses on learning technology, we ask students to produce instructional videos. Within 2 weeks, relative novices produce learning-relevant videos with more visual appeal and information than they could prepare given months of computer programming. Yet, despite the ease of camera use, the array of editing features, and the many video genres, we find it frustrating that the literature provides few resources that can help these students make even more effective use of video for learning. Excepting work on mass media (e.g., Fisherkeller, 2002), there are relatively few empirical evaluations on the use of video for learning, even when compared to computer-aided efforts, as suggested in Table 21.1. There are also few practical publications to help design video for learning.

Our frustration with the available literature has led us to write this chapter on designed video. We offer some suggestions for educational researchers and instructional designers alike. Within the field of the learning sciences, most practitioners fill both roles—they design activities for student learning, and they assess the effects of those activities so they can learn what works. Thus, in the following discussion, we consider applications of designed video for both instruction and assessment. To help organize our suggestions, particularly for beginners in the learning sciences, we present a framework for matching different genres of video with different types of learning. It would be a mistake on our part to delineate the many different genres of video, but then treat learning as a single thing. There are many types of learning, and different applications of video are more or less appropriate for each. To design an effective video, it

TABLE 21.1
Percentage of Journal Abstracts That Indicate Research on Video-Aided or Computer-Aided Instruction.

Journal Title (number of abstracts)	*Video-Aided Learning*	*Computer-Aided Learning*
Cognition and Instruction (n=31)	3.2%	25.8%
Educational Technology Research and Development (n=56)	3.6%	57.1%
Journal of Educational Psychology (n=159)	2.5%	9.4%
Journal of the Learning Sciences (n=31)	9.7%	61.3%
Learning and Instruction (n=48)	6.3%	16.6%

Note: As found in the ten issues prior to October 2005. Each journal has a strong peer-review system and accepts articles on learning interventions, at all ages, and on all topics.

is important to have a clear target, so in our discussion, we describe some important findings about learning and how to promote and measure it.

In the first section, we describe common learning outcomes, give examples of video genres that achieve those outcomes, and suggest methods for determining whether an outcome has been achieved. We do not provide technical details for creating videos. Instead, our goal is to help people consider the relation between video and learning. In particular, we suggest a number of ways to help assess learning (with and without using video), because our experience has been that creating learning assessments is very difficult, until one has seen many, many different examples.

In the second section, we offer some examples of how one might use digital video in a larger, multimedia context. Video does not have to be stand-alone, like a TV program. Video is a more forgiving and powerful learning medium when it is embedded within a larger context of use. Thinking of a larger context is particularly useful for repurposing the raw footage that is frequently collected by researchers. This footage rarely makes a self-contained video story, but when embedded within a multimedia environment, it can be used in many creative ways to encourage learning interactions.

FOUR COMMON LEARNING OUTCOMES

A significant challenge for any designer of learning involves deciding exactly what people are supposed to learn. This decision requires more than choosing a topic to teach. It also requires a consideration of how the subject matter will be used. Take the example of asking an American to learn the sport of cricket. Is the goal that the person will be able to (a) play; (b) explain the history of the game; (c) recognize a good play; or, (d) want to learn more? Clarifying the learning outcome is important, because one would probably use video differently for each outcome.

Figure 21.1 provides a map of different learning outcomes that will guide our discussion. The core shows four broad classes of learning outcomes. The inner ring refines each learning outcome into approaches one might take to achieve the learning outcome. For example, one might promote the outcome of being able to "say" by providing people with facts or explanations. The second ring indicates the types of behaviors people will exhibit if they have learned. We will describe these behaviors in some detail because they can help clarify the meaning of the learning outcomes, plus they provide the keys to successful assessments. Finally, the outer ring samples some relevant video genres for each outcome. We now take up each wedge of the learning pie in turn.

Seeing

A signature quality of video is that it can help people see things they could not see before. There is a continuum of seeing outcomes. On one end is a familiarity approach that introduces people to phenomena they are unlikely to have seen—a strange animal, an industrial process, a foreign land. The familiarity approach counts on people recognizing what is novel. At the other end is a discernment approach that helps people perceive details they might otherwise overlook—the balance point of a painting or the difference between a 5.6 and a 5.8 in a gymnastic routine. The discernment ap-

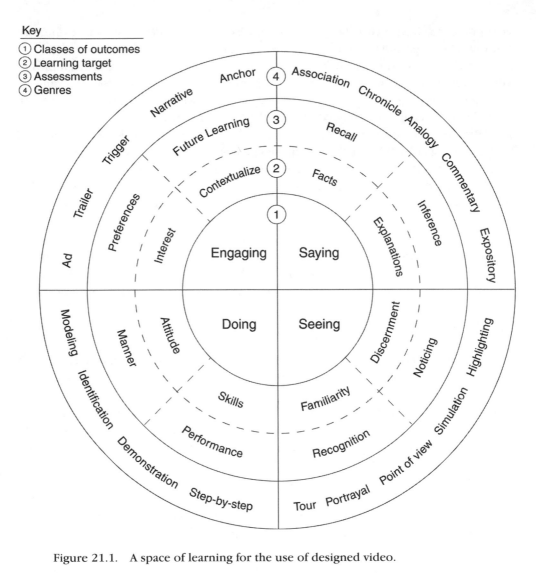

Figure 21.1.   A space of learning for the use of designed video.

proach presupposes that novices may not see what is significant even though it is "in plain sight." To help people develop an "enlightened eye" (Eisner, 1998), it is necessary to educate their perception because people tend to assimilate what is familiar rather than accommodate to new subtleties. Therefore, learning to discern often requires special provisions to help people notice (Sherin, this volume).

A common genre for exposing people to unfamiliar sights might be labeled tour videos. These include travelogues, nature shows, historical re-creations, and the like. They can include commentary to give a name and background to what is shown. There are also portrayals, as in the case of period pieces, where the novelty is woven into the dramatic narrative and setting. Moving toward the discernment side of the continuum, there are point of view videos that use camera angles, audio commentary, or interviews to give insight into new ways of seeing, for example, from the point of view of a character, a coach, or a hunted prey. Extreme points of view can create simulated experiences, as in the cases of video from the helmet of a skydiver or a highly empathetic character portrayal. Each of these approaches helps expose people to unfamiliar sights and perspectives. For the goal of helping people discern subtlety in the familiar, highlighting techniques are appropriate (Goodwin, 1994). Sport broadcasting has a sophisticated toolkit that contains instant replay, slow motion, zooming, and digital pens that commentators employ to highlight patterns of movement. Several chapters in this volume (e.g., Pea) describe digital video environments that help researchers notice and annotate. These environments can be re-purposed to support specific instructional goals, such as noticing aspects of one's performance that are difficult to observe in real time.

Given the range of seeing outcomes and possible video techniques, how can designers know whether their video has been successful? A recognition paradigm is the least demanding way to test for familiarity. One might show pictures or video clips and ask people to select the ones they recognize. People, however, are very good at recognizing images (Shepard, 1967), so this can be made a more sensitive measure by showing things at a different angle or setting. It is not trivial to recognize a plant in the wild after seeing a video of the plant in a laboratory. For videos that target discernment, noticing assessments are appropriate. One form of assessment might take a forced-choice approach where people have to select which of two pictures or videos is exemplary. Preservice teachers, for example, might watch two videos of teachers handling a student question and select which one shows the correct technique. A more open-ended approach provides the learner with a new video and has the learner describe what is important to notice about it. This is a difficult task because what is important is often buried under a hundred other details. In this case, people need to discern what is significant despite the surrounding "noise." In all cases, appropriate assessments of learners' abilities to see will require asking learners to look at something. Thus, video is not just useful for initial learning, it is also useful as an assessment tool (Derry et al., 2005).

Engaging

Engagement may be characterized as the pull that brings people to a situation or topic and keeps them involved. Engagement creates the mental context that prepares people to learn. Video is superior at creating engagement and setting the stage for learning, even though the video itself may not contain the new information people are supposed to learn. Video can help people bring to bear relevant knowledge to raise interest and make sense of subsequent instruction. One approach to engagement is to develop the learners' interest so they are more likely to take steps to learn. For example, a video might show examples of the enjoyment or money people attain once they have learned. A useful distinction distinguishes extrinsic and intrinsic motivations to learn (Lepper & Greene, 1978). Extrinsic motivators or rewards are often irrelevant to the knowledge gained through learning (e.g., receiving money for good grades). Intrinsic motivators, or what we labeled interest, depend on the target content itself being engaging. Piquing people's curiosity or showing the relevance of a topic to their own lives would be canonical examples of raising intrinsic motivation or interest. A similar form of engagement is to contextualize the information in ways that make it meaningful and relevant to the learners. One technique is to provide background information or activate prior knowledge that anchors the meaning of subsequent activities. For example, a video biography of an artist and her era can help students make more sense of a subsequent painting. Without relevant prior knowledge, people can have difficulty making sense of a lesson and often have no recourse but to memorize the content rather than understand it.

A familiar way to raise interest is through advertisements. Car and beer commercials use visual narratives of a desirable lifestyle. These advertisements have large production budgets, and thus, are out of reach for most designers. However, we recall a conversation with a venture capitalist whose hero creates cheesy infomercials to sell appliances, "The man can sell anything!" Evidently, production values need not be a stumbling block to persuasion. Trailers, like a tourist video or movie preview, are similar to advertisements but tend to sample the content of what is to come more realistically. Educational video often uses trailers by painting an initial picture of a domain (e.g., life in the sea) in preparation for a subsequent unit (e.g., on ocean ecosystems). Another method of engagement uses video as a trigger to set the stage for subsequent discussion. For example, television news magazines will often present an initial report to which experts and a call-in audience can react. Similarly, in a classroom, one might present a case study or example that is intended to stir discussion and raise a host of relevant issues and tacit beliefs. Anchor videos can be used to contextualize learning and problem solving. For example, *The Adventures of Jasper Woodbury* comprise 20 min video narratives that contextualize multistep mathematical problem solving (Cognition and Technology Group at Vanderbilt, 1997). The narrative includes a challenge (e.g., students need to plan how to save a wounded eagle in the video) and the information needed to solve the challenge (e.g., the relevant distances to safety and the speeds of different vehicles). The goal is for students to learn to solve complex mathematical problems. The contextualization of the video makes it possible to present a problem with a level of complexity that is hard to match with a word problem. The video makes the complex problem tractable by using a vignette that is easy to grasp and engages students' everyday knowledge.

Assessments of engagement should ideally target how well students are intellectually or motivationally prepared to learn. For interest, one can assess people's preferences toward learning-relevant behaviors. For example, do students exhibit enough interest that they spontaneously choose further investigation and conversation over another option, such as going outside and playing? One might also populate a database or web pages with information and record if students access those pages associated with the video. For contextualization, one might measure future learning from subsequent instruction (Bransford & Schwartz, 1999). For example, one might compare students who did and did not watch a biography of an artist. Afterward, both groups of students receive a subsequent lesson on a relevant painting. Ideally, students who watched the biography would learn more or different things from the lesson on the painting. So, rather than directly measuring what people learn from the video, one can measure what people learn from the lessons following the video, on the assumption the learners can engage the new lessons more fully.

Doing

Video is ideal for presenting human behaviors. There are two quite different subclasses of "doing outcomes"—those involving attitude and those involving skills. For attitudes, people readily learn by modeling other people's behaviors (Bandura, 1986). People can model other people so well that learning can be unintentional, which is one reason that violence on television is of concern. The second outcome is skill acquisition and typically involves intentional effort and practice on the learner's part. The number of skills, or procedural knowledge, an adult possesses is hard to fathom— brushing teeth, riding a bike, taking conversational turns, computing best buys, and so on. As with attitudes, people can learn skills by imitating behaviors shown in a video. Sometimes skills are quite complex so that replaying, zooming, and slowing the motion can be quite helpful. Other times, when it is too much to expect a learner to imitate an expert's fully integrated performance, it makes sense to decompose a task into subskills that are learned separately. Additionally, for some skills, it is important to help people see the critical components of the behavior. For example, novice tennis players may not see the key moves of a professional, in which case, they cannot possibly imitate it. Good procedural instruction makes sure that students can discern the behaviors of significance so they can imitate them. In this case, to achieve the outcome of doing, one also needs to target the outcome of *seeing*.

Video models can help to shape attitudes. The young children's program, *Mr. Rogers*, modeled politeness on the assumption that children would be polite. Another way to invoke attitudinal learning uses identification. Effective drama causes people to emulate a hero, and teens are well known for their identification with media stars. Simple demonstrations are a useful way to help people learn skills. Cooking and home repair shows are familiar examples. There are also training videos for learning sports and mathematical procedures where experts perform the task in a rather slow fashion (for them). When tasks get complex, videos often use step-by-step instruction that breaks a task into manageable chunks. It should be noted that the best skill instruction also includes a narrative that explains why a particular procedure takes the form it does. Without an explanatory overlay, people can learn a skill, but they may not have any flexibility when performing it. For example, if people merely imitated the behavior of pulling the parking-break lever without understanding why, they would be in trouble with a car that uses a floor pedal instead. Examples of children blindly copying mathematical procedures without understanding are legion, as are the consequences.

Ideally, "doing" evaluations require people to perform some action, rather than just say what they would do. For attitudes, assessments evaluate the learner's manner in context. For example, researchers can study if portrayals of violence affect children. Do young children exhibit a more aggressive manner after watching video violence, for example, by walking over the legs of children sitting on the ground instead of walking around them? This evaluation is notable, in part, because it looks for a proximal effect (immediate behavior) rather than the distal effect that people really care about (violent behaviors in later years). Although this is not completely satisfying, it is wise advice to first look for proximal outcomes. For example, after a video on sharing, one might put a learner in a relevant situation and see what happens. This is much easier than seeing if the children use sharing in their everyday habitats or waiting to see if children become adults who share. For skill acquisition, performance assessments are ideal because they directly test the relevant behavior. There are various genres of performance assessment. For routine skills that require efficiency (e.g., driving a car), it makes sense to require full-blown performance (e.g., a driving test) and evaluate the number of errors and time to execute the skills. These types of assessments can be made more revealing by throwing in variations the student has not fully practiced (e.g., parallel parking on a hill). For less mature skills, one might ask people to perform subcomponents of a task. This has the advantage of identifying what skill components require additional attention. Additionally, one can scaffold a performance for a novice by providing hints or supports to make the task easier. For example, with intellectual skills, like doing a science investigation, it is useful to create a set of supportive materials a novice can use, rather than just asking the learner to do an investigation cold. One interesting way to turn scaffolding itself into a dynamic assessment is to evaluate how many hints people need to complete a task (Feuerstein, 1979). For example, how many times does a student ask to review a video when trying to execute an assembly procedure shown on the video? Unfortunately, performance assessments can be impractical. In a sex education class, it might be best to simply ask students to recall the correct steps for safe sex. Of course, this does not mean students will execute the performance when the time comes. Talking the talk is not as good a measure as walking the walk, but sometimes, it is an acceptable proxy that can evaluate whether a video is moving people in the desired direction.

Saying

Whereas the other outcomes tap into the unique strengths of visual media, "saying" outcomes—verbal or "declarative" knowledge—can be achieved by many media. Nevertheless, video is also good in this realm. One type of verbal outcome is the acquisition of facts. Facts include things like the name of an animal, the average temperature of the sun, 2 + 2 = 4, and so on. Facts are often seen as the stuff of memorization, but the difference between a good and bad news report is often whether it includes critical facts to help viewers draw their own inferences. Despite common belief, repetition is not the best way to learn facts. Repeatedly subvocalizing a phone number (e.g., 451–9180) is less effective than making up a formula with the numbers (e.g., $4 + 5 \times 1 = 9 = 1 + 8 + 0$). Bransford, Franks, Vye, and Sherwood (1989) demonstrated that people remember facts better when those facts come as a solution to a problem an individual has attempted rather than as a bald assertion. Explanation is a second class of verbal knowledge. Explanations, like those provided in good science shows, provide the "why" and "how." They tie the facts together. A challenge for any video designer is estimating how far to move from facts to explanations. For example, a weather report might present the day's high temperature; it might explain the winds that cause the temperature; or, it might develop the science behind atmospheric phenomena. The further one moves from facts, the more important it becomes to create videos that make processes and explanations transparent; talking heads only work if viewers already have sufficient prior knowledge to understand what the heads say (Schwartz & Bransford, 1998).

A number of genres emphasize retaining factual knowledge. Quiz shows are a bad example, because they typically do not help viewers remember. A better example is *Sesame Street*, which uses association to pair entertaining images and names to help children memorize the letters and numerals. Chronicles, like news broadcasts and narratives, deliver facts embedded within the context of a larger story. Moving toward explicit explanation, video can use analogy to help people understand; for example, the trade winds are like rivers in an ocean of air. Typically, single analogies (and examples) are not nearly as effective as pairs of analogies in helping people cull the deep explanatory structure (people tend to focus on surface features of a single analogy or example, Gick & Holyoak, 1983). Commentary or interpretation is a powerful way to supplement video with verbal explanations as is often the case with news magazines and sports broadcasts. Expository videos explicitly develop a sustained account of some set of facts as in the case of the science show *Nova* or other documentaries.

To assess factual knowledge, a recall paradigm is effective. This can take the form of free recall where people simply say or write what they remember. It can also take the form of cued recall, as in the case of showing a numeral and asking a child to state the name or say what numeral comes next. A more difficult version of cued recall asks people to identify what is missing, for example, from a classroom lesson plan or video demonstration. Multiple-choice tests, although a staple of educational testing, are problematic because students can get the right answer by guessing, especially if the test has obviously wrong "foils." For explanatory outcomes, the standard approach requires learners to draw inferences; students need to go beyond the information given to show they have not simply memorized the words. Inferences include things like problem solving, applying ideas in a new situation, predicting, taking up a point of view, and constructing an argument. Open-ended formats (e.g., essays, making a video) are more difficult to score than right/wrong problems, but they also provide more latitude for students to exhibit what they have learned. In general, explanatory knowledge is trickier than the other outcomes, because there are so many possible levels of explanation and understanding. More than the other outcomes, it is useful to think of explanation assessments before designing a video because the assessments can help shape what is included in the video.

PUTTING VIDEO IN MULTIMEDIA CONTEXT

The history of visual media comprises the invention of techniques and formal features that guide people's thoughts to particular outcomes. The zoom mimics attention; the slow motion in *Raging Bull* emphasizes the poignancy of critical moments of life. Rembrandt's self-portraits are exceptionally sharp around the eyes to reflect where he is looking. Multimedia, however, is relatively new. There are fewer established techniques, and it is useful to capitalize on extant formal features. At the same time, it means the potential for innovation and exploration is extremely high. Here, we offer a sample of a single multimedia program that uses video in a variety of ways.

The examples come from work with undergraduate, preservice teachers in an educational psychology course. The course used a multimedia authoring shell, called *STAR.Legacy*, that was designed to help teachers use their local resources to create pedagogically sound instruction (Schwartz, Brophy, Lin, & Bransford, 1999). The main interface of the shell, shown in Figure 21.2, includes an explicit inquiry cycle to help teachers and students navigate through distinct phases of learning.

Typically, for any given inquiry topic, we populate the shell with initial videos and resources to seed the lesson. For example, each inquiry cycle begins with a challenge in the form of a trigger video to increase interest, contextualization, and discernment. For this Legacy, the trigger video was a recording of a local news segment on children building model rockets in "a mock mission to mars." The trigger set the stage for the challenge of creating instructional resources to help teachers improve learning from project-based activities (see Barron et al., 1998 for ways to improve project-based

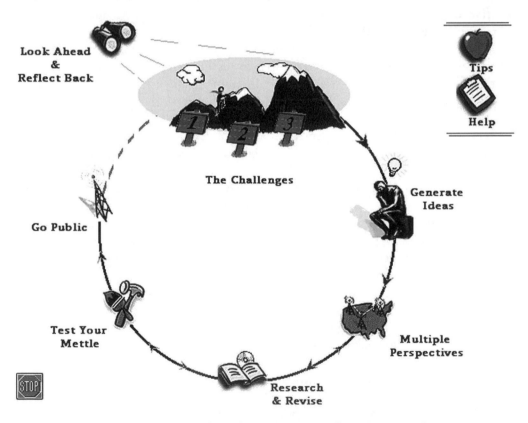

Figure 21.2.   The main interface of a multimedia shell, STAR.Legacy, for using designed video within a larger inquiry cycle. Each icon is a clickable button that leads students (and authors) to sections that support different aspects of complex learning. Each STAR.Legacy is typically populated by different uses of video that are either seeded by the instructional designer or developed by the teacher and students themselves.

learning). The pre-service teachers watched the news segment and then had to generate ideas about what they saw that was relevant to improving project-based learning. They typically noticed, in rather vague fashion, that the kids in the video were doing hands-on learning. After this, the students saw clips showing the multiple perspectives of six experts who had watched the same video. These videos helped students notice things they had missed, modeled an attitude of looking beneath the surface, and provided short explanations. (We had shown the trigger to a variety of experts and videotaped them briefly as they described what they noticed.) One expert observed that the children were collaborating quite well, which meant the teacher had done some activities on collaboration. Another expert observed that the children did not measure how high their rockets went, and therefore, had no way to do science on their designs. The inquiry cycle then led the students to explore further resources (e.g., papers) and to test their mettle (self-assess what they had learned). Ultimately, they had to go public with what they had learned; they had to design a video-based lesson directly into the multimedia shell. Their task was to leave a "legacy" for future cohorts of students who would use the software. They found this engaging, and for us, it was a performance assessment of their ability to "do" multimedia instruction based on what they had learned about learning. In the following paragraphs, we describe some of the innovative ways they used video in a multimedia context. We separate their efforts by the primary class of outcome they targeted.

Seeing

Some of the students had read a paper on classroom practices. These students asked the author for the classroom video footage on which the paper was based. They cut the video into segments that mapped into each of the concepts in the paper and turned them into a "seeing" lesson. Next to each clip on the page was an audio button where the student designers had recorded explanations of the critical concept. Users had to indicate where in the video clip the concept was put in play. These students also included an assessment page; users saw new clips and had to determine which concept was exhibited in each clip. Another group of students used contrasting cases, like wines side-by-side, that can help novices notice key features (see Schwartz & Bransford, 1998). These students had read an article on how different parents appreciate different types of teachers. They directed videos of their friends acting out different teachers. As shown in Figure 21.3, users had to click on the juxtaposed video portrayals, notice the differences, and then match each teacher portrayal with the appropriate parents who were described in text fields to the side.

Engaging

One team of students created their video instruction based on an article about problem- and case-based approaches to education. They interviewed law students, lawyers, deans, and law professors about their experiences with case-based legal education. The students organized the brief video clips to reflect the different points of view of the stakeholders and how their attitudes toward case-based instruction

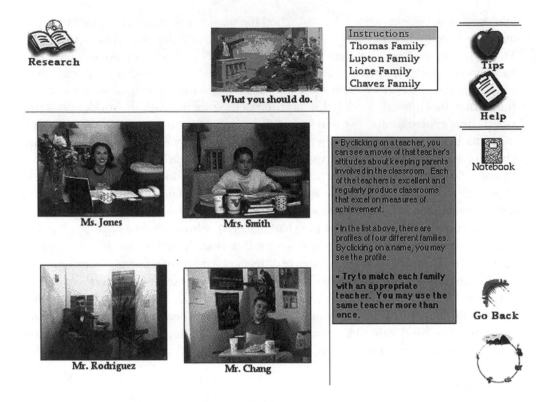

Figure 21.3.   An example of video-based instruction authored into STAR.Legacy by preservice teachers. See text for explanation.

changed with time. They included some simple audio with a button that explained how users should watch the video clips. The students' goal was to help contextualize case-based instruction to prepare learners to engage the associated article themselves.

Doing

Another team of students recreated a method of teaching simple mathematics using manipulatives. Users watched the demonstration video. Users then had to move simple images of manipulatives to set up instruction for a new problem. When done, the users could click to see the correct answer. Another group of students interested in collaborative learning created videos about activities that help to make more cohesive groups. They showed an example of a group of students that followed the activities correctly next to an example of a group that had such divisiveness they could not complete the activities. This latter video showed how the teacher intervened to make the activity successful.

Saying

One team of students indicated that users should first read a target article on project-based learning. They then showed a video clip of some children working on a science project. The video stopped and the students had to predict what would happen next and give a justification for the prediction. They wanted students to draw inferences based on the explanation in the article. Afterward, they showed videos of what different "experts" predicted would happen and their justifications. Yet, another team of students created a talking-head video to explain a visual framework they had invented to guide the design of project-based learning. Students could click on a part of their visual framework to get a short snippet of the explanation. Based on all of the foregoing video implementations, we were impressed by what the students had learned to do with video for learning and assessment and in a relatively short period of time.

CONCLUSION

Our modest goal for this chapter was engaging reader interest in the potential of video within multimedia for instruction and research. We observed that although digital media has put video in the hands of many, there has been surprisingly little work exploring its unique potentials for learning. To help people make some headway in light of the limited literature on video for learning, we tried to help people see that there are distinct learning outcomes, and we compared different techniques of assessment to help readers discern what makes these outcomes different. We also gave some examples of how to design videos that help learners learn and help learners assess if they have learned. Finally, we aimed for some saying outcomes by providing select facts and explanations about why certain forms of presentation affect different types of outcomes. Ultimately, we suspect this would have all been much more effective had we used video in an interactive multimedia context. Nevertheless, we hope that we have provided some impetus for people to consider the value of designing video for learning. The digital era is not just about computer programs, it is also about digital video.

REFERENCES

Bandura, A. (1986). *Social foundations of thought and action: A social cognitive theory*. Upper Saddle River, NJ: Prentice-Hall.

Barron, B. J., Schwartz, D. L., Vye, N. J., Moore, A., Petrosino, A., Zech, L., Bransford, J. D., & CTGV. (1998). Doing with understanding: Lessons from research on problem- and project-based learning. *Journal of the Learning Sciences, 7,* 271–312.

Bransford, J. D., Franks, J. J., Vye, N. J., & Sherwood, R. D. (1989). New approaches to instruction: Because wisdom can't be told. In S. Vosniadou & A. Ortony (Eds.), *Similarity and analogical reasoning* (pp. 470–497). NY: Cambridge University Press.

Bransford, J. D., & Schwartz, D. L. (1999). Rethinking transfer: A simple proposal with multiple implications. *Review of Research in Education , 24,* 61–101.

Cognition and Technology Group at Vanderbilt (CTGV). (1997). *The Jasper project: Lessons in curriculum, instruction, assessment, and professional development*. Mahwah, NJ: Lawrence Erlbaum Associates.

Derry, S. J., Hmelo-Silver, C. E., Feltovich, J., Nagarajan, A., Chernobilsky, E., & Halfpap, B. (2005). Making a mesh of it: A STELLAR approach to teacher professional development. *Proceedings of Computer Support for Collaborative Learning (CSCL) (pp. 105–114), Taipei, Taiwan*. Mahwah, NJ: Lawrence Erlbaum Associates.

Eisner, E. W. (1998). *The enlightened eye: Qualitative inquiry and the enhancement of educational practice*. Upper Saddle River, NJ: Merrill.

Feuerstein, R. (1979). *The dynamic assessment of retarded performers: The learning potential assessment device, theory, instruments, and techniques*. Baltimore, MD: University Park Press.

Fisherkeller, J. (2002). *Growing up with television: Everyday learning among young adolescents*. Philadelphia: Temple University Press.

Gick, M. L., & Holyoak, K. J. (1983). Schema induction and analogical transfer. *Cognitive Psychology, 15*, 1–38.

Goodwin, C. (1994). Professional vision. *American Anthropologist, 96*, 606–633.

Lepper, M. R., & Greene, D. (Eds.). (1978). *The hidden cost of reward: New perspectives on the psychology of human motivation*. Hillsdale, NJ: Lawrence Erlbaum Associates.

Schwartz, D. L., & Bransford, J. D. (1998). A time for telling. *Cognition & Instruction, 16*, 475–522.

Schwartz, D. L., Brophy, S., Lin, X. D., & Bransford, J. D. (1999). Software for managing complex learning: An example from an educational psychology course. *Educational Technology Research and Development, 47*, 39–59.

Seels, B., Fullerton, K., Berry, L., & Horn, L. J. (2004). Research on learning from television. In D. H. Jonassen (Ed.), *Handbook of research on educational communications and technology* (2nd ed., pp. 249–334). Mahwah, NJ: Lawrence Erlbaum Associates.

Shepard, R. N. (1967). Recognition memory for words, sentences, and pictures. *Journal of Verbal Learning and Verbal Behavior, 6*, 156–163.

Teachers' Gestures as a Means of Scaffolding Students' Understanding: Evidence From an Early Algebra Lesson

Martha W. Alibali
Mitchell J. Nathan
University of Wisconsin–Madison

During classroom instruction, teachers often attempt to scaffold students' understanding of lesson content. But how is this scaffolding achieved? One obvious possibility is that teachers adjust the ways in which they communicate information relevant to the lesson. Surprisingly, relatively little is known about how teachers vary their communicative behavior in order to scaffold student understanding. However, video technology has greatly increased the range of behaviors that can come under rigorous study. Using video analysis techniques, we examined a teacher's use of verbal and gestural forms of communication.

In this chapter, we consider the possibility that teachers use spontaneous hand and arm gestures along with their speech in an effort to scaffold students' understanding. Previous research has documented that teachers do indeed use gestures in classroom settings (Flevares & Perry, 2001; Neill & Caswell, 1993; Núñez, 2004; Zukow-Goldring, Romo, & Duncan, 1994), as well as in tutorial settings (Goldin-Meadow, Kim, & Singer, 1999; Wang, Bernas, & Eberhard, 2001). However, previous studies of teachers' gestures have not directly examined gesture as a form of scaffolding.

Studies conducted in noneducational settings have demonstrated that listeners do in fact glean information from speakers' gestures (see Kendon, 1994, for a review). Speakers' gestures facilitate listeners' comprehension of the accompanying speech, particularly when the verbal message is ambiguous (Thompson & Massaro, 1994), highly complex (Graham & Heywood, 1976; McNeil, Alibali, & Evans, 2000), or degraded in some way (Riseborough, 1981, Experiment 3). Based on this prior work, it seems likely that students' comprehension of lesson content may also be aided by teachers' gestures. Gestures may be particularly important in classroom settings because students' comprehension is often challenged by instructional discourse that presents new concepts and uses unfamiliar terms. In addition, classrooms are often noisy, with multiple individuals speaking at once. Under such circumstances, gesture may play a particularly important role in comprehension.

We hypothesize that teachers use gestures to "ground" (cf. Glenberg & Robertson, 1999; Lakoff & Núñez, 2001) their instructional language, that is, to link their words with real-world, physical referents such as objects, actions, diagrams, or other inscriptions. This grounding may make the information conveyed in the verbal channel more accessible to students. We suggest that, by providing gestural grounding where appropriate, teachers scaffold students' comprehension of instructional language, and in so doing, foster students' learning of lesson content. Thus, gestural grounding may be one means by which teachers scaffold students' understanding.

If teachers are sensitive to this grounding function of gesture, they should vary their use of gesture, using more gestures during parts of the lesson for which students need greater scaffolding. There are at least three types of circumstances in which greater scaffolding is likely to be needed. First, greater scaffolding is likely to be needed when new instructional material is introduced. As the material becomes more familiar, the scaffolding is no longer necessary, and it can "fade" away without consequences for students' understanding. Second, greater scaffolding is likely to be needed for material that is more complex or more abstract. For material that is uncomplicated and concrete, scaffolding may not be necessary. Third, scaffolding may increase in response to students' questions. Based on these ideas, we can derive three specific predictions about gesture frequency during instruction: First, teachers should use gestures more frequently when they introduce new material than when they cover familiar material. Second, teachers should use gesture more frequently when they speak about material that is more complex. Third, teachers should use gestures more frequently in response to students' utterances than prior to students' utterances.

The purpose of the present study was to investigate a teacher's use of gesture in naturalistic classroom communication, with a focus on the role of gestures in grounding verbal content. In so doing, we outline a technique for analyzing video of classroom instruction. We expected that the teacher would regularly use gesture to "ground" her verbal utterances. Further, we predicted that the teacher would vary her use of grounding in an effort to scaffold students' understanding. Specifically, we predicted that she would produce more gestures when introducing new material, when talking about aspects of the lesson content that are more complex, and when responding to students' questions and comments.

To address these issues, we selected a sixth-grade mathematics lesson that focused on algebraic relations. This lesson was chosen for several reasons. First, the lesson content was challenging and unfamiliar for the students. Second, the lesson provided ample opportunities for discussing abstract concepts. Third, the lesson included a long segment in which the teacher addressed the class as a whole, during which the teacher's gestures could be examined. Fourth, the teacher's use of the overhead projector during her instruction gave us the opportunity to observe her gestures simultaneously in two planes of motion, even though we used only one camera in the classroom. This rich visual display made the video data particularly rich and mitigated some of the ambiguity sometimes associated with gesture. Our analyses focus on the teacher's use of gestural grounding and how it varies throughout the lesson.

METHOD

Source of Data

The data for this study were drawn from a video recording of a sixth-grade mathematics lesson that focused on algebraic relations. The lesson was conducted by a regular classroom teacher in a suburban school that operated with a "middle school" philosophy. The student body was 86% Caucasian, 6% Asian, 5% Hispanic, 2% American Indian, and 1% African-American. Twelve percent of the student body received free or reduced lunch, and 13% were in need of special education services. The mathematics performance of the students in the classroom on the California Achievement Test (CAT) ranged from the 5th to the 99th (highest) percentile.

Students were algebra novices, although they had participated in a set of lessons on simple algebra story problem solving about 3 months earlier. Algebra was not a standard part of the sixth-grade curriculum for this school. However, the teacher was participating in an experimental program aimed at understanding early algebra learning and instruction. The content of this lesson was not specifically developed within this research collaboration, but was chosen by the classroom teacher, based on materials she had obtained earlier.

As is typical of many middle school mathematics instructors, the teacher was trained in elementary education rather than in mathematics. She had been teaching for over 10 years. She had a warm manner with her students and saw many of them for multiple subject areas (including language arts, foreign language, science, and homeroom). Throughout the class periods, the teacher used several forms of class organization, including small groups, individual seatwork, and whole class discussion.

The target lesson was designed to introduce students to the power of algebraic sentences (equations and inequalities) to model the physical world. The goal was to help students understand how algebraic relations can serve as mathematical models of physical systems, and thereby help to give meaning to systems of equations and inequalities, while providing a formal approach for determining the value of unknown quantities. The lesson involved illustrations of pan balance scales, like that seen in Figure 22.1, with various combinations of objects of unknown weights. As the teacher described it (video of 1/21/99), "We're going to translate some of these … pans into equations."

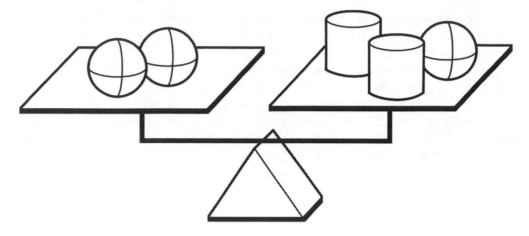

Figure 22.1.   Pan balance problem 1 ($2S = 2C + S$).

The teacher operated with several major objectives: First, she set out to show the parallel structure between the pan balance and important features of the algebraic sentences. In particular, she wished to highlight how balance related to equality (and to the equal sign), how lack of balance related to inequality (and to the greater than sign), how collections of the same object related to scalar multiplication, how collections of different objects related to addition (and to the "+" operator), and how the unknown weight of each object related to the use of variables. Second, the teacher supported students' construction of possible algebraic models of the pan balances. Third, the nature of the problems led to the construction of systems of interrelated equations or inequalities that could be simplified through meaningful operations, such as canceling out identical objects from opposite sides of the pan balance (or algebraic sentence), and substituting known relations (as when one sphere balances the weight of two cylinders).

The video analyses focus on instruction with three pan balance configurations, including two configurations that balanced (problem 1, $2s = 2c + s$, and problem 2, $2b + c = s$), and one that did not (problem 3, $3b + 2c > s + c$). Students constructed equations to symbolize the configurations of the pan balance (i.e., balanced or unbalanced scales), and attempted to simplify these equations, first working individually, then in small groups. The current analyses focus on a segment that follows the small group work when the teacher convenes a whole-class discussion on algebraic modeling. The analyses are based on 23 min and 40 sec from a 90-min class that also included other mathematical and administrative activities.

Video Analysis

The video excerpt was transcribed in two "passes." In the first pass, the teacher's speech was transcribed, and the verbal transcript was divided into meaningful idea units, following a procedure similar to Kintsch (1998). In this chapter, we refer to these units as utterances.

In the second pass, all of the observable gestures that the teacher produced along with her speech were identified and incorporated into the transcript. Gestures were classified into three categories. Pointing gestures were defined as gestures used to indicate objects, locations, inscriptions, or students. Most pointing gestures were produced with the fingers or hands; some were produced using a pen as a "pointer." For example, in Figure 22.2, a frame from the video shows that the teacher points with her pen to the sides of the unbalanced scale as she states, "This side is more than this side." Representational gestures were defined as gestures in which the hand shape or motion trajectory of the hand or arm represented some object, action, concept, or relation. The category of representational gesture as used here collapses across the categories of iconic and metaphoric gestures as described by McNeill (1992). It refers to gestures that pictorially "bear a close formal relationship to the semantic content of speech" (p. 12) by depicting either a concrete object or event, or an abstract, "invisible" idea (p. 14). For example, Figure 22.3 shows a sequence of still frames of the teacher (a) first miming plucking (pictures of) two spheres from either side of pan balance A, and then (b) "picking up" an *s* variable from both sides of an algebraic equation (note how her left hand has shifted down the screen from the pan to the equation area). (A transcript of the teacher's utterances leading up to and accompanying these gestures is provided in Appendix A.) Writing gestures were defined as writing that the teacher produced while speaking, and that was temporally integrated with speech in the same way that

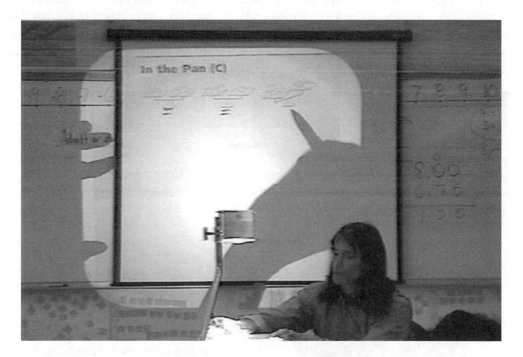

Figure 22.2.   Example pointing gesture. In this instance, the teacher points with her pen.

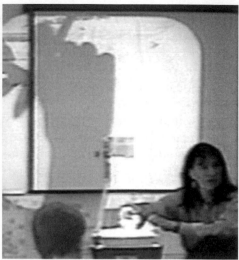

Figure 22.3.   Examples of representational gesture, accompanying the verbal utterance "I'm gonna take away a sphere from each side." (a) Teacher mimes picking up a sphere from each side of the pan balance. (b) Teacher mimes picking up the variable *s* from each side of the equation.

hand and arm gestures are temporally integrated with speech. For example, in Figure 22.4, the teacher used a writing gesture while talking about the effects of removing a sphere from one side of the pan balance and the equation. (Note that writing that was produced in the absence of speech was not coded as a form of gesture.) Pointing, writing, and representational gesture are all taken to be forms of grounding, as discussed earlier.

Figure 22.4.   Example writing gesture, accompanying the verbal utterance "Do I have a sphere anymore? No. Naa (sound like game show buzzer), alright?"

There were a total of 301 teacher utterances in the data set. For 29 of these utterances, the teacher was off camera, so the presence of grounding could not be assessed. Each of the remaining 272 utterances was classified as either speech alone or speech with grounding. For utterances with grounding, the method of grounding (e.g., pointing, representational gestures, writing, or some combination of these) was noted. In addition, the total number of grounding acts (i.e., distinct gestures) that accompanied each utterance was noted.

In addition to coding for gestural grounding, utterances were also coded for their referents. Three categories of referents were of greatest interest; (1) pans, defined as the concrete, pictorial representations of the pan balance; (2) mathematical relations, defined as symbolic equations or inequalities, in this case, algebraic sentences used to model the pan balance; and (3) links, defined as conceptual links or correspondences made between particular configurations of the pan balance and algebraic sentences. Of course, many of the teachers' utterances did not refer to one of these three categories of referents. These utterances were excluded from analyses that focus on referents.

These three categories of referents vary in level of abstraction. Pans as illustrations of physical objects are considered to be the most concrete of the three types of referents. Algebraic sentences as inscriptions of equalities and inequalities are less concrete, but still physically locatable, and are therefore considered to be of a moderate level of abstraction. Links, which lack physical and visual presence but exist conceptually, are considered to be the most abstract of the three types of referents.

RESULTS

The data analyses focus on the teacher's use of grounding; how it changes over the course of the lesson, how it varies depending on the referent of the utterance, and how it relates to student utterances.

How Many of the Teacher's Utterances Included Some Form of Gestural Grounding?

Overall, 56% of the teachers' utterances included some form of gestural grounding; 21% included at least one instance of pointing; 20% included at least one instance of representational gesture; and 15% included at least one instance of writing gesture. Some utterances included more than one form of grounding.

If the analysis is restricted to utterances that focus on the instructional task itself (i.e., utterances that focused on the pan, algebraic sentences, or links between the two, rather than on classroom management or other matters, $N = 158$), the teacher's use of gesture is even more striking. For this subset of utterances, 74% of the teachers' utterances included some form of gestural grounding. 34% included at least one instance of pointing, 22% included at least one instance of representational gesture, and 24% included at least one instance of writing gesture. Thus, grounding with gesture was pervasive in the teacher's instructional communication.

How Did the Teacher's Use of Grounding Change Over the Course of the Lesson?

Recall that we hypothesized that the teacher would produce more gestural grounding when new material was introduced than after that material had become familiar to the students. Based on this hypothesis, we predicted that the teacher would use more gestural grounding at the outset of the lesson, and that this scaffolding would "fade" from the first to the second problem, which were similar in structure (i.e., both simple equations). Further, one might expect that gestural scaffolding would "rebound" for the third problem, which was a new problem type, namely, an inequality.[1] As seen in Figure 22.5, this prediction was borne out. The proportion of utterances that included gestural grounding was much lower for the second problem than for the first or third problems, $\chi^2(2, N = 263) = 12.17, p < .01$.

Because many utterances included multiple grounding acts, we next examined the mean number of grounding acts that the teacher produced per utterance for each problem. This provides some insight into how heavily concentrated the teacher's use of gesture was at various points in the lesson. As seen in Figure 22.6 (total heights of the bars), the mean number of grounding acts per utterance decreased from the first problem ($M = 0.78, SE = 0.08$) to the second problem ($M = 0.40, SE = 0,07$), and then increased again for the third problem ($M = 0.67, SE = 0.08$). The number of grounding acts per utterance differed significantly across problems, $F(2, 260) = 5.35, p < .01$. Post hoc tests indicated that the teacher used significantly more grounding acts per utterance on the first problem and on the third problem than on the second problem. Part of the reason for this pattern is that the teacher was more likely to use multiple grounding acts for any given utterance on the first problem (32% of utterances) and the third problem (25% of utterances) than on the second problem (10% of utterances). Thus, it

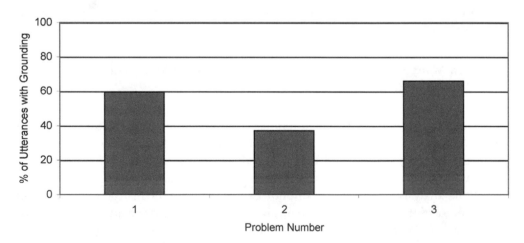

Figure 22.5.   Percentage of utterances that included some form of grounding for each of the three problems.

---

[1] Nine utterances were excluded from these analyses because they focused on helping one of the students find some missing pens and pencils rather than on any of the target problems.

seems clear that the teacher used gestural grounding most extensively when she introduced new material (i.e., on Problems 1 and 3) and this practice faded as the material became more familiar to students. Finally, we examined the mean number of each of the three types of grounding acts (pointing, representational gesture, writing) for each problem.

The data are presented in Figure 22.6. As seen in the figure, pointing was the most frequent type of grounding act overall, followed by representational gesture and then writing. Further, pointing was the most frequent method of grounding for each of the three problems. Representational gesture was used more frequently than writing on the first and second problems, and they were used similar amounts on the third problem.

How Did the Teacher Ground Utterances With Different Types of Referents?

Recall that we hypothesized that the teacher would produce more gestural grounding for aspects of the lesson content that were more abstract. Based on this hypothesis, we predicted that the teacher would produce the greatest proportion of grounding acts for utterances about the links between algebraic sentences and the pans, and the lowest proportion of grounding acts for utterances about the pans themselves. Figure 22.7 presents the proportion of utterances with each type of referent that included some form of grounding. As predicted, a greater proportion of utterances about links included some sort of grounding than utterances exclusively about algebraic relations or pans, $\chi^2(1, N = 158) = 4.13, p < .05$. The proportion of utterances that included grounding was lower and similar for utterances exclusively about algebraic relations and utterances exclusively about the pans

We next examined the mean number of grounding acts per utterance. As seen in Figure 22.8 (total heights of the bars), utterances that referred to links between the

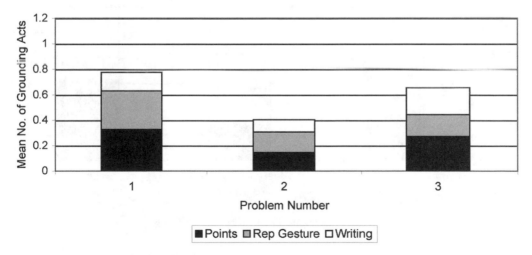

Figure 22.6.   Mean number of grounding acts of each type produced per utterance for each of the three problems.

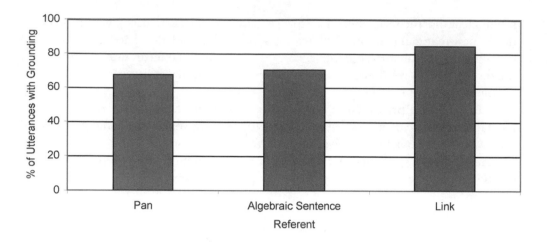

Figure 22.7.   Percentage of utterances with each referent that included some form of grounding.

pans and the algebraic sentences received the greatest number of grounding acts per utterance ($M = 1.18$, $SE = 0.12$). Utterances that referred exclusively to the algebraic sentences and utterances that referred exclusively to the pans received comparable numbers of grounding acts per utterance (algebraic sentences, $M = 0.83$, $SE = 0.09$; pans, $M = 0.84$, $SE = 0.12$). The number of grounding acts per utterance differed significantly across referents, $F(2, 155) = 3.60$, $p < .03$. Post hoc tests indicated that the teacher used significantly more grounding acts per utterance for utterances about links than for utterances about pans or algebraic sentences. Part of the reason for this pattern is that the teacher was especially likely to use multiple grounding acts for utter-

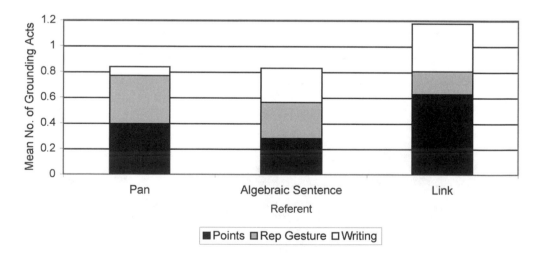

Figure 22.8.   Mean number of grounding acts of each type produced per utterance for each of the three referent categories.

ances about the links between pans and algebraic sentences. Indeed, 57% of utterances about links received multiple grounding acts, compared to only 27% of utterances about algebraic sentences and 28% of utterances about pans.

Utterances about links often involved a pointing gesture to some aspect of a pan, followed by a pointing or writing gesture to the corresponding aspect of an equation or inequality, or vice versa. For example, near the outset of the lesson, the teacher sought to establish the correspondence between the fulcrums of the pan balances (which she referred to as "wedges") and the relational symbols (equal signs) used in the algebraic sentences.

Speech: OK, um (pause) we want to think of these wedges as the equals sign, right?

Gesture: {1} {2} {3} { 4 }

Gesture descriptions:
1. points with pen to fulcrum in Problem A
2. points with pen to fulcrum in Problem B
3. points with pen to fulcrum in Problem A
4. writes equal sign under fulcrum in Problem A

As in this example, the teacher frequently used gestures to delineate the correspondences between different representations of the same mathematical information. In other cases, the teacher integrated aspects of multiple representations in a single utterance, as in the following example, where she highlights that the greater-than symbol can describe the relationship between the pans.

Speech: you'd have to say, well right now, this side's greater than that side

Gesture: {1} {2} { 3 }

Gesture descriptions:
1. points with pen to left pan in Problem C
2. points with pen to greater-than symbol below Problem C
3. points with pen to right pan in Problem C

We suggest that students may be more likely to attend to and understand the relationships among representations when these relationships are instantiated with gestures.

Finally, we examined the mean number of each of the three types of grounding (pointing, representational gesture, writing) for each referent. The data are presented in Figure 22.8. As seen in the figure, pointing was especially frequent for utterances about the links between pans and algebraic sentences. This is due in part to the fact that pointing is adept at highlighting correspondences, such as the correspondence between a written variable symbol in an equation and its referent object on the pan balance. Representational gesture was especially likely to be used for utterances exclusively about the pan, in part because actions such as removing objects from the pan could readily be described using representational gestures. Writing gestures were highly likely to be used for utterances about the algebraic relations, in part because these mathematical relations could be produced or elaborated with writing (e.g., underlining each side of an equation) as it was being described in her speech.

Was the Teacher's Use of Grounding Influenced by Student Utterances?

Another way to evaluate whether the teacher uses grounding adaptively is to examine whether she uses gesture more in response to students' questions and comments than otherwise. To address this issue, we compared her gesture rate in utterances that respond to student utterances to her gesture rate in utterances that immediately precede student utterances. These two categories of teacher utterances should be similar in terms of topic, level of engagement, and the point at which they occur during the class period. For extended back-and-forth sequences between student and teacher, we focused on the first student utterance of the sequence, and included the teacher utterances that immediately preceded and followed that student utterance in the data set.

The differences in the teacher's behavior were clear. The teacher was more likely to gesture after a student utterance than before (56% vs. 31%). In cases when the teacher gestured only before or after a student utterance ($N = 15$), she was much more likely to do so only after, rather than only before (binomial test, $p < .02$) Further, the mean number of gestures per teacher utterance was significantly greater in utterances that immediately followed student utterances than in utterances that immediately preceded student utterances [0.69 vs. 0.33, $t(35) = 3.17, p < .01$]. Thus, it appears that student utterances trigger more elaborate, grounded responses on the part of the teacher.

DISCUSSION

The present results indicate that gesture is pervasive in instructional communication. Furthermore, the results suggest that gesture serves a scaffolding function. As predicted by the scaffolding hypothesis, gesture was used most frequently for new material, for referents that were highly abstract, and in response to students' questions and comments. Thus, gesture appears to be one means by which this teacher attempted to scaffold student comprehension.

Of course, some limitations of the present work must be acknowledged. The study focuses on a single lesson from a single teacher. Differences in lesson content, teaching style, classroom context, and the population of students may influence how gesture is used to scaffold new mathematical concepts. The larger body of gesture research suggests that some of the patterns we observed in this study should also appear across different lessons, teachers, settings, and populations, but this must be established empirically. In addition, we did not examine whether individual gestures reinforced or complemented the accompanying speech. Future analyses of this lesson may reveal even more about the ways that information conveyed in gesture relates to the accompanying speech. Finally, it remains to be seen whether the grounding of abstract ideas through gesture aids student comprehension of lesson content and leads to greater learning. Some recent experiments using videotapes of teachers carrying out instruction suggest that this may be the case, at least for lessons that focus on procedural knowledge (e.g., how to use a compass, Glenberg & Robertson, 1999) and visuospatial concepts (e.g., symmetry, Valenzeno, Alibali, & Klatzky, 2003; Piagetian conservation, Church, Ayman-Nolley, & Mahootian, 2004). However, studies of the impact of gesture on student learning in naturalistic, instructional settings are needed. We suspect that such studies will show that teachers' gestures do indeed influence students' comprehension and learning.

We have argued that the evidence supports a view of teacher gesture as a form of scaffolding. However, at least one alternative possibility must be acknowledged—namely, the teacher's gestures may index her own cognitive state. Past research has shown that gestures are not solely communicative; they also serve a cognitive function for the speaker, helping to support the reasoning process (e.g., Alibali, Kita, & Young, 2000; Goldin-Meadow, Nusbaum, Kelly, & Wagner, 2001). This perspective suggests that the teacher might decrease her use of gesture as she becomes more familiar with the mathematical content in the lesson (e.g., from the first to the second problem). When new cognitive demands arise (e.g., on the third problem, or in response to a student question or comment), she may gesture more as means to support her own reasoning processes. Future research will be needed to explore whether the cognitive account or the scaffolding account better explains teachers' use of gesture in classroom settings, or whether both accounts are viable. It is worth noting that, even if the teacher produces gesture to support her own reasoning processes, those gestures may still help student learning by helping her articulate her thoughts more clearly.

Regardless of which account ultimately proves correct, it is clear that teachers' gestures reveal aspects of their thinking. Indeed, an extensive body of past research has shown that gestures are a unique and revealing window on speakers' thought processes (e.g., Crowder, 1996; Goldin-Meadow, Alibali, & Church, 1993; Schwartz & Black, 1996). We suggest that teachers' gestures reveal aspects of their thinking, not only about lesson content but also about students' abilities. If teachers produce more gestures when they believe students need greater scaffolding, then teachers' gestures may reflect their implicit models of students' knowledge and potential areas of difficulty. It is possible that teachers may use gesture differently when explaining the same content to different students, depending on their beliefs about the students. We plan to explore this issue in future work.

This research emphasizes the communicative nature of teaching (e.g., Flevares & Perry, 2001; Valenzeno et al., 2003), and underscores the importance of gesture research for teacher education and professional development. In addition, this work highlights the many tasks that teachers face in communicating abstract material to students for the first time. Because gesture works in tandem with speech, it can provide conceptual grounding for new and abstract ideas in a visual and holistic manner. Teachers can use the verbal modality to articulate abstract concepts, while using gesture to direct students' attention to the referents of their speech, and to ground those abstract concepts to ideas that are concrete and familiar.

The use of gesture as a form of grounding bears on one of the fundamental challenges in the field of cognitive science, namely, the "symbol grounding problem"

(Harnad, 1990). This is the problem of how meaning can be explained in terms of arbitrary symbols (such as words) that have behavior governed exclusively by syntactic rules (or grammars). Embodied accounts of cognition (e.g., Barsalou, 1999; Glenberg, 1997; Glenberg & Kaschak, 2002) offer one solution to this problem, namely, that symbolic representations are grounded from the "bottom up," via their links to bodily experiences and actions. We suggest that gestures may be one manifestation of the embodiment of cognition. The teacher's gestures reveal a natural, spontaneous way to provide the grounding that is necessary for meaning to be attached to the objects of instruction.

According to Harnad (1990), grounding can be achieved either through "iconic representations," which are analogs of sensory information, or through "categorical representations," which encode invariant features of object and event categories. This perspective could help explain the teacher's adaptive use of gesture across different referents. As seen in Figure 22.8, the teacher often used representational gestures, such as iconic representations of physical actions, when she spoke about the pan balances. Further, she often used pointing to highlight object features when she spoke about the links between the pans and the algebraic sentences. Thus, different referents afforded different types of gesture. These findings suggest that lessons that involve different types of representational material will likely elicit different types of gestural grounding. Future research on this issue is needed.

The prevalence of gesture and its role in instruction raises an additional issue fundamental to classical cognitive theory. In its canonical form, information processing is the core of the cognitive system. Outside of the "central processing unit" are the "peripherals"—the sensory inputs that provide information from the perceptual systems, and the actuators that are output systems for acting on the world. As Wilson (2002) characterizes it, "Perceptual and motor systems, though reasonable objects of inquiry in their own right, were not considered relevant to understanding 'central' cognitive processes. Instead, they were thought to serve merely as peripheral input and output devices" (p. 625). Thus, a great deal of cognitive research assumed that very narrow forms of data—reaction times, accuracy, and verbal responses—were sufficient to characterize complex behavior. The symbol processing view, seemingly fundamental to classical cognitive science, has recently come into question. As Eisenberg (2002) muses,

> As I write this sentence, I am glancing over at the color printer sitting beside my screen. In the popular jargon of the computer industry, that printer is called a "peripheral"—which, upon reflection, is a rather odd way to describe it. What, precisely, is it peripheral to? If the ultimate goal of my activity is to produce a physical artifact, then one would have to conclude that the printer is a central—maybe the central—technological device in sight. (p. 1)

From our perspective, physical gesture appears to be essential to teachers' ability to conduct their practice, and its place in the complex, cognitive activity of real-time instruction is not peripheral, but instead quite central. For this reason, we expect to see a growing appreciation for the importance of gesture research and the use of video data as part of the scientific study of instruction and teacher cognition.

The current study sheds some light on instructional practices that are necessarily overlooked in analyses of classroom speech alone. By broadening the scope of discourse analysis to include body motions, referent objects, and inscriptions, as well as the language that accompanies them, one can achieve a more nuanced view of the communicative processes involved in instruction and learning. Although gesture often reinforces the information expressed in the verbal channel, it sometimes reveals information that is not expressed in speech (e.g., Church & Goldin-Meadow, 1986; Perry, Church, & Goldin-Meadow, 1988). Further, listeners may glean different messages from speech, depending on whether or not that speech is accompanied by gesture, and depending on whether gesture provides reinforcing or complementary information (e.g., Goldin-Meadow & Sandhofer, 1999; McNeil et al., 2000; Singer & Goldin-Meadow, 2005). In some situations, such as the lesson analyzed here, gesture may serve as the "glue" that helps to forge the links among various other representations (in this case, the illustrations of the pan balances, and the symbolic representations) for listeners. A deeper understanding of the communicative aspects of instruction will require serious attention to the multi-modal nature of instructional communication. This goal can be achieved only with the use of video data, and with the continuing development of methods for video analysis.

ACKNOWLEDGMENTS

We are grateful to Amy French for her participation in this study of her teaching practices. We also thank Kate Masarik for assistance with videotaping and field-based research, and Emily Coleman for assistance with transcription and coding. This research was funded in part by a grant award "Understanding and Cultivating the Transition from Arithmetic to Algebraic Reasoning" from the Interagency Educational Research Initiative, an alliance of the National Science Foundation, the Department of Education Institute of Educational Sciences, and the National Institute of Child Health and Human Development within the National Institutes of Health.

REFERENCES

Alibali, M. W., Kita, S., & Young, A. (2000). Gesture and the process of speech production: We think, therefore we gesture. *Language and Cognitive Processes, 15*, 593–613.

Barsalou, L. W. (1999). Perceptual symbol systems. *Behavioral and Brain Sciences, 22*, 577–660.

Church, R. B., Ayman-Nolley, S., & Mahootian, S. (2004). The role of gesture in bilingual education: Does gesture enhance learning? *International Journal of Bilingual Education and Bilingualism, 7*, 303–319.

Church, R. B., & Goldin-Meadow, S. (1986). The mismatch between gesture and speech as an index of transitional knowledge. *Cognition, 23*, 43–71.

Crowder, E. (1996). Gestures at work in sense-making science talk. *Journal of the Learning Sciences, 5*, 173–208.

Eisenberg, M. (2002). Output devices, computation, and the future of mathematical crafts. *International Journal of Computers for Mathematical Learning, 7*, 1–44.

Flevares, L. M., & Perry, M. (2001). How many do you see? The use of nonspoken representations in first-grade mathematics lessons. *Journal of Educational Psychology, 93*, 330–345.

Glenberg, A. (1997). What memory is for. *Behavioral and Brain Sciences, 20,* 1–55.

Glenberg, A. M., & Kaschak, M. P. (2002). Grounding language in action. *Psychonomic Bulletin and Review, 9,* 558–565.

Glenberg, A. M., & Robertson, D. A. (1999). Indexical understanding of instructions. *Discourse Processes, 28,* 1–26.

Goldin-Meadow, S., Alibali, M. W., & Church, R. B. (1993). Transitions in concept acquisition: Using the hand to read the mind. *Psychological Review, 100,* 279–297.

Goldin-Meadow, S., Kim, S., & Singer, M. (1999). What the teachers' hands tell the students' minds about math. *Journal of Educational Psychology, 91,* 720–730.

Goldin-Meadow, S., Nusbaum, H., Kelly, S. D., & Wagner, S. M. (2001). Explaining math: Gesturing lightens the load. *Psychological Science, 12,* 516–522.

Goldin-Meadow, S., & Sandhofer, C. M. (1999). Gesture conveys substantive information to ordinary listeners. *Developmental Science, 2,* 67–74.

Graham, J. A., & Heywood, S. (1976). The effects of elimination of hand gesture and of verbal codability on speech performance. *European Journal of Social Psychology, 5,* 189–195.

Harnad, S. (1990). The symbol grounding problem. *Physica. D., 42,* 335–346.

Kendon, A. (1994). Do gestures communicate? A review. *Research on Language and Social Interaction, 27,* 175–200.

Kintsch, W. (1998). *Comprehension: A paradigm for cognition.* New York: Cambridge University Press.

Lakoff, G., & Núñez, R. (2001). *Where mathematics comes from: How the embodied mind brings mathematics into being.* New York: Basic Books.

McNeill, D. (1992). *Hand and mind: What gestures reveal about thought.* Chicago: University of Chicago Press.

McNeil, N. M., Alibali, M. W., & Evans, J. L. (2000). The role of gesture in children's comprehension of spoken language: Now they need it, now they don't. *Journal of Nonverbal Behavior, 24,* 131–150.

Neill, S., & Caswell, C. (1993). *Body language for competent teachers.* London: Routledge.

Núñez, R. (2004). Do *real* numbers really move? Language, thought, and gesture: The embodied cognitive foundations of mathematics. In F. Iida, R. Pfeifer, L. Steels, & Y. Kuniyoshi (Eds.), *Embodied artificial intelligence* (pp. 54–73). Berlin: Springer-Verlag.

Perry, M., Church, R. B., & Goldin-Meadow, S. (1988). Transitional knowledge in the acquisition of concepts. *Cognitive Development, 3,* 359–400.

Riseborough, M. G. (1981). Physiographic gestures as decoding facilitators: Three experiments exploring a neglected facet of communication. *Journal of Nonverbal Behavior, 5,* 172–183.

Schwartz, D., & Black, J. B. (1996). Shuttling between depictive models and abstract rules: Induction and fallback. *Cognitive Science, 20,* 457–497.

Singer, M. A., & Goldin-Meadow, S. (2005). Children learn when their teacher's gestures and speech differ. *Psychological Science, 16,* 85–89.

Thompson, L. A., & Massaro, D. W. (1994). Children's integration of speech and pointing gestures in comprehension. *Journal of Experimental Child Psychology, 57,* 327–354.

Valenzeno, L., Alibali, M. W., & Klatzky, R. L. (2003). Teachers' gestures facilitate students' learning: A lesson in symmetry. *Contemporary Educational Psychology, 28,* 187–204.

Wang, X.-L., Bernas, R., & Eberhard, P. (2001). Effects of teachers' verbal and non-verbal scaffolding on everyday classroom performances of students with Down Syndrome. *International Journal of Early Years Education, 9,* 71–80.

Wilson, M. (2002). Six views of embodied cognition. *Psychonomic Bulletin & Review, 9,* 625–636.

Zukow-Goldring, P., Romo, L., & Duncan, K. R. (1994). Gestures speak louder than words: Achieving consensus in Latino classrooms. In A. Alvarez & P. d. Rio (Eds.), *Education as cultural construction: Exploration in socio-cultural studies* (Vol. 4, pp. 227–239). Madrid, Spain: Fundacio Infancia y Aprendizage.

APPENDIX

Transcript

If I take away a sphere on each side / does this still balance like Robbie said? / Yeah, it still balances / Doesn't it? / OK! So a way that you can notate that down in your equation, down here is you can say "OK, now I am gonna take away an S" / "I am gonna take away a sphere from each side" / "instead of taking it off the pans" / "I am going to take it away from this equation" / So, I'm gonna take away an S here / which is like crossing that one off / Are you with me? / (Yeah) / and it's like taking away an S over here / Follow me? / ... / Just teaching you a (short pause) way to notate this / and a way to think about this / Ok, so now what happens if I take a sphere / if I have a sphere and I take one away / do I have a sphcrc anymore? / No. / Naa (sound like game show buzzer), alright? (Video of 1/21/99, 5:35–6:15)

Epistemic Mediation: Video Data as Filters for the Objectification of Teaching by Teachers

Wolff-Michael Roth
University of Victoria

> *Video provides a shared resource to overcome gaps between what people say they do and what they, in fact, do. Video provides optimal data when we are interested in what "really" happened rather than in accounts of what happened.*
>
> —*Jordan and Henderson (1995, p. 50)*

> *It was really neat with the videotape, to be able to see how there are things happening in the classroom that I can't see, that I don't know that they are happening. I really noticed that there is not enough teacher movement away from the chalkboard when I am not using it. So that was one thing that I was struck with.*
>
> —*Christina (fourth-grade teacher)*

Learning science researchers attempting to understand situated human practices traditionally have relied on ethnographic observation and field notes recorded after the events have occurred. However, as Jordan and Henderson articulated in the opening quote, they are faced with the gap between accounts of action and (situated) actions themselves. The problem is heightened when learning science researchers become themselves participants in the setting under study. Thus, a number of learning science researchers—including Magdalene Lampert, Jim Minstrell, David Hammer, and myself—conducted research on cognition and instruction all the while teaching the lessons that are the focus of their studies. Furthermore, an increasing number of teachers continue their formal education and become learning science researchers and teach at elementary and secondary schools. Teacher-researchers are confronted with particular challenges arising from the fact that they are participants in rather than onlookers to the situation to be analyzed and theorized. They are interested rather than disinterested participants, and therefore have something at stake, which harbors particular dangers for the quality of the analyses of learning and instruction in their classrooms that accompany the analytic advantages that derive from their insider role (Roth & Tobin, 2002). Video, as the second quote shows, provides opportunities to teacher-researchers to see themselves and their experiences differently, even pertaining to their own actions. In the second quote, Christina described how watching herself on videotape allowed her notice that she was standing a lot next to the chalkboard even when it was not used during interactions with students. That is, by means of the video, she became aware of her own actions in a different way.

When teachers become researchers, the bias may even be more serious because of the different intentionality and experiential horizon distinguishing teaching and researching (observing teaching). Whereas the ethnographic observer focuses on events subsequently recorded in the form of field notes, the teacher is absorbed in the act of teaching rather than the recording of events—they have, as all practitioners, a particular perspective that radically differs from those any onlooker has (Bourdieu, 1990). Later, recollections are seen not only through the filter of the different intentionality but also through a filter that changes, imperceptibly and with time. Experience is not recorded in the human body and mind as an indelible trace but takes the character of cinder (Derrida, 1995).

I have used video since the beginning of my career, which I spent as a teacher-researcher. I was a science teacher and a department head. Although video allowed me to make my teaching and myself the object of reflection, that is, video allowed me to objectify myself, I thought that others in the learning sciences community might find the distance between participant and analyst too small. To protect myself from charges of subjectivism, I sought the collaboration of a university-based researcher, which allowed me to gain some distance from the data that I had collected, transcribed, and constructed initial analyses of. This chapter is concerned with epistemological issues of video as a tool in the (self-) study of learning and instruction, and in application of learning sciences to teacher development. I exemplify an empirical approach for teacher-researchers in the learning sciences striving for high levels of objectivity of their analyses and theorize the process of objectification.

BACKGROUND

Some individuals in the learning sciences not only research but also teach the lessons that constitute the object of their studies. Some investigations are interested not only in their students' cognition of but also in the cognition of teaching itself (Lampert,

1998). Conducting research in the teacher-as-researcher mode harbors dangers during the data analysis phase, particularly when the analyses are conducted without recourse to others who do not have stakes in the project. For example, the teacher's retrospective accounts of what he has done in some situation recorded on tape may not accurately render the plans, goals, and beliefs that motivate his utterances (van Zee & Minstrell, 1997). At best, the teacher's reflections are plausible (rather than to be privileged) accounts of the events experienced earlier (e.g., Hammer, 1996). Being an insider also comes with the potential danger that we simply reify our preconceptions about how some place works rather than pushing the analysis to get to the deep structures of meaning and seeking configurational validity that transcends our own experiences (Goldman-Seagall, 1995).

Investigations of teaching as thinking practice involve reflection on practice rely on teachers' and researcher-teachers' introspection and self-reports. These data are central to the practice of reflection on action, one of the most salient practices not only in the study of teaching but also for developing it. A recently published critical review of reflection in teaching listed confessional journals and autobiographical narratives as two common ways of achieving the reflexive turn (Fendler, 2003). A central problem with introspection is that in most instances they constitute flawed data in the study of cognition and consciousness. At issue here is not the inherent subjectivity of the viewpoint any human being has on the world; I both acknowledge and value it. Rather, at issue are surface accounts of experience, which necessarily differ from person to person, whereas deep accounts get at the structures that allow particular experiences to emerge. On the other hand, stimulated recall using videotaped episodes constitutes an important but underused means in the process of this form of data production. The resulting texts can be viewed as participants' own a posteriori analyses of events and therefore provide researchers with their views of the world, which may be substantially different from the outside observer (Jordan & Henderson, 1995; Stevens & Toro Martell, 2003).

Other cognitive science researchers do not view some forms of introspection as something negative (e.g., Gallagher, 2002). They recognize that to access an individual's thinking and objects of consciousness in the process of teaching, we have to rely on his or her self-report produced after the event. Donald Schön (1987) considered reflection on action to be an important tool in professional development. We cannot do without such self-reports of teaching experience, because human beings do not merely react to some abstract, scientifically described conditions but act in the face of what they perceive to be the case. To understand what is or was perceptually salient, we require access to these perceptions and the related experiences. But we also need to control or at least to understand the possible effects of a subjective perspective arising from the dual roles of teachers who are also learning sciences researchers.

EPISTEMOLOGICAL ISSUES

How and why do analyses become more objective, as I claimed earlier, when teacher-researchers draw on video? Here, I address some of the fundamental epistemological issues involved in the interpretation of, and reflection about, experience.

Subject and Object in Activity

Knowledge necessarily arises from experience, but all experience is suffused with preunderstanding. Most crucially, human experience is shaped by the current activity, which provides a particular motive and intentional horizontal character to what and how we perceive the current situation and the object of actions (Leont'ev, 1978). To understand activity and the actions that bring it forth, we cannot separate the knowing subject and the object of its conscious actions (including plans and intentions). The question of who knows cannot be separated from the question of what is known: The subject and object of activity mutually presuppose one another; they are dialectically related. Therefore, the object of reflection on some event is very different from the way the event originally appeared in the consciousness of the individual involved.

Experience and Traces

On the surface, the original experience can be said to leave something like a trace—aspects of which can now be represented with the help of video-based software tools (Stevens & Hall, 1997). When we revisit some event for the purpose of learning from it, for example, we present it again to ourselves, that is, we re-present it. In other words, the trace of the original event presents itself to our consciousness during reflection on action. We no longer have direct access to the event but only through the mediation of signs. More so, each time we reflect, the object changes, as it is mediated by all reflections that have occurred between the original event and the present day (Roth, 2002). The motive of activity is different now, learning from past events, and so is the object of the activity—the representation of events as they occur to us at the present moment. These representations have become the object of our activity; we have objectified ourselves.

We need to understand that experience does not just leave a stable trace. It does not even leave a trace that slowly fades while keeping its original structure, just a little less vivid in color and intensity. Rather, experience is like cinder, "something that remains without remaining, which is neither present nor absent, which destroys itself, which is totally consumed" (Derrida, 1995, p. 208). Every new experience transforms how past experiences present themselves to us. That is, a past event never remains the same but is transformed through new experience over and over again. More generally, experience integrates over itself; experience not only transforms experience but also incinerates and obliterates how it originally had appeared to us. In every experience there is this incineration, this experience of incineration, which is experience itself (Roth, 2004). However, we do not just perceive or envision events but also believe that something will occur or has occurred, and this belief will transform what we represent.

Practical Understanding and Explanation

The use of video changes the contributions teacher-researchers can make to the learning sciences. Although the difference between teaching and reflecting on teaching (research) remains, the object of reflection has changed. When teacher-researchers rely on videotape, they still face the effects of their preunderstanding and beliefs. However, the videotape provides them with a stable material basis for their analytic work rather than the ever-changing and even disappearing memory traces. Recorded on the videotape, teacher-researchers see themselves (or rather, a representation of themselves) literally objectified, become material objects standing in dialectical opposition to the subject. They perceive events not from their own position—coming with all the affordances and constraints of perspective—but from the position of another.

Whereas reflection mediates and turns experience into an object of inquiry, which distances us from our experience, it also brings us closer to it, makes us increasingly familiar with the events and allows us to better understand them. This seemingly contradictory movement occurs because practical understanding and explanation stand in a dialectical relationship. We cannot articulate deep structures of, and therefore explanations of, experience unless we already have a practical understanding of how the world works (Garfinkel, 1967); but our understanding of how the world works increases through explanation-seeking analysis (Ricœur, 1991).

Recollected events appear to teacher-researchers from the same perspective that they originally experienced them, although transformed through and disappearing in subsequent experiences (Roth, 2002). Even though the form of the experiential trace changes, the perspective has not changed. Video changes the perspective, allowing us to see ourselves in situations as other may or indeed have seen us. This change in perspective from the experienced I–You relation in situation to the objectified I–It relation is crucial in constructing explanations of behavior that can critically deal with the dangers of solipsism.

Reflection and Collectivity

Consciousness is based on and requires social mediation (Mikhailov, 1980). We are human exactly because we have the means to see ourselves as other to another. Video is a means of perceiving our (past) selves as bodies among other bodies. Collectively conducted interaction analysis (Jordan & Henderson, 1995) accentuates the objectification because we not only come to see ourselves from the outside, but we also come face to face with other individuals' perception of the recorded events. In collective analysis, we come to confront not only our own representations of past events but also those of others—collective analysis is a step toward reaching configurational validity (Goldman-Seagall, 1998). For this reason, teacher-researchers engaging in the forms of research presented here are in a position to make unique contributions to research on teacher cognition, because they are in the position to provide plausible accounts from an insider (I–You) as well as from an outsider (I–It) perspective.

Extending Reflexivity

Video is an important if underused tool not only in the construction of knowledge in the learning sciences but also in the reproduction of the field. How do we become better researchers? Video provides opportunities for collaborative analysis with novice researchers, who are, in the process, enculturated to the learning sciences community largely through tacit modes of learning (Jordan & Henderson, 1995). How can our graduate students learn not only through tacit modes of learning by participating in our research but also through reflection on their experiences? I have recorded data analyses sessions used for many years now, thereby allowing us to get a handle on the cognitive processes of analyzing data and generating theory. Graduate students use these tapes to access the observations made by different participants of the earlier analysis sessions, a process that has been referred to as "cannibalizing the [video]tapes" (Jordan & Henderson, 1995, p. 46). That is, used in a secondary way, video not only records the analysis session and therefore constitutes a form of external, objective, and collective memory, but also affords the objectification of scientific analysis itself.

EXEMPLIFYING THE ISSUES

In the foregoing section, I articulated some epistemological issues, specifically the increasing objectification with the use and fine-grained analysis of videotapes, especially when it is conducted collectively.

As part of my research on learning, I studied and theorized not only students' cognition but also the cognitive processes of teaching and change in these processes (learning, development) over time and with increasing experience. Videotapes of my teaching constituted a major tool for objectifying the classroom events. The following materials come from a study of the differential participation of students and teachers in classroom science conversations, where I had been teaching a science unit on simple machines to a split sixth- and seventh-grade class comprising 26 students.

One day, I had literally and figuratively rigged a tug of war using a block and tackle. The point was to win a tug of war against several students and then make their loss the focus of a discussion that would explain the events and theorize the workings of pulleys and block and tackles. Making a conservative estimate, I had initially invited eight students to compete against me, but as the events unfolded, more and more students joined in until about 20 of them were pulling on the other side. They still lost.

Reflection on Action

Immediately after class, I am reflecting about the lesson that had just come to a completion. I am thinking particularly about the effect of my challenge, and I am satisfied about having been able to get so many students interested in participating, first in the tug of war and then in the discussion. But I also notice that details of the event were not available to me: Exactly how many students contributed, who were these contributors, and how much did each actually contribute? I further note that I always see the events from the position that I had taken in the events; I do not see myself through the eyes of another [as Piaget (e.g., 1970) suggested], from the outside so to speak. Not only the events but also my changing recollections and understandings are always from a first-person perspective. From experience, I know that if any initial ways of seeing are not actually fixed, in terms of field notes or a reflective journal, a particular way of seeing that differs considerably from what can be seen on a mechanical record of the events may actually emerge (Goldman-Seagall et al., 1993).

Early in my career as a teacher-researcher, I had come to recognize the limitations of reflection on action. This had become even clearer after attending teacher meetings organized by Gaalen Erickson, a champion of reflection on action (e.g.,

Erickson & MacKinnon, 1991). There were many teachers proudly explaining that they are constructivists, all the while talking about some teaching strategy that they had used for more than 20 years, that is, way before there was any constructivist movement in education.

Debriefing—Collective Reflection

We are in the car, driving back to the university. I am talking to my team about the tug of war, in particular about not having expected that students would be so keen so that the subsequent discussion would go on for as long as it had. I was thinking aloud about a crucial moment I had experienced in the classroom, perhaps 5 min into the entire discussion, where I was getting ready to move on to the next part of the lesson. I had just provided a summary when Shamir (a seventh-grade student), in his habitually provocative way, made some comment involving a different configuration. I do not know why I picked up on the issue, but it turned out to be productive, leading into another 12 min of further discussion. Perhaps it was a professional sense.

Another team member tells me that Ian, a physically handicapped student moving about in a wheelchair, had not only participated but had also slam shut the brakes of his chair. I realize that I had barely noticed him at the end of the line, hidden by the 20 other students who had eventually lined up in front of him, and I had not noticed at all that the brakes were on. I become aware of the fact that as time went on. I am seeing the events through the filter of my initial articulations of what had happened. When I look at the video today, I cannot see it other than through the filter of knowing that Ian had his brakes slam shut.

Such debriefing sessions, while they allowed me (as all members of the research team) to gain insight as to how others had perceived and remembered some event, still have serious drawbacks. These arise from the fact that any differences in assertions, if there were any, were simply confronting one another. We had no touchstone that we could use to test the different assertions. All we had were our always-positioned recollections of the events, which began to change during the debriefing sessions as others contributed seemingly incontrovertible facts that had escaped our own attention and perception. The videotapes provided us with such a touchstone.

Rough Transcripts and Macroanalysis

A member of the research team subsequently prepared a rough transcript, which was used in our analyses that occurred about 4 weeks later. Our initial transcription looked like this:

Shamir:	((erases diagram)) Banister, long string, Roth, pull here
MR:	OK
Shamir:	then there is a pulley
MR:	and where do you pull?
Shamir:	and then there is another banister, here, and then we pull (.) here ((Design 5)) 11:18 (2)

MR: thank you very much, can you ((sign to sit down))

Aslam: but then Mr Roth doesn't have anything to pull at ((MR applauds him))

Shamir: that's just the thing

Daniel: that's the point

Initially, we did not use the videotape other than to extract what had been said or drawn—in part because we still thought about knowledge in terms of mental models and conceptions that could be recovered or inferred by the analysis of the spoken text. Thus, for example, we were interested in Shamir's pulley-rope-support configuration (Fig. 23.1), attributing it to his conception. We were (as many researchers) not interested in the microlevel details of the utterances or rather constituted noise behind which we wanted to find the underlying knowledge; nor were we interested in understanding the configuration as a situated and interactional achievement.

The videotape allowed me to recover certain facts in an objective way—that 13 of the 26 students in the class had directly contributed to the discussion or the nature of Shamir's configuration. These put my memory traces of the events into relief, now appearing against objectively knowable facts. In the end, I come to the conclusion that Shamir attempted to convince others, but especially me, that his pulley configuration would not disadvantage the class. My sense was that having provided opportunities to talking about and arguing with the diagrams at the chalkboard was an important aspect of the emerging pulley-related discourse. The diagrams had functions similar to their use in scientific discussions and debates. Access to the chalkboard allowed students to engage in multimodal science talk characteristic for scientists, that is, "thinking with

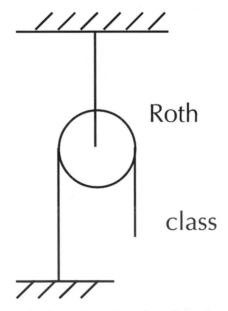

Figure 23.1.   One student's configuration of a pulley system that was to take away from the teacher (Roth) the advantage that he gained from rigging the tug of war.

hands, eyes, and signs" (Roth, 1996) that facilitated the emergence of a pulley discourse in this class. Discussions such as this were also important for introducing students to another aspect of pulley talk; the use of diagrams to represent key aspects that are not as evident in actual models or real pulley systems.

Debriefing and ethnographic field notes, while capturing events in a global manner, have the shortcoming that their contents cannot be verified or replayed; video provides ethnographers with tools to enhance, and sometimes overcome the shortcomings of, other ethnographic means (Spindler & Spindler, 1992). Quite the contrary, because they both rely on recollections, they are subject to transformation and decay of initial memory traces. Here, video and macrolevel analysis provide a means of establishing with some accuracy what has been said, how many individuals have contributed, what the spatial configuration of the speakers was, and so forth (Goldman-Seagall, 1998). However, video allows us to push our analyses to deeper and more accurate levels.

Video and the Analysis of Interactional Turns

Working with video allows me to distance myself from the original experience (or its trace) and thereby enter a whole new level of analysis of the events that I had originally experienced.

To allow the close analysis exemplified later, I digitize the video using iMovie, which I also use to create a Quicktime movie file. I open this file in Peak DV, a software package that researchers can use to work with the audio track (e.g., increasing the volume) and to represent the volume in graphical form. This graphical form aids the precise measurement of the length of words, phonemes, and pauses. Relevant video off prints are created within iMovie, converted into grayscale images, and their size is changed so that the pixel density increases from 72 to 300 points per in for printing purposes. The resulting transcript is evidence for the opportunities that arise from working with video.

During one part of the discussion, I was standing next to the chalkboard filled with representations of pulleys, interacting with Shamir who was proposing a pulley configuration that would have allowed the class to win rather than lose the tug of war against me.

The transcript figures the event in all its material detail that others can (or could) verify against the videotape. Readers will immediately notice that there are pauses throughout the event. Thus, rather than saying "And there is another banister here and we pull here," (Lines 19–25) as if he was simply reading from a mental screen to his audience something that already existed in his mind; Shamir produces the sentences in real time, constrained and enabled by the requirements for writing and drawing as well as gesturing (e.g., Line 28) over those parts already drawn. We see that Shamir is not just articulating a description but in fact gesturing, requiring the operation of muscles, and therefore the enaction of sensorimotor schema and the sensation of forces [e.g., inertial forces and gravity when accelerating the arm upward (line 28)]. In fact, Shamir also engaged his sensorimotor schema to draw the "long string" (Line 04) itself, producing a force in the direction of those that operate in the pulley system and a form of representation in the body that differs from those operating while writing "Roth."

Transcript

01	Shamir:	B*a*nister.
02	Teacher:	Okay.
03		(0.64) ((A student chuckles.))
04	Shamir:	Long string. *
05	Teacher:	Okay. Pulley?
06		(0.30) ((A student chuckles.)) ((Shamir writes "Roth".))
07	Shamir:	*ROth::*
08		(0.52)
09	Teacher:	Okay.
10	Shamir:	(0.30) Pull he::re. *
11		(1.23)
12		↑Ok[ay?
13	Teacher:	[Okay.
14		(0.20)
15	Shamir:	And then there is a pulley. *
16		(0.67)
17	Teacher:	And where do you pull?
18		(1.58) ((Several Ss chuckle.))
19	Shamir:	And there is another banister
20		(1.05) ((Several Ss chuckle.))
21		here
22		(0.61)
23		and we pull
24		(0.81) ((Writes "class."))
25		here. *
26		(2.71) ((Several Ss clap.))
27	Student:	Oh yeah.
28	Shamir:	°You khh- is-° *
29	Student:	Uhhh.
30	Teacher:	Okay. Thank you very much. [Can you?*]
31	Shamir:	[T's just-]
32	Student:	Uhhhm.
33		(0.57)
34	Aslam:	But then, Mister- Doctor Roth has nothing to *pull* at.
35	Shamir:	[That's just it.
36	Teacher:	[((Claps.))
37	Daniel:	*It*'s ↑the point.

Transcript 23.1.

We note that Shamir repeatedly did not answer my query. In Line 05, I asked for the pulley but the student articulated and then wrote "Roth" next to the "long string" that he had just drawn (Lines 07–10). The pause of 0.30 sec (Line 10) was longer than would be expected from the literature on turn taking at talk (ten Have, 1999). However, Shamir was writing. If we do not a priori privilege verbal over other forms of actions, then there is no contradiction, for the action of producing an inscription is still continuing. In Line 17, I asked where the class would pull, but, after a conversationally long pause of 1.58 sec, during which he was drawing another rope, Shamir talked about "another banister" (where the second rope was to be fixed). Ordinary turn taking in everyday conversation "requires" co-participants to respond to the other person. The allotment of turns is particularly structured in formal educational settings where the teacher officially is in charge of turn allocation (Jordan & Henderson, 1995). If Shamir does not respond, he challenges these ordinary practices, and with it, the institutional hierarchy that place student and teacher differently. His challenge is further expressed in Line 28, where he changed his bodily orientation from the chalkboard to face me while gesturally articulating my "weakened" situation in the scenario that he had just finished articulating. Claims that the student challenged and resisted the institutional power hierarchy is further evidenced when he continued gesturing toward the diagram even though I thanked him (Lines 29–30), and even though he began another utterance while pointing to the diagram (Line 31). However, at this point, I explicitly invited him to return ("Can you?" accompanied by a pointing gesture to his seat).

In this situation, rather than playing a passive role, Shamir played an active role in the constitution of the enacted curriculum (Roth & Tobin, 2002). My research in very different classrooms and different countries showed that—consistent with an ethnomethodological framework and opposite claims to the contrary (e.g., Jordan & Henderson, 1995)—students do play active roles in the constitution of school situations. My deeper understanding of the classroom events, which implies and comes with greater familiarity, emerged from my continued engagement with the transcript in increasingly distancing ways. That is, reflection had led to a better practical understanding and better theories at the same time.

By itself, the transcript is available to the teacher-researcher in the way it is to other researchers. Unlike my recollections and my after-the-fact reflections (even ethnographic field notes when I am not teaching), which highlight particular macroevents while omitting everything else, especially microlevel details, the video camera records interactions as they occur. Whereas there is an unavoidable bias in the way the camera always operates from some position and features events under some angle, this bias is relatively constant—although zooming can vary the detail actually available for analysis (e.g., Harel, 1991). Furthermore, whereas our ways of seeing and understanding the recorded video may change—even at the level of physically hearing what is really being said-the record of the event itself is rather constant in contrast to recollection.

This form of analysis comes with the advantage that the interactional details rather than the always hidden intentions and mental content, which constitute the objects of many learning science researchers, come into focus. Studying interactional details makes sense, because these are the same details interaction participants make available to one another, and which therefore are the grounds for subsequent actions. Explanations are constructed for the ways in which interaction participants react to what others make available to them, which is also available to the analyst. Such explanations in my view are more powerful, as they rather than hidden intentions explain the making of classrooms in particular and society more generally. This turns the teacher-researchers' attention away from their own (recollected) intentions to how others use their (material, verbal) actions as resource in their response, and how they dealt with the actions of others as this is revealed on the tape. Collective analysis further objectifies a teacher-researchers' understanding of what has happened in their classrooms.

Collective (Interaction) Analysis

Participation in the collective analysis of videotaped lesson allows me to step back even further, because now I am confronted not only with the microlevel details of the tape but also with the way other analysts see them.

Collective analysis proceeded in the following way. The person at the control ran the video until someone requested to stop to talk about a feature or episode— usually the episodes have a duration of somewhere between several hundred milliseconds to the order of 10 sec. The person requesting the stop pointed out what was salient to him, described and interpreted the episode, and generated hypotheses to be tested in the remainder of the same tape and in the remainder of the database. The other persons also provided their descriptions and interpretations. The episode is discussed until all participants feel that there is nothing more to say about it at the moment—although subsequent periods of writing often turn up additional features, which are discussed during the next meeting on the following day. In this way, we worked image-by-image through the video and, correspondingly line-by-line through the transcript.

Analyzing the tapes from a first-time through perspective assists us in avoiding teleological interpretations, whereby the outcomes of interactions are used to attribute sense to historically earlier interactions. This prevents us from providing sense to the moment of interaction that none of the participants could have, because at the moment of their interaction, they cannot see what the ultimate outcome would be. For example, after viewing the events in Lines 04–05, someone might request a stop and wonder what the sequence means and what its implications are. Initially, I responded to Shamir saying "Okay," thereby acknowledging what had been said and shown and also that I was still following. The utterance of "Pulley?" with a rising inflection is heard as a question. Now one hypothesis might be that I, the teacher, made a move to scaffold the student's construction of a representation. Someone else might suggest that whether a discursive move can be evaluated in terms of the notion of scaffolding, one needs to see how the interactional partner responded—a decision whether a move was scaffolding can then be established only by analyzing the response. Here, Shamir wrote the word "Roth" next to the end of his "long string." Rather than drawing or talking about the pulley, as I had requested, Shamir labeled a part of the diagram. Utterances are designed for recipients and there always exists the expectation that the addressee responds in an appropriate fashion. Shamir did not respond as we assumed. The implications are (a) that my actions did not constitute a scaffold and (b) that Shamir had not submitted to my request, which can be hypothesized to be an instance that the hierarchical relationship between teacher and student has been put into question.

Such hypotheses are noted and subsequently tested in the unfolding events or, retroactively, by referring to earlier events that had not yet been interpreted along these dimensions.

Collective microanalysis of interactions addresses many of the problems that plague other forms of data and analysis. We even videotape these sessions, thereby objectifying the analyses even further. These new tapes are not only records of the different perspectives that have been articulated within the research team, but also they document how scientific explanations emerge and are refined over time. This, then, objectifies the process of theory construction, providing an audit trail, and making it inspectable by peers.

When I look at the experience now, I no longer see how it might have appeared to me at the moment but through all the conversations about and analyses of the events of that day. Reflections not only add to but transform the way the original experience appears to me. I do not remember the moments when I had set up the block and tackle part that was part of the rope system and when I had challenged the students to the tug of war or how I had set it up. I had reflected on the teaching sequence repeatedly and analyzed it in a number of articles and chapters, allowing me, among others, to understand the role of and access to representational tools, interactions with others, and individual and collective learning (e.g., Roth, 1996). It is through these publications that I know that I had set up the ropes and pulleys for the tug of war during recess, when the children had been in the schoolyard on the other side of the building, which allowed me to use a railing just outside a door that led directly from the classroom into the parking lot as a support for the block and tackle.

CODA

The early analyses of this and other episodes from my design experiment allowed me to develop better models of whole-class conversations, particularly the effects differential access to the inscriptions themselves had on the content and form of discursive contributions (e.g., Roth, McGinn, Woszczyna, & Boutonné, 1999). More so, these improved explanations gave me immediate but distancing and objectifying access to my teaching practice, which I immediately adapted, as one might expect from design experiments, to build on the affordances of inscriptions to student–student and student–teacher transactions. The frequently pervasive gap between research (theory) and practice was bridged to the extent that this is possible at all. In this particular study, I increased the use of inscriptions for mediating small-group and whole-class discussions.

Temporal distance to the events appears to assist in creating more objective analyses. Today, 9 years after the episodes, I have forgotten what I felt in the situation. I only identify with the teacher featured on the tapes in so far as he was another instantiation of "me." But at that point I had different ideas, theories, and social relations and therefore was a different person. Today, having no stakes, no face to lose, I find it easy to conduct analyses as if it was any other teacher.

Teacher-researchers who use particular metaphors as referents for their being in the classroom often find these metaphors confirmed when they subsequently reflect on their actions, particularly when they only draw on their recollections of the events (Roth, 2002). Video analysis—for example, in the form of conversation analysis—provides a means of circumventing the negative effects of introspection, for example, the reification of preconceptions and myths. By using multiple data sources, teacher-researchers can indeed arrive at principled analyses of knowing and learning that occur in their classroom, including their own knowing and learning.

The use of video in reflection for understanding teaching practice introduces some real dangers—despite apparently facilitating changes in teacher classroom actions—because teacher-researchers may merely reify discourses without changing beliefs or practices (confirmation bias). Teacher-researchers interested in understanding, explaining, and changing practice require conceptual and methodological tools that allow them to critically engage with their own presuppositions. Because of my interest in the interaction between teacher and students, I chose to take a close look at our conversations at the level of individual utterances to construct an understanding of how participants managed them. Such an understanding was exactly what our previous analysis could not provide.

Using video, teachers as learning-sciences researchers can objectify their practices, that is, create external representations of what they have done that can be shared with and agreed on when they are the objects of collective inquiry. In this way, they can get a handle on the most pressing of scientific problems, taking "as one's object the social work of construction of the pre-constructed object" (Bourdieu, 1992, p. 229) and thereby get a better handle on their own preconceptions that potentially interfere with the construction of sound theories. At the same time, this distancing also develops our practical understanding of how the world of the classroom works. A science that fails to question itself does not know what it does. Video used to record events and their analyses allows us to objectify research and the research process, and therefore develop a more reflexive science of the learning sciences. Because they also experienced the situation as insiders, teachers have unique contributions to make to the theorizing of teaching. I do not suggest that these contributions are to be privileged, because the possibilities and likelihood for confirmatory bias lie particularly near. I understand teachers' contributions to be among many viewpoints, stories, or voices on teaching, learning, and education. Disciplined and principled first-person methodologies that deal with the bias coming from preconceptions, which, as with all methodologies, require substantial training, will allow teachers to gain the next level of depth in their analyses.

ACKNOWLEDGMENTS

The work presented in this chapter was made possible in part by a grant from the Social Sciences and Humanities Research Council of Canada. I am grateful to Sylvie Boutonné, Michelle K. McGinn, and Carolyn Woszczyna who contributed to data collection and transcription.

NOTES

1. I use the following transcription conventions that are based on the work of Gale Jefferson (e.g., ten Have, 1999).

    *—denotes the moment in the transcript that corresponds to the video offprint on the right;

    ((Nods.))—Transcribers comments including descriptions of salient and relevant actions are noted are enclosed in parentheses;

    (0.79)—pause in one-hundredths of a second;

    °Compounds°—words between degree signs are spoken with very low, almost unnoticeable voice;

    [ ]—Square brackets in consecutive lines indicate beginning of overlapping speech;

    What?—Words spoken with clearly hearable emphasis are italicized;

    (?)—Question mark in parentheses indicates inaudible utterance(s);

    (No?)—Indicates uncertain hearing of the utterance "No";

    —— n-dash marks sudden stop in utterance;

    lo::ng—Lengthening of a phoneme is indicated by colon;

    , . ! ?—punctuation marks are used to mark intonations rather than as grammatical markers;

    Right?—The arrow indicates a rise in the pitch much sharper and more clearly noticeable than normally occur in questions.

REFERENCES

Bourdieu, P. (1990). *The logic of practice*. Cambridge, England: Polity Press.

Bourdieu, P. (1992). The practice of reflexive sociology (The Paris workshop). In P. Bourdieu & L. J. D. Wacquant (Eds.), *An invitation to reflexive sociology* (pp. 216–260). Chicago: University of Chicago Press.

Derrida, J. (1995). *Points ... Interviews 1974–1994*. Stanford, CA: Stanford University Press.

Erickson, G. L., & MacKinnon, A. (1991). Seeing classrooms in new ways: On becoming a science teacher. In D. Schön (Ed.), *The reflective turn: Case studies in and on educational practice* (pp. 15–36). New York: Teachers College Press.

Fendler, L. (2003). Teacher reflection in a hall of mirrors: Historical influences and political reverberations. *Educational Researcher, 32*(3), 16–25.

Gallagher, S. (2002). Experimenting with introspection. *Trends in Cognitive Sciences, 6*, 374–375.

Garfinkel, H. (1967). *Studies in ethnomethodology*. Englewood Cliffs, NJ: Prentice-Hall.

Goldman-Segall, R. (1995). Configurational validity: A proposal for analyzing multimedia ethnographic narratives. *Journal for Educational Multimedia and Hypermedia, 4*, 163–182.

Goldman-Segall, R. (1998). *Points of viewing children's thinking: A digital ethnographer's journey*. Mahwah, NJ: Lawrence Erlbaum Associates.

Goldman-Segall, R., Riecken, T., Adams, A., Taaffe, K. D., Isles, J., & van Osterhout, W. (1993). Growth of a multimedia school culture: A multi-voiced narrative. *Arachnet Electronic Journal on Virtual Culture, 1*(7).

Hammer, D. (1996). Misconceptions or p-prims: How may alternative perspectives of cognitive structure influence instructional perceptions and intentions? *The Journal of the Learning Sciences, 5*, 97–127.

Harel, I. (1991). The silent observer and holistic note taker: Using video for documenting a research project. In I. Harel & S. Papert (Eds.), *Constructionism: Research reports and essays, 1985–1990* (pp. 449–465). Norwood, NJ: Ablex.

Jordan, B., & Henderson, A. (1995). Interaction analysis: Foundations and practice. *The Journal of the Learning Sciences, 4*, 39–103.

Lampert, M. (1998). Studying teaching as a thinking practice. In J. Greeno & S. V. Goldman (Eds.), *Thinking practices in mathematics and science learning* (pp. 53–78). Mahwah, NJ: Lawrence Erlbaum Associates.

Leont'ev, A. N. (1978). *Activity, consciousness and personality.* Englewood Cliffs, NJ: Prentice Hall.

Mikhailov, F. (1980). *The riddle of self.* Moscow: Progress.

Piaget, J. (1970). *Genetic epistemology.* New York: Norton.

Ricœur, P. (1991). *From text to action: Essays in hermeneutics, II.* Evanston, IL: Northwestern University Press.

Roth, W.-M. (1996). Thinking with hands, eyes, and signs: Multimodal science talk in a grade 6/7 unit on simple machines. *Interactive Learning Environments, 4,* 170–187.

Roth, W.-M. (2002). *Being and becoming in the classroom.* Westport, CT: Ablex.

Roth, W.-M. (2004). Das Video als Mittel der Reflektion über die Unterrichtspraxis [Video as a means for reflecting on teacher practice]. In M. Welzel & H. Stadler (Eds.), *Nimm doch mal die Kamera: Zur Nutzung von Videos in der Lehrerausbildung-Beispiele und Empfehlungen aus den Naturwissenschaften* [Why don't you take the camera: On the use of videos in teacher training—Examples and recommendations from the natural sciences] (pp. 4–21). Münster: Waxman-Verlag.

Roth, W.-M., McGinn, M. K., Woszczyna, C., & Boutonné, S. (1999). Differential participation during science conversations: The interaction of focal artifacts, social configuration, and physical arrangements. *The Journal of the Learning Sciences, 8,* 293–347.

Roth, W.-M., & Tobin, K. G. (2002). *At the elbow of another: Learning to teach by coteaching.* New York: Peter Lang.

Schön, D. A. (1987). *Educating the reflective practitioner.* San Francisco: Jossey-Bass.

Spindler, G., & Spindler, L. (1992). Cultural process and ethnography: An anthropological perspective. In M. LeCompte, W. Millroy, & J. Preissle (Eds.), *The handbook of qualitative research in education* (pp. 53–92). New York: Academic Press/Harcourt Brace.

Stevens, R., & Hall, R. (1997). Seeing tornado: How video traces mediate visitor understandings of (natural?) spectacles in a science museum. *Science Education, 81,* 735–748.

Stevens, R., & Toro-Martell, S. (2003). Leaving a trace: Digital meta-exhibits for supporting visitors to represent and exchange their ideas about museum exhibits. Supporting museum visitor interpretation and interaction with digital media annotation system. *Journal of Museum Education, 28*(2), 25–31.

ten Have, P. (1999). *Doing conversation analysis: A practical guide.* London: Sage.

van Zee, E., & Minstrell, J. (1997). Using questioning to guide student thinking. *The Journal of the Learning Sciences, 6,* 227–269.

24

The Development of Teachers' Professional Vision in Video Clubs

Miriam Gamoran Sherin
Northwestern University

What is it that enables the archeologist to see a collection of stones as part of a larger structure that once existed? Why is it that a meteorologist can look at the sky and recognize patterns in the shape and coloring of clouds? Goodwin (1994) provides one answer by introducing the notion of professional vision. As Goodwin explains, professional vision involves "socially organized ways of seeing and understanding events that are answerable to the distinctive interests of a particular social group," (p. 606). In other words, members of a professional group develop specific ways to interpret the phenomena that are the focus of their work.

This chapter is concerned with teachers' professional vision (Sherin, 2001). Although professional vision is only one component of teaching expertise, it is of particular importance today in light of the demands that mathematics education reform places on teachers. Traditionally, the structure of a mathematics lesson was determined by the teacher prior to instruction. Reform efforts, however, call for teachers to base their instruction, at least in part, on the lesson as it unfolds in the classroom with close attention given to the ideas that students raise (National Council of Teachers of Mathematics, 2000; Smith, 1996). Thus, teachers need to be able to "see" how a lesson is going and to interpret students' ideas in the midst of instruction. Even veteran teachers who are already skilled at interpreting classroom events may need to develop a new kind of professional vision. That is, they may need to learn to see different kinds of events and in different ways. An important challenge then, is to provide teachers with experiences in which they can develop this new kind of professional vision.

In this chapter, I examine the development of professional vision as teachers participate in a professional development program called video clubs. In video clubs, groups of teachers watch and discuss excerpts of videos from their classrooms. In many ways, the use of video may seem like an obvious choice for helping teachers to develop professional vision. Video appears to be able to capture much of the richness of classroom interactions and it can be used in contexts that allow teachers time to reflect on these interactions. Yet despite a wealth of video-based professional development programs, there is little empirical evidence to support the claim that viewing videotapes of classroom instruction can be an effective context for teacher learning (Fuller & Manning, 1973; McIntyre, Byrd, & Foxx, 1996; Sherin, 2004). As a result, the goals of this research are two-fold. First, I want to propose that professional vision offers a productive way to conceptualize what it means to examine the role of video in teacher learning. Second, I want to explore the ways in which participating in video clubs can support the development of teachers' professional vision.

In the remainder of this chapter, I first define more explicitly what it means for teachers to have professional vision. I then describe the results of a year-long study in which four middle-school mathematics teachers participated in monthly video club meetings with a facilitator.

WHAT DOES IT MEAN FOR TEACHERS TO HAVE PROFESSIONAL VISION?

Unlike archeologists who examine stones and sand, the phenomena that are of interest to teachers are classroom events. Thus, for teachers, professional vision involves the ability to make sense of what is happening in their classrooms. And like all perceptual processes, professional vision is not a simple passive observation of the world. Instead, it involves a dynamic interplay of top-down and bottom-up processes. As a teacher observes a classroom, he or she is constantly reasoning about what is seen, and this drives where and how the teacher will look in the future. Although this dynamic interplay is complex, here I simplify somewhat and describe professional vision as consisting of two distinct subprocesses; (a) selective attention, and (b) knowledge-based reasoning.

Classrooms are complex environments, with much happening simultaneously. A teacher cannot pay attention to everything with equal weight; instead certain things will stand out to the teacher. It is this process that I refer to as selective attention. For instance, a teacher's attention may be drawn to the mathematical ideas that are being discussed in class. Alternatively, a teacher may focus on the ways in which students interact with each other—who is speaking to whom and in what manner. In still another case, a teacher may pay particular attention to the level of noise in the classroom as a way of keeping track of how a lesson is proceeding. In each of these examples, the teacher's attention is tuned to particular kinds of events that take place in the classroom.

Other researchers describe related phenomena. Specifically, Frederiksen (1992) defines the notion of a "call-out" in which teachers literally call out what appears noteworthy to them in a videotaped mathematics lesson. Similarly, Jacobs and Morita

(2002) describe "stopping points" in which a teacher pauses a videotape of a mathematics lesson in order to comment on the instruction. Finally, van Es and Sherin (2002) use the term "noticing" to describe the process through which a teacher identifies what is important in a classroom episode. Common to all of these is the idea that as a teacher views instruction, whether live or via video, certain aspects of the classroom interactions receive more attention from the teacher than do others.

The second process is what I refer to as knowledge-based reasoning. The idea is that once the teacher's attention is drawn to a particular event, next the teacher will begin to reason about that event based on his or her knowledge and understanding. Some examples of this reasoning process that are particularly important for teachers are described in van Es and Sherin (2002). These include using what one knows about the subject matter to make sense of an idea that has been raised, using what one knows about the classroom context to understand why a person spoke or acted in a particular way, and making connections between the specifics of the classroom and broader principles of teaching and learning.

Although I have suggested here that a teacher will first notice a classroom event and will then begin to reason about it, the situation is, in practice, much more complicated. To reiterate, selective attention and knowledge-based reasoning interact in a dynamic manner. On the one hand, what stands out to a teacher will certainly influence the reasoning that takes place. But in addition, a teacher's expectations and knowledge also drive what a teacher perceives.

In this chapter, I examine the development of teachers' professional vision by looking at changes in each of these processes as teachers participated in a series of video club meetings. Specifically, I argue that, over time, there was a shift in the teachers' selective attention as they began to focus on different kinds of events that were visible in the video excerpts. In addition, I claim that the teachers developed new techniques for reasoning about the phenomena that they viewed on video. I will also briefly consider the interaction between these two processes.

RESEARCH DESIGN

Four middle-school mathematics teachers participated in a monthly video club with a facilitator across one school year. All of the teachers taught at the same public school outside a large U.S. city. The group of teachers included one female and three male teachers, and their teaching experience ranged from 1 to 28 years.

The video clubs took place after school once a month for approximately 40 min each. Prior to each meeting, the facilitator videotaped in one of the teacher's classroom and later met with that teacher to review potential excerpts to show in the video club. The selected video excerpts were generally from whole-class discussions and lasted between 5 and 7 min. At the beginning of each meeting, the teacher whose video was being viewed provided background information on the lesson that appeared in the video. Transcripts of the excerpt along with any handouts used in class were then passed out to the participants and the video was played. After viewing the video excerpt, the facilitator typically began by asking the teachers "What did you notice?" or "Any comments?"

In all, seven video club meetings took place across the school year.[1] Each of these meetings was videotaped and transcribed. The teachers also participated in an individual interview following the final meeting in which they discussed their experiences in the video clubs.

Data Analysis

Data analysis consisted of two main components. First, to examine to what the teachers attended in the video, I chose to focus on the topics that they raised for discussion. Clearly, topic alone does not come close to capturing all of the complexity that is involved in teachers' selective attention. Nonetheless, I felt that topic could serve as a useful indicator of changes occurring to the teachers' selective attention.

To study the topics raised, two researchers independently examined the transcripts of the video clubs and noted where there was a change in the topic of conversation. This process is similar to what Jacobs and Morita (2002) describe as dividing up a transcript into "idea units." Initial agreement among the researchers on the resulting discussion segments was 90.8%. Points of disagreement were then reviewed together and consensus was reached.

Next, select segments from each of the video club meetings were used to identify the different topics that were discussed by the teachers. Five topics were identified; (a) pedagogy, (b) student conceptions, (c) classroom discourse, (d) mathematics, and (e) other. Segments relating to pedagogy concern what the teacher in the video is doing and saying. Segments about student conceptions concern what the students understand about mathematics in a lesson. The third category, classroom discourse, has to do with the ways that the teacher and students talk with each other during class, that is, how many students participate in a discussion or whether students' comments are directed to the teacher or to other students. Next, mathematics involves the teachers' own ideas about the mathematics in a lesson. The final category, other, includes comments that do not fit into any of other four categories, for instance, comments about technical aspects of the video that was viewed.

Using these five categories, the researchers coded the discussion segments that had been identified previously. This involved independently assigning one of the five categories to each segment. Initial agreement among the researchers was 86.6%. After reviewing the points of disagreement, consensus was reached on all segments. Following this, the amount of time spent discussing each topic in the different video club meetings was calculated. It was also noted whether each segment was initiated by one of the teachers or by the facilitator.

The second component of analysis focused on the ways in which the teachers reasoned about what they noticed in the video. Again, I did not try to capture all of the complexity involved in teachers' knowledge-based reasoning. Instead, I focused on the ways in which the teachers reasoned about the two topics that were discussed most

---

[1]The group actually met 10 times during the school year, although the final three meetings had a different format than the initial seven video club meetings.

frequently in the video clubs, student conceptions (42% of the total time) and pedagogy (35% of the total time). Specifically, all discussion segments related to these two topics were examined for differences in the ways that the teachers approached these topics over time.

RESULTS

During the course of the video clubs, the teachers' professional vision developed in important and interesting ways. First, in terms of selective attention, the teachers began to pay attention to new aspects of classroom events. Second, with respect to knowledge-based reasoning, the teachers developed new techniques for thinking about what they noticed. In addition, the data provides evidence of the complex relationship between these two processes, specifically, that changes in selective attention influenced the teachers' knowledge-based reasoning and vice versa.

The Development of Selective Attention

There are many ways in which a teacher could begin to attend to different aspects of classroom events. Here, I examine this issue by investigating the topics that the teachers raised for discussion across the seven video clubs. As described in Sherin and Han (2004), in the first video club meeting, the comments raised by the teachers focused primarily on pedagogical issues—on what the teacher in the video was doing. For example, the teacher whose video was shown to the group asked if he had made the right decision by discussing a particular issue with the class. "What I wanted to [ask] was, how do you decide? [Because] you know, as teachers you make decisions right on the spot about explore it or don't explore it." Later another participant asked about the goals of the teacher in the video. "Did you actually have a plan or did you want to [see what would happen]?"

In contrast, in the seventh video club, the teachers' comments focused much less on pedagogy. Instead, the teachers appeared to be primarily concerned with issues relating to student conceptions—to what the students in the video understood about the mathematical ideas raised in class. Specifically, early in this video club, one of the teachers mentioned different approaches that students seemed to be using to examine two related sets of data. "A lot of them were comparing the [data sets] instead of just looking at one list of numbers." This prompted a lengthy discussion concerning which of the two data sets appeared to be most relevant to students and whether students were "looking at the relationship between the two lists." The teachers also initiated an extended discussion of one student's thinking, trying to make sense of what this student, Brenda, understood about the data presented. Later, the teachers considered the different kinds of comments students made relating to the idea of correlation. "[Are students talking about] the difference between correlation and cause and effect?" Finally, prior to the conclusion of the video club, one of the teachers raised a pedagogical issue, asking about the teacher's goals for the lesson and whether the ways in which the students had discussed the data aligned with what the teacher hoped to achieve in class that day.

The teachers' attention in these two video clubs was drawn to very different kinds of events. In the first video club, the teachers were mainly concerned with what the teacher in the video was doing and saying. To be clear, there was some discussion of student conceptions during the first video club, but it was almost exclusively the facilitator who raised the topic. On the contrary, in the seventh video club, the teachers themselves initiated a great deal of discussion about student conceptions, and the facilitator played a much more minor role.[2] Issues of pedagogy were still of interest to the teachers, but in addition, what the students were doing and saying about mathematics had become more visible to the teachers—their selective attention had become attuned to an additional topic.

This pattern of increasing attention to student conceptions over time was confirmed through a detailed analysis of all seven video club meetings. Table 24.1 summarizes some of what was found. Specifically, Table 24.1 shows the percent of teacher-initiated segments of conversation by topic across the seven video clubs. First, note that student conceptions and pedagogy were in fact the two most commonly raised topics by the teacher.[3] Second, looking at the data for these two topics illustrates that while pedagogy remained an important focus of attention for teachers throughout the video clubs, the teachers also began to generate discussions of student conceptions. Specifically, in video club meetings 1 and 3, less than 15% of the issues raised by the teachers concerned student conceptions and 50% or more related to pedagogy.[4] In video clubs 4, 5, and 6, student conceptions were raised, on average, 40% of the time, whereas pedagogical issues were brought up 36% of the time. In the final meeting, video club 7, the teachers initiated discussion of student conceptions 86% of the time, whereas pedagogical issues were raised only once, 14% of the time. It is not clear whether video club 7 is representative of a trend that would have continued had there been additional video clubs meetings. Nonetheless, the data suggest that the teachers came to pay greater attention to student conceptions over the course of the video clubs.[5]

To be clear, on its own, the data presented in Table 24.1 are not intended to establish, with any confidence, changes in the selective attention of teachers. My primary argument for changes in selective attention relies on the detailed qualitative

---

[2] In video club 1 the facilitator initiated 89% of the discussion segments related to student conceptions. In video club 7, the facilitator initiated only 25% of the discussion segments related to student conceptions. In general, the facilitator's participation in the video clubs decreased over time.

[3] Segments coded as other comprised only 4% of the teacher-initiated discussion segments and are not included in Table 24.1.

[4] Video club 2 is somewhat of an anomaly in that the teachers initiated almost twice as many discussion segments overall than in video clubs 1 and 3, and more of these segments concerned student conceptions than pedagogy. I believe, however, that in video club 2, the teachers' goal was simply to raise a variety of issues for discussion and that they were less concerned at this point with whether or not an issue raised was particulary noteworthy. In fact seven of the fifteen discussion segments initiated by the teachers in video club 2 were discussed for less than 20 seconds. This included three of the five segments concerning student conceptions.

[5] Here the unit of analysis is the group of teachers. Although not the focus of the current study, there are indications in the data to suggest that the trends reported here hold across the individual teachers as well.

TABLE 24.1
Percent (and Number) of Teacher-Initiated Segments of Discussion Per Topic

	Video Club 1	*Video Club 2*	*Video Club 3*	*Video Club 4*	*Video Club 5*	*Video Club 6*	*Video Club 7*
Student Conceptions	14%(1)	33%(5)	13%(1)	50%(4)	40%(4)	30%(3)	86%(6)
Pedagogy	57%(4)	27%(4)	50%(4)	38%(3)	30%(3)	40%(4)	14%(1)
Discourse	14%(1)	27%(4)	25%(2)	13%(1)	10%(1)	20%(2)	0%(0)
Mathematics	14%(1)	13%(2)	13%(1)	0%(0)	20%(2)	10%(1)	0%(0)

Note. Due to rounding, some of the percent totals may add up to more than 100%.

analysis described in Sherin and Han (2004). Table 24.1 is intended only to provide a summary for the reader of what was observed, and to point to the plausibility of my overall conclusions.

The Development of Knowledge-Based Reasoning

In using the phrase knowledge-based reasoning, I mean to encompass a broad range of cognitive processes. In this section, however, I focus on the teachers' reasoning about issues related to student conceptions. As stated previously, over the course of the video clubs, the teachers began to raise issues of student conceptions with increasing frequency. Here I make a different claim—that the teachers developed new ways to reason about student conceptions

As described in Sherin and Han (2004), when exploring student conceptions, the teachers engaged in three different levels of analysis, each representing an increasingly complex way to explore the ideas that students raised in the video. First, comments at Level 1 generally involved simply reading what a student had said directly from the transcript of the video. For example, in video club 1 the facilitator asked the teachers what the students had said about a particular graph and the teachers responded by listing statements from the transcript. "[Amy] says 'It's not very realistic.'" "Ben says, 'I goofed.'" At Level 2, the teachers' comments involved beginning to analyze students' ideas. Thus, they went beyond simply restating what a student had said and tried to make sense of the meaning of a students' comment or method. For example, in video club 3, the teachers tried to understand what one student meant when she said that a line was "straighter" and that a straight line is "like a 90 degree angle." Level 3 comments consisted of generalization and synthesis of the students' ideas. At this level, the teachers looked at connections across the ideas of several students, or at how a specific idea related to other mathematical concepts that had been explored by the students in the video. An example of Level 3 analysis occurred during video club 7 when the teachers discussed the issue of correlation, and tried to understand the different ways that students addressed this issue in the video.

Although over time the teachers examined student conceptions at all three levels, in the early video club meetings the teachers engaged heavily in Level 1 analysis. In addition, during the first video club, there were two occasions in which the teachers did not follow up on the facilitator's suggestion that they consider issues related to student thinking. Instead, the teachers proceeded to change the topic of discussion. Beginning with video club 4, however, the teachers began primarily to discuss student conceptions at Level 2. Table 24.2 illustrates this pattern by indicating the highest level of analysis achieved for each discussion segment that was coded as relating to student conceptions.

To be clear, there were instances early on in which the teachers engaged in Level 2 and Level 3 analyses. However, these were heavily scaffolded by the facilitator. In contrast, in the later video clubs, the facilitator played a less central role and the level of analysis was determined instead by the participating teachers. I return to this issue of the decreasing influence of the facilitator in the next section of this chapter.

Thus, during the course of the video clubs, the teachers came to think about issues of student conceptions in new ways. Yet, these changes raise questions concerning what was actually learned by the teachers. Are we seeing deep changes in the reasoning that the teachers were capable of doing? Or are these changes representative merely of changes in inclination or in how the teachers saw the task as presented by the facilitator in the video club? To make progress on these questions, I believe that it is necessary to look at the interaction between selective attention and knowledge-based reasoning.

Exploring the Relationship Between Selective Attention and Knowledge-Based Reasoning

Thus far, I have examined the development of teachers' selective attention and knowledge-based reasoning independently, and have pointed to possible changes in

TABLE 24.2
Teachers' Analysis of Student Conceptions

Discussion segments concerning student conceptions	Video Club 1	Video Club 2	Video Club 3	Video Club 4	Video Club 5	Video Club 6	Video Club 7
No response to facilitator's prompt to discuss student ideas	2	0	0	0	0	0	0
Level 1: Teachers quote student statements	4	4	2	1	0	1	0
Level 2: Teachers explore meaning of student statements	1	3	1	4	4	1	4
Level 3: Teachers synthesize student ideas	2	1	1	1	1	1	4

each of these areas. Yet as described earlier, in practice, there exists a dynamic relationship between these two processes. Here I examine this interaction at two time scales.

Interactions Between Selective Attention and Knowledge-Based Reasoning at a Broad Time Scale. I first consider the interaction between selective attention and knowledge-based reasoning at a broad time scale. Across the hours and weeks during which the teachers participated in the video clubs, their new focus on student conceptions prompted the development of new reasoning processes in two ways. First, as described already, paying more attention to events related to student conceptions drove the development of increasingly sophisticated strategies for reasoning about student conceptions. Second, this new focus on student conceptions also influenced how the teachers reasoned about pedagogical issues.

When the teachers discussed pedagogical issues, they tended either to explore alternative pedagogical strategies that the teacher in the video might have used or to offer explanations for why the teacher in the video used a particular approach. In either case, student thinking was initially not a factor in trying to understand the teachers' actions. Later however, discussions of pedagogical issues were closely tied to student thinking (Table 24.3). Consider, for example, the pedagogical issue that came up at the end of video club 7. The group's discussion of whether the teacher's goals for the lesson had been achieved was based heavily on what the group had come to recognize concerning student understanding of the data. For instance, the teacher whose video was viewed explained that he "wasn't asking [students] to decide if they thought there was [a] correlation" and that having students raise a variety of ideas about the data matched his goals for the day. This stands in sharp contrast to the group's discussion, during video club 1, of the teacher's goals for the lesson. In that case, the group focused on the teacher's plan independent of students and asked specifically about the teacher's decision to introduce the lesson in a particular way.

In sum, I claim that the teachers' increased attention to student conceptions had a powerful influence on the way that they reasoned about what they noticed. Not only did the teachers develop new techniques for analyzing student thinking, they also began to reason about pedagogical issues in terms of student conceptions. Moreover, I suggest that these developments in the teachers' knowledge-based reasoning ultimately influenced their selective attention. To explain this, consider once again the different levels at which the teachers engaged in discussions of student conceptions.

TABLE 24.3
Connecting Issues of Pedagogy and Student Thinking

Teacher-Initiated Discussion Segments Concerning Pedagogy	Video Club 1	Video Club 2	Video Club 3	Video Club 4	Video Club 5	Video Club 6	Video Club 7
Discussed independent of student thinking	4	4	4	1	0	0	0
Discussed in light of student thinking	0	0	0	2	3	4	1

Table 24.4 is similar to Table 24.2 except that it distinguishes between those discussion segments that were initiated by the teachers and those that were initiated by the facilitator. What I want to point out is that in video clubs 1–3, all Level 3 analysis involved discussion segments that were initiated by the facilitator. Thus, it was the facilitator who selected the issues to be analyzed at this complex level. Beginning in video club 4, however, the teachers initiated discussion segments that involved Level 3 analysis. Therefore, in addition to my earlier claim that the teachers began to reason about student thinking in more complex ways over time, here I argue that they also began to "see" more complex issues concerning student thinking in the video excerpts—that what they noticed about student thinking in the video was itself more complex as a result of their experiences in reasoning about these types of events.

Interactions Between Selective Attention and Knowledge-Based Reasoning at a Narrow Time Scale. Selective attention and knowledge-based reasoning also interact at a much more narrow time scale. Within a single "discussion segment," the teachers were likely to engage in several cycles of attending to and reasoning about an event. If our goal is to understand the development of professional vision, then it is critical to examine the interaction at this level. A teacher may be able to learn to "see" something interesting or to engage in a certain kind of analysis, but being able to use these processes fluidly and in support of each other indicates an important level of attainment in the development of professional vision. Here I present one example of this kind of interaction taken from video club 7.

As described briefly earlier in this chapter, in video club 7, the teachers viewed a video clip in which students were examining the relationship between two sets of data. After discussing some of the different ways in which the students were exploring the data, one of the teachers asked "What is Brenda talking about?" Given what Brenda had said, the teacher was unsure whether Brenda was analyzing only one or both sets of data. The teachers then worked together to understand the meaning of Brenda's idea,

TABLE 24.4
Comparison of Teacher- and Facilitator-Initiated Analyses of Student Thinking

		Video Club 1	Video Club 2	Video Club 3	Video Club 4	Video Club 5	Video Club 6	Video Club 7
Teacher-initiated segments	Level 1	1	3	0	0	0	1	0
	Level 2	0	2	1	3	3	1	3
	Level 3	0	0	0	1	1	1	3
Facilitator-initiated segments	No response	2	0	0	0	0	0	0
	Level 1	3	1	2	1	0	0	0
	Level 2	1	1	0	1	1	0	1
	Level 3	2	1	1	0	0	0	1
Total		9	8	4	6	5	3	8

paying particular attention to Brenda's claim that "there's a medium correlation." The teachers discussed the meaning of the term correlation and what Brenda might understand about this idea. They then came up with two different ideas for what Brenda might have been trying to say. One idea was that Brenda was looking at a single set of data and was informally correlating the data to the order of the numbers. A second idea was that Brenda was in fact considering both sets of data and "realizes that the numbers … in the left column are going up … [but] she sees some inconsistency." Unable to reach consensus, the teachers decided to watch more of the video from the class. They did so, paying close attention to Brenda's participation. Using this new information, the teachers continued their discussion of Brenda's idea, finally coming to consensus concerning the idea that she had been trying to share in class.

This example illustrates a complex interaction between selective attention and knowledge-based reasoning. First, selective attention is at the fore as one of the teachers noticed that what Brenda said did not make sense to him and he raised this issue with the group. In response, the group as a whole examined Brenda's statements and tried to understand their meaning—they drew on their knowledge-based reasoning skills to make sense of Brenda's ideas. In doing so, the group pointed to Brenda's use of the term "medium correlation" as particularly important. Thus, in the midst of their reasoning about Brenda's ideas, selective attention was still in play as they chose which part of Brenda's statement required special attention. With this in mind, the teachers reasoned further about Brenda's idea but were unable to come to consensus. At this point, the teachers turned once again to selective attention, looking for more information about Brenda's thinking. Based on what they now noticed, the teachers reasoned additionally and were able to come to consensus.

I now return to my question concerning the depth of the changes that occurred in the teachers' attention to and reasoning about student conceptions. Although I believe that the changes described earlier in this chapter are impressive and nontrivial, the extended interactions of the type that I describe earlier are more indicative of deep change. Thus, I suspect that at some level, it may be a rather simple matter to get teachers to, in some cases, attend to and reason about student conceptions. In contrast, it will turn out to be much more difficult, through professional development, to have teachers engage in the sort of extended reasoning illustrated earlier. Firmly establishing this hypothesis is a matter for future work.

CONCLUSIONS AND IMPLICATIONS

In this chapter, I have tried to show that while participating in a series of video clubs, teachers developed professional vision. In particular, they began to pay close attention to student thinking and began to reason about what they noticed in new ways. Furthermore, these processes interacted in powerful ways as teachers tried to make sense of what they viewed on the video. The teachers learned through their participation in the video clubs—and learned in ways that are likely to help support their efforts to implement mathematics education reform. As described in prior research, a focus on student thinking is critical to the successful implementation of reform (Ball, 1997; Fennema et al., 1996; Sherin, 2002) and connecting ideas about pedagogy with student thinking is also a powerful component of reform-based instruction (Franke, Carpenter, Levi, & Fennema, 2001).

What I have presented here is more than a story of teacher learning. In introducing the notion of teachers' professional vision, I have extended current approaches for conceptualizing teacher cognition. In addition, this research establishes the importance of examining professional vision, in particular, in order to better understand the affordances of video for teacher education. Finally, if we agree that the development of teachers' professional vision is an important goal, then this research suggests that we pay close attention to the design and implementation of video-based professional development, for such contexts may be uniquely situated to helping develop teachers' professional vision.

Many questions remain to be answered in future work, two of which I mention here. First, what is the relationship between teachers' professional vision as it is used in video clubs and teachers' professional vision during instruction? In other words, how do changes in professional vision that occur in a video club influence teachers' instructional practices? Second, how might video clubs be designed to support the development of professional vision in particular ways? Here I claimed that teachers learned as a result of participating in a series of video clubs, but I did not look closely at the design of the video clubs themselves. For example, how did the facilitator's role influence the learning that took place? How did the particular video clips influence what teachers were able to "see?"

ACKNOWLEDGMENTS

This research was supported by a post-doctoral fellowship from the National Academy of Education and the Spencer Foundation and by the National Science Foundation under Grant Nos. 0133900. The opinions expressed are those of the author and do not necessarily reflect the views of the supporting agencies. The author wishes to thank Bruce Sherin for his thoughtful comments on an earlier version of this manuscript.

REFERENCES

Ball, D. (1997). From the general to the particular: Knowing our own students as learners of mathematics. *The Mathematics Teacher, 90*(9), 732–737.

Fennema, E., Carpenter, T. P., Franke, M. L., Levi, L., Jacobs, V. R., & Empson, S. B. (1996). A longitudinal study of learning to use children's thinking in mathematics instruction. *Journal for Research in Mathematics Education, 27*, 458–477.

Franke, M. L., Carpenter, T. P., Levi, L., & Fennema, E. (2001). Capturing teachers' generative change: A follow-up study of professional development in mathematics. *American Educational Research Journal, 38*(3), 653–689.

Frederiksen, J. R. (1992, April). *Learning to "see": Scoring video portfolios or "beyond the hunter-gatherer in performance assessment.* Paper presented at the annual meeting of the American Educational Research Association, San Francisco.

Fuller, F. F., & Manning, B. A. (1973). Self-confrontation reviewed: A conceptualization for video playback in teacher education. *Review of Educational Research, 43*(4), 469–528.

Goodwin, C. (1994). Professional vision. *American Anthropologist, 96*, 606–633.

Jacobs, J. K., & Morita, E. (2002). Japanese and American teachers' evaluations of videotaped mathematics lessons. *Journal for Research in Mathematics Education, 33*(3), 154–175.

McIntyre, D. J., Byrd, D. M., & Foxx, S. M. (1996). Field and laboratory experiences. In J. Sikula (Ed.), *Handbook of research on teacher education* (pp. 171–193). New York: Simon & Schuster.

National Council of Teachers of Mathematics. (2000). *Principles and standards for school mathematics.* Reston, VA: Author.

Sherin, M. G. (2001). Developing a professional vision of classroom events. In T. Wood, B. S. Nelson, & J. Warfield (Eds.), *Beyond classical pedagogy: Teaching elementary school mathematics* (pp. 75–93). Mahwah, NJ: Lawrence Erlbaum Associates.

Sherin, M. G. (2002). When teaching becomes learning. *Cognition and Instruction, 20*(2), 119–150.

Sherin, M. G. (2004). New perspectives on the role of video in teacher education. In J. Brophy (Ed.), *Using video in teacher education* (pp. 1–27). New York: Elsevier.

Sherin, M. G., & Han, S. (2004). Teacher learning in the context of a video club. *Teaching and Teacher Education, 20,* 163–183.

Smith, J. P. (1996). Efficacy and teaching mathematics by telling: A challenge for reform. *Journal for Research in Mathematics Education, 27*(4), 458–477.

van Es, E. A., & Sherin, M. G. (2002). Learning to notice: Scaffolding new teachers' interpretations of classroom interactions. *Journal of Technology and Teacher Education, 10*(4), 571–596.

Teaching in and Teaching From the Classroom: Using Video and Other Media to Represent the Scholarship of Teaching and Learning

Desiree H. Pointer Mace
The Carnegie Foundation for the Advancement of Teaching

Thomas Hatch
The Carnegie Foundation for the Advancement of Teaching

Toru Iiyoshi
The Carnegie Foundation for the Advancement of Teaching

PROLOGUE: SHARING TEACHING EXPERTISE

All day long, Yvonne Divans Hutchinson demonstrated, encouraged, celebrated, and guided students through an active and critical reading process that undercut the common perception that reading simply involved the decoding of words, that print had single, basic meanings that students had to decipher quietly and store away. She had students write in a "reading journal" a dialogue between themselves and the author of whatever book they were currently reading, agreeing, disagreeing, sympathizing, questioning—engaging the ideas in the pages.

—Mike Rose (1995, p. 8)

Yvonne Divans Hutchinson has worked throughout her 38 years teaching middle and high school English in central Los Angeles to address the gap she has observed between the considerable oral language talents that her students bring and the literary, academic discourse that she expects of them. Her passion, drive, and focus are anchored in a deep commitment to social justice, and have had positive impact on the students, student teachers, and fellow colleagues with whom she's worked over the years. Although books like *Possible Lives* inspire us with descriptions of her outstanding classroom practice, few have had an opportunity to see what she does in her work with students or to analyze the strengths and weaknesses in her practice. As Hutchinson's work has become known through *Possible Lives*, and her involvement in the National Board for Professional Teaching Standards and the National Writing Project, in fact, the interest in learning more about her work has led a number of individuals to ask for copies of the videotapes that served as the basis for her entries in her National Board Portfolio. Beyond those homemade videos, however, lies a slew of other materials and artifacts, including Hutchinson's own writings, classroom assignments, and student work, that can help to bring her teaching into focus.

Like exploring a new frontier, developing the means to use video and other media to capture the complexity of teaching and to examine the expertise that goes into teaching well can open up entirely new avenues for research, professional development, and teacher preparation. By using video and other new media to represent their practice and knowledge, scholarly, innovative, and experienced teachers like Hutchinson can simultaneously teach in their classrooms and teach from their classrooms. Through websites and other forms of materials distribution, such teachers can contribute to the development of insights, resources, and strategies for education, bring forth new opportunities for public and scholarly discourse, and contribute to the creation of a professional knowledge base.

If this were an issue of depicting the work of a few master teachers like Hutchinson, then skilled videographers and documentarians could do the work themselves. However, a few isolated examples of teaching cannot adequately represent the issues and opportunities that teachers encounter in many different contexts. Learning from teaching can benefit from developing large numbers of representations of teaching in many different contexts, making them available for examination in research as well as in practical settings like teacher education courses and professional development activities. In that sense, Hutchinson's teaching, like that of thousands of teachers around the country, serves as a vast, untapped resource that could be used to support improvements in teaching on a wide scale.

In this chapter, we describe the evolving efforts of the Carnegie Knowledge Media Lab to develop multimedia websites that can be used to capture and convey the expertise that goes into the work of many different teachers around the country. In the process, we identify two key challenges that confront those who seek to represent and examine a complex activity like teaching; provide examples of new formats or genres that can help to "compress" videos of classroom interactions and a wide variety of other artifacts of teaching into productive web-based representations; outline one effort to enable larger numbers of faculty to produce these kinds of representations; and explore implications for the future.

THE WORK OF THE CARNEGIE KNOWLEDGE MEDIA LAB

The ideas discussed in this chapter grow out of collaborations among members of the Carnegie Knowledge Media Lab (KML) and a variety of faculty members from K–12 and higher education who have participated in the Carnegie Academy for the Scholarship of Teaching (CASTL). These collaborations have yielded a variety of multimedia websites that document faculty members' examinations of their teaching practice, as well as tools and resources that can enable many faculty members to make their teaching practice and inquiries public. The K–12 faculty who have produced these websites have incorporated diverse media—reflections on practice, video excerpts of classroom events, interviews with students, and examples of class documents and student work—into representations that convey to others their central concerns, key learning issues, and useful pedagogical and curricular strategies.

CASTL has encompassed a large initiative since 1999 to render K–12 teaching public, to subject teaching to critical evaluation and peer review, and to enable others to study and build upon work of teacher-scholars selected into the program. The programs have provided 1–2 year fellowships in which the selected faculty are invited to carry out an investigation into their own teaching and to produce a final project to

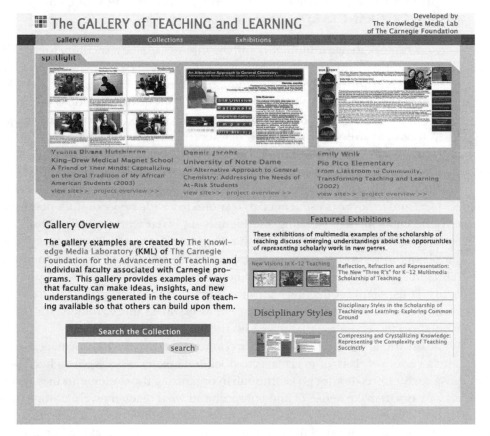

Figure 25.1.   The Carnegie Foundation Multimedia Gallery of Teaching and Learning.

share their results with others. In recent years, CASTL scholars have increasingly made use of multimedia and online technologies in creating and sharing their final products (Hatch et al., 2005).

The Knowledge Media Lab has worked closely with CASTL since its inception to help faculty imagine ways to use multimedia to make work and expertise public and accessible to others. In addition, the KML has developed tools and resources to support faculty pursuit of their scholarship, to enable them to share their expertise, and to build on one another's work. Beginning in 1998, the KML worked very closely with a few CASTL scholars who were interested in making their work public online, and developed several pilot websites showing some different possibilities for sharing scholarship online (for information on both CASTL programs, see (http://www.carnegiefoundation.org/CASTL). Because of the limited number of scholars interested in this early experimentation, the KML was able to provide one-on-one support and technological professional development to enable scholars to document and represent their scholarship online. As will be described later, this model quickly became unsustainable as more scholars became interested in putting their work online, and led to the invention and development of tools and technologies to enable scholars to work more independently.

REPRESENTING TEACHING THROUGH NEW MEDIA: PROBLEMS OF SCOPE, GRANULARITY, AND GENRE

Those who seek to use multimedia and the internet as a new medium for the production, examination, and exchange of ideas about teaching face a two-fold problem. First, teaching is an enormously complex human endeavor. (Lampert, 1985, 2001; McDonald, 1992; Shulman, 1983) The sources of that complexity include the fact that teaching is highly situated, requiring considerable contextual knowledge and access to the perceptions of a variety of participants in order to unpack and understand it. Teaching is also highly ambiguous with both the actions and the outcomes undetermined and open to interpretation (Eisner, 1998; Lampert & Ball, 1998). The incomplete and evolving nature of subject-matter knowledge, the multiple and shifting demands teachers face, and the ever-present need to respond to a wide range of students all contribute to the challenges of isolating and identifying "what works" (Ball, 1996).

Second, although new media and the internet offer possibilities for capturing that complexity in ways that traditional texts cannot, they also come with their own constraints that affect whether or not and how viewers can come to understand that complexity. Thus, simply collecting many hours of video and numerous materials from teachers like Hutchinson does not mean that viewers will be able to understand, examine, or build on the expertise that goes into them.

While the initial websites developed in conjunction with the KML managed to use video and a variety of classroom artifacts to document many aspects of teachers' practice, those early efforts did not go far enough in organizing those elements in ways that enabled viewers to make sense of and comprehend what teachers were doing. These experiences suggested that any effort to capture the complexity of teaching depends on developing a conception of the key aspects of practice and on taking into account the constraints that viewers bring, including limited attention spans and current knowledge and conceptual frameworks. In particular, in subsequent work, we have tried to address two fundamental challenges:

- The challenge of scope: What aspects of a teacher's practice and thinking need to be made explicit and accessible in order for others to understand what that teacher is doing and why? Lesson plans or actual classroom interactions? Drafts and preparation or simply major assessments? What the teacher hoped would happen or what the teacher makes of what happened?

- The challenge of granularity: At what level of depth should those aspects be described? [In a sense, granularity is the "flipside" of scope: If a representation describes too many aspects of a teachers' practice or does so at too general a level, viewers may fail to grasp the specific ideas and strategies that make that practice effective (or ineffective).]

In addressing these challenges, we were influenced by the belief that providing opportunities to revisit the same material at different times, for different purposes, and from different perspectives is one way to develop an understanding of a complex activity like teaching (Spiro, Feltovich, & Coulson, 1992). At the same time, we were also concerned that many people have very limited conceptions of what teaching is or what it entails and they might lack the prior knowledge and conceptual structures needed to make sense of such a complex activity and such complex representations (Bransford, Brown, & Cocking, 2000). The likelihood that viewers would quickly skim only a portion of the many materials that could be provided in a web-based environment also contributed to concerns that, whatever the possibilities, few people would spend the time it usually takes to develop robust understandings of such complex material (Krug, 2000).

In response to these concerns, we sought to experiment with the development of both different genres and forms of web-based representations that resolve the issues of scope and granularity in different ways. These genres provide viewers with some common frameworks that can facilitate their comprehension of particular aspects of a teacher's practice while still leaving the teachers' work open to interpretation and useful for a variety of purposes. In particular, by drawing on genres of representation (like the format of traditional research reports, newspaper articles, or even art exhibitions) and playing off images and ideas with which viewers might be familiar, we hoped to "compress" the large amounts of video, curriculum artifacts, and other teaching materials needed to represent teaching into arrangements that viewers can make meaning out of relatively quickly and easily. This process of compression is central to many scholarly disciplines, where methods and genres have evolved to enable scholars to turn large amounts of data and information into forms that others can understand and examine. Furthermore, by focusing viewers' attention on a relatively small number of video clips and other resources but also providing access to much longer videos and larger collections of artifacts of teaching, we sought to engage viewers in key ideas quickly, encouraging them to return for much more in-depth (and ideally collaborative) investigations of the teacher's work.

Genre in the Making: Yvonne Hutchinson's "Class Anatomy"

Although there are numerous ways to conceptualize teaching practice, building on the work of Shulman (1999) suggests six key aspects that may enable viewers to gain a broad impression of a faculty member's teaching:

- Context: Where, with whom, and under what conditions teaching and learning take place.
- Vision: Ideals and beliefs; "images of the possible"; the place of teaching in a larger context.
- Design: Goals, time line for activities, key activities, and assignments.
- Interaction: Interactions and exchanges between and among students and teachers in class and out.
- Assessment: The formal and informal means teachers use to judge students' learning.
- Reflection: Teacher's impressions and analysis of their teaching and the design, interactions, assessments, and events involved.

However, each of these elements of a teacher's practice could be examined at a variety of levels of detail over relatively short or long periods of time (including a part of a lesson, a "class," a course, or a career, among others.) Yvonne Hutchinson's website resolves these issues of scope and granularity by inventing the "class anatomy." Thus, the website focuses on one 2-hr instructional block in her English course for 9th–10th grade students to document how she supports the development of abilities to analyze literary texts and the ways in which she orchestrates classroom discussions that involve all students.

By using the first page to highlight a series of video clips that outline the evolution of one class period, Hutchinson makes available ways to examine both the design of her group discussions and the kinds of interaction that go on in them. At the same time, the first page also provides a quick summary of the context in which Hutchinson teaches and offers a link to Hutchinson's own description of her teaching context. The site provides access to Hutchinson's vision and her personal historical context by linking to a narrative about her own educational history and how it connects to her goals for her students as well.

While the site focuses on one class period, the first page of Hutchinson's website also includes links to some of her key assignments and strategies and a link to an hour-long reflective interview that situates her pedagogical choices shown in the documented class within the context of her overall goals and year-long instructional design. Similarly, viewers can make their own assessment of the quality of students' discussion and analyses in the video of the single class period, but links to videos of the same student's projects and presentations the following year make it possible to consider whether or how students' abilities may have developed over the longer term. Finally, the site also includes materials she has authored for professional development settings in which she reflects on aspects of her practice, provides rationales for her requirements for student oral participation, and describes how she sets the tone for the classroom in the beginning of the year.

Figure 25.2. Yvonne Divans Hutchinson's class anatomy.

Toward a Range of Representations of Teaching

Hutchinson's class anatomy has served as a kind of prototype or "template" for other faculty working with the KML to examine key issues in their practice. At the same time, the invention of any genre or format also creates limits and constraints that may or may not be appropriate depending on what teachers want to communicate about their practice (Hatch, Bass, Iiyoshi, & Pointer Mace, 2004). A class anatomy like Hutchinson's makes it possible to get an in-depth look at the interactions and design of a single instructional period, but that comes at the expense of learning about the design of her course, her "full" vision for her students' development, and the hundreds of other aspects of teaching and learning that she deals with during the year.

Websites developed in conjunction with Irma Lyons, Heidi Lyne, and Emily Wolk illustrate some of the other ways that the representational challenges of scope and granularity can be addressed and suggest some of the other formats that might be worth further exploration. These websites reflect decisions about which aspects of their practice they wanted to foreground. Each also designed a representational interface (in collaboration with the KML) that would advance the central themes of their inquiry and the ways in which they chose to share videos of their teaching.

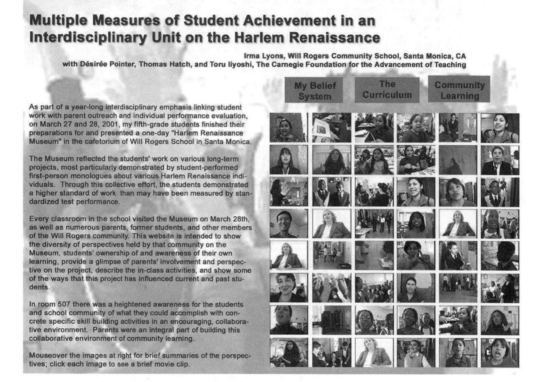

Multiple Measures of Student Achievement in an Interdisciplinary Unit on the Harlem Renaissance

Irma Lyons, Will Rogers Community School, Santa Monica, CA
with Désirée Pointer, Thomas Hatch, and Toru Iiyoshi, The Carnegie Foundation for the Advancement of Teaching

| My Belief System | The Curriculum | Community Learning |

As part of a year-long interdisciplinary emphasis linking student work with parent outreach and individual performance evaluation, on March 27 and 28, 2001, my fifth-grade students finished their preparations for and presented a one-day "Harlem Renaissance Museum" in the cafetorium of Will Rogers School in Santa Monica.

The Museum reflected the students' work on various long-term projects, most particularly demonstrated by student-performed first-person monologues about various Harlem Renaissance individuals. Through this collective effort, the students demonstrated a higher standard of work than may have been measured by standardized test performance.

Every classroom in the school visited the Museum on March 28th, as well as numerous parents, former students, and other members of the Will Rogers community. This website is intended to show the diversity of perspectives held by that community on the Museum, students' ownership of and awareness of their own learning, provide a glimpse of parents' involvement and perspective on the project, describe the in-class activities, and show some of the ways that this project has influenced current and past students.

In room 507 there was a heightened awareness for the students and school community of what they could accomplish with concrete specific skill building activities in an encouraging, collaborative environment. Parents were an integral part of building this collaborative environment of community learning.

Mouseover the images at right for brief summaries of the perspectives; click each image to see a brief movie clip.

Figure 25.3.   Irma Lyons' community collage.

Community Collage.   Irma Lyons' site is designed as an interactive photo montage of stills from the video clips that she assembled to describe her annual "Harlem Renaissance Museum," a one-day culminating event in which her fourth-and fifth-grade students demonstrated a range of projects and performances developed over several months around the theme of the music, art, literature, and historical events of the Harlem Renaissance period. To document the "Museum," Lyons invited a teacher colleague and a member of the KML to visit the school and videotape interviews with the parents, students, and community members in attendance at the event. On her website, each interview clip is represented by a photo still. When each still is moused over, it reveals more about the clip, as well as connections to other related materials and videos. For example, clicking on an image of a young girl from Lyons' class who studied Langston Hughes brings up links to videos interviewing her father about

How do we help kids know what we mean by high standards and good work? The most important part of the portfolio and graduation process that we're using here... Is It helps make visible to kids the values of the school, and its definition of what an educated person that age should look like and be able to do."

-Deborah Meier, Mission Hill School founder and co-principal

Mission Hill School
Heidi Lyne, Mission Hill School

with Desiree Pointer, Thomas Hatch, and Toru Iiyoshi
The Carnegie Foundation for the Advancement of Teaching

During the critical transition years between childhood and high school, young people need schools that share certain characteristics. First of all, they should be schools where young people work in small groups with a small number of adults who know them well. Second, students need to experience high and rigorous academic expectations. Third, they need to see the connections between academic work and the work that takes place outside of the classroom. Fourth, they need opportunities to explore the world in authentic and engaged ways – through music, dance, theater, visual arts and outdoor education. And finally, students should have opportunities to make their own positive individual and collective mark on their environment – to be useful to others. Graduation from our school is a process that incorporates all of these opportunities and expectations. But how do we measure students' accomplishments and gauge their knowledge?

This website is built around a central documentary video about the portfolio and graduation process at Mission Hill School. Click on the photos above to jump directly to those sections of the film. To see an outline of the project and associated materials, and to view the video in its entirety, **click here**.

Portfolio Process Overview (PDF) • Examples of Student Work • "Redemption:" Narrative by Heidi Lyne (PDF)
Supplementary Videos: Akwasi's Portfolio Presentation • Rebeca's Initial Presentation Judging • Rebeca's Redo

Figure 25.4.   Heidi Lyne's longitudinal documentary.

what she learned in the process of researching Hughes' life, as well as a link to a former student who'd studied Hughes and who had returned to the Museum to "check up on" the current Langston's performance. The nature of the montage presents the website's audiences with an immediate visual experience of the community diversity in attendance at the Museum.

Longitudinal Documentary.   Heidi Lyne's site is designed to share her documentary video showing the year-long preparation process undergone by her 8th grade students at Mission Hill School in preparation for their graduation portfolio presentations. Lyne divides her website into the four stages of her documentary—overview, preparation, presentation, and reflection—each of which is juxtaposed with student work samples (in multiple drafts), documents from the school, and Lyne's own reflections on her learning about students' experiences of the authentic assessment of their "habits of mind."

Multimedia "Bulletin Board."   Emily Wolk's website, a project description, relates her work with a multiage, extracurricular group of student "Participatory Action Researchers" to take proactive steps to address problems in their school community. Although Wolk has written about her work in theoretical articles related to her gradu-

Figure 25.5. Emily Wolk's multimedia bulletin board.

ate studies, she was nervous about the idea of creating a website—she didn't consider herself technologically savvy. However, when she was given the suggestion that she think about it as if she were creating a bulletin board in her school's hallway to communicate her students' achievements, she found the task much easier to think about. She immediately designed a format that played off of images of a stop light—a symbol of the ultimate goal of the students' research on pedestrian safety in their neighborhood, which conclusively demonstrated the high rate of pedestrian injuries and fatalities and the great impact that installing a stoplight would have at one particularly perilous intersection. Her students' work is emphasized by the inclusion of video clips from a local news program that highlighted the student-researchers' work. Once the task was put into terms that overlapped with existing areas of Wolk's teacher expertise, she found that designing an engaging interface for her website was something she could do well.

Supporting Faculty Scholarship With Tools and Templates.   To make public the expertise of thousands of teachers and foster the production, critical examination, and advancement of knowledge of teaching on a large scale, faculty need new kinds of support and resources. Faculty need support in order to participate in the process of "going public" as Hutchinson, Lyons, Lyne, and Wolk have. These supports must be designed to be seamlessly and easily integrated into the teaching lives of involved faculty, supporting them in the intellectual and technological "heavy lifting" required to make their teaching practice public online.

Over the past several years, the KML has responded to this need by developing representational and organizational tools to enable faculty to work with significantly less direct technological support than has been available for CASTL scholars. Because the technological challenges can be daunting for many teachers, the KML has worked over the past few years to develop several support documents and tools that enable faculty to quickly make their work public on the web, to link relevant documents and artifacts of practice, and to arrange web pages in a way that will engage their intended audiences.

The KEEP toolkit—Knowledge Exhibition, Exchange, and Presentation Toolkit—is an example of the kinds of resources that facilitate the technical process and support the intellectual process of creating powerful representations of knowledge. KEEP is a suite of online functions that allow faculty to quickly upload text, images, videos, and other teaching materials , dynamically rearrange them to develop different representations, and seamlessly connect their work to that of other scholars. KEEP is envisioned as a way to allow scholars to easily contribute their work to a growing online network, a "living archive" of multimedia representations of teaching. The interface is flexible, so

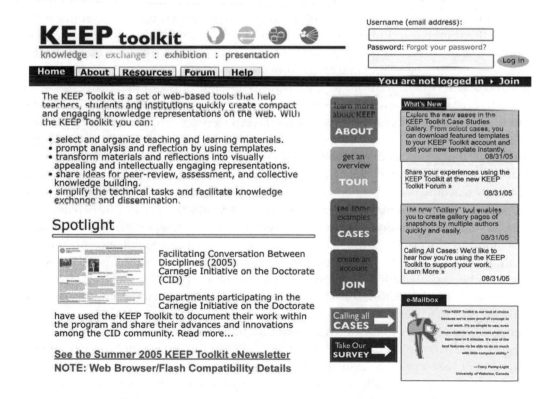

Figure 25.6. The Carnegie Foundation's "KEEP toolkit"(http://www.cfkeep.org).

that scholars can easily add multimedia data, classroom documents, and links to rich descriptions of practice and inquiry, as well as archive their websites as templates for future projects. Each template is intended to prompt the scholars to consider the choices about how to share their inquiry with their intended audiences. The KEEP tools and the evolving archive of teaching practice have been designed for open-source use by individual teacher-scholars as well as by institutions, professional development groups, and reform organizations.

New Directions: Transforming Professional Development and Preservice Teacher Education

Scholars working to document and represent their inquiry is an important first step; gaining access to tools that will make the scholarship of teaching increasingly available is another. Once teaching is made public, however, education communities have an incredible new opportunity to use those multimedia examples of scholarly teaching in professional development, in preservice teacher education, and in national reform networks. Traditionally, teacher educators and professional developers have relied on exemplary articles and educational theory, supplemented by anecdotal sharing of classroom experiences, or narrative cases of teaching. Occasionally, teacher educators also incorporate videos of exemplary lessons into their courses (Brophy, 2004). However, in order to learn new teaching practices, novices and practitioners need opportunities both to analyze the complex moves that go into a successful lesson, and to access the thinking and planning of the teacher that informs those moves. Images of the complexity of teaching—how teachers create goals for their students, the strategies that they use, and outcomes that result from these strategies—are precisely the materials that teachers need in order to learn how to teach throughout their careers.

If teachers and their colleagues in schools and universities can produce relatively large numbers of powerful representations of teaching practice in short periods of time, then a critical new resource for preparation and professional development will emerge. Already, teacher educators who have begun to use these websites as texts in preservice teacher education courses are beginning to see how using multimedia exemplars of scholarly teaching opens up new conversations with their students. Teacher educators commonly face the dilemma of talking about pedagogical strategies and curricular innovations that may be implemented very differently across the diasporas of their students' teaching placements. By having a rich set of resources that can be reflected on and explored in community, the teacher educators are finding that their students are developing a richer vocabulary to express educational ideas. Pam Grossman of Stanford University has used Yvonne Hutchinson's website in her curriculum and instruction class for preservice secondary English teachers. In fall 2003, Grossman assigned her students to work in pairs to answer particular questions about Hutchinson's practice and inquiry. One pair explored the role of Hutchinson's "anticipation guide" task before and during her classroom discussion; these students appropriated the use of the anticipation guide with their fellow students in leading a discussion about their findings. Grossman observed that having access to a common set of materials allows students to experiment with new pedagogical and curricular strategies in a smaller setting, before trying things out in their student teaching placements. Anna Richert, professor of education at Mills College, has found similar outcomes in her use of Hutchinson's materials in a course on adolescent development. The students in Richert's class are able to draw parallels between their own teaching experiences and the ones they encounter on the websites; the problems of teaching described online are incorporated into the classroom vocabulary. Because Richert is now able to assign the materials of Yvonne Hutchinson along with John Dewey and Lev Vygotsky, she commented that the materials, "privilege teaching in a new way."

Richert and Grossman are among several teacher-educator scholars who are currently participating in a new project of the Carnegie Foundation—the Quest Project—intended to explore the use of these multimedia materials of K–12 teachers as alternative texts for the preparation of new teachers. Over the 2 years of the Quest Project, participating teacher educators will adapt their courses to incorporate the multimedia websites most relevant to their preservice teachers' needs (for example, math methods teachers have long used Deborah Ball's "Shay's Numbers" video to unpack the subtleties of student thinking). The teacher educators will document their own teaching of the multimedia websites, and in some cases, will videotape their students in their field placements trying out the strategies demonstrated by experienced teachers. In so doing, the Quest Project will show the ways in which the documentation of K–12 teaching practice can augment teacher educators' attempts to create accomplished novices (Stylianides & Ball, 2004), and make visible the transfer of the wisdom of practice (Shulman, 2004). The research agenda of the Quest Project will track the use of these websites in teacher education, with a preliminary focus on the significant work undertaken by the teacher educators to adapt their courses to embrace these alternative materials. Subsequent directions for research include following graduates of the Quest Fellows' programs into their beginning years of teaching to measure the impact of the "virtual apprenticeships" afforded by the Quest and CASTL websites. We also plan to explore how teacher educators learn from each other's practice, and if and how the documentations of K–12 and teacher educator practice contribute to advancement of knowledge in the discipline.

There are exciting new initiatives being pursued by many different institutions to document and represent teaching and learning. We are also collaborating with the National Writing Project to share teacher knowledge, videos of classroom practice, and innovative curricular documents for a NWP Reading Initiative. By encouraging the teachers involved in the initiative to record their classroom practice and make their innovations in reading public, the NWP / CASTL partnership will disseminate the practices of multimedia scholarship to hundreds, if not thousands, of new teachers.

Envisioning a future in which the practice of scholarly teaching can become the content of conversations about teaching is transformative. Using these new materials has the potential to significantly alter the landscape of teacher education as well as provide more general conceptions of what "counts" as accepted texts on teaching. By bringing teachers' scholarly voices from margin to center, and by making their teaching lives public, by allowing them to teach in and teach from their classrooms, we have the potential to reshape widely held assumptions about the work of teaching and the lives of teacher-scholars.

REFERENCES

Ball, D. L. (1996). Teacher learning and the mathematics reforms: What do we think we know and what do we need to learn? *Phi Delta Kappan, 77*(7), 500–508

Bransford, J., Brown, A., & Cocking, R. (Eds.). (2000). *How people learn: Brain, mind, experience, and school.* Washington, DC: National Academy Press.

Brophy, J. (Ed.). (2004). *Advances in research on teaching: Vol. 10. Using video in teacher education.* New York: Elsevier Science.

Eisner, E. W. (1998). *The kind of schools we need : Personal essays.* Portsmouth, NH: Heinemann.

Hatch, T., Bass, R., Iiyoshi, T., & Pointer Mace, D. (2004, September/ October). Building knowledge for teaching and learning: The promise of scholarship in a networked environment. *Change Magazine,* 42–49.

Hatch, T., Ahmed, R., Lieberman, A., Faigenbaum, D., Eiler White, M., & Pointer Mace, D. (2005). *Going public with our teaching: An anthology of practice.* New York: Teachers College Press.

Krug, S. (2000). *Don't make me think: A common-sense approach to web usability.* Berkeley, CA: New Riders Press.

Lampert, M. (1985). How do teachers manage to teach? *Harvard Educational Review, 55*(2), 178–194.

Lampert, M. (2001). *Teaching problems and the problems of teaching.* New Haven, CT: Yale University Press.

Lampert, M., & Ball, D. L. (1998). *Mathematics, teaching, and multimedia: Investigations of real practice.* New York: Teachers College Press.

McDonald, J. P. (1992). *Teaching: Making sense of an uncertain craft.* New York: Teachers College Press.

Rose, M. (1995). *Possible lives: The promise of public education in America.* New York: Houghton Mifflin Company.

Shulman, L. S. (1983). Autonomy and obligation: The remote control of teaching. In L. S. Shulman & G. Sykes (Eds.), *Handbook of teaching and policy* (pp. 484–504). New York: Longman.

Shulman, L. (1999, September/October). The scholarship of teaching: New elaborations, new developments. *Change Magazine,* 10–15.

Shulman, L. (2004). The wisdom of practice: Essays on teaching, learning, and learning to teach. San Francisco: Jossey-Bass.

Spiro, R. J., Feltovich, P. J., & Coulson, R. L. (1992). Cognitive flexibility, constructivism, and hypertext: Random access instruction for advanced knowledge acquisition in ill-structured domains. In T. M. D. D. H. Jonassen (Ed.), *Constructivism and the technology of instruction: A conversation* (pp. 57–75). Hillsdale, NJ: Lawrence Erlbaum Associates.

Stylianides, A. J., & Ball, D. L. (2004). *Studying the mathematical knowledge needed for teaching: The case of teachers' knowledge of reasoning and proof.* Paper prepared for the 2004 meeting of the American Educational research Association, San Diego, CA, April 14, 2004.

WEBSITES

The Carnegie Academy for the Scholarship of Teaching and Learning (CASTL): http://www.carnegiefoundation.org/castl/

The Carnegie Foundation Multimedia Gallery: http://gallery.carnegiefoundation.org

Yvonne Hutchinson: http://gallery.carnegiefoundation.org/yhutchinson

Irma Lyons: http://gallery.carnegiefoundation.org/ilyons

Heidi Lyne: http://gallery.carnegiefoundation.org/hlyne

Emily Wolk: http://gallery.carnegiefoundation.org/ewolk

Bill Cerbin: http://gallery.carnegiefoundation.org/bcerbin

Elizabeth Barkley: http://gallery.carnegiefoundation.org/ebarkley

Dennis Jacobs: http://gallery.carnegiefoundation.org/djacobs

Bruce Cooperstein: http://gallery.carnegiefoundation.org/bcooperstein

The KEEP tool: http://www.cfkeep.org

Teachers as Designers: Pre and In-Service Teachers' Authoring of Anchor Video as a Means to Professional Development

Anthony J. Petrosino
The University of Texas at Austin

Matthew J. Koehler
Michigan State University

In their influential article, Brown, Collins, and Duguid (1989) emphasized the importance to education of looking carefully at authentic cognition and of creating cognitive apprenticeships based on authentic tasks, defined most simply as the "ordinary practices of the culture" (p. 34). Our focus on the apprenticeship of pre and in-service teachers (i.e., teacher-learners) has led us toward a consideration of the authentic tasks that would support continued and long-term thinking about classroom cultures, specific problems of practice, and conceptual knowledge. But working within an apprenticeship model inevitably introduces issues of feasibility. For instance, how can we engage teacher-learners in authentic real-world tasks and experiences while staying in the confines of the traditional college and university classroom walls? How can experiences and mentors be provided consistently? Is it possible to provide opportunities to apprentice in a scalable manner?

One approach to address these challenges is anchored instruction [Cognition and Technology Group at Vanderbilt (CTGV), 1990]. Anchored instruction is a technology-based learning approach that stresses the importance of placing learning within a meaningful, problem-solving context. Anchored instruction uses context as a learning device. The anchor refers to the bonding of the content within a realistic and authentic context. In anchored instruction, learners are encouraged to not only solve problems, but also to think about the processes involved. Essential to this approach is the use of interactive video within a narrative format. The narrative provides a story and a context while the video provides an essential element to deliver real-world complexity.

The approach taken by anchored instruction (CTGV, 1990) uses video-based "adventures" to initiate sustained exploration by students and teachers into understanding authentic problems. In addition to being more engaging than text (CTGV, 1990), video helps situate learners in a variety of contexts. Students can be placed at the feet of a mathematician, at an important moment in history, or even on a field trip to Mars. Bringing students to the corresponding real-life experiences would be difficult at best (if not impossible).

The prototypical incarnation of anchored instruction theory into practice is the creation of "The Adventures of Jasper Woodbury" (a.k.a. "Jasper"). "The Adventures of Jasper Woodbury" consists of 12 videodisc-based adventures (plus video-based analogs, extensions, and teaching tips) that focus on mathematical problem finding and problem solving. Each adventure is designed from the perspective of the standards recommended by the National Council of Teachers of Mathematics (NCTM). Specifically, each adventure provides multiple opportunities for problem solving, reasoning, communication, and making connections to other areas such as science, social studies, literature, and history (CTGV, 1997). The success of the anchored-instruction approach has been well documented in Jasper classrooms (CTGV, 1990, 1997; Hickey, Moore, & Pellegrino, 2001) as well as in the area of specific disciplines such as reading (Sharp et al., 1995) , special education (Glaser, Rieth, Kinzer, Colburn, & Peter, 2000), and science (Goldman et al., 1996; Sherwood, Kinzer, Bransford, & Franks, 1987).

In constructing an anchored approach to the development of teacher-learners, we have also sought to use video as a means of connecting teachers learning to authentic contexts. By connecting knowledge to the contexts in which it is to be used, we are seeking to improve the likelihood that the experiences teacher-learners will be able to apply their knowledge to real classroom situations, and transfer it to other situations and contexts when appropriate (Bransford & Schwartz, 1999). It is worth noting the conceptual and theoretical overlaps between anchored instruction and case-based approaches insofar as teacher-learners are concerned. In both approaches, authentic contexts are used to engage teacher-learners in the types of thinking they will use in future classrooms. In both, they are asked to analyze and reflect about situations, knowledge, theories, and problems. Accordingly, learning in these authentic situations is more likely to bridge the gap between the theory and the practice of teaching (Shulman, 1986, 1992).

Historically, anchor videos were prepackaged for teachers and presented as finished products for practical reasons: (a) Most teachers do not have the time and resources to develop video anchors; (b) most teachers do not have the collaborative support structure needed to take an idea from concept to actualization; and (c) producing video was technically challenging and cost prohibitive. But prepackaging also made key ideas such as curriculum design, coordination with state and national standards, and finding engaging multistep problems unproblematic to the teacher-learners. Recent advances in hardware and software have removed most of the technical barriers to teacher-created video. Also, engaging teacher-learners enrolled in teacher education and master's-level courses can remove the time and support barriers.

But if teacher-learners are not taught and encouraged to design their own curricula, they can have little or no ownership, responsibility, or voice in the curricula process. Traditionally, in colleges of education across the country, curriculum and teaching have been separated. Teacher-learners are not given opportunity to create curriculum, only instruction on how to enact it. The role of design, so prevalent in the development of learning environments, has rarely been accessible to teacher-learners in any meaningful manner (Goldman-Segall, 1998).

Our work is about changing the authentic cultural practice of teaching and teacher education, creating new forms of teaching practice that do not yet exist on a broad scale. By building on the strengths of anchored instruction and the creative possibilities afforded by new technology, we have to look seriously at a reinterpretation of the notion of apprenticeship and teacher education. What can happen if these powerful learning tools are put into the hands of teacher-learners in order to look at their own teaching or to create their own learning environments?

We will detail our efforts to have teachers author their own anchor videos. We present three examples, drawn from our own teaching experiences with pre and in-service teachers. In the first example, preservice teachers in mathematics and science education create their own anchor videos as they develop 4- to 5-week project-based learning curricula. The second example describes preservice teachers' efforts to create best-practice teaching video cases from a library of footage in a K–8 literacy methods class. The third example documents the experiences of master's students (in-service teachers) as they create iVideos (movies that exemplify a big idea in education) as a means of situating learning in educational technology. After describing these examples, we summarize the learning afforded by the teacher as designer approach.

EXAMPLE 1: CREATING ANCHORS FOR PROJECT-BASED INSTRUCTION

Krajcik and his colleagues at the University of Michigan (Krajcik, Czerniak, & Berger, 2002; Krajcik, Soloway, Blumenfeld, & Marx, 1998) have described project-based instruction as engaging learners in exploring authentic, important, and meaningful questions of genuine concern to students. Through a process of scaffolded investigation and collaboration, using the same processes and technologies that experts use, students formulate questions, make predictions, design investigations, collect and analyze data, make products, and share ideas. As in other complex learning environments, well-designed and scaffolded project-based investigation has the potential to help all students—regardless of culture, race, or gender—engage in learning.

In work at the University of Texas at Austin with teacher-learners, we have combined attributes of both anchored instruction and project-based instruction. Within our program, teacher-learners are supported in creating an initial challenge, or anchor video, and then in placing that anchor within a 4- to 5-week, student-created, project-based unit.

In our version of project-based instruction, students use a wide variety of software to develop project-based curricular units that are infused with technology. Software applications allow Web authoring, video editing, concept mapping, and modeling. Units produced by students are posted to the Web and copied onto a class CD so that students have access to a library of projects (see http://www.edb.utexas.edu/anchorvideo/).

A major hurdle in creating project-based curricula is that the process requires simultaneous changes in curriculum, instruction, and assessment practices—changes often foreign to the students as well as to the teachers. In our program, teacher-learners design, implement, and evaluate project-based curricula in collaboration with their cooperating K–12 teachers and university researchers. Previous work has identified four important design principles for this type of instruction; (a) defining learning appropriate goals that lead to deep understanding; (b) providing scaffolds such as beginning with problem-based learning activities before completing projects and using "embedded teaching," "teaching tools," and sets of "contrasting cases"; (c) including multiple opportunities for formative self-assessment; and (d) developing social structures that promote participation and revision (Barron et al., 1998).

Although all four goals are important, the development of a quality anchor video most satisfies the first design principle and also paves the way for the other three design principles. This course has many innovative aspects, but the most salient for the immediate issue at hand is the design, development, and incorporation of teacher-learner created video anchors for their project-based units.

Over the past 4 years, project-based units have been developed in such diverse areas as energy expenditure of muscles during exercise, oil spills, habitats of Austin area bats, chemical bonding, virus transmission, and mathematical modeling. All cases have incorporated a set of design principles for creating a motivating question. These design principles have been informed by the work of Krajcik as well as the CTGV. Criteria for a quality "driving" question (Krajcik et al., 2002) include issues of whether the question is (a) worthwhile (i.e., promotes higher order thinking), (b) feasible (i.e., students can design and perform investigations to answer the question), (c) contextualized (i.e., related to real-world problems), (d) meaningful (relevant to learners' lives), and (e) open ended (a complex problem with multiple solution paths). Design principles for the creation of the anchor video include a narrative structure to the story, a generative design to the story that allows the user to develop their own problem-solving strategy, embedded data, a complex problem involving multiple steps to mimic real-world problem solving, and the use of digital video to make the complexity manageable (CTGV, 1990; Goldman et al., 1996).

We now take an example of one student-created video and show how the students employed these design principles. The project we look at examines how viruses spread. In the words of one student group, their project addressed the design principles of a driving question in the following way:

The students will be able to perform and sustain an experiment where they are controlling the factors so that they can observe and report their data. The investigation allows the students to infect the plants with the tobacco mosaic virus. It allows the students to perform and sustain an experiment where they are controlling the factors, so that they can be the ones to report and observe their data. This driving question is anchored to what is a real world situation. Virus spread is a topic that is found in every state, nation, country and continent. It is something that does affect our daily lives…Meaning is also important to a successful driving question … Understanding the importance of vaccines can become interesting for those students who know of people who are not vaccinating their children. Learning about past epidemics can lead to great insight of our technological advances. The driving question has sustained inquiry. The students are constantly challenged on learning new material and being able to present it as well. The investigations we have included in the unit require collaboration from the students so that they have success in their investigation. The students will be able to create their artifacts so that their knowledge on this subject does sustain. The investigation leads to a great sustainable driving question. The students can go into great details and pursue the answers over a long period of time.

An examination of their anchor video shows the incorporation of a narrative. Story—various clips of people addressing the impact of harmful viruses in their lives, including farmers, sharecroppers, and lovers. Multiple pictures of viruses are embedded in the video along with text fields that pose problems to the students. The video presents the problem in such a way as to make it obvious that solving a problem like the spread of a virus is very complex because it mimics other diseases and is often disguised or invisible to the novice. The use of digital video assists in managing the complexity of the problem.

Students not only design, film, and edit their own anchor video, but also develop a 4- to 5-week curriculum consistent with state and district standards. Over a single semester, students are exposed to and eventually produce their own project-based units. Furthermore, during their 10-week student-teaching semester, students are given the opportunity to enact the project-based units they developed with their partners. We believe this gives our teacher-learners an amazing head start to reflect on their own understanding and ability to enact complex instructional pedagogy while very early in their careers.

EXAMPLE 2: CASES OF LITERACY INSTRUCTION

Michigan State University has a long history of researching, advocating, and implementing case-based approaches to the professional development of future teachers (e.g., Lampert & Ball, 1998; Sykes & Bird, 1992). In TE 401, an elementary education methods course, Dr. Cheryl Rosaen had used case materials and videos of classroom teaching (e.g., Lampert & Ball). Together, Rosaen and Koehler worked to engineer some classroom experiences that would:

1. Help the teacher education students connect what they were reading about to actual classroom practices.
2. Make students' thinking visible and on the table.
3. Help students learn and understand the content area (early grade literacy instruction).
4. Help students learn to be thoughtful, reflective, and analytical about classroom teaching.

The approach combines elements of a case-based approach, exploring video anchors, and student authoring. Using IVAN (2003), a system designed at Michigan State University, students are presented with a library of video clips and linked commentary to a case representing a wide variety of examples of literacy instruction for a single teacher (multiple IVAN cases cover different teachers in different grades). Students can view any segment of the library, read materials that go with the video segments, and even follow Web links to external sources (see Fig. 26.1). Students also can mark portions of any video they see and easily place that segment on the time line, in essence allowing them to make their own movie, case, or video anchor out of the larger video library. They also can write commentary that is linked to specific portions of their time line (video anchor).

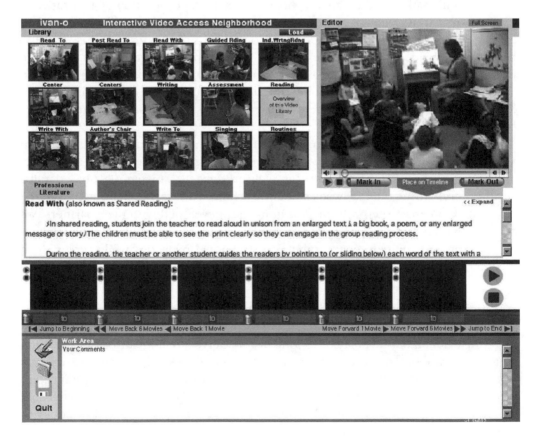

Figure 26.1.   Example of IVAN interface for teacher-authored videos.

Using this interface, students receive task-specific instructions for their video authoring, in groups or as individuals (depending on the constraints of time, computer resources, and task for the day). In one example, students were asked "to explore how reading skills and strategies are taught and learned at that grade level ... to create a timeline that includes at least three excerpts." In addition, students were asked to write linked text to connect their video exploration with concepts and terminology used in class and in the reading assignments. Specifically, for each segment they chose, they were to explain (a) the stage of the reading process, (b) specific reading skills and strategies being taught, and (c) any questions they had about this clip. Students also were asked to document any "aha! experiences" as they completed the assignment.

This approach has several advantages. The instructor has the opportunity to see how well students are making connections between theory (their readings) and practice (the videos). The class shares the multiple (and varied) understandings and perspectives brought by different groups. It also has been our experience that the activities are very engaging to students, and they report the activity to be a valuable source of learning not available through the text alone. Fruitful class discussions also follow, as groups show the segments chosen by each group, discuss their thinking, explore any questions, and discuss any commonalities and differences among the groups' works. This allows students to think about and discuss video from grades their specific group did not see and to discuss how (or if) literacy instruction differs between grade levels.

Anchors are used in two ways. First, the library and task structure provide a form of anchored instruction very similar to that provided by Jasper. Students receive an authentic context, work on an open-ended and semistructured problem, are situated in video, work collaboratively to reach answers, and engage in extended inquiry. Second, the task makes use of anchors because in the process of creating their own time line, the students design their own form of anchored instruction for their fellow classmates. The student-authored time lines serve as "mini-anchors" for the ensuing classroom discussions (and other possible uses of the products we have not explored yet).

This approach also connects strongly to the literature on case-based instruction. By combining the best elements (and benefits) of case discussions (see Lundeberg, Levin, & Harrington, 1999, for a review of the topic), case authoring (e.g., Shulman, 1992), and case analyses (e.g., Harrington, 1999), this approach asks teachers to engage in the kinds of reasoned decision making (Toulmin, Rieke, & Janik, 1984) that they will be asked to do as teachers.

Additionally, students ask questions that they might not have voiced without the opportunity to explore activities through video. For example, one teacher education student wrote that one advantage was recognizing that second-grade students are making different connections than the one you are trying to teach. I can see how it would be easy to overlook statements that may seem to be disruptive or not pertinent to a discussion.

That is, what this student took as good teaching—a clear transmission of ideas to students—was challenged by her work in the IVAN cases. Initially she saw a teacher doing this kind of instruction, and saw children's responses as disruptive behavior. Later, she saw that it was possible to see children's responses as indicators of possible connections to alternative understandings children were constructing (some of them were misunderstandings). Aha! moments are perhaps the most rewarding experience in this approach—they represent times in which the material suddenly came together, suggesting it was not well understood before using the IVAN case. We are currently conducting studies exploring the effectiveness of this approach compared to other models of instruction.

EXAMPLE 3: TEACHERS AS FILMMAKERS: USING iVIDEOS TO LEARN ABOUT EDUCATION AND TECHNOLOGY

At Michigan State University, practicing teachers come to the master's program in educational technology hoping to learn technology skills that will enrich their teaching. Teaching educational technology is no easy task, as it requires careful identification of what teachers need to know about technology (see Zhao, 2003; or Mishra & Koehler, 2003, for a full discussion). The challenge is not to teach specific skills that are soon obsolete as hardware and software change. Moreover, students in our program cut across every conceivable grade level and content area. So, it is not simply a matter of helping teachers teach with technology in, for example, high-school mathematics. Instead, we argue that the goals of educational technology courses should be the following:

1. Learn technology skills and concepts that are likely to persevere beyond the newest version of hardware and software (e.g., the concept of a file format—gif, .jpg, etc.).

2. Appreciate the reciprocal relationship between teaching and technology: Technology changes what students can teach and how they can teach it. Likewise what (and how) students teach impacts the technology used.

3. Teach students "how to learn" about technology. To relieve teachers from becoming constantly updated (and retrained) via workshops, teachers would do well to learn how to learn a new piece of technology on their own (or with colleagues). This means learning that failures, trying out new ideas, exploring interfaces, and consulting other resources are part of the process.

4. Encounter technology in authentic contexts in which multiple technological and educational concepts are tied together.

5. Encounter new ideas in education. What is the point of an educational technology program in which students learn nothing about education?

One way we have tried to design learning environments that foster these kinds of educational technology learning is a series of design tasks (see Mishra & Koehler, 2003, for a fuller description of the design philosophy). In one of these design activities, we have teachers collaborate in groups to produce iVideos (idea-based videos) that communicate an idea of education importance to a wider audience (Wong, Mishra, Koehler, & Siebenthal, in press). The videos should inspire others with passion for the idea. For example, in a capstone course, we asked teachers to complete a 2-min iVideo that completes the following sentence: "Teaching is _____." Instead of teaching teachers how to do digital video, the teachers had to learn the technology in the context of communicating a broader educational idea.

We typically devote a good portion of the semester to working on iVideos, as groups work to organize their thinking, storyboard, and idea; begin to produce and edit their video; get feedback on the message they are trying to convey; and then cycle back through this process. In making an iVideo, students confront a lot of issues:

1. Learning about technology. Students learn a lot of key technology concepts and ideas, including file formats, client and servers (as they move files around to share pieces), Web pages and servers (as they share their work), the ins and outs of video editing and formatting, and the ability to play with technology (they are given very little explicit instruction on how to use cameras or editing equipment).

2. Learning about education. In producing an iVideo that completes the sentence "Teaching is _____," teachers have to wrestle with what constitutes good teaching, what their teaching philosophies are, what that philosophy looks like in practice, and the extent to which these values are shared and encompassing.

3. Learning to learn. Teachers are used to learning about educational technology by attending workshops. In the iVideo approach, students learn skills in context, as needed, by themselves (or with the help of other teachers). They learn that learning with technology means unexpected troubles (technology can drive you nuts), but more importantly, how to work through those troubles. They soon appreciate that they do not need a learned other (the instructor) every time they have a question. Instead they can figure things out by themselves, from a classmate, or from a Web site. In short, the process tames the fear of learning about technology.

Working with iVideos has, of course, other advantages (cooperative learning, learning to express ideas, etc.) that cannot be fully described here (see Wong et al., in press, for further details).

In essence, asking teachers to create an iVideo is asking them to create a video anchor. Doing so requires a lot of learning, collaboration, organization of knowledge, vision, and hard work. It has been our observation (Wong et al., in press) that the process of making this anchor alone is worth the experience. Furthermore, teachers absolutely love the experience and are highly engaged and extremely proud of their work (they will show their video to just about anyone who will watch). We have found the anchor itself is worthwhile as well. As students preview their work in progress to the class, we have found very engaging classroom discussions to be the norm—the ideas raised in the video are fruitful grounds for discussion that we seldom can evoke through a good reading. We have also found these anchors to be very useful in other courses as catalysts for discussion. For instance, as appropriate topics arise in our PhD courses, we sometimes use the master's students' iVideos to initiate discussions.

THE LEARNING AFFORDED BY TEACHER-AUTHORED VIDEO ANCHORS

Although we have presented three different examples, there are common themes that arise across our approaches to the use of teacher-authored video as a means of learning for our teacher preparation candidates. Each example is consistent with the goals of anchored instruction. That is, each uses video stories to situate learning in an authentic context to serve as a bridge between the theoretical knowledge (that is often covered in a methods class) and the practical knowledge of expert teachers. Furthermore, these anchors serve as a point from which to learn content (i.e., subject matter). For example, to author IVAN time lines about best practice literacy instruction, students learn about literacy concepts and how to apply them. In all three examples, the anchor also embeds opportunities to talk about pedagogy, and what makes for good instruction.

The three approaches we have presented represent promising approaches to the development of in-service teachers. Although we are currently in the process of conducting more formal research studies, we are encouraged by our anecdotal observations of the classrooms, and the results presented by the work of similar scholars. For example, in one study, teacher candidates who used digital video tools to augment their classroom placement observations outperformed control groups in measures that required them to identify, to interpret, and to analyze evidence of exemplary teaching (Beck, King, & Marshall, 2002). We see the use of developing and designing anchor video as an effective means to help preprofessional students organize their knowledge.

Because teacher education students are the authors of the video anchors, we have also seen a re-energized teacher-education classroom. We feel that in a more traditionally designed curriculum, teacher education students have lost ownership, responsibility, and voice in the curricula process. However, in all of our examples, and many that we have experienced but not reported in this chapter, prospective teachers have all been highly engaged and motivated in curricula design, in their own teaching, and in student learning. All students want copies of their iVideos and video anchors to show their family, friends, and colleagues. This desire or pride by the students is not evident in any other assignment we give to the students throughout the academic year.

Unleashing the power to be creative as well as a thoughtful and reflective professional seems to be a great leverage point for their professional education, leading to a sense of community, shared goals, and the development of collaborative skills (Petrosino & Dickinson, 2003). Also, there has long been a call for teachers to integrate technology into their own teaching (Mishra & Koehler, in press; PCAST, 1997; Zhao, 2003). The use of anchor video is a specific example of integrating technology into their own learning, which, when combined with its motivating effects, makes it much more likely to extend into their own teaching.

SUMMARY

This chapter presented three promising approaches to integrating digital video tools with pre and in-service teachers to support their development as instructional designers and reflective practitioners. Despite the differences in goals and implementations inherent in these three examples, together they afford opportunities to accomplish many of the goals of anchored instruction—that is, they encourage future teachers to pose and solve complex, realistic problems that bridge the gap between theory and practice. To be sure, teaching structured around teacher-designed video comes with challenges: Technology comes with its own set of costs in time, money, and potential frustrations; faculty have to buy in to this different model of teaching; and time spent doing these activities is time taken away from other instruction.

Extensive assessment and evaluation must be undertaken to see how this approach impacts in- and preservice teachers' cognition and (more importantly) the understanding of their students. However, the process we have begun has emerged from a sound research base, and such studies should provide insight to the instructional design community as well as to the learning sciences.

ACKNOWLEDGMENTS

We would like to thank a number of people who in different ways have helped us conceptualize these ideas. In particular, Matthew would like to thank David Wong, Punya Mishra, and Sharman Siebenthal for their collaboration and ideas for iVideos. Matthew would also like to thank Dr. Cheryl Rosaen for her collaborative work in the TE 401 course, none of which would be possible without her classroom, ideas, and rich supply of classroom video. Anthony would like to thank Gail Dickinson for her collaborative work with EDC 365-E. He would also like to thank Ann Cunningham of Wake Forest University for her insightful, practical, and theoretical discussions on teachers' creation of videos. Finally, he would also like to thank the students of The University of Texas at Austin's UTeach-Natural Sciences Program. Both authors would like to thank Sharon Derry and Ricki Goldman for their help and support.

This work has been partially funded by the Joe L. Byers and Lucy Bates-Byers Endowment awarded to Matthew J. Koehler from the College of Education at Michigan State University.

REFERENCES

Beck, R. J., King, A., & Marshall, S. K. (2002). Effects of video case construction on preservice teachers' observations of teaching. *Journal of Experimental Education, 70*(4), 345–355.

Bransford, J. D., & Schwartz, D. (1999). Rethinking transfer: A simple proposal with multiple implications. In A. Iran-Nejad & P. D. Pearson (Eds.), *Review of research in education* (Vol. 24, pp. 61–100). Washington, DC: American Educational Research Association.

Brown, J. S., Collins, A., & Duguid, S. (1989). Situated cognition and the culture of learning. *Educational Researcher, 18*(1), 32–42.

Cognition and Technology Group at Vanderbilt (CTGV). (1990). Anchored instruction and its relationship to situated cognition. *Educational Researcher, 19*(6), 2–10.

Cognition and Technology Group at Vanderbilt. (1997). *The Jasper project: Lessons in curriculum, instruction, assessment, and professional development.* Mahwah, NJ: Lawrence Erlbaum Associates.

Glaser, C. W., Rieth, H. J., Kinzer, C. K., Colburn, L. K., & Peter, J. (2000). A description of the impact of multimedia anchored instruction on classroom interactions. *Journal of Special Education Technology, 14*(2), 27–43.

Goldman, S. R., Petrosino, A. J., Sherwood, R. D., Garrison, S., Hickey, D., Bransford, J. D., & Pellegrino, J. W. (1996). Multimedia environments for enhancing science instruction. In S. Vosniadou, E. De Corte, R. Glaser, & H. Mandl (Eds.), *International perspectives on the psychological foundations of technology-based learning environments* (pp. 257–284). New York, NY: Springer-Verlag.

Goldman-Segall, R. (1998). *Points of viewing children's thinking: A digital ethnograper's journey*. Mahwah, NJ: Lawrence Erlbaum Associates.

Harrington, H. L. (1999). Case analyses as a performance of thought. In M. A. Lundeberg, B. B. Levin, & H. L. Harrington (Eds.), *Who learns what from cases and how: The research base for teaching and learning with cases* (pp. 29–48). Mahwah, NJ: Lawrence Erlbaum Associates.

Hickey, D. T., Moore, A. L., & Pellegrino, J. W. (2001). The motivational and academic consequences of elementary mathematics environments: Do constructivist innovations and reforms make a difference? *American Educational Research Journal, 38*(3), 611–652.

Krajcik, J. S., Soloway, E., Blumenfeld, P., & Marx, R. W. (1998). Scaffolded technology tools to promote teaching and learning in science. *1998 ACSD yearbook: Learning and technology.*

Krajcik, J. S., Czerniak, C., & Berger, C. (2002). *Teaching science in elementary and middle school classrooms: A project-based approach* (2nd ed.). Boston: McGraw-Hill.

Lampert, M., & Ball, D. L. (1998). *Teaching, multimedia, and mathematics: Investigations of real practice*. New York: Teachers College Press.

Lundeberg, M. A., Levin, B. B., & Harrington, H. L. (Eds.). (1999). *Who learns what from cases and how: The research base for teaching and learning with cases.* Mahwah, NJ: Lawrence Erlbaum Associates.

Mishra, P., & Koehler, M. J. (2003). Not "what" but "how": Becoming design-wise about educational technology. In Y. Zhao (Ed.), *What teachers should know about technology: Perspectives and practices* (pp. 99–122). Greenwich, CT: Information Age.

Mishra, P., & Koehler, M. J. (in press). Technological pedagogical content knowledge: A framework for integrating technology in teacher knowledge. *Teachers College Record.*

Petrosino, A. J., & Dickinson, G. (2003). Integrating technology with meaningful content and faculty research: The UTeach Natural Sciences Program. *Contemporary Issues in Technology and Teacher Education* [Online serial], *3*(1). Retrieved from http://www.citejournal.org/vol3/iss1/general/article7.cfm

President's Committee of Advisors on Science and Technology, Panel on Educational Technology. (1997). *Report to the president on the use of technology to strengthen K–12 education in the United States.* Retrieved from http://www.whitehouse.gov/WH/EOP/OSTP/NSTC/PCAST/K-12ed.html#exec

Sharp, D. L. M., Bransford, J. D., Goldman, S. R., Risko, V. J., Kinzer, C. K., & Vye, N. J. (1995). Dynamic visual support for story comprehension and mental model building by young, at-risk children. *Educational Technology Research & Development, 43*(4), 25–42.

Sherwood, R. D., Kinzer, C. K., Bransford, J. D., & Franks, J. J. (1987). Some benefits of creating macro-contexts for science instruction: Initial findings. *Journal of Research in Science Teaching, 24*(5), 417–435.

Shulman, J. H. (Ed.). (1992). *Methods in teacher education*. New York: Teachers College Press.

Shulman, L. S. (1986). Those who understand: Knowledge growth in teaching. *Educational Researcher, 15*(2), 4–14.

Shulman, L. S. (1992). Toward a pedagogy of cases. In J. H. Shulman (Ed.), *Case methods in teacher education* (pp. 1–30). New York: Teachers College Press.

Sykes, G., & Bird, T. (1992). Teacher education and the case idea. In G. Grant (Ed.), *Review of research in education* (Vol. 18, pp. 457–521). Washington, DC: American Educational Research Association.

Toulmin, S., Rieke, R., & Janik, A. (1984). *An introduction to reasoning* (2nd ed.). New York: Macmillan.

Wong, D., Mishra, P., Koehler, M. J., & Siebenthal, S. (in press). Teacher as filmmaker: iVideos, technology education, and professional development. In M. Girod & J. Steed (Eds.), *Technology in the college classroom*. Stillwater, OK: New Forums Press.

Zhao, Y. (Ed.). (2003). *What teachers should know about technology: Perspectives and practices*. Greenwich, CT: Information Age.

Part Four

Video Collaboratories and Technological Futures

27

Video Workflow in the Learning Sciences: Prospects of Emerging Technologies for Augmenting Work Practices

Roy Pca
Stanford University

Eric Hoffert
Versatility Software

VIDEO WORKFLOW AND PROCESSES

The aim of our chapter is to provide our readers with a comprehensive model of the stages of video workflow and their affiliated work practices, and with a road map of present and future technologies that support these practices. We situate the contributions provided by the chapters of this section of our volume within this workflow framework. Armed with this orienteering guide for video workflow, the reader should have a sense of the sociotechnical context of digital video and its affiliated technologies that they will be able to leverage today and anticipate in the years ahead. It is important to understand not only the epistemological and representational issues involving research video, in applications of video research on peer, family, and informal learning, and on classroom and teacher learning—but to recognize and to use productively the advances that are enabling and transforming video workflow practices for the work that we do as learning scientists.

We were inspired to sketch out this framework by a talk from Carl Rosendahl, Executive Producer of *Antz* and Founder of Pacific Data Images, the company that produced *Shrek*'s digital effects for the DreamWorks studio. He highlighted how computer software and networks are transforming every stage of the filmmaking process, from development to preproduction including storyboarding, production, and postproduction, including nonlinear editing, cinematography, visual effects, and distribution. We also foresee computer software and networks transforming every stage of the video workflow for the learning sciences, and we review later with illustrative examples how these transformations are beginning to surface in core technologies for each video workflow area. These advances, exciting as they are, are nonetheless at an early stage compared to the rapid development of digital video workflow in the film industry.

The chapters of this volume together help illuminate the extraordinary workflow complexity of video research in the learning sciences, a multistage and iterative process that, as Hay and Kim argue in their chapter (this volume), is beset with too much "friction" today—in which the researcher's needs and desires to do particular things with video such as share it or open it up to collaborative commentary are slowed down with the present state-of-the-art. And as Stevens highlights in this volume, capturing ideas in digital things and structuring learning around them using new video tools is a new version of a solution to the longstanding problem of inert knowledge in education.

Figure 27.1 provides a top-level view of the video workflow framework that transitions from video capture, to analysis, sharing, and collaboration. It begins in the upper lefthand corner with strategy and planning for video record capture, and moves quickly into the tactics of preproduction: Where, when, and how will you capture the video data that you seek? Our chapter does not treat these facets of workflow as they are addressed elsewhere in this volume. Then you are on-site, capturing video records with whatever devices suit your aims. Depending on the nature of your device, encoding of your video record may happen at the same time as capture—witness the advent of consumer digital video recorders that save video to computer hard disks that are part of the recorders. The video researcher then begins the processes of pulling these records into some kind of order, from the simple act of labeling them to easily find them later to the much more intricate activities that add the value of interpretation to these records. The researcher may chunk the video record into segments defined by event boundaries, time markers, or a variety of semiotic considerations. And marking video segments of interest, creating transcripts at different levels of detail, developing and using categories that the researcher considers useful for the aims of their research works in a recursive manner with both the deepening analysis of the video records and the never-ending tracking and finding of the rapidly growing population of data through searching and browsing. The researcher marks, transcribes, and categorizes a little, analyzes and reflects a little, needs to search and find a little, and so on, in the recursive loops that define such knowledge building activities (analogously to the writing process). In essence, there are close interdependencies between the activities of video record de-composition (e.g., segmenting, naming, coding) and re-composition (e.g., making case reports, collections of instances of commonly categorized phenomena, statistical comparisons of chunked episodes). Then the workflow moves on to presenting and sharing video analyses, in a variety of formats, and such sharing may be formative as one collaboratively develops and/or comments on a developing video analysis, or a summative account as the video analysis is published (e.g., on the web or a DVD) and commented on by others in the community. To close the loop, the substantive insights from specific video research workflow activities have the prospects of influencing the next cycles of video research workflow in the field.

VIDEO TECHNOLOGIES

Video Capture, Standards, Storage, Input/Output, and Display

Video Capture: Formats for Inputs

Video input to a computer may be in a digital or analog format. Many "legacy" video sources are still in an analog form including a wide variety of VCRs, TVs, and a previous generation of analog video camcorders; these devices use NTSC (525 lines of resolution), PAL (625 lines), or SECAM standards—each adopted in different regions of the world (e.g., North America and Japan for NTSC; Europe for PAL; Eastern Europe for SECAM). The devices all require a process of analog to digital video conversion. Classic analog video formats include VHS (250 lines of resolution), S-VHS (400 lines), Hi8, Betacam, and BetaSP. Video data on tape is in YCrCb format (one luminance/brightness channel and two chrominance/color channels), and converted to RGB (three color channels, stored as 8 bits per pixel/channel) on a computer. There are a

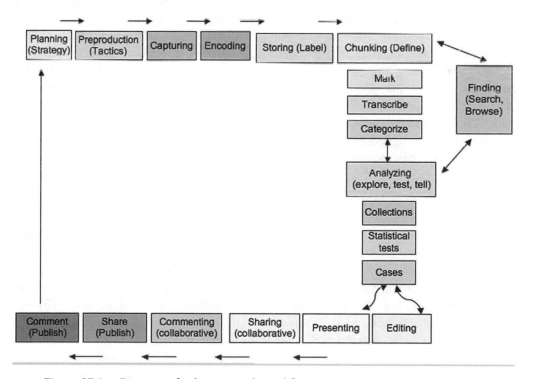

Figure 27.1. Diagram of video research workflow processes.

broad array of devices specialized in the analog to digital video conversion process. From lowest to highest quality, analog video signals span composite, S-video, and RGB component video. The serial control protocol RS-422 is often used to control a video source from a computer to allow computer-controlled video commands (i.e., rewind, fast-forward, play, jump to time code X).

Video imagery has an aspect ratio that is expressed as "x:y," where *x* represents its width and *y* its height. For digital video, pixels may be square or nonsquare; aspect ratios are typically 4:3 (1.33:1) for traditional video (the traditional TV screen, as well as IMAX), but newer widescreen formats include high definition, or HDTV (16:9, or 1.78:1), flat (1.85:1, American Theatrical Standard), and anamorphic scope (2.35:1). These widescreen video formats will gravitate from home theatre systems to the research laboratory. A proliferation of digital video formats includes D1, DV, DV25, DVCam, Digital8, DVCPro, and DVCPro50. Digital video device types include digital camcorders with video content recorded using mini-DV and other formats; direct-to-flash memory as in Nokia Series 60 3G phones; and direct to hard disk for other larger devices. PDAs, digital cameras, and cell phones now have integrated cameras with direct digital video input; video is captured using variants of digital video standards such as H.263, motion JPEG, or MPEG-4 (see later). These video-aware devices are exploding in popularity; 500 million camera phones were sold in 2005, and 63 million digital cameras were sold in 2004, with a projected 100 million digital cameras to be sold in 2008.

FireWire cables (using the IEEE 1394a standard) are rapidly becoming the prevalent high-bandwidth input mechanism that work across computer platforms to import video into a personal computer through serial bus ports. This standard is integral to a large number of popular digital camcorders for video I/O and runs at a rapid 400 megabits per sec. There are also FireWire 800 products, based on the IEEE 1394b version multimedia standard, that deliver speeds starting at 800 megabits/second, scalable to 3.2 gigabits/second. The new ultra high bit-rate standard also extends the distance that FireWire-equipped devices can send video and audio to more than 100 meters over CAT-5, plastic fiber, and other media.

The tools using these video inputs to the workflow process include digital camcorders, media enabled PDAs, and video-rich cell phones, web-cams, wireless video-on-IP (Internet Protocol) devices, and analog video cameras with video digitizers.

Video Standards

A great number of the revolutionary advances in digital video available to learning sciences researchers have developed due to world-wide MPEG standards (see http://www.mpeg.org for pointers), and because the kinds of functions available to a researcher are dependent on these standards and their evolution, we provide a brief account of them here. Whereas initially MPEG standards made for advances in video compression that were essential to reducing the costs of storing and transmitting video records, newer MPEG standards delve into the semantic content of videos, and enable new interactive capabilities with that video for authors and consumers that use such standards. In addition to the MPEG standards, several other important video standards will be reviewed; 3GPP and 3GPP2 for mobile video, and SMIL—a World Wide Web consortium standard for describing multimedia presentations.

MPEG

MPEG (http://www.chiariglione.org/mpeg) is a series of broadly adopted world-wide audio and video coding standards. Major standards developed from MPEG include:

- MPEG-1: Coding for digital storage media (1992).
- MPEG-2: Coding for digital TV and DVD (1994).
- MPEG-4: Interactive Multimedia Audiovisual Objects (1998).
- MPEG-7: Content Description Interface (2001).
- MPEG-21: Multimedia Framework (2003).

These standards are reviewed later, with the table providing a quick summary of key aspects of the primary MPEG standards today, MPEG-1, MPEG-2, and MPEG-4.

MPEG-1 (1992): Coding for Digital Storage Media

MPEG-1 represents the first major standard from the MPEG family of video codecs, designed for coding of moving pictures and associated audio for digital storage media at up to about 1.5 Mbit/s. The term "codec" combines the terms "compressor-decompressor" to characterize either hardware or software that can perform transformations on a data stream at both ends of its use in telecommunications. Codecs compress video for such purposes as storage, transmission, or encryption, and then decompress it for its uses including display, as in videoconferencing. In MPEG-1, each frame of video is decomposed into macroblocks—regions of 16 x 16 pixels. The macroblocks contain brightness and color samples called YUV, where U and V (chrominance information) are sampled at one quarter of the rate of Y (luminance information). Each 8×8 block of pixels is converted from a two-dimensional to a one-dimensional representation of 1×64 (in a zigzag sequence from the upper left to the lower right corner of the pixel block). A transform algorithm (discrete co-

TABLE 27.1
MPEG-x Characteristics Including Video and Audio Resolution,
and Application Domain

Video Codec	Video Resolution	Audio Resolution	Applications
MPEG-1	352 × 240 for NTSC at 29.97 fps	224 to 384 kbps including MP3	CDROM, Internet, Video conferencing including MP3
MPEG-2	720 × 480 for NTSC at 29.97 fps	32 to 912 kbps including MP3	DVD, Digital-TV, HDTV
MPEG-4	Variable	Variable including MP3, AAC	Cell phone, PDA, PC, variable rate Internet delivery

sine transform / DCT) converts the content from the spatial to the frequency domain as a series of numeric values, where the result is further coded, with frequent values replaced by short codes and infrequent values replaced by longer codes.

"Motion compensation" is an algorithmic technique common across the MPEG-x standards, and used to predict the movement of pixel blocks (macroblocks) from frame to frame; the prediction error for a macroblock is then stored and quantized and is typically smaller than storage of the actual pixel values, increasing the compression rate further. Motion compensation works by searching for "matching" macroblocks in adjacent video frames. Compressed sequences of MPEG video are constructed from groups of pictures (GOP). There are three classes of frame types in the MPEG standard that comprise a GOP. These are:

- I-Frames (intra-coded frames) compress a single frame of video without any reference to other frames in the video sequence. For random access in an MPEG video sequence, decoding starts from an I-frame. I-frames are included every 12 to 15 frames. These frames are also used for fast forward and reverse.

- P-Frames (Predicted frames) are coded as differentials from a prior I-Frame or a prior P-Frame. The prior P-or I-frame is used to predict the values of each new pixel in order to create a new predicted P-frame. P-frames provide a compression ratio superior to the I-Frames although this is a function of the degree of motion; small amounts of motion produce better compression for P-frames.

- B-Frames (Bi-directional frames) are coded as differentials from the prior or next I-or P-frame. B-frames use prediction similar to P-frames but for each block in the image, the prior P-frame or prior I-frame is used or the next P-frame or the next I-frame is used. Because the encoder can select which I-frame or P-frame to select, the encoder can select the bi-directionally predicted frame that produces the highest possible level of compression.

A coding sequence for MPEG-1 for NTSC video is I B B P B B P B B P B B P B B I, where I-Frames may be spaced 15 video frames apart, and two B-Frames precede each P-Frame (or I-frame). MPEG-1 standards vary for NTSC video (National Television System Committee, a 525-line/60 Hz, 30 fps system, principally used in the USA/Japan) and for PAL video (phase alternation by line, a 625-line/50 Hz, 25 fps system, used principally in Europe). For PAL video, the sequence is I B B P B B P B B P B B I and is typically 12 frames long. MPEG-1 resolution is 352×240 for NTSC at 29.97 fps and 352×288 for PAL/SECAM at 25 fps. Audio bit rates are typically 224 kbps for MPEG-1 layer II audio where 384 kbps is the typical rate utilized. MPEG-1 layer III audio coding represents the well-known MP3 audio standard, with typical rates of 128kbps and 192kbps.

Importantly for research, because MPEG-1 combines intraframe and interframe encoding, the co-dependence of certain frames makes this codec inappropriate for editing and other image postproduction applications. MPEG-1 displays progressive scan images, noninterlaced frames that cannot be used for broadcast, but it can achieve three times or more the compression factors of JPEG. It is good for playback only on applications such as games, distribution, publishing, VCD, and CD-ROM although it is occasionally used for desktop-based rough cut editing applications.

MPEG-2 (1994): Generic Coding for Digital TV and DVD (Moving Pictures and Associated Audio Information)

The MPEG-2 specification was designed for broadcast television using interlaced images. It provides superior picture quality compared to MPEG-1 with a higher data rate. At lower bit rates, MPEG-1 has the advantage over MPEG-2. At bit rates greater than about 4 Mbits/s, MPEG-2 is recommended over MPEG-1. MPEG-2 includes support for high quality audio and full surround sound with 5.1 channels, representing left, center, right front, right rear, and left rear audio channels. The audio can be extended to 7.1 with left center and right center channels. Audio bit rates range from 32 kbps up to 912 kbps where 384 kbps is the typical rate utilized and the sampling rate is fixed at 48 kHz.

MPEG-2 supports variable video bit rate and broadcast applications; MPEG-2 tends to be encoded at 6 to 8 Mb/s fixed data rate. For high-end production, typically the highest bit rates are used, such as 50 Mbps. This is called master quality MPEG-2 video encoding. Component ITU-R 601 format video running at 270 Mbits/sec will run at 2–50 Mbits/sec when transcoded into MPEG-2. MPEG-2 can also support both 4:3 and 16:9 image aspect ratios. MPEG 2 is used for DVD, digital TV, and HDTV.

MPEG-2 uses a group of pictures (GOP) at 12 (PAL) or 15 (NTSC) frames in length where each frame is constructed of two interlaced fields. A coding sequence for MPEG-2 for NTSC video is I P B P B P B P B P B P B P B I, where I-Frames may be spaced 15 video frames apart, and a P-Frame precedes each B-Frame. For PAL video, the sequence is I P B P B P B P B P B P I and is typically 12 frames long. MPEG-2 resolution is 720 × 480 for NTSC at 29.97 fps and 720 × 576 for PAL/SECAM at 25 fps.

Like MPEG-1, the I-frames in MPEG-2 are encoded independently and are the only independent frames in an MPEG-2 sequence. Only the I-frames can be edited when working with MPEG-2. MPEG-2 has been proven to be a good video standard to handle the use of transcripts along with standard (and noisy) classroom interactions. MPEG-2 consisting only of I-frames at high bit rates is often used for video editing and/or production applications due to its high picture quality and flexible random access support. In the case of MPEG-2 where only I-frames are used, production quality MPEG2 at 50 Mbps is also referred to as IMX; this format is frequently utilized with equipment such as AVID editing stations and storage subsystems.

MPEG-4 (1998): Coding of Interactive Multimedia Audiovisual Objects

MPEG-4 resulted from a new international effort incorporating and extending MPEG-1 and MPEG-2 and involving hundreds of researchers and engineers. MPEG-4 builds on three fields; digital television; interactive graphics applications (synthetic content); and interactive multimedia (distribution of and access to content on the Web). MPEG-4 provides the standardized technological elements for integrating of the production, distribution, and content access paradigms for the three fields.

Unlike its predecessors, MPEG-4 is an object-based video standard. Audiovisual scenes can be composed of objects, where a compositor within a decoder places video objects into a scene using the optimal encoding process for each object. An objective is to go beyond the typical start/stop/rewind/fast-forward level of interaction common to video content; with MPEG-4 the objective is to allow for interactivity with video objects directly embedded within a scene. Relevant to computer graphics practitioners, the standard is targeted for the combination of natural and synthetic objects in a scene. Audiovisual objects can include 2D/3D computer graphics, natural video, synthetic speech, text, synthetic audio, images, and textures. MPEG-4 streaming delivers the same quality video streaming as MPEG-2, the current industry standard, but MPEG-4 uses only one third of the MPEG-2 bit rate. This bit rate reduction at the same quality level is quite substantial and yields significant speedups in transmission time. MPEG-4 video provides very high quality across the bandwidth spectrum—from cell phones up to high bit rate broadband—that rivals the best proprietary compression algorithms available today. MPEG-4 was designed to be a scalable Codec that could support a broad array of delivery devices (PDA, PC, Set-top box, etc.) and it has delivered on that promise.

At the core of the MPEG-4 standard is the audio codec—AAC (Advanced Audio Codec). AAC offers support for multichannel audio, up to 48 channels; high resolution audio with sampling rates up to 96 KHz; decoding efficiency for faster and more efficient decoding; and compression with smaller file sizes. Multilingual support is also provided. AAC is used for audio coding at 32 kbps per channel and higher. The standard is targeted for audio coding in 3G wireless phone handsets and is used in the Apple iTunes Music Store. Apple Computer strongly supports MPEG-4 (Apple Quick-Time7/MPEG, 2005). MPEG-4 is an integral element of QuickTime 7 (and beyond) and Real Networks has adopted the standard as well. However, and in notable fashion, Microsoft has yet to embrace the standard and has provided an alternative scheme in Windows Media 9 and 10. Most recently, a flavor of MPEG-4 referred to as MPEG-4 Part 10, which is also known as H.264, is rapidly coming into place as a favored standard for high-quality video compression. H.264 is being used to store video as a "broadband master" at bit rates from 3 to 6 Mbps from which the content can be further transcoded into a variety of lower bit rates for broadband distribution. Only the fastest PC and Macintosh computers can decompress H.264 at acceptable playback speeds and resolutions; as a result, this nascent format is expected to take an extended time period to come into widespread consumer usage.

3GPP and 3GPP2

Launched in 2003 as consumer services, 3GPP (the Third Generation Partnership Project) defines Mobile Video Codecs, with capability to download or stream video for mobile media devices, and often to capture video as well. The similar 3GPP and 3GPP2 are based on Mobile Video and Audio Codec Standards and are primarily targeted for ultra-low bandwidth downloadable and streaming video for cell phones and mobile devices (3GPP works for GSM networks-Global System for Mobile Communication; 3GPP2 for CDMA networks-Code Division Multiple Access). Variants of key standards are used such as H.263 and MPEG-4, with video download rates defined at 64 kbps. Streaming rates range from 25–45 kbps with AMR audio spanning 4–12 kbps and AMR-WB spanning 6–25 kbps. Image resolution and frame rates include Sub QCIF (128 × 96) and QCIF

(176 × 144 for PAL, 176 × 120 for NTSC) using 7.5, 10, and 15 frames per sec. Transport Mechanisms are designated as GPRS (General Packet Radio Service) for Internet access, WAP (wireless access protocol), MMS (multimedia messaging service) for e-mail exchange, and MMC (multimedia memory card) for memory to PC synchronization. This mobile multimedia standard resolution is 128 × 96 at 15 fps but may scale higher. Stanford's DIVER Project has been experimenting with Nokia cell phone short video capture (e.g., one min clips). A movie thus captured is sent to the DIVER software web server as an MMS e-mail attachment, and is then transcoded into Flash video format for research analyses, commentary, and remixing over standard web browsers using DIVER (see later in this chapter). As cell phone video cameras increase in resolution and storage media on phones allow capture of longer movies, this approach could enable a flexible and ever-present component of video research technology.

MPEG-7 (2001): Multimedia Content Description Interface. MPEG-7 is a standard focused on video and rich content metadata. The metadata for video includes semantic characterizations of video and interactivity. Once powerful mechanisms for video object detection and segmentation are in place and validated as a reliable capability, MPEG-7 can support these advanced functions with an end-user ability to edit out objects, people, and scenes. The main elements of the MPEG-7 standard include: (a) description tools, or descriptors (D), that define syntax and semantics of each feature (metadata element); and description schemes (DS), that specify structure and semantics of the relationships between their components, (b) a description definition language (DDL) that is used to define the syntax of the MPEG-7 description tools and allow creation of new description schemes, and (c) system tools, used to support binary representation for efficient storage and transmission, transmission mechanisms (both for text and binary formats), multiplexing of descriptions, synchronization of descriptions with content, and management and protection of intellectual property in MPEG-7 descriptions. MPEG-7 descriptions of content may include information on:

- Creation and production of the content.
  - Director, title, or short feature movie.
- Usage of the content.
  - Copyright pointers, usage history, and broadcast schedule.
- Storage features of the content.
  - Storage format, encoding.
- Spatial, temporal, or spatiotemporal components of the content.
  - Scene cuts, segmentation in regions, region motion tracking.
- Low level features in the content.
  - Colors, textures, sound timbres, and melody description.
- Reality captured by the content.
  - Objects and events, interactions among objects.
- How to browse content in an efficient way.
  - Summaries, variations, spatial, and frequency sub bands.
- Collections of objects.

- The interaction of the user with the content.
  - User preferences and usage history.

MPEG-21 (2003): Multimedia Framework

The major aim of MPEG-21 has been to establish a transparent multimedia framework for all ways in which one user interacts with another user, and the object of that interaction is a fundamental unit of distribution and transaction called the "digital item" or "resource" (where "user" has the technical sense of any entity that interacts in the MPEG-21 environment or makes use of a digital item). Digital items are the "whats" and users are the "whos" of the MPEG-21 framework. The standard defines a "resource" as an individually identifiable asset, such as video clip, audio clip, an image, and text. Interactions concerning resources include creation, production, provision, delivering, modification, archiving, rating, aggregating, syndicating, retail selling, consuming, subscribing, and facilitating as well as regulating transactions that occur from any of such kinds of interactions. The goal of MPEG-21 has been characterized as defining the technology needed to support users to access, consume, trade, and otherwise manipulate digital items in efficient, transparent, and interoperable ways. So for example, MPEG-21 includes an XML-based standard "rights expression language" for sharing digital rights, restrictions, and permissions for digital resources between creators and consumers, and for communicating ubiquitous and secure machine-readable license information (Wang, 2004).

Video Interaction: SMIL

SMIL (synchronized multimedia interaction language) is a Web Consortium standard for describing multimedia presentations (http://www.w3.org/AudioVideo). SMIL can be used to create time sequential and time parallel composited layers of image, text, and video within a single, synchronized multimedia presentation. Graphical regions of the screen are defined and temporal events can be mapped into the graphical regions. SMIL is compatible with both QuickTime (QuickTime/SMIL, 2005) and RealMedia. SMIL is based on an XML representation and allows for the integration of distributed web resources into a unified end-user experience. Examples of SMIL usage include starting one video clip after another video clip completes, or triggering a demographic trend graphic to appear beside a video news clip. In addition, a completely new user experience—such as launching a new browser window with a new user input form—can be triggered from a user mouse click in a particular graphical region or visual icon.

SMIL data files are typically comprised of links, media content, spatial and temporal layouts, semantic annotations, and alternative content (for varying bandwidths, tasks, and user characteristics). SMIL uses the concept of "layout adaptation," where SMIL documents can adapt to browsers and/or playback devices with different characteristics such as screen sizes, bit depths, language characteristics, and so forth. Adaptation can be based on environment, user, and purpose. Selected SMIL "dialects" may also be skipped using a "skip-content" flag. For example, a full color video could be represented in black and white on a monochrome cell phone display; a text region could be shown in French rather than in English based on the location of the end-user. SMIL is rich in hierarchy—regions can be hierarchical spatially—and time-based constructs can be nested such as parallel and sequential time-based media playback (also called "temporal hierarchy"). SMIL can be adapted to the needs of the education community because of its flexibility, features for rich media and interactivity, and ability to support curriculum tool building and delivery.

Video Storage and Archives

Video storage of sufficient scale and reliability to deliver rich media to multiple users is a key requirement. Video storage capability is rising dramatically (from GB to TB) while costs are falling quickly (Langberg, 2004). In the 50 years since IBM invented computer hard disk storage, the density of information that can be recorded per square inch has increased 50 million times, from 2K bits to 100 Gigabits (Walter, 2005), with ultra-high densities achieved of 50 terabits per square inch with SeaGate's labs (McDaniel, 2005). A gigabyte of storage today costs on the order of $1 (Gilheany, 2004; Napier, 2006), with terabyte storage for $1,000; this is remarkable when contrasted with storage pricing in the year 2000, when a terabyte of industrial grade storage might cost as much as $1,000,000. Yet, there is a direct correlation between the cost of the storage and its inherent reliability. Storage with ultra high levels of redundancy and reliability can be costly, usually 10–100 times more than standard SCSI storage on a PC.

Video storage systems, when used for ongoing production and archiving, may include any of the following storage approaches; *online, nearline,* and *offline*. The access time and amount of storage utilized for each "tier" of storage increases as one progressively transitions from online to offline. Likewise, the relative expense of storage declines as one moves from online to offline. Video storage often follows a scheme similar to traditional hierarchical storage management (HSM; e.g., see Front Porch Digital, 2002; IDC, 2005).

Online Storage. Online storage is the fastest storage media and is used for all production-level work. Online provides near instantaneous access to video material and content. The access time for online storage is on the order of 10 to 15 ms measured as the time it actually takes for the disk read/write head to locate a data sector on the disk drive. Online content typically ranges from gigabytes to terabytes. Data transfer rates can range from 10 to 1,000 Mbytes/sec or more, with higher speeds using special disk arrays.

Nearline Archiving (Hierarchical Archiving). With nearline archiving, an archiving mechanism can be used for content that has not been requested for an extended period of time. Under this scenario, file data is normally stored on a server so that it can be accessed quickly as needed. When a particular event occurs, as when files are not accessed for a specified period of time, a nearline archiving system automatically transfers files to an external removable tape device, providing additional disk

space for online work. When it is necessary to open a file whose data has been transferred to a nearline archive, the data is automatically recalled from remote storage.
Nearline automated tape libraries are the primary mechanism used for archival and
short-term storage. Access time for nearline archives is on the order of 100s of ms or
single digit seconds. Storage cost is on the order of 10¢ per GByte or less (Gray, 2004).
Data transfer rates range from 10 to 250 Mbytes/sec or more. Nearline content typically
ranges from gigabytes to terabytes.

Offline Archiving.   With an offline archiving approach, the selected content
will be moved to offline storage when it is no longer required and the content will not
be available without manual content restoration. Offline storage is based on tape. Tape
devices access data in sequence; this means that accessing files from tape devices can
require significant time even if the tape is already on site and loaded in the tape drive.
When storage costs drop, the need for offline and nearline archives is reduced. Offline
archives are typically on high-density tape; for example DLT tape contains 600 GBytes
of data on a single tape and it is rated to last 30 years. Access time for offline archives is
on the order of 10s or 100s of seconds or more. Storage cost is on the order of 1¢ per
GByte or less. Offline content can range from petabytes to exabytes, particularly for organizations with very large-scale data backup and retention requirements over large
time scales. Data transfer rates range from 1 to 100 Mbytes/sec or more. Holographic
memory is a new optical media contribution to superior low-cost offline archiving. In
late 2006, Maxell is releasing new optical storage media with 3-D holographic recording technology so that a single 5¼" diameter optical disc has a 1.6 terabyte capacity, offering a 50-year media archive life and random data access with data rates as high as 120
mbytes/sec.
    Tape archival libraries provide affordable mass storage, which often handles tape
library management. An archive manager is a middleware software solution serving as
an abstraction layer and bridge between an online storage system and an automated
tape library's tape drive and robotics mechanisms for physical tape movements to and
from use. Archive managers must be compatible with automation systems (e.g.,
Sundance, Encode, Probus), content management systems (e.g., Documentum,
Artesia), and video servers (i.e., QuickTime, Real Networks, Quantel, Avid). Archive
managers hide the complexity of interfacing to disparate tape library systems, which
tend to have complex logic and proprietary interfaces. Archive operations require
physical retrieval of content from remote devices, mounting of tapes and extracting of
video from the tape back to an online storage facility; video files may be distributed
across multiple tapes. Video is also routed to the correct video server when more than
one server is utilized. Key archive manage operations include "archive," "delete," and
"restore."
    Online storage, especially the type that is used on high performance media serv-
ers, will often include support for RAID (Redundant Array of Independent/Inexpensive
Disks), a category of disk drives that employ two or more drives in combination for
fault tolerance and performance. RAID supports six tiered levels and can handle data
striping (where data resides across multiple disks for faster access), and data mirroring
(where data resides on multiple disks for fault tolerance). Storage Area Networks

(SANs) are becoming more widespread now, especially in networked or hosted environments. A SAN (see SAN, 2005) is a high-speed subnetwork of shared storage devices—machines that contain only disks for storing data. SAN architecture enables all storage devices to be accessible to servers on a local or wide-area network. Because data stored on disk does not reside on network servers, server power is utilized for applications only and disk servers handle data access only. Alternative and lower cost solutions are available including newer systems such as Mirra, where a specialized multimedia/data/backup server is connected to a PC and allows for external Internet access to stored content. Such storage appliance devices cost only a few hundred dollars but allow for automated backup of content, integrated with web-friendly access for storage sharing.

As storage costs drop substantially and as interest rises in the archiving of video material, a number of major video archives are rising in prominence. Notable among public archives are the Internet Moving Image Archive (Internet Archive, 2005, http://www.archive.org/movies/movies.php) and the Shoah Visual History Foundation (Shoah, 2005). The Internet Moving Image Archive is a repository of video that is searchable, indexed, and available to the public as part of the Internet Archive Project, which contains 400 terabytes of indexed content, growing at the rate of 12 terabytes per month. The video archives provide a collaborative environment (user votes, ratings, most popular videos, most popular categories, and number of downloads) and multiformat delivery including MPEG-1, MPEG-2, and MPEG-4 (standard playback and editable). Internet Archive research is underway to develop the "Petabox," a petabyte archive and processing matrix using 800 PCs to process the archived data. The Shoah Visual History Foundation has captured 120,000 hours of video testimony from holocaust survivors. This content is stored on a 400 terabyte digital library system where robotics are used to retrieve the appropriate tape matching user requests for video access; a high-speed fiber optic network connects a number of major universities to the archive and can deliver full resolution video to web browsers at these institutions. One could imagine these concepts extended to the learning sciences—for example, an archive could be developed with hundreds of thousands of hours of video captured from teachers and classroom interactions providing a valuable resource for the education community, subject to appropriate human subjects approvals from institutional research boards (IRBs) required by federally funded research in the United States.

Video Clients and Servers

Video clients are ubiquitous as digital media players, and are now used widely across the Internet for multimedia delivery. The three primary client players are Microsoft Windows Media, Real Networks, and QuickTime. These players are dominant on the Mac and PC platforms. While technically not a client video player, Macromedia Flash format streaming video files are growing in usage because Flash is installed in over 98% of the world's computers and plays through web browsers. Recent projections indicate that streaming Flash video could grow to a market share of as much as 35% or more in the next 5 years. Linux and SUN Microsystem's Solaris operating systems are not as directly connected to the media player ecosystem although there are options available. Recent co-operations announced between SUN Microsystems and Microsoft could lead to advances for media players across the Solaris/Linux and Microsoft worlds—this would be helpful for greater standardization of media delivery, but only time will tell. The installed base of QuickTime players is now several hundred million worldwide. The media players include support for graceful degradation where CPU power, networked bandwidth limitations, or other issues that reduce throughput are adaptively handled by the players, which will reduce frame rate, picture resolution, or audio sampling rate if and where possible. Although the concept of scalable video has been highlighted for many years—where video content can be adaptively modified in real-time based on constraints of bandwidth and CPU power, alternative concepts are used for desktop delivery such as graceful degradation and multitrack reference movies (discussed later). Many video capture systems are now configured, through the use of multiple video input cards, so that video can be digitized and transcoded (converted into new media formats) in parallel into multiple downloadable and streaming media formats. Once the video content has been encoded into any of the key media formats, it can then be played back using a variety of the video server platforms. A subset of video servers is described in the next section.

Video servers store and deliver streaming and downloadable media. Streaming allows for random access at the client, no placement of the media file on the user's local hard drive, and supports delivery of live media streams. Storage for video may also be provided as a service on the Internet. Servers may be open source or proprietary. Representative video server platforms include: (a) the Real Networks Helix Server, the first major open-source streaming media server (Helix Server, 2006), which supports a large variety of video codecs (i.e., QuickTime, MPEG-2, MPEG-4, Windows Media, Real Media, etc.) as well as provides access to an open source code base for enhancing and extending the media server itself. This server is particularly useful when developing new streaming media algorithms and protocols; it can be used to build customized encoders and players for solutions requiring new codecs. A Helix web site provides source code, documentation, and useful reference information; and (b) the QuickTime Streaming Server is server technology for sending streaming Quick-Time content to clients across the Internet using the industry standard RTP (Real-Time Transport Protocol) and RTSP (Real-Time Streaming Protocol) protocols. The streaming server has a number of key features including skip protection, which uses excess bandwidth to buffer ahead data faster than real time on the client machine. When packets are lost, communication between client and server results in retransmission of only the lost packets (not all of the data in a block containing lost packets, as is typically used), reducing impact to network traffic. ISO (International Standards Organization) compliant MPEG-4 files can be delivered to any ISO-compliant MPEG-4 client, including any MPEG-4 enabled device that supports playback of MPEG-4 streams over IP. QuickTime supports the concept of reference movies that allow for storing tracks of movies at alternate data rates and delivery methods. For example, you can store movies at data rates of 56 Kbits/sec, 384 kbits/sec, and 1.5 Mbits/sec all in the same QuickTime movie file and select the appropriate movie depending on connection speed. Similarly, it is feasible to include both HTTP FastStart and Streaming media versions of the same content as separate tracks within the same QuickTime movie to allow selection based on the connection scenario for a user.

QuickTime FastStart is a delivery mechanism for QuickTime Movies that works with web browsers over HTTP, that is, after a preset amount of content has been delivered to a client over HTTP, the movie will begin to playback. Concurrent with playback, the HTTP process of downloading proceeds in parallel. A movie can thus play while the next set of content is being downloaded. The process emulates real streaming but is only an approximation. The ubiquity of HTTP, thanks to browsers, makes this a real advantage as no special server side or streaming server software is required and there are rarely any issues with traversing corporate or university firewalls or network translation tables. However, live content cannot be provided, as one must wait until an entire movie is downloaded before true random access is available, and a copy of the content is placed locally on a user's hard drive, which the content provider (commercial or researcher) may not desire.

Darwin is a related platform for the QuickTime Streaming Server. A key metric for streaming servers is the number of concurrent streams that may be delivered by a server. With Darwin streaming server, up to 4,000 simultaneous streams can be served from a single server, and resources can be scaled up to meet increased traffic by adding multiple servers. Darwin is based on the same code as Apple's QuickTime streaming server product (available for Mac OS X) and is available as open source, with support for Solaris, Windows NT/2000, Linux, and an ability to be ported to additional platforms (http://developer.apple.com/opensource/server/streaming).

Video Resolution, I/O, and Display

Megapixel image resolution and frames per second (fps) parameters continue to improve as per unit cost drops rapidly. Many digital camcorders provide much better imagery in using 520 lines of horizontal resolution versus 240 lines with VHS analog camcorders. Hi-resolution formats are becoming increasingly important due to the widespread proliferation of HDTV compatible displays and plasma screens. Based on high demand, costs are expected to drop rapidly over the next few years for HD (High-Definition) technology. Key standards for high-resolution video (SGI HDTV, 2005) include 1920 × 1080 @ 60i (an analog video interlaced display at 60 fps, i.e., SMPTE 274M), and 1280 × 720 @ 60p (a digital video progressive scan display at 60 fps, i.e., SMPTE 296M). Frame rates per second for HDTV content can include, for progressive scan, 24, 25, 30, 50, or 60 fps, and for interleaved scan, 50 or 60 fps. An important consideration for the next few years in the transition to more pervasive HDTV is the adverse effects of displaying conventional 4:3 aspect ratio interlaced video on HD displays. A standard cable feed, or video footage shot on DVD can be stretched, distorted, or aliased in appearance. This is unfortunate, because traditional analog TV displays will look better when displaying this content than would more expensive HDTV displays. Addressing these issues will be a slow process; making 4:3 aspect ratio video higher quality in appearance on 16:9 displays, and waiting for more content to be produced as original in widescreen format.

New high-resolution production systems are emerging from vendors such as Silicon Graphics, Apple, and Discreet; these environments include an ability to capture, compress, store, and manage HDTV data streams.

The Japanese broadcaster NHK has developed a possible successor to HDTV that uses the same 16:9 wide screen aspect ratio, Ultra High Definition Video (UHDV), but with an immersive field of view that is four times as wide and four times as high at HDTV, yielding a picture size of 7680 × 4320 pixels. UHDV also refreshes 60 frames per second, twice conventional video. Because HDTV took 40 years from its development as standard in 1964 to its consumer growth today, UHDV may be a long time coming.

To move from the large screen to the palm-size video display, we note that in 2003, the standard camera phone resolution was 300 kilopixels. By 2005, 2-megapixel camera phones are commonplace from Nokia and other companies, with even 7- to 8-megapixel camera phones available from Samsung. The next horizon will come from much higher resolution and much smaller form factor cameras using CMOS rather than CCD technology. CMOS (pronounced "see-moss") stands for complementary metal oxide semiconductor, and CMOS integrated circuits are very low in power consumption and heat production and thus allow for very dense packing of logic functions on a chip, resulting in greater, cheaper video functionality in a smaller package.

Video input and output capabilities span a broad range of bandwidths and form factors. Transfer mechanisms can be analog or digital. Video can be readily transferred to and from cell phones, cameras, PDAs, TVs, VTRs, and HDTV systems. For video I/O, one can transfer directly to computer from capture devices using USB or FireWire cabling, or by removable storage media like Compact Flash, Memory Stick Pro, Smart Media, Secure Digital, XD Photo Card, or tiny CD-R discs. In 2005, tape-free camcorders emerged with a tiny removable 4-gigabyte MicroDrive hard disk.

The primary worldwide standard for digital video is ITU 601, with video at a resolution of 720 × 480 pixels (SMPTE 601 @ 270 Mbps via serial digital interface, or SDI). HDTV video is typically managed using fiber-channel disk arrays with digital video transfer via the SMPTE 274M and 296M standards. However, certain classes of systems (Sony HDTV, 2005) allow HDTV signals to be encoded using the pervasive 601 standard so that they can be easily imported and then subsequently manipulated back in the HDTV format region.

The area of handheld and mobile devices continues to advance at a dramatic pace, with new models of handhelds and cell phones offering color screens, higher memory, increased network bandwidth (via WiFi and 3G) and enhanced removable storage. It is logical to consider the use of these devices as a platform on which to distribute and display rich media. For example, data storage cards such as CompactFlash, SD Memory, Memory Stick, MemPlug, and others, offer storage on the order of a gigabyte or more. This level of storage is well suited to handling compressed digital video files with duration of one hour or more. Content authors can create and store standard 4:3 or panoramic video content on this new class of data storage for mobile devices.

Kinoma (2005) has provided a strong solution for displaying high-quality digital video on handhelds. Kinoma Producer provides an authoring environment for PCs or Macs that enables the user to convert a movie into a specialized media format suitable for playback and interaction on a handheld device, using Kinoma Player software.

Kinoma video is compatible for playback on the many handhelds and cell phones running PalmOS. The Kinoma video codec supports full screen, full motion, full color, high-resolution video for Palm Powered handhelds plus VR objects, VR panoramas, animation, and still images with synchronized audio. For output, Kinoma can generate video, audio, still images, and with proper setup, can transform a PowerPoint presentation into a format for handheld display. In 2005, Apple introduced a video capable iPod which provides similar capabilities as Kinoma but for a much broader audience on the iPod, where video is more of an integral part of the product design; the video iPod is based on the MPEG-4 standard described earlier. Apple has also integrated its digital rights management technology (FairPlay) into the MPEG-4 video content used on the iPod to manage distribution of protected intellectual property including content that is for sale.

As Fishman (this volume) notes, multimedia handheld PCs and smart phones have the potential to dramatically improve teachers' ability to access multimedia records for their uses in professional development, or for making notes during instruction that can be synchronized with subsequent reflections on video for their practices, or to anchor mentoring dialogues.

Media phones, PCs, PDAs, TVs, and HDTVs can all serve now as video display devices. Users must be able to author in a multimedia multidevice world. Display technology is advancing rapidly and future advances must be anticipated now. High-resolution display can be crucial for "seeing" for analysis on larger displays like HDTV (e.g., Microsoft Windows Media 9 support for HDTV; development of Mark Cuban's HDNet HDTV channels and broadcasting) and already 13 million U.S. HDTVs compatible monitors were in the United States at the end of 2004, with projections of 74 million by 2010 (Chanko. Wiqkcr, & Scevak, 2005). The use of non-PC and non-TV platforms can provide nondesktop opportunities for reviewing and analyzing videos, for example, cell phones (with 320 × 240 pixel displays).

Video Editing, Indexing, and Analysis

Video Editing

Nonlinear editing is a key method to identifying and prioritizing video streams to produce final output. Nonlinear editing tools are now mainstream and available in simpler form factors than ever before, and without data loss during digital editing and copying, unlike analog videotape copying. Computer-based, nonlinear editing systems have radically changed the editing paradigm and have become the standard tools in both the film and the video industry (Hoffert & Waite, 2003), so much so that all the major television stations are dismantling their linear video editing suites and many of the youngest generation of editors have never edited linear video.

Nonlinear editing systems range from high-end, professional systems (such as the Avid Media Composer and Film Composer) to industrial grade or "prosumer" systems (such as Adobe Premiere and Apple Final Cut Pro) to the most basic consumer variants (such as Apple's iMovie or QuickTime Pro, or Pinnacle's Studio MediaSuite). All editing tools share random-access capabilities for retrieving digitized video and sound material and most utilize the concept of the time line as a working tool. In various systems, a low-resolution and more highly compressed digital video proxy of the broadcast grade content is utilized to speed the nonlinear editing process. An edit decision list (EDL) is employed with time code pairs and pointers to the original material. EDLs are then applied to the high-resolution content to generate the final edited video content. In higher end systems (e.g., Sony HDTV, 2005), working images are often stored in uncompressed format at full resolution and with R:G:B as 4:4:4 (12 bits) where possible to avoid any degradation of image quality during multigenerational image manipulations.

The time line is a graphical representation of the edited program, which allows an overview of the linear flow of the program. It shows representations of the clips assembled to create the master edit using the length of the clips to represent their durations and vertical lines to represent the locations of edits between clips. The clip names are displayed at each edit point. Optional thumbnail image representations can be displayed, as well as symbols for segment and transition effects present in the program. The time line consists of tracks that represent separate video and audio streams. A basic master edit has three tracks—one video track and two audio tracks for stereo sound (see Fig. 27.2). More tracks are possible in the more advanced systems.

The time line allows the editor to move through the master edit without having to scroll through the program. This is achieved by providing a clear overview of the location of the various elements within the program. The position locator, represented by the long vertical line and yellow arrowhead to which the arrow is pointing in the figure is moved to any location with one click of the mouse.

Multichannel video editing is also on the horizon not only as a new art and media form, but as a legitimate tool for presentation of multiple video channels; there is a relevance to studying distributed learning with video capture of interactions at each node of collaboration. Commercial editing tools have principally been developed for creating single-channel programs, for example, a linear film or video with one image stream accompanied by simultaneous audio in mono, stereo, or multiphonic variants. Although the semiprofessional and professional systems are capable of multiple video and audio tracks, these are intended as intermediary steps in creating the single-channel master. Layered video tracks are intended for creating composites, keys, and other image effects that will ultimately be reduced to one image track via rendering. The multiple audio tracks are likewise an aid in working with different sound elements such as speech, sound effects, and music that will be mixed down to the finished stereo audio

Figure 27.2.   A typical time line. This example is taken from Apple Final Cut Pro. The position locator is represented by the long vertical line and yellow arrowhead near the center.

tracks. Nevertheless, these existing tools can be used to other ends for creating non-standard, multichannel filmic environments. These include synchronized, multichannel films, nonlinear hypernarratives, and stereoscopic films.

Automated editing is a next generation area where certain decision areas can be delegated to smart video editing software. Video metadata standards such as MPEG-7 (see earlier) can be employed initially to segment video into discrete scenes and objects. Such segmentation and objectification of video can support the premise of automated editing schemes. Content can then be organized thematically and semantically based on business rules with automated editing algorithms applied. In the future, one may also employ scripting languages (a "film grammar" that allows for "when X and Y show up zoom into Z") to automate the video editing process for large corpuses of content. Davis (2003) argues that with metadata and media reuse, consumers could more readily become daily media producers through automated mass customization of media.

Video Indexing: Object and Scene Detection

Video indexing allows video to be segmented and deconstructed into component elements suitable for browsing, indexing, search, and retrieval. Tools to handle this can be manual or automated. Manual tools typically allow marking of relevant scenes, frames, or sots (sounds on tape, such as interview clips) using time codes or time code pairs. Users are also able to add annotations or links to related content. These requirements are evident in the learning sciences as many research tools employ them in the feature sets they provide to their user communities. Commercial tools such as the Virage Videologger (Virage, 2005) provide support for both manual and automated indexing of audio and video content, for both stored and lived media. Automated indexing tools seek to index media with little or no human intervention. Video object and scene detection allows for the detection of objects, scenes, key frames, and scene changes in well-understood visual domains, and is enhanced when multimodal information can be used (Snoek & Worring, 2005). Complementary algorithms for speech recognition support speaker identification, speech-to-text conversion, and transcript creation. More advanced models for automated indexing are in a research mode for event detection such as determining there is an event where two people are coming together; this work goes beyond traditional object and scene detection. Panoramic cameras are also being used to capture full 360°degree scenes (e.g., Pea et al., 2004; Sun, Foote, Kimber, & Manjunath, 2001) and the captured content may be used in conjunction with indexing to identify speakers (via audio) and to locate individuals or objects (via video).

Video Analysis

As the chapters of this volume illustrate, "video analysis" circumscribes an extremely diverse set of theoretical underpinnings, researcher objectives, and affiliated work practices concerning what is done with video when it is analyzed. Video analysis includes at least two broad and complementary categories of research; one bottom up from observations and the second top down from theory. Both forms of analysis often rely on having ready-to-hand some form of transcripts of the talk—from levels ranging from coarse grained to phonological—and possibly annotations regarding gestures, body orientations, actions on artifacts and documents, visual regard—all depending on the purpose of the video analysis.

In the first case of bottom-up inquiries, the researcher is viewing video and building up category definitions and exemplars inductively, from watching video and noting features of activities that appear worthy of designating with a name as a category, and additional exemplars are sought out to render the utility of the category more evident. In their classic paper on interaction analysis using video recordings of human activities, Jordan and Henderson (1995) articulate an accumulating body of wisdom concerning productive interaction analysis work practices, as well as providing an exposition of a number of "analytic foci," or "ways into a tape" that are orienting strategies for the theoretical issues of special interest to interaction analysts and that help in identifying video segments for collaborative group analytic work.

In the second case, following such inductive work, video "coding" is the major video analysis activity, and it depends on having a set of categories, definitions, and exemplars of the category to guide coding practice (e.g., Barron, 2003). For the purpose of conversational analysis of video records of human interactions, researchers tend to use either commercial (e.g., Atlas/ti, HyperResearch, Qualrus), or open source software and analytic tools (e.g., CLAN, Transana, see later) as key enablers to interpret conversations in video interactions.

Current video analysis tools are strong individually in various aspects of annotation and coding of time segments of video (such as Anvil: Kipp, 2001; CAVA: Brugman & Kita, 1998, a replacement for MediaTagger: Brugman & Kita, 1995; ELAN, 2005; Silver: Myers et al., 2001; Signstream: Neidle, Sclaroff, & Athitsos, 2001), editing of video (such as Silver and Transana, 2005), or producing and analyzing transcripts (such as the CLAN programs used in the Child Language Data Exchange System/CHILDES, for studying conversational interactions: MacWhinney, this volume), and multiple points of view on video with attribute significance ratings and visualizations (Goldman's Orion, this volume). Yet none of these tools has directly tackled the core challenges of supporting the broader use, sharing, publishing, commentary, criticism, hyperlinking and XML standardized referencing of the multimedia data produced and output by the tools. This area is a key remaining challenge where support for critical collaborative commentary and cross-referencing—a process that allows researchers to make XML standardized, accessible, and direct contact with competing analyses of video and audio data—is a fundamental advance still needed for scientific disciplines that depend on video data analyses. The integration of video analysis for the work of a community of researchers and practitioners poses technical and design issues that go beyond those inherent in developing video analysis tools, such as Transana and SILVER, which are more focused on specific tasks like video editing or transcribing than on providing a generalinteroperable and global XML standards-based infrastructure for collaboration.

Our research community also faces the challenges of preserving human subject anonymity where this is required by informed consent protocols, while also desiring to develop a cumulative knowledge base where multiple perspectives and competitive research argumentation can be brought to bear using video data of learning interactions. In order to address privacy concerns, the future of video analysis in the learning sciences may include the possibility of automatically anonymizing video using face and voice recognition (Kitahara, Kogure, & Hagita, 2004) and then transforming faces and voices. Research is also making progress in detecting emotions from facial expressions and contextual information (Picard, 2000), and mapping facial gestures onto computer-animated 3-D facelike surfaces may eventually be used to obscure identity otherwise revealed in the video source recordings.

Video Sharing

Video Asset Management

Specialized content and digital asset management systems allow video to be tagged with metadata, stored in multiple versions, transcoded into alternate formats for delivery (e.g., MPEG-2 and MPEG-4), and automated for generation of hierarchical low-bandwidth media previews and visual proxies for rapid access. E-mail notifications with hyperlinks to video and enabling video for public Web site access fit into this class of system. Metadata schema allow a variety of descriptive and rights-related parameters to be associated with the multimedia content, including but not limited to copyright information, production data, educational topic, level K–12 educational appropriateness, contractual usage restrictions, time codes, scripts or transcripts connected to the content, associations between assets, and composition structure for layered video (i.e., effects, titles, independent tracks, etc.). For example, the Corporation for Public Broadcasting has released *PBCore* (the Public Broadcasting Metadata Dictionary, http://www.utah.edu/cpbmetadata/) as a standard metadata vocabulary of 48 categories for describing and using media including video, audio, text, images, and interactive learning objects to enable content to be more easily retrieved and shared across developers, institutions, educators, and software systems. To provide a bit more detail, 13 different elements describe the intellectual content of a media resource, 7 describe the intellectual property elements that relate to the creation, creators and usage of a media resource, and 28 describe the instantiation elements that identify the nature of the media resource as it exists in some form/format in the physical/digital worlds.

Digital video asset management systems typically include a search, retrieval, and indexing engine as a core component of their design. Database indices often include indexing of free-form text as well as of structured metadata. To provide value to external systems, an increasing trend is to enable an export capability where video assets may be transmitted to external systems via an XML representation of metadata with pointers to related assets in databases. Enterprise class video asset management systems are based on multitier architecture with a canonical Web server, application server, and database server. Enterprise scale asset management systems (e.g., Artesia, Documentum, North Plains, Oracle's Intermedia) start at $50K and can range into the $500K and multimillion dollar levels when deployed for thousands or tens of thousands of users. Low-cost asset management systems as alternatives (i.e., Canto's Cumulus, Extensis, etc.) share a number of similar traits with enterprise scale systems but with streamlined functionalities. These systems can start at much lower price points in the hundreds to low thousands of dollars for individual use, or in the $10K–$50K range, depending on the number of clients in one's workgroup. More recently, video-centric digital asset management systems such as Venaca and Ardendo have begun to generate significant interest because they embed the functionality of video logging, annotation, rough-cut video editing, transcript searching, and other video features, directly into the core digital asset management architecture.

"Web Services" have provided a new method of abstraction by establishing a globally recognized language and computer-platform independent API (application programming interface) and messaging mechanism by means of a set of definitions of the ways one piece of computer software can communicate with another. Web services are likely to become more prevalent soon for use in video development, including API access for video capture, playback, transformations and so on (http://www.w3.org/TR/ws-arch/). Standards such as XML[1] and SOAP[2] are likely to be used to create a new level of standardization for accessing rich media and video functionality across the Internet; emerging content management standards such as JSR-170 and the nascent JSR-283 should be tracked; progress on these matters also depends on resolving key issues for security, billing, and provisioning. Web services directories (e.g., GrandCentral Communications) and web services interface builders (e.g., Dreamfactory, Curl, Laszlo Systems), are expected to be integral to the advancement of Web services usage. These trends should be watched closely for learning sciences video research support infrastructure.

Video Security

Video security middleware is increasingly required to ensure the security and privacy of video content. Authentication and authorization for media access, roles, and permissions is required. Digital media files can typically be copied and distributed freely across open networks. This approach, while promoting content access and usage, provides limited protection and no direct compensation to copyright holders of media content or protection of data records required by IRB (Institutional Review Board) human subjects protocols. Digital Rights Management (DRM) systems, designed to address such issues, restrict the use of digital files in order to protect the interests of copyright holders, to monetize content delivery, and to allow consumers to legitimately access vast libraries of copyrighted multimedia material. DRM technologies control file access (number of views, length of views, timeframe during which viewing is allowable), as well as file altering, sharing, copying, printing, and saving. DRM technologies can be made available within the operating system, within dedicated software, or in the actual hardware of media capable devices. DRM systems are

---

[1]XML stands for the global standard and general purpose Extensible Markup Language, which makes it possible for groups to create markup languages for describing data (thus, metadata) to support sharing of data across Internet-connected systems.

[2]Simple Object Access Protocol is an XML messaging protocol that encodes information in Web service request and response messages before they are sent over a network. SOAP messages are independent of any operating system or protocol and can be transported using many Internet protocols, such as HTTP, MIME, and SMTP.

now widespread, with close to a billion media players in computers enabled for DRM support. Many content authors and consumers are not aware of the availability of DRM platforms relative to the ubiquity with which these systems are now utilized.

DRM solutions take two distinct approaches to securing content. The first approach is "containment," an approach where the content is encrypted in a package so that it can only be accessed by authorized users. This limits access to content where a user had a valid license to interact with the media. The second approach is "watermarking," the practice of placing a watermark on content as a signal to a device that the file is copy protected. Our focus here is on containment methods. A number of DRM systems are currently used in high-profile media on-demand commercial services to secure content and to generate content revenues for content providers. Note that DRM is not yet available as a capability across all digital media formats. Sample media on-demand services include iTunes Music Store, RealNetworks' Rhapsody Digital Music Service, and for movies—MovieLink, PressPlay, Akimbo, and LaunchMedia (part of Yahoo!).

The component of the DRM system used to package the content is often called a "License Server." DRM systems typically secure content to a server platform and require users to be authenticated for content access through use of a license key. License server platforms package media files and issue licenses. License servers encrypt a given media file, lock it with a license key, and incorporate additional information from the content provider. This results in a packaged file that can only be played by the person who has obtained a license. The license itself may be distributed together or separately from the content in a conventional or encrypted format.

When a user requests playback or access to content that is secured, they must enter the license for the content (or are redirected to a page where they can learn how to obtain a license including payment details), or there must be a communication mechanism with the server to exchange a license key with the server to enable playback. License management allows users to make a specified number of local copies of the content, and to restore media files on a secondary computer in case of a hardware failure on a primary system. Users may also transfer files to secure portable devices, to portable media, and can burn content onto CDROM; however, rules must be set by the content owner to allow each of these types of operations.

The encrypted content may be placed on a Web site, streaming media server, CD/DVD, or e-mailed. Strong encryption is used to protect the content using cryptographic and antipiracy mechanisms. A number of the algorithms are based on published ciphers that have undergone intense review from the cryptographic community. Major commercial DRM systems include Windows Media and Office DRM, RealNetworks DRM, and Apple's FairPlay system used with iTunes (Salkever, 2004). The highest revenues generated to date for digital rights management targeted at pure consumer delivery of digital music have come from Apple Computer and the iTunes Music Store. FairPlay, with many of the aforementioned security characteristics, was able to achieve critical buy-in from the content providers to enable their media for distribution and purchase. It provided a strong DRM solution, along with a networked-based metadata service that can be updated dynamically (such CDDBs, or CD Data Bases, include the open-source sites FreeDB and MusicBrainz, and the commercial encoding CDDB platform from Gracenote used by tens of millions of digital music users, Copeland, 2004).

The significance of DRM solutions for video was underscored in autumn 2005 when Apple Computer began providing major network television shows, music videos, and video podcasting capabilities for its iPod portable media players.

Video Gateways and Media Appliances

Broadband video gateways and media appliances are starting to make major headway into the home. This trend and the component technologies are powerful enough and moving to a more open architecture so that they can be considered and leveraged for research purposes. In 1999, TiVo and ReplayTV launched the first personal video recorders (PVRs), which have since then revolutionized time shifting of consumer TV viewing experiences by providing hard-disk storage for digital video recording. TiVo, the leading PVR company, has an installed base as of September 2005 of 3.6 million units, with estimates of more than 10 million PVRs across all suppliers, a trend expected to accelerate as satellite and cable companies such as DirecTV and Comcast incorporate DVR functions into their set-top boxes. In addition, TiVo is making major advances with their PC to TV connection —based on their home media option. From a price perspective, compared to a media PC—this is very low cost—because as of November 2005, a 40-hour TiVo PVR costs $50 with 12 months prepaid service at a $12.95 per month rate (or a lifetime use fee of $299). TiVo supports TV to PC linkages with TiVoToGo so video can be watched on PCs or on the road. Fishman (this volume) sees TiVo and other PVR devices' capabilities to record live content in a buffer of "constant recording" and save prior events on command as a viable direction for teachers to collect video assets as records of their practice in classrooms.

The TiVo platform today is still a relatively closed architecture, but by the time this book is published, the linkage from PC to TV is expected to strengthen with additional support for a more open programming and extension environment (as indicated by the announced partnership of TiVo and NetFlix to deliver movies-on-demand). Related to TiVo and the home video space, Happauge Digital serves as a very low-cost stand-alone bridge between the PC and the TV, and Digital Blue and Mattel's Vidstar each provide low-cost capture and movie-making appliances for kids.

Consumers and researchers have alternate platforms considerably more open in design yet way more costly than TiVo—since 2002, Microsoft "Home Media" PC and debuting in 2005, Apple Computer's "Front Row," each allow for direct interconnect to a TV with a specialized interface for remote control and for viewing media rich information at a distance. As prices on such products drop in the coming year, they effectively cross an "affordability chasm" for using these systems in a consumer living room context as well as in research labs. With ultra fast 64-bit chips such as Intel's "Prescott" to power the next generation of Home Media Center PCs (speeds from 2.4 to 3.4 GHz) and the new Viiv initiative, the gaps between computing and high quality home video experiences are disappearing.

Video Publishing—DVD Recorders, Video Web Sites, Hybrid Models

Digital Video Disc (DVD) has rapidly become a common distribution format for video material with low-cost authoring platforms widely available. In addition to the digitization of video and archival protection relative to analog videotape, interactivity can be added easily with chapter markers for key scenes and shots. Typical storage is on the order of 4.5 gigabytes with discs ranging from one to 2 hours or more, depending on the video bit rate selected. Typical bit rates for MPEG-2 on DVD range from a modest picture quality video stream at 4 Mbps (1.8 GBytes per hour) up to a high quality video picture stream at 10 Mbps (4.5 GBytes per hour). An alternative to reducing the bit rate of MPEG-2 to store more video is to utilize the MPEG-4 format on DVD, which achieves a much higher level of compression (MPEG-4 compression can be 2x to 4x higher than MPEG-2). However, this would require a new generation of consumer players and as a result, is unlikely to occur soon. To further simplify the process of conversion from analog to digital video disc conversion, DVD service bureaus are now broadly available to convert archival videotape into digital video format—the typical cost per tape conversion is now in the range of $15 to $25 per hour of analog video onto DVDs.

The next generation DVD standards are also now on the horizon; key contenders for the next 10 year's of relevance in a world of high-definition video and high-quality audio include Blu-Ray and HD-DVD. A consumer electronic industry coalition led by Sony is supporting Blu-Ray (Belson, 2004)—a new intermediate format for DVDs with 8.5 gigabytes of storage and support for 4 hours or more of video at high quality, and it is called a "double layer disc" (Sharma, 2004a). Double-layer DVD discs are single sided with two data layers that can be independently recorded to and read from, where both layers can be accessed from the same side of the disc. Blu-Ray uses blue lasers instead of the red lasers typically used in optical drives to read data off discs, and supports 50 gigabytes of storage capacity with standards development backed by Dell, Hitachi, Hewlett-Packard, Sony, Samsung, Panasonic, Sharp, and so forth. The use of blue lasers allows storage of more data for the same surface area of the disc. HD-DVD uses a single-lens optical head that integrates both red and blue laser diodes, and supports 30 gigabytes of storage capacity with standards development backed by leaders Toshiba and NEC, 200 other companies in the DVD forum, and supported by Microsoft, and now Intel (in November, 2005). Although it has less storage than Blu-Ray, its backers consider HD-DVD more reliable as a storage medium (Sharma, 2004b). As of early 2006, these competing standards are playing out in a drama, with a standoff in Hollywood (Belson, 2005), as to whether consumer electronics/TV (Blu-Ray) or computer companies (HD-DVD) will rule the future of digital video disk-based technologies.

Based on the high bandwidth required for TV resolution video, DVD stands as a superior publishing medium relative to the Web. However, as bandwidth of the Internet rises overall, and broadband to the home, office and schools rise, expect that TV quality video will start to migrate to the Web. Even today, video Web sites that allow posting of indexed and searchable video with commentary can form the basis for new formats of e-publications of video material.

For example, in this volume, Beardsley, Cogan-Drew, and Olivero describe the VideoPaperBuilder system, software that enables teachers and researchers to work together to build multimedia Web-page documents called VideoPapers that closely link video, text, and still images from classroom practices. Authors may annotate segments of digital video of teaching or learners, with text comments or scanned records of student work or teacher handouts on paper or whiteboards. The completed document uses JavaScript menus, html links and frame sets, and QuickTime image slide shows to interweave the authors' video, slides, and text into a single multimedia presentation that can be interactively experienced by users using Web browsers.

A frequently noted example of this video-sharing trend can be found in the Open Video project (Geisler, 2004). The Open Video site (http://www.open-video.org) showcases effective video sharing across the Internet, metadata use, and how to make video more broadly accessible across a range of research and public user communities. The site houses a variety of video collections comprising over 3,000 videos, such as the CMU Informedia Project, the Howard Hughes Medical Institute, and the Prelinger Archives. All video content can be easily searched and browsed; metadata (with rights information) has been tagged for all video clips, and each video segment has a short preview (7 sec), a multiframe storyboard representation, and access to the original high-resolution video source material in digital format (i.e., MPEG-x). The next stage of the project will add more video formats, genre characteristics (student television, anthropological footage, technology demonstrations), and more collections for the video community site. During 2005, a plethora of Web sites was also launched that enable the general public to upload their videos and tag them with categories and brief commentaries, including Google, YouTube, Vimeo, Clipshack, OurMedia, VideoEgg, and so forth. Secure peer-to-peer group sharing of video and audio recordings is the focus of other commercial ventures such as Grouper and Veoh.

There are also commercial Web sites, such as Teachscape, LessonLab, CaseNex, and TeachFirst, all of which incorporate high-quality digital video as an integral element of their service. In this case, the focus is on teacher professional development and/or preservice education. Some of these companies, such as Teachscape (2005), use hybrid models that utilize the connectivity and interactivity of the web in conjunction with the high-bandwidth, high-speed media delivery platforms such as DVD in combination. A video program can be authored in tandem so that users interact with a Web browser or application constructed on the Mac or PC, with high-bandwidth media accessed rapidly from a local DVD drive. Users of hybrid model Web sites are shipped physical video discs for the video. Such hybrid systems will be able to provide very rich media soon with the advent of the DVD dual layer drive, HD-DVD, and Blue-Ray standards described earlier. This model was used frequently in the past as well with CD-ROM discs; however, as video delivery over the web has improved, this mode of interaction makes less sense. The higher density DVDs make the model viable yet again, but there will always be a race between high-bandwidth optical media and high-speed Internet connectivity to the home, school, university, and office. Raul Zaritsky's stimulating chapter (this volume) works in a related vein, but with several surprises, advancing high quality video case studies as what he designates "educational research visualizations" to serve as scaffolding for teachers seeking to understand and emulate the rationale and situated practices of a reform-oriented mathematics curriculum. If effect, he argues that these visualizations serve as warrants in an argument for the appropriation of new teaching practices, and how such a "workshop in a box" as a new media form could accelerate the adoption curve for theory-driven designs into education. The results he reports are sobering for innovators seeking to advance new media grammars that exploit multiple camera angles, multiple audio tracks, 3-D graphics, and other new affordances of digital video for educaion, as they can be perceived as complex and disorienting distractions rather than helpers without more teacher familiarity.

VIDEO COLLABORATION

Digital Video Collaboratories

Video collaboration systems will support key elements of the learning sciences video research workflow, enabling end-users to analyze, share, and collaborate around video records. Such systems will form the backbone of a video research framework. Earlier work enabled real-time and/or asynchronous text messaging among multiple participants as they are watching video broadcasts or video archives (e.g., Bargeron et al., 2002; White et al., 2000), but our work practices in the learning sciences require far more than that. Despite consumer-driven advances in video capture technology, and a sharp rise in the use of video for analysis purposes by solo researchers, video circulates sluggishly, if at all, within research communities. The same researchers who use video for analysis typically rely exclusively on text to present results. Researchers default to text because they cannot readily ensure that an audience can view video as source data, much less in a form that integrates an argument with video evidence.

This promise–reality gap for digital video has serious consequences for researchers. Connections between evidence and argument are obscured, the development of shared examples of exemplary analyses using video that can serve training and socialization functions for researchers is impeded, and sharing of video data among scientists is discouraged in favor of an isolated and inefficient approach to gathering and analyzing primary data. Research communities will not make full use of video data so long as significant obstacles remain at any of the key points of video capture, encoding, storage, retrieval, analysis, sharing, and commentary. Enabling research communities to build knowledge through sharing video data and analyses would constitute an important enhancement to the global research and education infrastructure. We see this emphasis shared throughout the chapters of this section of our volume.

In the Digital Video Collaboratory Project, where the DIVER team at Stanford has teamed with Brian MacWhinney's TalkBank team housed at CMU and the University of Pennsylvania, we have been addressing these critical issues and enabling communities of researchers and practitioners to collaborate in producing, analyzing, and commenting on an evolving corpus of video records in diverse disciplines studying learning and human interactions. Our project is centered on creating highly accessible tools for video analysis, sharing, and collaboration. We seek to establish a strong basis for broader impact across multiple disciplines and applications with our focus on accessibility, ease of use, core technical advances, and metadata/API standards. Achieving these goals requires leveraging information technology advances and innovations in Web-based computing, video analysis and collaboration tools, and video compression and streaming. To achieve the primary goal and validate our tools, we have been conducting our research initially as a multi-institution collaboration between Stanford University and Carnegie Mellon University, but we plan to develop and use our enabling infrastructure as a unified Digital Video Collaboratory for broad accessibility to researchers in a range of disciplines studying interactions in classrooms and in other contexts of human activity.

Enabling the free flow of video data and analyses within research communities requires three capacities that are currently lacking, each of them addressed by our Digital Video Collaboratory Project. First, video data has to be universally accessible without regard to its physical location. Currently, video data is available, if at all, in heterogeneous repositories with idiosyncratic access control, search and retrieval interfaces, and metadata structures. We are addressing this obstacle by developing a virtual video data repository and video analysis community portal, implementing a metadata scheme designed to support research use of video-as-data.

Second, research communities require video analysis tools that support the full range of scientific activities from inductive development of categories for interpretation to coding analysis and through collaboration, critique, and publishing. Currently, video analysis tools maroon data on islands of incompatible file formats, making it difficult to share data among applications, much less among other researchers in a free flow of data and argumentation. We address this obstacle by developing both generic and discipline-specific XML-based schema for video analysis to facilitate application interoperability, and flexible desktop and Web-based video analysis tools that directly support sharing, critique, and output of video analyses. The use of XML will facilitate development of specialized XML extensions to represent discipline-specific metadata for use by such components, and also make possible data exchange with other video analysis tools using XML, such as Atlas.ti and SignStream tools.

Finally, if video is to be a primary communications medium, other researchers must be able to respond to a video analysis using the medium of video itself. Our DIVER team at Stanford in this project has developed an approach for enabling distributed video analysis that allows random space-time access into compressed video streams, while not requiring the downloading of video into local computer storage for authoring new video clips (see Pea, 2006).

The DIVER system uniquely enables "point of view" authoring of video analyses in a manner that supports sharing, collaboration, and knowledge building around specific references into video records. We do this by enabling users to easily create an infinite variety of new digital video clips from a video record. This process is called "diving," and the author a "diver," because the DIVER user "dives" into a video record by controlling—with a mouse, joystick, or other input device—a virtual camera viewfinder (see yellow rectangle, Fig. 27.3) used to mark snapshots of specific moments, or to record multiframe video "pathways" through a video to create their "dive." The use of DIVER to focus the attention of an observer of one's dive on a video resource is what we call "guided noticing." Guided noticing is a two-part act for a visual scene that has been a vital part of cultural learning episodes, long before computers existed: First, a person points to, marks out, or otherwise highlights specific aspects of that scene. Second, a person names, categorizes, comments upon, or otherwise provides a cultural interpretation of the topical aspects of the scene upon which attention is focused. In the case of DIVER, such guided noticing is time-shifted and shareable by means of recording and display technologies. Diving creates a persistent act of reference with dynamic media—which can then be experienced by others remote in time and space, and which can additionally serve as a focus of commentary and re-interpretation. Why is guided noticing important? Because achieving "common ground" (e.g., Clark, 1996) in referential practices can be difficult to achieve, and yet is instrumental to the acquisition of cultural categories generally, and for making sense of novel experiences in the context of learning and instruction especially.

As illustrated in Figure 27.3, a dive is made up of a collection of "panels" on the right side of the web page, each containing a small video key frame representing a mark or video clip, as well as a text field containing an accompanying annotation, code, or other interpretation. Both the annotations and the space–time coordinates of a user's dive on video records are represented by the DIVER software system as XML metadata, so that one is not literally creating new video clips, but simply views into parts of one or more video files through a dive.

DIVER is designed to serve the purposes of both the video researcher who captured the video records, and his or her research collaborator or colleague who desires to have conversational exchanges anchored in specific moments that matter to them in the video segments. First, the video researcher uploads video data in any one of a typical range of formats to the DIVER server. Once DIVER software services automatically transcode the video into a streaming format, the researcher then may use a client-side Web browser to mark and record space–time segments of videos with the virtual camera, and to make text annotations about them as they build up their "dive" for analyzing the video record. (To provide security to video records, streaming video files in the Macromedia Flash, or .flv format, are made accessible through a Web server over the Internet, so that video files will not be downloaded to personal computers.)

Figure 27.3.   WebDIVER: Streaming media interface for Web-based diving.

The researcher's dive may then serve as the multimedia base medium for pro-cesses of scientific interchange, supporting collaboration and elaboration, as well as a critique of scientific argument and research evidence. The databases of primary video records and secondary analyses, or "dives" then become available to approved users through a browsable, text-searchable community-based Web site. Other researchers viewing the originating researcher's dive can respond to that diver's annotations by posting their own textual comments linked to the video in question (as in the first panel of Fig. 27.3), which could then be viewed by the diver and by other researchers, and be developed further in a Web-enabled video-anchored dialog. Dive respondents may also make their own dives on video records, referencing segments from any of the terabytes of video files available through the DIVER servers. All of these activities gener-ate searchable metadata, and support finding analysis-relevant clips and analyses in a video research community of practice.

Our DIVER team is extending these developments to realize our vision of the Digi-tal Video Collaboration (DVC) for robust access control, group formation, e-mail notifi-cations of changes in dives one has authored or subscribed to, and so on. We have also integrated WebDIVER with the DVC virtual data repository concept, so WebDIVER users can store and retrieve video data and analyses without regard to their underlying physi-cal storage locations. WebDIVER users can make dives into videos stored and served from distributed web servers or content delivery networks (CDNs), and play back dives as 'remixes' that reference only the spatio-temporal segments pointed to within the dive.

MacWhinney (this volume) characterizes progress toward building tools to facili-tate this new process, which he calls "collaborative commentary," and defines as the in-volvement of a research community in the interpretive annotation of electronic records, with the goal of evaluating competing theoretical claims. The collaborative commentary process involves linking comments and related evidence to specific segments of digital video, transcripts, or other media. He describes seven spoken language database pro-jects that have reached a level of Web-based publication that makes them good candi-dates as targets and beneficiaries of collaborative commentary technology.

Goldman (this volume) has been pioneering for many years in her video ethnog-raphy software systems the importance of "points of viewing theory," most recently software in her work on the Web-based Orion video analysis system, in which the in-sights generated by diverse participants on a given video record are valued as ethno-graphic contributions. Reed Stevens' desktop VideoTraces software (this volume) is oriented to reflection and presentation—enabling users to lay down a "trace" on top of a "base" video record (playable at variable speeds). The trace consists of voice annota-tion and a gesture depicted as a pointed hand cursor. When a VideoTraces file is played, one hears the audio trace overlay and sees its gestural focus. Stevens and colleagues have used this system in science education museums and in higher education courses such as rowing and dance composition. VideoTraces' uses of virtual pointing and voice-recorded commenting on video provide a complementary mechanism to our use in DIVER of guided noticing for achieving common ground in a referring act in the complexity of a video record.

We hope these new capabilities to establish digital video collaboratories will ac-celerate scientific advances across a range of disciplines. Beyond the bounds of academia, the availability of a fluid and reliable mechanism for publishing video data and analyses can support research dissemination by providing a means for public access to research results and enabling community commentary (see Pea, 1999).

MacWhinney (this volume) reviews a variety of approaches for enabling collaborative commentary among research video user communities—collaboration around the core elements of video research including video clips, video, transcripts, coding schemes, and annotations. Such community environments allow users to view and annotate material from one another, and include tools such as Zope, Zannot, Annotea, B2Evolution, Blogger.com, Blogging.com, and RSS (Really Simple Syndication).

Baecker, Fono, and Wolf (this volume; also Baecker, Moore & Zijdemans, 2003) have been developing a system called ePresence, designed to enable global broadcasts over the Web—of video and slides and presentations and live software demos, real-time interactive access to broadcasts by remote viewers who can have public or private text-based chats and submit questions to presenters, and postevent access to presentation archives. Although not initially conceived as a collaboratory for video data sharing in the learning sciences, Baecker et al. highlight how the video channel in ePresence can be used not only for video of lecturers, but for collaborating researchers to share video data access and to have text-based chats and threaded discussions about such resources. Because these discussions can incorporate Web links to other video sources and documents, ePresence provides a potentially powerful infrastructure for digital video collaboratory activities among learning science communities.

Communities of Interest Networks (COIN)

Communities of interest networks have emerged in recent years, thanks to blogging, RSS Web feeds (e.g., pubsub, newstrove, rocketinfo), and Web-based community platforms that enable participants to specify topics of interest so that they are regularly notified of results of searches (e.g., Google Alert), other news streams (RSS) culled from millions of Web sites, or new citations of articles published (e.g., Ingenta provides 20 million online articles from nearly 30,000 publications). We believe that as video resources become more widely available on the Web and in communities of practice, such as learning sciences research, and interest "grows" around them, that COIN infrastructure services will become available and widely used. One attractor to such COIN services is that they come to provide a form of "social information filtering" in which highly used, highly rated, or highly cited resources bubble to salience through patterns and levels of use of these resources by participants in the networks who are using these resources. In this way, video collaboratories can come to leverage the network effects (Katz & Shapiro, 1985; Rohls, 1974) seen in other Web spheres from e-commerce to communications, in which the value of a network grows exponentially with the number of nodes attached to it (see Barabasi, 2002).

CONCLUSIONS

It is noteworthy that we have observed a great proliferation of genres in the past few years that incorporate interactive multimedia and differing levels and kinds of affordances for collaboration. These include systems such as Orion's "constellations," DIVER's "dives," VideoPapers, VideoTraces documents, ePresence "archives," Talk-Bank video transcripts, and so on. The multiplicity of such systems raises a number of important issues, and dimensions for comparative analysis as researchers seek to find the best fit to their desired work practices. These issues include ease of use, embedability in Web-commentary layers, access/security/IP regarding content, search capabilities, and virtual access to video stored across multiple distrivuted servers. As we have indicated, video researchers studying learning and teaching have a great deal to look forward to as the converging advances of computing and media communication technologies make formerly advanced technologies into everyday consumer and research tools.

ACKNOWLEDGMENTS

The National Science Foundation has provided support at Stanford University that we gratefully acknowledge for work reported in this chapter (#0216334, #0234456, #0326497, #0354453). The opinions and findings expressed are our own. We'd like to thank Joe Rosen for his many contributions as DIVER software architect and engineer for advancing the agenda of open-standard digital video collaboratories.

REFERENCES

Apple QuickTime 7/MPEG. (2005). Retrieved October 30, 2005 from http://www.apple.com/mpeg4

Baecker, R. M., Moore, G., & Zijdemans, A. (2003). Reinventing the lecture: Webcasting made interactive. *Proceeding of HCI International 2003* (Vol 1, pp. 896–900). Mahwah, NJ: Lawrence Erlbaum Associates.

Barabasi, A.-L. (2002). *Linked: The new science of networks*. Cambridge, MA: Perseus.

Bargeron, D., Grudin, J., Gupta, A., Sanocki, E., Li, F., & LeeTiernan, S. (2002). Asynchronous collaboration around multimedia applied to on-demand education. *Journal of Management Information Systems. 18*(4), 117–145.

Barron, B. (2003). When smart groups fail. *The Journal of the Learning Sciences, 12*(3), 307–359.

Belson, K. (2004, September 20). New economy: Format wars, Part 3—Sony and its allies battle 200 companies over the next generation of digital videodiscs. *New York Times, Section C, p. 3.*

Belson, K. (2005, July 11). A DVD standoff in Hollywood. *New York Times, Section C, p. 1.*

Brugman, H., & Kita, S. (1998, February). CAVA: Using a relational database system for a fully multimedial gesture corpus. *Workshop: Constructing and Accessing Multi-media Corpora: Developments in and around the Netherlands*. Nijmegen, The Netherlands.

Brugman, H., & Kita, S. (1995). Impact of digital video technology on transcription: A case of spontaneous gesture transcription. *KODIKA/CODE Ars Semeiotica, An international journal of semiotics, 18,* 95–112.

Chanko, T., Wiqder, Z. D., & Scevak, N. (2005, July 19). *US DTV and iTV Forecast, 2005 to 2010.* Report may be purchased from http://www.jupiterresearch.com/bin/item.pl/research:vision/1211/id=96505/

Clark, H. H. (1996). *Using language.* Cambridge, UK: Cambridge University Press.

Copeland, M. (2004, March). The magic behind the music. *Business 2.0, 5*(2), 40.

Davis, M. (2003). Editing out video editing. *IEEE Multimedia, 10*(2), 2–12.

ELAN. (2005). *EUDICO linguistic annotator.* Software downloaded October 30, 2005 from http://www.mpi.nl/tools/elan.html

Front Porch Digital. (2002). *An overview of digital video archives in broadcast: A white paper for the media and entertainment industries.* Retrieved October 30, 2005 from http://www.fpdigital.com/uploads/1115225972.pdf

Geisler, G. (2004, February). *The open video project: Redesigning a digital video digital library.* Paper presented to the American Society for Information Science and Technology Information Architecture Summit, Austin, TX.

Gilheany, S. (2004). *Projecting the cost of magnetic disk storage over ten years.* Retrieved November 5, 2005 from http://www.aiim.org/documents/costmagstorage.pdf

Gray, J. (2004). *The five minute rule.* Microsoft Research Presentation on PetaByte Server Infrastructure. Retrieved October 30, 2005 from http://research.microsoft.com/~Gray/talks/FiveMinuteRule.ppt

Helix Server. (2006). Retrieved December 28, 2006 from http://www.realnetworks.com/products/media-delivery.html ; http://www.helixcommunity.org

Helix Community. (2006). Retrieved December 28, 2006 from http://www.helixcommunity.org

Hoffert, E. M., & Waite, C. (2003, August). *Post-Linear Video: Editing, Transcoding, and Distribution.* Paper presented at the Conference Proceedings of the ACM SIGGRAPH 2003, San Diego, CA.

IDC. (2005, January). *WGBH digital asset management prototype lays foundation for lower costs, increased efficiencies, and enhanced services: An IDC eBusiness case study.* [IDC Report FE2016-0]. Retrieved October 30, 2005 from http://www.artesia.com/pdf/IDC_CaseStudy_WGBH.pdf

IEEE 1394b. (2005). *About IEEE 1394b technology.* Retrieved from http://www.1394ta.org/Technology/About/1394b.htm

Internet Archive. (2005). *Internet moving image archive.* Retrieved October 30, 2005 from http://www.archive.org/movies/movies.php

Jordan, B., & Henderson, A. (1995). Interaction analysis: Foundations and practice. *The Journal of the Learning Sciences, 4*(1), 39–103.

Katz, M. L. & Shapiro, C. (1985). Network externalities, competition, and compatibility. *The American Economic Review, 75*(3), 424–440.

Kinoma. (2005). Kinoma player, Kinoma producer. Retrieved October 30, 2005 from http://www.kinoma.com

Kipp, M. (2001). ANVIL: A Generic Annotation Tool for Multimodal Dialogue, *Proceedings of Eurospeech* (pp. 1367–1370). Aalborg, Denmark, September, 2001.

Kitahara, I., Kogure, K., & Hagita, N. (2004, August). Stealth vision for protecting privacy. *Proceedings of the 17th International Conference on Pattern Recognition, 4,* 404–407.

Langberg, M. (2004, March 26). New hard drives may turn handhelds into tiny TiVos. *Forbes.* Retrieved October 30, 2005 from http://www.forbes.com/technology/2004/03/0326harddrivespinnacor_ii.html

McDaniel, T. W. (2005). Ultimate limits to thermally assisted magnetic recording. *Journal of Physics: Condensed Matter, 17,* R315–R332.

Myers, B., Casares, J. P., Stevens, S., Dabbish, L., Yocum, D., & Corbett, A. (2001). A multiview intelligent editor for digital video libraries. *Proceedings of the 1st ACM/IEEECS Joint Conference on Digital Libraries* (pp. 106–115). Roanoke, VA.

Napier, D. (2006, January). Build a home terabyte backup system using Linux. *Linux Journal, 141,* p. 3.

Neidle, C., Sclaroff, S., & Athitsos, V. (2001). A tool for linguistic and computer vision research on visual–gestural language data. *Behavior Research Methods, Instruments, and Computers, 33*(3), 311–320.

Pea, R. D. (2006). Video-as-data and digital video manipulation techniques for transforming learning sciences research, education and other cultural practices. In J. Weiss, J. Nolan, & P. Trifonas (Eds.), *International handbook of virtual learning environments* (pp. 1321–1393). Dordrecht: Kluwer.

Pea, R. D. (1999). New media communication forums for improving education research and practice. In E. C. Lagemann & L. S. Shulman (Eds.), *Issues in education research: Problems and possibilities* (pp. 336–370). San Francisco: Jossey Bass.

Pea, R., Mills, M., Rosen, J., Dauber, K., Effelsberg, W., & Hoffert, E. (2004, January/March). The DIVER™ Project: Interactive digital video repurposing. *IEEE Multimedia, 11*(1), 54–61.

Picard, R. W. (2000). Toward computers that recognize and respond to user emotion. *IBM Systems Journal, 39*(3&4), 705–719.

QuickTime/SMIL. (2005). *Usage of the synchronized multimedia interface language in the QuickTime multimedia standard.* Cupertino, CA: Apple Developer Technical Specification.

Rohlfs, J. (1974). A theory of interdependent demand for a communications service. *The Bell Journal of Economics and Management Science, 5*(1), 16–37.

Salkever, A. (2004, March 24). Digital music: Apple shouldn't sing solo. *Business Week Online*.

SAN. (2005). Retrieved from November 2, 2006 from http://www.webopedia.com/TERM/S/SAN.html

SGI/HDTV. (2004). Real-Time, Full-Bandwidth HDTV I/O with SGI HD XIO. Discreet / Silicon Graphics Technical Specification.

Sharma, D. (2004a, March 17). SONY debuts double-layer DVD drive. *News.com Article.* Retrieved October 30, 2005 from http://news.com.com/2100–1041–3–5174122.html

Sharma, D. (2004b, January 7). Toshiba spotlights high-definition DVD player, *News.com Article.* Retrieved October 30, 2005 from http://news.com.com/2100–1041–5136601.html

Shoah. (2005). *Survivors of the Shoah Visual History Foundation* Retrieved October 30, 2005 from http:// www.vhf.org

Snoek, C. G. M., & Worring, M. (2005). Multimodal video indexing: A review of the state-of-the-art. *Multimedia Tools and Applications, 25*, 5–35.

SONY/HDTV. (2005). *Real-time, HDTV I/O with the SONY HDCAM Software Codec. Discreet / SONY Technical Specification, 2005.*

Sun, X., Foote, J., Kimber, D., & Manjunath, B. S. (2001). Panoramic video capturing and compressed domain virtual camera control. *Proceedings of the 9th ACM International Conference on Multimedia,* pp. 329–338.

Teachscape. (2005). Teachscape web site and service. Retrieved October 25, 2005 from http:// www.teachscape.com

Transana. (2005). *Qualitative analysis software for video and audio data.* Retrieved from http://www.transana.org

Virage. (2005). Retrieved October 25, 2005 from http://www.virage.com/

Walter, C. (2005, August). Kryder's law. *Scientific American, 293*(2), 32–33.

Wang, X. (2004). MPEG-21 rights expression language: Enabling interoperable digital rights management. *IEEE MultiMedia, 11*(4), 84–87.

White, S. A., Gupta, A., Grudin, J., Chesley, H., Kimberly, G., & Sanocki, E. (2000). Evolving use of a system for education at a distance. *Proceedings of the 33rd Hawaii International Conference on System Sciences* (Vol. 3, p. 3047).

28

Toward a Video Collaboratory

Ronald M. Baecker
University of Toronto

David Fono
University of Toronto

Peter Wolf
University of Toronto

Wulf (1989) defined a collaboratory as a "'center without walls,' in which the nation's researchers can perform their research without regard to geographical location—interacting with colleagues, accessing instrumentation, sharing data and computational resources, [and] accessing information in digital libraries." We shall present the ePresence system, an open source interactive real-time webcasting and multimedia archiving solution. We sketch how it could be extended and applied to enable research and collaborative use of video-as-data by a worldwide community of educational researchers, teachers, and learners. Particular emphasis will be placed on how the system supports representation, reflection, interaction and collaboration.

FACILITATING THE USE OF VIDEO IN THE LEARNING SCIENCES

Goldman, Pea, Barron, and Derry (this volume) describe the learning sciences as "a distinctive branch of the multidisciplinary cognitive sciences, with distinctive emphases on the problems of education and learning," and assert that an understanding of human learning requires insights from multiple disciplines including cognition, developmental psychology, educational psychology, linguistics, anthropology, education, and computer science. A central goal of the learning sciences is to produce enhanced descriptions and understandings of education and learning, using various kinds of data and various methods of deriving meaning from the data.

Learning science descriptions are often expressed in text, such as the teacher was "eloquent," the student was "puzzled," or the class was "disruptive." Descriptions may also be represented mathematically, such as, the student made "two errors." Yet, such abstractions are reductionist abstractions of what one or more observers have concluded is the meaning of the data; they do not have the richness of video records of the actual behavior. As this volume demonstrates, there is therefore much interest and activity in the use of video-as-data.

This chapter begins by reviewing past and current research on tools to facilitate the use of video-as-data, with particular interest in the concept of the "collaboratory" and in the support of video data in collaboratories. We then introduce ePresence Interactive Media—an open source interactive real-time webcasting and multimedia archiving solution, present some sample uses of ePresence, and discuss its architecture and implementation. Although not originally designed for use as a video collaboratory, the system is modular and malleable enough to allow modest extensions that enable such use. We present these planned extensions in terms of support for representation, reflection, interaction, and collaboration.

RELATED RESEARCH

Since the publication of Wolf's paper, many collaboratory initiatives have been undertaken, primarily by organizations studying the physical sciences (e.g., Berman, 2000; Caspar et al., 1998; Van Buren, Curtis, Nichols, & Brundage, 1995). Much of the existing research on collaboratories describes lessons learned from these ventures, focusing in particular on the sociological conditions necessary to facilitate successful collaboration (e.g., Olson, Finholt, & Teasley, 2000; Schunn, Crowley, & Okada, 2002).

The National Research Council (1993) identified key technological needs that must be satisfied in order to achieve a fully functional collaboratory:

- Data sharing (including electronic libraries, accessible archives, and a comprehensive retrieval system).
- Software sharing (including interoperability of local software with remote data, and network-accessible storage of results).
- Communication (including voice, video, text, data, and images in both synchronous and asynchronous modes).
- The ability to control remote instruments.

In particular, the system must integrate these functionalities so that the borders between them are transparent to the collaboratory participant, who works with data, tools, and colleagues in unison. These requirements were conceived with the physical sciences in mind, but most are typically applicable to learning science collaboratories, although it would be rare to need to control remote instruments. However, most projects have been developed without any guiding plan, using off-the-shelf technologies not designed to work well together (Finholt, 2002).

The prospect of a video collaboratory faces additional barriers to an effective implementation. Video can be a primary data source for observational inquiry, permitting qualitative analyses of behaviors and processes (Smith & Blankinship, 1999). However, video-as-data is a novel application requiring novel tools (Nardi, Kuchinsky, Whittaker, Leichner, & Schwarz, 1996).

Video can also be used to share real-time depictions of shared work objects, and thus bring complex objects at one physical location into a virtual shared workspace to coordinate distributed teams (Whittaker, 1995). For example, a study by Gaver, Sellen, Heath, and Luff (1993) emphasizes that in many cases, real-time views of an object under study in a shared workspace is preferable to conversational views among fellow collaborators. Use of video in this manner can help to establish common ground among collaborators, a shared physical context that adds meaning to indexical utterances (Clark & Brennan, 1991). Thus, the development of WYSIWIS (What You See Is What I See) interfaces is one of strong imperatives in collaboratory design (Finholt, 2002).

Pea (2006) summarizes the challenges facing a video collaboratory as follows: Video data and analyses must be universally accessible, collaborators must have access to video analysis tools that support discipline-specific analytic practices and that are interoperable, and analyses and commentary must be available for public participation and collaboration.

Several ongoing and past projects have aimed to make video data universally accessible by developing large, public corpora of digital multimedia recordings. The Open Video Project (www.open-video.org <http://www.open-video.org>; Marchionini & Geisler, 2002) is in the process of accumulating a shared digital video repository to serve as a test bed for research into information retrieval in multimedia libraries. The Informedia Digital Video Library (Hauptmann, 2005) and the CAETI Internet Multimedia Library (Wolf et al., 1997) are similar test bed repositories, together having archived over 1,000 hours of educational video for K–12 students as part of ongoing research into constructing digital video libraries; the Informedia collection has since been added to the Open Video Project. The Talkbank site (www.talkbank.org) maintains a large body of video and audio data, as well as transcription, coding, and annotation tools, designed to help researchers studying human and animal communication.

Many projects also aim to provide comprehensive retrieval functionality for existing libraries. For example, the Informedia Project uses combined speech, language, and image understanding to provide intelligent search and selective retrieval within its database of videos (Christel, Papernick, Huang, & Moraveji, 2004; Wactlar, Christel, Gong, & Hauptmann, 1999). Users can search for "video paragraphs" based on text extracted from the soundtrack or captioning, and rapidly browse the results using a "video skimming" technique. Other projects explore the automatic capture and archival of live content (Brotherton & Abowd, 2004; Chiu, Kapuskar, Reitmeier, & Wilcox, 2000; Hurst, Maass, Muller, & Ottmann, 2001; Kientz, Boring, Abowd, & Hayes, 2005; Moran et al., 1997; Mukhopadhyay & Smith, 1999).

Tools for video analysis and interoperability of these tools with the corpus of networked data form complementary challenges. Pea and Hay (2003) conducted a workshop with video researchers in the learning sciences to identify the functions that analytical tools should support; acquisition, chunking, transcription, way finding, organization/asset management, commentary, coding/annotation, reflection, sharing/publication, and presentation.

VideoPapers (Beardsley, Cogan-Drew, & Olivero, this volume) are online multimedia publications designed for educational researchers and practitioners. Using the VideoPaper Builder, users can synchronize videos with textual analysis and images, and publish the presentation to the web.

DIVER (Digital Interactive Video Exploration and Reflection) allows researchers in the behavioral sciences to easily edit and annotate collected footage, thus creating lightweight "dives" that illustrate some specific point or piece of evidence (Pea et al., 2004). Dives can subsequently be exported to a Web site, so that other researchers have a chance to observe and comment on them.

VideoTraces (Stevens, this volume) offer a unique form of "show and tell" multimedia representations for conveying embodied knowledge. Users augment video with an interpretive layer of audio and visual gestures to create traces, which can then be exchanged with others or used for future reflection.

The importance of point of view in interpreting and constructing thick descriptions of video data has been particularly emphasized in the video ethnographic work of Ricki Goldman (Goldman-Segall, 1998). Goldman has also for the past two decades been designing a variety of interesting systems for video annotation and analysis (Goldman, this volume; see also Harrison & Baecker, 1992).

Numerous technologies have also been developed to facilitate synchronous and asynchronous collaboration at a distance. Fishman's KNOW (Knowledge Networks on the Web) system (Fishman, this volume) envisions a collaborative teacher professional development environment that combines hyperlinked curriculum documents, student work with teacher feedback, tools for personal Web logging and discussion, and several types of instructional video.

Several systems developed by Microsoft Research allow real-time interaction among individuals who are collaboratively watching a video webcast (Cadiz, et al., 2000; Jancke, Grudin, & Gupta, 2000; Rui, Gupta, & Grudin, 2003; White, et al., 2000). Finally, Grudin and Bargeron (2005) present technology to allow asynchronous collaboration around video archives, using a shared annotation system integrated with e-mail.

The system we are about to describe was based around the dual goals of enabling the real-time transmission through the Internet of synchronized multimedia including video, and of enabling the efficient archiving of such presentations to allow flexible browsing, navigation, and searching of the material. We shall introduce the system, then sketch how it could be extended to enable the use of video-as-data in a video collaboratory for the learning sciences.

ePRESENCE INTERACTIVE MEDIA

ePresence Interactive Media (Baecker, 2003; Baecker et al., 2006; Baecker, Moore, & Zijdemans, 2003; Baecker, Wolf, & Rankin, 2004; Rankin, Baecker, & Wolf,

2004) is a Web-based streaming (webcasting) and collaboration tool for the large-scale broadcast of events over the Internet—from university lectures to demonstrations by master teachers to public health briefings to annual meetings to rock concerts. Events are streamed live and can later be easily deployed as browsable, searchable archives accessed through a customizable Web portal. Web casting itself is noninteractive, which is overcome by combining it with interactive features. For example, ePresence currently employs text chat as a mechanism for allowing interaction among remote participants, and between these individuals and the speaker via a moderator, and has also recently added VoIP support for voice questions and discussion.

For both live and archived events, ePresence provides a rich and engaging multi-media experience for viewers connecting over the Internet using desktop and mobile clients. During a live event, end-users have access to an audio–video feed, navigable slide images, and a text chat system. Live events can be quickly and easily archived, and made available to users via the portal. Archives are full-text searchable, and provide an interactive time line and two-level table of contents for easy browsing and navigation. The concept of hierarchically structured video is based in part on work described in Baecker, et al. (1996) and in Baecker and Smith (2003).

More specifically, ePresence currently includes support for:

- Video, audio, slide, and live desktop demos.
- Slide review.
- Moderated chat, private messages, and the submission of questions via text and voice.
- The automated creation of structured, navigable, searchable event archives.

ePresence also allows configurable live and archive interfaces through tailorable "skins," which allow site-specific control over the layout and typography of both interfaces, and the inclusion of corporate logos for purposes of "branding." The server-side software runs under Linux and Windows; media capturing and streaming engines run under Windows; client viewers exist for the IBM PC, the Macintosh, and Linux platforms. Media may be transmitted using Windows Media, Real Media, and MPEG-4. Web casts may be received with bandwidths as low as a 56Kbits/second. The software is implemented with .NET and Mono technologies, is highly modular, and has been released open source and community source (Baecker, 2005, see also http://epresence.tv).

User Interface

One interface to access live webcasts is illustrated by the screen snapshot in Figure 28.1. The video window and its controls are in the upper left; the slide window and its controls are in the upper right; the chat system is at the bottom. Slide controls allow a remote viewer to review any slide already presented by the speaker. The chat system supports public chat, private messages, and questions to the speaker. Web links can also be sent by the speaker and synchronized with the video. A "live demo" feature enables transmission of live 600×800 screen captured streams of live demos from the

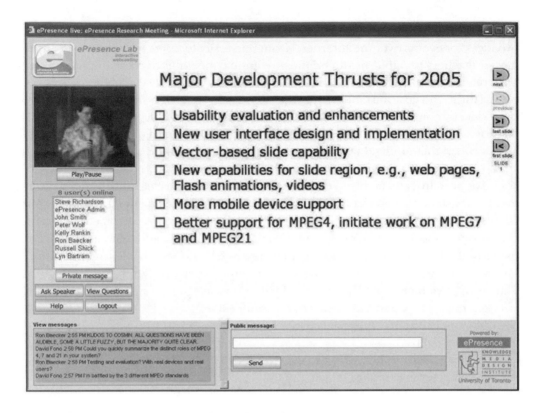

Figure 28.1.   A screen shot from a live webcast.

presenter's computer. There is an integrated registration and systems check procedure so that potential viewers can ensure technology compatibility in advance.

The archives interface allows retrospective navigation and browsing through a webcast using a two-level outline of the logical structure of the talk and its slides and live demo sessions (Fig. 28.2, right side). Slide titles are picked up automatically from PowerPoint in case it is used; the outline is input by the moderator during the talk and if need be, updated afterward using the ePresence Producer (see later). Archive viewers can also navigate by a timeline (Fig. 28.2, bottom). We also allow searching based on key words in the slides when PowerPoint is used. (Both dependencies on Power-Point are to be removed in the February 2006 release of ePresence Version 3.1.) Chapter titles appear darker in the table of contents, and as the upper tick marks on the time line. Slides appear lighter in the table of contents, and as the lower tick marks on the timeline.

Example Uses of ePresence

An interesting case study of ePresence (Zijdemans, Moore, Baecker, & Keating, 2006) has been its use by the Millennium Dialogue on Early Child Development (MDECD) project, part of the Atkinson Centre for Society and Child Development's

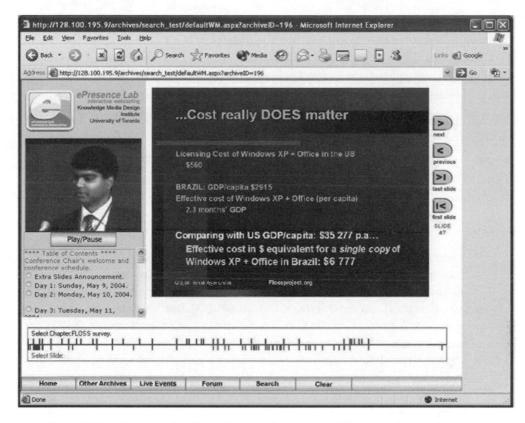

Figure 28.2. A screen shot from the interface to an archive constituting the multimedia proceedings of a conference on open source and free software.

steps toward establishing a learning community for child development based on a theoretical model for developing a learning society network (Keating & Hertzman, 1999; Matthews & Zijdemans, 2001; http://www.webforum2001.net/; http://www.acscd.ca).

MDECD brought together eight experts from different areas in child development for a two-day conference in November, 2001. The meeting was attended by roughly 200 local participants and was webcast using ePresence to 20 remote North American groups. Over 600 public and private chat messages among the remote groups were exchanged. Table 28.1 shows how the composition of the chat messages changed over the 2 days. Note the increase in the percentage of messages related to the content of the sessions, from an average of 8% on Day 1 to 15% on Day 2, and in the percentage of social messages, from 15% on Day 1 to 29% on Day 2.[1]

Since the conference, the ePresence multimedia archive of the scientist presentations has served as a knowledge base and nurtured the learning community through ongoing activities:

_____

[1]White, et al. (2000) similarly report that text exchanges went from 27%:62%:11% content:technology:social messages to 60%:14%:26% over the last three sessions of an eLearning course.

TABLE 28.1
Categorizing Chat Messages Over the 4 Half-Days of the WebForum 2001 Webcast

	A.M. Day1	*P.M. Day1*	*A.M. Day2*	*P.M. Day2*
Content-related	11	5	13	16
Technology-related	116	112	44	41
Administrative	38	21	13	10
Social	30	1	28	30
Other	18	19	13	14

- Ongoing curriculum development and incorporation of the knowledge base into courses for graduate students and professionals.
- Creation of *Conversations on Society & Child Development* (see http://www.cscd.ca), an interactive CD/Web ePublication that provides an environment for accessing the knowledge and supporting exchange among researchers and those who want to apply the findings.
- Plans to translate the knowledge for use by parents, educators, and policy makers.

A Second Example

On July 16, 2003, North Network, with the assistance of Videotelephony Inc., used their videoconferencing network and the ePresence system to webcast a talk entitled "West Nile Virus: First Canadian Experiences." Simultaneous talks were given by two regional infectious disease experts, and the North videoconference network was used to connect remote communities with these experts. The videoconferencing feed was bridged into ePresence, and used as the basis for a simultaneous webcast. The talks were recorded, digitized, and mounted on NORTH Network's server for archive access (www.northnetwork.com). ePresence technology was also used to create a knowledge product (on Web and CD) "Just-in-Time' Clinical Education During SARS and West Nile," which was awarded "best innovation in use of technologies in health education" at the Canadian Society for Telehealth Conference.

A Third Example

May 9–11 2004 saw the Knowledge Media Design Institute hosting a major international conference entitled "Open Source and Free Software: Concepts, Controversies, and Solutions."

There were roughly 20 hours of lectures, panel discussions, and question and answer dialogues between the 250 local audience members, 25 remote audience members, and 30 speakers. Graduate student editors reviewed the proceedings in detail and added hundreds of chapter titles that, together with the slide titles, provide a rich table of contents into the multimedia proceedings (see Fig. 28.2 and also http://www.epresence.tv/website_archived.aspx?dir=7).

SYSTEM ARCHITECTURE AND IMPLEMENTATION

The system is implemented using .NET and Mono technologies. The server software runs under Windows or Linux. webcasts can be viewed on client personal computers running the Linux, Windows 98/2000/2003/XP, and Mac 9.x or OS/X operating systems, and the Internet Explorer, Netscape Navigator, Mozilla Firefox, Opera and Safari 1.2+ browsers, and using either Real Media or Windows Media live streaming. Archives may be produced in Real Media, Windows Media, and MPEG-4 formats.

The architecture of our highly modular system may be portrayed as in Figures 28.3 and 28.4. For further details, see Baecker, Wolf, and Rankin. (2004) and Rankin et al. (2004).

Interactive Live Web Casting

An ePresence live Web cast is created by a speaker, an operator, and a moderator (Fig 28.3). These can be different individuals or the same person depending upon the scale of the event. The ePresence Mobile Station (4) includes several live media encoding and capturing software applications (e.g., Windows or Real Media) controlled by the operator (3) or speaker (7) via a single unified remote control interface. The remote control interface has been developed for different internet-connected devices (Laptops, tablet PCs, and PDAs). The operator can perform the following operations either locally or remotely; initiate live broadcast, start or stop archiving session, control slides transmission, submit URLs, and initiate multiple live software demo sessions. The speaker may give a talk to a local audience or remotely via a telephone, VoIP, or

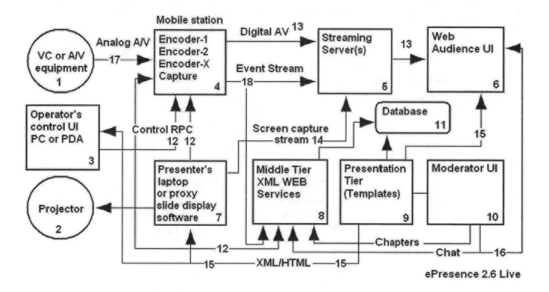

Figure 28.3. Current ePresence system architecture for live webcast.

videoconferencing (1). This allows us to webcast a meeting that is being held via video-conference. Web-based slide controlling and projecting (7) software allows having multiple distributed audiences listening and following the slide presentation in real time. The moderator interface (10) supports a local moderator who is watching the webcast, sending public announcements to a web audience, and submitting notes (chapter titles) to the archiving application. The moderator works as a communication "bridge" between the speaker and web audience transferring questions and comments on behalf of remote participants.

An ePresence webcast is typically viewed by both a local audience (2) and a live web audience (6). The web audience receives video and audio (13) of the speaker(s) from the streaming servers (5), a synchronized slide presentation stream (18) or a screen capture stream (14) from the presenter's computer, and web URLs (15). Remote viewers can also submit questions to the speaker (directly via text or voice or indirectly via the moderator), have public or private text-based dialogs (16), and review the slides that have been already presented. The live interface (9) has been developed as a set of templates ("UI skins") that support different layouts, media formats, video resolutions, and other features. The operator can choose the most suitable template depending on the content of the talk. Adopters of the ePresence system can easily develop their own skins using XML, HTML, and a choice of several scripting technologies.

Archiving and Publishing a Webcast

The webcast data (4) such as video (1), slides (3), and event streams (2) is automatically captured during the live webcast (Fig 28.4). The events stream data includes

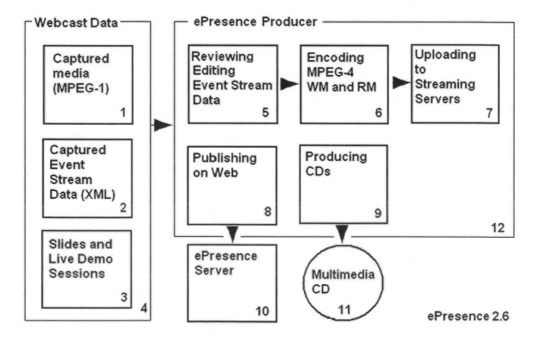

Figure 28.4.   Producing an archive of an ePresence webcast.

time stamp information of slides and chapters submitted during the live Web cast. Event streams can be updated (5) after the Web cast using the ePresence Producer application (12). The operator can add additional keywords to enhance search, update slide synchronization data, edit chapter and slide titles, and replay the event with all synchronized materials before publishing the archive. The ePresence Producer software also allows encoding the captured video in different popular streaming formats (6), automatic uploading to a streaming server (7), automatic creation and publishing of web archives (8), and production of multimedia CDs (9). The software provides a selection of archive templates. The published archive becomes automatically available on the ePresence website (10). It includes video player, slide frame, interactive time line component, search tool, interactive table of contents, and threaded discussion board. Every archive exposes its keywords through the XML web services. This makes it easy to integrate the archives into different document repositories, "learning object" banks, and other searchable data storage systems.

ePRESENCE AS A COLLABORATORY ENABLING THE USE OF VIDEO IN THE LEARNING SCIENCES

ePresence was originally developed to enable the worldwide broadcast of presentations, interactive access to these broadcasts in real time, and flexible retrospective access to structured archives of the presentations. Yet, there is nothing in the technology that restricts the video channel to "talking heads" presenting lectures, or to a presentation in which a small video image is portrayed as an adjunct to a large slide image.

In Figure 28.5,[2] we see a screen shot from an experimental video collaboratory version of ePresence developed by Russ Shick (2005) as part of his MSc work. The application is collaborative video viewing (Cadiz et al., 2000) over the Internet of a structured archived video of a classic Canada—U.S. hockey match. Video structure consists of the periods of the game and within these periods, interesting events such as goals, penalties, and "near goals." Multiple viewers are able to converse in two modalities (Schick, Baecker, & Scheffel-Dunand, 2005) using both ePresence text chat and an experimental spatial audio voice-over Internet system known as Vocal Village (Kilgore, Chignell, & Smith, 2003). Control over video playback is distributed among all viewers of the video.

More generally, this system could contain any video stream of relevance to educators and to the learning sciences, such as a teacher demonstrating a difficult concept in the analysis of an English text, a group of students discussing an ethical quandary, an animation of a law of physics, or the movement of organisms seen under a microscope.

This implementation is encouraging, but we need to go further. If ePresence is to serve as a collaboratory for the learning sciences, then its capabilities for representation, reflection, interaction, and collaboration need to be enhanced. We shall now address each of these in turn.

---

[2]This image is included for illustration purposes only, and implies nothing about current or planned product or service offerings in the Bell Sympatico portal.

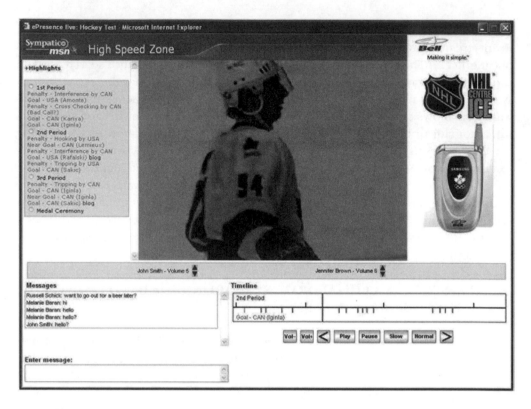

Figure 28.5.   A consumer prototype of an ePresence-based video collaboratory.

Representation

ePresence represents video as a structured document with a two-level hierarchy. The upper level encodes logical sections of a video, which we typically term "chapters." The lower level is usually used for representing associated slides or live demonstrations. But these are totally general indices into a video, so they can be used in whatever way is desired by the producer of the multimedia event. For example, in the multimedia proceedings of the open source conference discussed earlier, the upper level was used to provide detailed topic references to talks or panels at a rate of once every minute or two, and, in the hockey game collaboratory, the lower level was used to encode interesting events.

For simplicity, we resisted the temptation to provide arbitrarily linked hypervideo. Yet, links from anywhere in one video to anywhere in any other ePresence multimedia archive can be introduced. We can currently reference slides, screen capture videos, and HTML pages. Other media formats (like Flash movies) will be added soon. We can reference the entire repository as a single object, which is used in our "repository search" feature. We can also reference arbitrary entry points or arbitrary sections of any video in order to enable chunks of material to function as "learning objects."[3] We plan to extend this feature to allow referencing synchronized presentation media from other applications, such as email and online learning environments.

The system does currently impose the limitation that there be only one video active at a single time, although lifting this restriction through use of the open source code would be possible. This would be required to realize the notion of multiple video points of view of an event as proposed by Goldman (this volume).

Reflection

ePresence currently provides the ability for viewers to chat over a live event both publicly and privately and to send questions to the speaker. Reflection and note taking could be realized by sending private messages to oneself, which could function as bookmarks or more generally as notes to oneself. The system stores all messages, although they have not as yet been made public in the archive. Doing this would be trivial, although we would need some way of distinguishing between those messages that are intended to be persistent, and those that are not, and some lightweight method to control access to certain messages. We need to allow notes to include hyperlinks as well as raw text.

More interesting than textual notes is the ability to annotate or scribble on the videos. This can easily be added to ePresence, leveraging insights from work such as Brotherton and Abowd (2004).

Interaction

We are currently investigating ways to extend interaction among users beyond basic live chat at the time of the event.[4] ePresence archives already have a rudimentary threaded discussion capability, but the interface lacks elegance and the feature has rarely been used. The discussion boards are fully separate and inaccessible from the live chat, which may explain the scarcity of usage. We intend to replace this functionality with a more robust discussion interface that tightly integrates synchronous and asynchronous communication, both during and after the event.

The extended interface that we are currently building consists of a single discussion view that supports chat as well as threading and basic annotation. The threading model is based in part on the threaded chat system (Smith, Cadiz, & Burkhalter, 2000). Users may contribute to discussions either by sending a chat message, or by posting a message in response to a previous message. When a response is posted, it also appears as a new chat message with a link to the corresponding thread, so that discussants only need to pay attention to a single stream of new content. The combination of "chatlike" and "message boardlike" functionality within a single interface means that participants can adapt the style of their conversations to the circumstances at hand, including the

---

[3]"A learning object ... [is] any digital resource that can be reused to support learning." (Wiley, 2002, p. 7).

[4]This work began with the now completed M.Sc. thesis of David Fono (Fono & Baecker, 2006).

number of discussants and the complexity of the topics involved, as well the circumstances of any concurrent webcast.

Messages may also be annotated using user-defined tags, as well as priority indicators. Tagging has been found to be a useful technique for collaboratively organizing a variety of media, and we expect that it will have a similar effect in the chat context. Using the interface's customizable visualization of chat history, users can navigate content marked with specific priorities and specific tags. Thus, participants can engage in discussions over a variety of topics within a single archive, even as that archive grows indefinitely.

The interface's appearance and behavior will be the same for both live and archived events. This will serve to narrow the distinction between the two modes of operation, and thus encourage sustained interaction. Discussion among users should start with the event, move into the archives, and continue well beyond. We are also exploring means for participants to export discussions to third-party servers and interfaces, in order to further facilitate varied forms of collaboration around video archives. We expect that the final version of our chat interface will serve not only as an interface for communication among participants, but also as a portal for tracking the various discussions about the video that develop elsewhere.

Collaboration

Collaboration in ePresence is enabled by communication using the various chat and discussion capabilities. But one can only interact with individuals if you are aware of their existence as viewers of an event, or as possible viewers of an archive. ePresence currently provides a rudimentary display of all individuals watching a live event (Fig. 28.1), but no display of all individuals who have watched an archive or who are currently watching it.

In order to enable any kind of collaboration, we need to have awareness of who is potentially available to chat, to brainstorm, or to do focused work of some kind (Kraut, Egido, & Galegher, 1998). There is much current research on mechanisms for group awareness (mechanisms to enrich group awareness (see, for example, Elliot & Greenberg, 2004; Gutwin & Greenberg, 2004; Rounding & Greenberg, 2000). Work is underway on how to integrate such capabilities into ePresence (Baecker et al., 2006).

SUMMARY AND CONCLUSIONS

We have presented a scalable, modular, extensible architecture for interactive multimedia webcasting and for providing access to structured archives of these webcasts. Various features to enable communication and collaboration over live and archived events have also been proposed and discussed. These features can be implemented and need to be implemented if the environment is to function well as a video collaboratory for educational researchers.

To facilitate this happening, as well as to allow ePresence to be molded by its adopters in many different directions, we have decided to release our software open source (Baecker, 2005; Benkler, 2002; DiBona, Ockman, & Stone, 1999; Weber, 2004;). Applications of ePresence to date, mostly in the eLearning space, have all been carried out with differing goals, modes of usage, and measures of success. An open source implementation that puts maximum control in the hands of adopters should also enable a rich set of new ePresence applications in the learning sciences.

ACKNOWLEDGMENTS

Guidance of early ePresence definition and scoping was shared with Dr. Gale Moore, Director of KMDI. Kelly Rankin has very capably anchored the business, marketing, and production side of ePresence, working side by side with system architect and developer Peter Wolf. Maciek Kozlowski of Videotelephony Inc. has provided many technical insights that have informed these developments. We are thankful for research support to the Natural Sciences and Engineering Research Council of Canada (NSERC) especially through the NECTAR (Network for Effective Collaboration Technologies through Advanced Research) NSERC Research Network and to the Bell University Laboratories.

REFERENCES

Baecker, R .M. (2003). A principled design for scalable Internet visual communications with rich media, interactivity, and structured archives. *Proceedings of the 2003 Conference of the Center for Advanced Studies on Collaborative Research.* IBM Press, 83–96.

Baecker, R. M. (2005, June 27–July 2). Open source strategies for educational multimedia. *Proceedings of World Conference on Educational Multimedia, Hypermedia and Telecommunication 2005*(1). AACE, 133–137.

Baecker, R. M., Baran, M., Birnholtz, J., Chan, C., Laszlo, J., Rankin, K., Schick, R., & Wolf, P. (2006). Enhancing interactivity in webcasts using VoIP. *CHI '06 Extended Abstracts on Human Factors in Computing Systems.* ACM Press, 235–238.

Baecker, R. M., Moore, G., & Zijdemans, A. (2003). Reinventing the lecture: Webcasting made interactive. *Proceedings of HCI International 2003*, Vol. 1, Lawrence Erlbaum Associates, New Jersey, 896–900.

Baecker, R. M., Rosenthal, A., Friedlander, N., Smith, E., & Cohen, A. (1996). A multimedia system for authoring motion pictures. *Proceedings of the Fourth ACM International Conference on Multimedia,* ACM Press, 31–42.

Baecker, R. M., & Smith, E. (2003). Modularity and hierarchical structure in the digital video lifecycle. *Proceedings of Graphics Interface 2003*, A. K. Peters Ltd., 217–224.

Baecker, R. M., Wolf, P., & Rankin, K. (2004). The ePresence interactive Webcasting system: Technology overview and current research issues. *Proceedings of World Conference on E-Learning in Corporate, Government, Healthcare, and Higher Education, 2004*(1), AACE, 2532–2537.

Benkler, Y. (2002). Coase's Penguin, or, Linux and the nature of the firm. *112 Yale Law Journal, 369*, 1–79.

Berman, H. M. (2000). Research collaboratory for structural bioinformatics. *Protein Data Bank Annual Report, 1999–2000.* New Brunswick, NJ: Rutgers University.

Brotherton, J. A., & Abowd, G. D. (2004). Lessons learned from eClass: assessing automated capture and access in the classroom. *ACM Transactions on Computer-Human Interaction, 11*(2), 121–155.

Cadiz, J. J., Balachandran, A., Sanocki, E., Gupta, A., Grudin, J., & Jancke, G. (2000). Distance learning through distributed collaborative video viewing. *Proceedings of the 2000 ACM Conference on Computer Supported Cooperative Work*, ACM Press, 135–144.

Caspar, T. A., Meyer, W. M., Moiler, J. M., Henline, P., Keith, K., McHarg, B., Davis, S., & Greenwood, D. (1998). Collaboratory operations in magnetic fusion research. *ACM Interactions, 5*(3), 56–64.

Causey, R., Brinholtz, J. B., & Baecker, R. M. (2006). Increasing awareness of remote audiences in webcasts. *Proceedings of the 2006 20th Anniversary Conference on Computer Supported Cooperative Work*, Conference Supplement, November 2006, 59–60.

Chiu, P., Kapuskar, A., Reitmeier, S., & Wilcox, L. (2000). Room with a rear view: Meeting capture in a multimedia conference room. *IEEE Multimedia, 7*(4), IEEE Computer Society Press, 48–54.

Christel, M., Papernick, N., Huang, C., & Moraveji, N. (2004). Exploiting multiple modalities for interactive video retrieval. *Proceedings of the IEEE International Conference on Acoustics, Speech and Signal Processing*, Volume 3, 1032–1035.

Clark, H. H., & Brennan, S. E. (1991). Grounding in communication. In L. B. Resnick, J. M. Levine, & S. D. Teasley (Eds.), *Perspectives on Socially-Shared Cognition* (pp. 127–149). American Psychological Association, Washington, DC.

DiBona, C., Ockman, S., & Stone, M. (Eds.). (1999). *Open sources: Voices from the open source revolution*. O'Reilly & Associates, http://www.oreilly.com/catalog/opensources/book/toc.html (last accessed: Feb. 5, 2007).

Elliot, K., & Greenberg, S. (2004, September). Building flexible displays for awareness and interaction. *Video Proceedings and Proceedings Supplement of the UBICOMP 2004 Conference* (Sept. 7–10, Nottingham, England) [6 min video, two-page paper].

Finholt, T. A. (2002). Collaboratories. In B. Cronin (Ed.), *Annual Review of Information Science and Technology*, Vol. 36. American Society for Information Science and Technology, Medford, NJ, 73–107.

Fono, D., & Baecker, R. M. (2006). Structuring and supporting persistent chat conversations. *Proceedings of the 2006 20th Anniversary Conference on Computer Supported Cooperative Work*, ACM Press, 455–458.

Gaver, W., Sellen, A., Heath, C., & Luff, P. (1993). One is not enough: multiple views in a media space. *Proceedings of the SIGCHI Conference on Human Factors in Computing Systems*, ACM Press, 335–341.

Goldman-Segall, R. (1998). *Points of viewing children's thinking: A digital ethnographer's journey.* Mahwah, NJ: Lawrence Erlbaum Associates.

Grudin, J., & Bargeron, D. (2005). Multimedia annotation: An unsuccessful tool becomes a successful framework. In K. Okada, T. Hoshi, & T. Inoue (Eds.), *Communication and collaboration support systems*. Tokyo: Ohmsha.

Gutwin, C., & Greenberg, S. (2004). The importance of awareness for team cognition in distributed collaboration. In E. Salas & S. M. Fiore (Eds.), *Team cognition: Understanding the factors that drive process and performance* (pp. 177–201). Washington, DC: American Psychological Association.

Harrison, B. L., & Baecker, R. M. (1992, May). Designing video annotation and analysis systems. *Proceedings of the conference on Graphics Interface '92*, 157–166. Vancouver, British Columbia: Morgan Kaufmann Publishers.

Hauptmann, A. (2005). Lessons for the future from a decade of Informedia video analysis research. *Lecture Notes in Computer Science, 3568*, 1–10.

Hurst, W., Maass, G., Muller, R., & Ottmann, T. (2001). The "authoring on the fly" system for automatic presentation recording. *CHI '01 Extended Abstracts on Human Factors in Computing Systems*, ACM Press, 5–6.

Jancke, G., Grudin, J., & Gupta, A. (2000). Presenting to local and remote audiences: Design and use of the TELEP system. *Proceedings of the SIGCHI Conference on Human Factors in Computing Systems*, ACM Press, 384–391.

Keating, D. P., & Hertzman, C. (Eds.). (1999). *Developmental health and the wealth of nations: Social, biological, and educational dynamics*. New York: Guilford.

Kientz, J. A., Boring, S., Abowd, G. D., & Hayes, G. R. (2005). Abaris: Evaluating automated capture applied to structured autism interventions. *Proceedings of the Seventh International Conference of Ubiquitous Computing (UbiComp)*, Springer, 323–329.

Kilgore, R., Chignell, M., & Smith, P. (2003). Spatialized audioconferencing: What are the benefits? *Proceedings of the 2003 Conference of the Centre for Advanced Studies on Collaborative Research*, IBM Press, 135–144.

Kraut, R., Egido, C., & Galegher, J. (1988). Patterns of contact and communication in scientific collaboration. *Proeeding, s of the 1988 ACM Conference on Somputer-Supported Cooperative Work*, ACM Press, 1–12.

Marchionini, G., & Geisler, G. (2002). The open video digital library. *D-Lib Magazine, 8*(12).

Matthews, D., & Zijdemans, A. S. (2001) Toward a learning society network: How being one's brother's keeper is in everyone's self interest. *Orbit V31, N4*, 50–54.

Moran, T. P., Palen, L., Harrison, S., Chiu, P., Kimber, D., Minneman, S., van Melle, W., & Zelweger, P. (1997). I'll get that off the audio: A case study of salvaging multimedia meeting records. *Proceedings of the SIGCHI Conference on Human Factors in Computing Systems,* ACM Press, 202–209.

Mukhopadhyay, S., & Smith, B. (1999). Passive capture and structuring of lectures. *Proceedings of the 7th ACM International Conference on Multimedia (Part 1),* ACM Press, 477–487.

Nardi, B. A., Kuchinsky, A., Whittaker, S., Leichner, R., & Schwarz, H. (1996). Video-as-data: Technical and social aspects of a collaborative multimedia application. *Computer Supported Cooperative Work, 4*, 73–100.

National Research Council Committee on a National Collaboratory. (1993). *National collaboratories: Applying information technology for scientific research*. Washington, DC: National Academy Press.

Olson, G. M., Finholt, T. A., & Teasley, S. D. (2000). Behavioral aspects of collaboratories. In S. H. Koslow & M. F. Huerta (Eds.), *Electronic collaboration in science*. Mahwah, NJ: Lawrence Erlbaum Associates.

Pea, R. D. (2006). Video-as-data and digital video manipulation techniques for transforming learning sciences research, education and other cultural practices. In J. Weiss, J. Nolan, & P. Trifonas (Eds.), *International handbook of virtual learning environments*. Dordrecht, The Netherlands: Kluwer Academic Publishing, 1321–1393.

Pea, R. D., & Hay, K. (2003). *Report to the National Science Foundation: CILT Workshop on Digital Video Inquiry in Learning and Education, November 25–26, 2002*. Stanford Center for Innovations in Learning, Stanford University, CA.

Pea, R. D., Mills, M., Rosen, J., Dauber, K., Effelsberg, W., & Hoffert, E. (2004). The Diver project: Interactive digital video repurposing. *IEEE Multimedia, 11*(1), IEEE Computer Society Press, 54–61.

Rankin, K., Baecker, R. M., & Wolf, P. (2004). ePresence: An open source interactive Webcasting and archiving system for eLearning, *Proceedings of the World Conference on E-Learning in Corporate, Government, Healthcare, and Higher Education 2004,* AACE, 2888–2893.

Rounding, M., & Greenberg, S. (2000, December). Using the notification collage for casual interaction. *ACM CSCW 2000: Workshop on Shared Environments to Support Face-to-Face Collaboration.* Philadelphia, PA. (http://www.edgelab.ca/CSCW/Workshop2000/workshop_papers.html, last accessed Feb. 5, 2007).

Rui, Y., Gupta, A., & Grudin, J. (2003). Videography for telepresentations. *Proceedings of the SIGCHI Conference on Human Factors in Computing Systems,* ACM Press, 457–464.

Schick, R., Baecker, R. M., & Scheffel-Dunand, D. (2005, June 27–July 2). Bimodal text and speech conversation during on-line lectures. *Proceedings of ED-MEDIA 2005*, 822–829, Montreal, Canada.

Schunn, C., Crowley, K., & Okada, T. (2002). What makes collaborations across a distance succeed? The case of the cognitive science community. In P. Hinds & S. B. Kiesler (Eds.), *Distributed work* (pp. 407–432). MIT Press.

Shick, R. (2005). A Study of Student Conversation in Text and Audio During Webcast Lectures, M.Sc. thesis, Dept. of Computer Science, University of Toronto, July 2005.

Smith, B. K., & Blankinship, E. (1999). Imagery as data: Structures for visual model building. *Proceeding of Computer Support for Collaborative Learning, 99*, 549–557.

Smith, M., Cadiz, J. J., & Burkhalter, B. (2000). Conversation trees and threaded chat. *Proceeding ACM CSCW 2000*, 97–105.

van Buren, D., Curtis, P., Nichols, D. A., & Brundage, M. (1995). The AstroVR collaboratory: An on-line multi-user environment for research in astrophysics. In R. A. Shaw, H. E. Payne, & J. J. E. Hayes (Eds.), *Astronomical Data Analysis Software and Systems*

IV (ASP Conference Series, Vol. 77). Astronomical Society of the Pacific, San Francisco, CA, 99.

Wactlar, H. D., Christel, M. G., Gong, Y., & Hauptmann, A. G. (1999, February). Lessons learned from building a terabyte digital video library. *IEEE Computer, 32*(2), IEEE Computer Society Press, 66–73.

Weber, S. (2004). *The success of open source.* Boston, MA: Harvard University Press.

White, S. A., Gupta, A., Grudin, J., Chesley, H., Kimberly, G., & Sanocki, E. (2000). Evolving use of a system for education at a distance. *Proceedings of the 33rd Hawaii International Conference on System Sciences, Vol. 3*, IEEE Computer Society Press, 3047.

Whittaker, S. (1995). Rethinking video as a technology for interpersonal communications: Theory and design implications. *International Journal of Human-Computer Studies, 42*(5), 501–529.

Wiley, D. A. (2002). Connecting learning objects to instructional design theory: A definition, a metaphor, and a taxonomy. *The Instructional Use of Learning Objects,* Section 1.1. Retrieved from http://www.reusability.org/read/ (last accessed Feb. 5, 2007).

Wolf, W., Liang, Y., Kozuch, M., Yu, H., Phillips, M., Weekes, M., & Debruyne, A. (1997). A digital video library on the World Wide Web. *Proceedings of the Fourth ACM International Conference on Multimedia*, ACM Press, 433–444.

Wulf, W.A. (1989). The national collaboratory: A white paper. In J. Lederberg & K. Uncaphar (Eds.), *Towards a national collaboratory: Report of an invitational workshop at the Rockefeller University* (Appendix A). Washington, DC: National Science Foundation.

Zijdemans, A., Moore, G., Baecker, R. M., & Keating, D. P. (2006). ePresence interactive media and WebForum 2001: An accidental case study on the use of Webcasting as a VLE for early child development. In J. Weiss, J. Nolan, & P. Trifonas (Eds.), *International Handbook of Virtual Learning Environments* (Vol. 14, pp. 1395–1428). New York: Springer.

VideoPaper: Bridging Research and Practice for Preservice and Experienced Teachers

Linda V. Beardsley
Tufts University

Dan Cogan-Drew
Tufts University

Federica Olivero
University of Bristol

Professional preparation and continuing professional development for classroom teachers are fields that are challenged by their reliance on two distinct traditions: the tradition of academic research, which seeks to provide scholarly and theoretical foundations for effective pedagogy; and the tradition of clinical practice, which supports practitioners to value their classroom experiences and use those experiences as a text to study and analyze in order to better understand their craft. Each of these two distinct traditions relies on a distinct discourse for articulating the schemas, scholarship, and terminology for communicating its values and perspectives. Many of the professionals whom these discourse traditions seek to engage find each of the traditions somewhat lacking. For some, academic discourse lacks the vitality and engagement of the classroom. For others, classroom-based research provides little opportunity to explore broad themes that inspire intellectual growth. In an effort to bridge the gulf between these two discourses, educators have been exploring the potential of new technologies and software. This chapter will discuss one of these innovative products, VideoPaper technology. A VideoPaper is a multimedia document representing analysis of video data. VideoPaper offers opportunities for integrating educational theory and academic research with the excitement of classroom practice and, thereby, transforming teacher education and professional development.

> Real teaching, I learned in time, happens inside a wild triangle of relations—among teacher, students, subject—and the points of the triangle shift continuously … what shall I teach? How can I grasp it myself … ? What are [my students] thinking and feeling? … Inside the triangle, clear evidence is very rare … Yet out of uncertainty, craft emerges. The wildness of the triangle provokes it. (McDonald, 1992, p. 1)

The intensely complex nature of teaching often overwhelms those who are learning the craft for the first time. Because teaching is such a fast and rapidly developing process, there needs to be a way to "freeze" those moments that intrigue us in order to deconstruct that text and learn from it to improve our practice. In their professional preparation programs, preservice teachers participate in discussions of educational and developmental theory in the scholarly tradition of the academy. However, as they encounter a broad range of challenges in their teaching, they become increasingly aware that the theories they learned in their academic program may not prepare them to answer the kinds of questions that McDonald poses as he replays the day in the classroom over and over again in his mind. New teachers find themselves moving away from the contact they once had with the research that posed the provocative ideas that stirred their interest in teaching in the first place. As they attempt to hold these two worlds in balance, they experience the tension of dueling discourses, the parallel trajectories of theory and practice.

This dilemma has been the topic of significant research in teacher education and professional development. It is a critical mission of teacher educators that we prepare teachers to accept that uncertainty is part of teaching and that there are ways to learn from the aspects of our craft that seem to challenge us at our very core. It is likewise essential that in professional development, we work with in-service teachers to sustain the intellectual vibrancy that first brought them to the profession. Two effective practices for building and supporting lasting intellectual curiosity—in teaching, as in other professions—are the development of reflective practice and the exposure to theory-driven research. Reflective practice enables one to learn from one's own experiences; engagement with research introduces clarifying lenses through which to reflect on these experiences. When reflection becomes an essential element of their craft, teachers may return to it as a way to embrace the uncertainties of classroom teaching and to develop strategies to learn from them.

How can we teach reflection (Schön, 1987) to preservice teachers? How can we enliven the professional conversations of veteran teachers who want to remain engaged in the intellectual work of teaching which they began as students in teacher education programs? One answer to these challenges is to keep teachers engaging in the two lines of discourse rather than allow one to predominate. VideoPaper and its creator, the VideoPaper Builder software (Nemirovsky, Galvis, Kaplan, Cogan-Drew,

& DiMattia, 2005) represent a new genre for the production, use, and dissemination of reflective practice and educational research. VideoPaper allows professionals to integrate and synchronize different forms of representation, such as text, video, and images, in one cohesive document. The essential features of VideoPaper (as described below) incorporate textual features that intrinsically belong to teachers' discourse, as opposed to the discourse of traditional academic papers. In this way, VideoPaper may fulfill the promise of the NSF program that funded its initial version in 2001, to "bridge research and practice." This chapter explores two ways in which VideoPaper has been used as a tool that integrates the discourse communities of teachers and researchers, incorporating both a theoretical basis and pedagogical principles in the ongoing intellectual work of teaching.

WHAT IS A VIDEOPAPER

Originally conceived in 1998 through a National Science Foundation "Bridging Research and Practice" project,[1] VideoPaper was designed to enable researchers to collaborate with practitioners through video. The format of VideoPaper is a presentation of text and video side by side, allowing authors to annotate digital video. The video footage is supplemented by the use of still images captured either from the video itself (e.g., notes the teacher has written on the blackboard, a student's facial expression), scanned content (e.g,. student work, teacher handout), or other digital images of interest (e.g,. explanatory diagrams, graphics).

VideoPaper is interactive in two directions: a reader can choose either to read the text and click on the author's "PLAY" buttons installed within the text that trigger selected clips of the video, or to watch the video and click on the "link-to-text" buttons that appear within the video, directing the reader to selected pages of related text. The ability to link raw data and video with text analysis and observations enables the "reader" to interact with the content in a way that is very different from reading a traditional linear text. The reader becomes a participant who can control how the text is read. A reader can select pages to view, watch and analyze pieces of video data, pause and expand time, experiment with other interactive content (available in the form of Java applets), and conduct further research by following the hyperlinks. Authors create html documents that become text sections, gif or jpg files that are used as slides, and a Quicktime or MPEG file for the video. VideoPaper Builder generates javascript menus, html links, frame sets, and Quicktime image slide shows in order to interconnect the author's video, slides, and text into a single multimedia presentation. The final product (see Fig. 29.1) is viewable in Mac or Windows, displayed using an Internet browser such as Internet Explorer or Mozilla Firefox.

Background

University-based teacher preparation programs introduce their preservice teachers to a range of research-based literature to provide a theoretical understanding of

---

[1]NSF # 9805289.

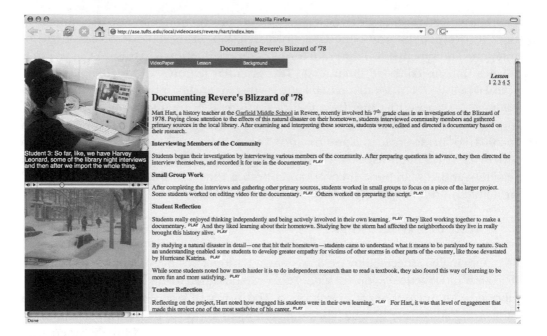

Figure 29.1. A VideoPaper.

pedagogy and cognitive development. Once these preservice teachers become professionals within a school district, they find that academic articles often ignore classroom realities and do not reflect the teacher's concerns. Therefore, their reliance on academic research to understand their practice more deeply seems to wane.

> Within the academic community, research knowledge and research genres are often viewed as relatively neutral or generic, and thus usable for those who engage in a wide variety of practices. Thus, teacher education programs often use journal articles to impart novice teachers with the professional knowledge and skills they need, with the hope that they will be able to use this professional knowledge in the particular professional situations they find themselves in. (Bartels, 2003, p. 737)

However, studies that investigate teachers and their experiences with academic research illustrate that research articles do not function well as a way to disseminate research to teachers (see e.g., McDonough & McDonough, 1990; Zeuli, 1994). Research articles tend to be organized in a way that speaks more directly to the practices of researchers. Bartels' (2003) study uses Gee's definition of discourses and addresses the problem of understanding the extent to which the discourse of teachers and researchers differ:

> A Discourse is a socially accepted association among ways of using language, of thinking, feeling, believing, valuing, and of acting that can be used to identify

oneself as a member of a socially meaningful group or 'social network', or to signal (that one is playing) a socially meaningful role. (Gee, 1990, p. 143)

One of Bartels' key findings indicates that teachers and academics have two different discourses and his data suggest that each party has difficulties in accepting and understanding the discourse of the other.

John (2003) shows that a key factor inhibiting teachers' engagement with research is the language barrier. Academic discourse, in particular, with its specialized tropes, schemas, and terminology was seen as a major obstruction to both reading and understanding. Of central importance, therefore, is the idea that teachers' knowledge is represented in the language of classrooms (Eraut, 1995; Hoyle & John, 1995) rather than in sets of distinct propositions and prescriptions (Day, 1999; Richardson, 1990). In fact, teachers define their professional lives through the sights, sounds, and interactive features of the classroom. The visual, aural, oral, and physical cues that are part of the natural world of communication of teachers are often lost in the "stream of words" that still dominate research dissemination, as Bartels (2003) concludes: "if academics want to write for teachers, they need to be careful to incorporate the kind of textual features and qualities that are central to teachers' Discourse" (p. 751).

Video has often been used in teacher education and professional development in different forms since its introduction in the 1960s (Sherin, 2004). Digital video is now emerging as a means to engage teachers in analyzing and improving their practice through studying frames of classroom activity. This provides an essential link for teachers to help them improve their understanding and interpretation of their practice (Carraher, Schliemann, & Brizuela, 2000; Derry, Siegel, Stampen, & STEP, 2002; Pea, 2003; Teachscape, n.d.).

Reflective Practice

Since the 1980s, reflection and reflective practice have increasingly gained recognition as essential components of a strong program of teacher education. The term "reflection" itself has come to mean so many things in so many different contexts that it has become confusing and vague. We see reflection as an ongoing collaborative process, a way to focus on particular aspects of one's teaching and the learning of one's students that makes the tacit explicit (Shulman, 1988). The "reflective practitioner" (Schön, 1983, 1987) takes a situation that is intriguing and/or perplexing and breaks it apart in order to reassemble it with a new and informed perspective. This intellectual endeavor helps the practitioner to recognize beliefs and assumptions that they bring to the classroom and how these shape their understanding of their role in the school and classroom.

Toward achieving this reflection, writing a VideoPaper helps a teacher to perform research on their practice. As Tochon (in this volume) writes, "[v]ideo editing is research process." The teacher chooses an episode that merits further discussion and reconstructs the moments that comprise that episode in order to understand those moments anew. This process mirrors Dewey's (1933) description of the process of reflective thought as helping one to break apart elements of a puzzling or enigmatic event in order to learn from it. We argue that developing a VideoPaper may be a valuable method for helping preservice and in-service teachers to reconsider moments in their teaching that might otherwise escape investigation. The VideoPaper is not only a review of scenes from a lesson, but an exercise in editing and annotating that promotes an intellectual, analytical involvement with the classroom and leads to a more profound understanding of the relationships essential to effective teaching (McDonald, 1992). From this understanding, teachers can learn to teach content more effectively, to refine their pedagogical skills, and to identify keen habits of mind that provide the foundation for professional acumen and expertise.

THE INTERACTIVE EDUCATION PROJECT: VIDEOPAPER AS A TOOL TO BRIDGE RESEARCH AND PRACTICE

"The InterActive Education Project: Teaching and Learning in the Information Age" [2] carried out at the University of Bristol (UK) was a collaboration between researchers and teachers aimed at exploring how information and communication technology (ICT) can be used most effectively to enhance teaching and learning across the curriculum and with all phases from primary to post-16 (Sutherland et al., 2004). One of the core strands of the project is concerned with issues related to knowledge transformation among practitioners and researchers in order to develop a new community of practice that engages classroom teachers (Triggs & John, 2004). The strand aims to transform the project's research results into forms of representation and communication that relate to teachers' discourse and may possibly develop into new professional development models (Olivero, Sutherland, & John, 2004).

One Teacher's Story

Dan Sutch was an English primary school teacher who began teaching in 1999 and joined the project when it began in 2000. In January 2003, he chose to create a VideoPaper in order to communicate to his colleagues how he had re-approached the challenge of teaching spelling to his fourth graders using a theoretical understanding of etymology and morphology that was proposed to him by a researcher from the InterActive Education project. In creating the VideoPaper, he had two goals: 1) he wanted to demonstrate his successful teaching of critical thinking and inductive reasoning through spelling lessons; 2) he wanted to share with his teacher colleagues the research that made this success possible. That his breakthrough had its origins in research was very important to Dan.

After the creation of his first VideoPaper, Dan reflected on the creation process and on the tool itself in the context of communicating research to teachers. After receiving feedback from both teachers and researchers, Dan developed a second VideoPaper specifically directed to teachers.

---

[2]For more information about the project see www.interactiveeducation.ac.uk

[C]ommunicating findings to others plays a much different role in the teachers' Discourse than in the academics' Discourse. For academics, the primary purpose of reading and communicating findings is to build a public base of abstract, generalizable knowledge. [...] For teachers, on the other hand, the primary purpose in reading and writing up research is to expand their personal, context-specific bases of knowledge about their teaching and their students' learning. (Bartels, 2003, p. 751)

Dan was encouraged by the structural features of VideoPaper as a way to integrate the two discourses (researcher's and teacher's) and, in fact, to bridge these two essential aspects of teacher development for practitioners.

Dan found the most appealing features of VideoPaper to be the way the video is hyperlinked in "a whole." In fact, Dan came to believe there was no adequate technical alternative to this "whole" to accomplish what he hoped to share with his colleagues. One of the main issues for Dan was the relationship between video and text. Which should come first? The teachers pressed the video play button before attending to anything else, which is not what Dan would have expected or wanted. He decided to remove the video control buttons so that the reader would not have control of the video independent of reading the VideoPaper's text. The reader's "reading" of the VideoPaper would then be driven by the text; viewing the video could be enabled solely by the reader activating a play button within the text.

When talking about rewriting his VideoPaper, Dan explicitly marked the fact that it was going to be a paper for teachers only, aiming to give them a window into what he had been doing over his three years with the Interactive Education project. He realized that he had to shift from being a "traditional writer," as he described himself in terms of the writing of his first VideoPaper, to being a "VideoPaper writer." In fact, his first VideoPaper contained lots of text and incorporated more features that are typical of the researcher's discourse. In creating the VideoPaper, he used the video to illustrate the text. Dan further described the significance of the video in providing the visual, oral, and aural dimensions of classroom teaching:

The video is really important because it shows that [the lesson being discussed] can be done in a real classroom ... I think with research it's very easy [for teachers to say] we can't do that ... While here you can see it's one classroom, one computer and it can be done. (D. Sutch, personal communication, May 10, 2003)

The reality of the classroom shown by the video is contrasted with the "made to measure" representations that often accompany official curriculum guides. These contrived demonstrations do not really incorporate elements of teachers' discourse, as the action and practices that characterizes real classrooms are not there. John (2003) reports that many teachers do not believe many of the research findings they encounter because of the overt ideological assumptions that appeared to dominate many of the studies. Dan's vision of the impact of his VideoPaper on the professional development of colleagues shapes how he incorporated the theory in a way that related well to teachers' discourse. His work demonstrates a way in which VideoPapers can be used to encourage teachers to explore the theoretical basis of their practice.

Dan clearly knew how he would like the VideoPaper to be read, and his idea resonates with the researchers' discourse concerning an issue of great importance: the role of evidence. Dan also recognized that one of the hardest parts of the process was choosing the video excerpts to insert into the VideoPaper because "[The video] didn't always show what I wanted to show. I was trying to say something and I didn't have the evidence to prove it …" This statement raises important questions about Dan's position within the teachers' and researchers' discourses. His language demonstrated a struggle to adapt the academic discourse of "proof" and "evidence" as a way to analyze learning and teaching in the classroom. This was stimulated by the features of the VideoPaper itself, suggesting that his medium does allow a practitioner to bridge the two discourses in an innovative way.

Toward a New Model for Teachers' Professional Development

> Watching video affords the opportunity to develop a different kind of knowledge for teaching—knowledge not of "what to do next," but rather, "knowledge of how to interpret and reflect on classroom practices." (Sherin, 2004, p. 17)

Dan's vision of the impact of his VideoPaper on the professional development of colleagues shaped how he incorporated the theory in a way that related well to teachers' discourse. As Dan explained, a VideoPaper communicates the classroom activity as well as the theoretical foundation and intellectual "hook" on which the lesson is constructed. This allows colleagues to teach a similar lesson based on a theoretical understanding rather than just copying the lessons as written, as he explains below:

> [I]t's got a lesson plan that teachers can follow, if they want to follow mine. But it's also got the theory behind it so they can change it themselves. Instead of taking my work and doing it, they'll be able to adapt it to their own class … And you can only adapt it when you understand it. (D. Satch, personal communication, May 10, 2003)

In this regard, Dan observed that the power of video needed to be substantiated by the text; the video did not just stand on its own. There was a need to incorporate both theory and practice and be more than just "tips for teachers."

Thus far, the use of video in teacher education and continuing professional development has focused mainly on the development of teaching skills (Sherin, 2004), which does not necessarily help teachers understand the intellectual dimensions of their profession. VideoPapers become a bridge, connecting the "academic theory, personal theory, and everyday planning" (Griffiths & Tann, 1992) of teaching. They connect it by linking the power of the video, which represents practice, to narrative and theory in a way that connects the discourses of the researchers and the teachers. Dan considered research important to understand what he had done in the classroom and to make it possible for other teachers to teach his lessons. But at the same time, seeing his own classroom was considered essential, too. VideoPaper allowed him to incorporate both dimensions without needing to have one necessarily dominate. We argue that VideoPapers "incorporate the kind of textual features and qualities that are central to teachers' Discourse" (Bartels, 2003, p. 751) and therefore may prove to be a way for novice and experienced teachers alike to develop an appreciation for both the values and practices of academics and the values and practices of teachers. Early experimentation with VideoPaper as a tool in teacher mentoring (Cogan-Drew, 2003) indicates that it may also prove useful in facilitating classroom study by mentors and protégés.

TUFTS UNIVERSITY: VIDEOPAPER AS A TOOL FOR REFLECTION IN TEACHER PREPARATION

The Tufts University Department of Education offers a one-year program of study for a Master of Arts in Teaching (MAT) degree. There are approximately 40 preservice teachers in the middle/high school certification program, divided into two cohorts; full-time interns, who begin the school year working 3 to 5 days a week in urban school placements; and traditional preservice teachers, who are in their schools only one day a week in the fall. All of the MAT candidates teach full-time in the spring semester. With few individual exceptions, preservice teachers arrive at the project with a basic understanding of digital video—conceptually what it takes to shoot and edit a film—and how meaning is conveyed through moving pictures. As a cohort, the interns may have more teaching experience than the traditional preservice teachers when they enter the program. From 2001 to 2004, both groups completed the VideoPaper project at the end of the fall semester as part of a required academic course.[3]

In creating reflective VideoPapers, the preservice teachers deconstruct a visual narrative of their experience, leading them to new thinking about their teaching (Bruner, 1996; Sarason, 1999;). They consider McDonald's (1992) notions of "text-making" and "gripping," which allow a teacher to hold critical moments in the classroom "steady." The faculty hypothesize that by finding meaning in these moments, preservice teachers tease out elements that puzzle and disturb them, making these the objects of their attention. Tufts faculty introduce and model reflection in an introductory teaching course. One of the foundation texts of the course is David Hawkins' essay, "I, Thou, It" (1974). Through his essay, preservice teachers encounter that triangle that McDonald (1992) terms the "wild triangle," a notion based in practice as well as philosophy. The preservice teachers learn that in each classroom this triangle—the teacher, the student, and the curriculum—must be connected in order for authentic learning to take place. They also learn to focus on the construction of relationships that will enable them to become effective practitioners.

Preservice teachers may choose to film any particular lesson in which they are featured as either the solo teacher or as a co-teacher. After they have completed their filming, they are asked to view the footage in its entirety, without taking notes or performing any written analysis of what they see. When they later watch the film inter-

---

[3]Starting in the 2004–2005 academic year, we moved the assignment to the end of the spring semester to allow all preservice teachers the opportunity to gain more classroom teaching experience before creating the VideoPaper. Initial evidence indicates that this move has had a positive effect on the quality of the assignment.

pretatively, they are asked either to actively view their footage for examples of a prede-termined topic, or to use the video as an investigative tool to inquire into an existing area of interest. The preservice teachers are encouraged to select clips that intrigue them for their unexpected or puzzling qualities. As with a literary "reader-response" approach to textual interpretation (see e.g., Andrasick, 1990), these seemingly random selections reveal the writer's salient inquiries.

The preservice teachers edit their film down to 2 ½ min of continuous or seg-mented footage. [Other programs in teacher education, such as Performance Assess-ment for California Teachers (PACT), and in teacher professional development, such as the process of National Board Certification (NBC), require two longer, contiguous clips of approximately 10 min each. We find it adds a useful rigor to the preservice teacher's inquiry to lower the limit.] When they have completed their initial editing, they convene in a video study group (Tochon, 1999) where they present their edited clips to peers for a discussion facilitated by a faculty member. Here the questions and comments by peers often require the author to reconsider their original vision of the classroom events. In the case where the preservice teacher is unsure of what themes emerge in the collection of clips they have chosen to present, the facilitator can assist in stimulating "an individual reorganization of apparently disorganized components" (Tochon, 1999, p. 62).

What matters most to Tufts faculty is the quality of contemplation of the teaching and learning portrayed in the video evidence rather than video evidence of achieve-ments or challenges. For example, Nakia, a preservice teacher, asked his mentor teacher to film on the day his sixth-grade students were presenting what they had learned about patterns of immigration to Boston from interviewing their families and neighbors. For a month, Nakia had been working particularly diligently with a sixth grader who needed a great deal of support. Nakia hoped that he could showcase this young man's presentation as a positive example of working well within the "wild trian-gle." However, as the tape rolled, a different story unfolded. Nakia called on the stu-dent and the student boldly announced, "I don't have my work ready to present." The videotape captures Nakia's momentary disbelief and disequilibrium. What do I do now? Nakia slowly repeats what the student has announced and uses this as a moment to calmly but clearly establish his expectations.

> You're not ready to present? I expected you to be ready. You and I had an agreement that today would be your turn to present. Your classmates are look-ing forward to hearing what you have learned. After class, you and I will meet again to go over what you have left to do and I want you to be ready to present tomorrow.

Out of the 50 min of presentations captured on that tape, his mentor teacher pressed him to work with those anxious few minutes and to use them as a point of de-parture for a discussion of his professional growth over the course of the semester. She saw the incident as evidence of how he was integrating his knowledge of developmen-tal issues and the theoretical "triangle" to become an authority figure who could deal with the unexpected. Nakia's response to his student allowed both teacher and student to maintain a positive relationship beyond the incident. The following day, the student completed his presentation effectively.

Faculty and Preservice Teacher Reflection

Videopaper projects like Nakia's have offered education faculty unique feedback about how preservice teachers are thinking about their classrooms. They use the textual features of the VideoPaper to express their thoughts on the "messiness" of the "wild triangle," just as Dan Sutch used the text to convey to his colleagues his teaching of etymology to develop critical thinking. As one faculty member related:

> The VideoPaper is not just offering you a glimpse into the classroom at that site; it's offering you a connection with the person who taught that lesson. It's much more sophisticated than just "here's my classroom and here's what we learned on Tuesday." It's "here's my classroom, here's what went on on Tuesday, and this is the essential strand that I'm taking out of that lesson. This is what I'm really thinking about as a pre-professional and I'm wondering where that is going to take me." (M. T. Tucker, personal communication, December 23, 2003)

As students, preservice teachers grapple with both the discourse of theory through their university program and the realities of practice in their practicum classroom. They can reconcile the conflicts and contractions between these two discourse communities by finding their own means of expressing strategies for integration of the two, as Dan Sutch has done through his VideoPaper on spelling. In both cases, the VideoPaper facilitates reflection at the same time that it enables expression. When the first group of Tufts preservice teachers completed their VideoPapers, the faculty noticed evidence of deeper self-reflection than on previous traditional video assignments.

The intellectual work the VideoPaper assignment demands arises from the fact that video, text, and slides must be connected in order for the narrative to emerge. This interconnectedness pushes the author to closely examine the relationship between the images and their text, to think carefully about exactly how to generate meaning from their media. The exactness of the medium demands that one make precise choices in editing and concentrate on discrete themes in the video. This close attention to the video has also been a two-edged sword. While on the one hand the preservice teachers are held accountable for their observations, the assignment's foregrounding of video has produced a kind of "video reality"—similar to what Dan Sutch encountered—that limits the extent to which comments that are not supported by video evidence can be perceived as persuasive. Dan could not comment on what he could not find video evidence to support. What is captured on the video is not entirely within the control of the preservice teacher, either. In many cases, they need simultaneously to be the actors in the video as well as the directors, instructing the cameraperson (usually a colleague) before class on the evidence they want to see on film. Prior to filming, we ask them to consider what questions this video may help them to answer. If they do not prepare to direct the camera's eye, what is captured may not be so useful to them as they reflect on their practice.

As faculty learn more about what is involved in asking preservice teachers to create video cases of their teaching, opportunities for deepening the professional relationships emerge. When the preservice teacher speaks with their mentor teacher or university supervisor about what constitutes video "evidence" of classroom learning, there is an opportunity for an exchange of ideas around "professional vision" (Sherin, 2001) which may lead to stronger connections between research-based theory and classroom practice.

Audience Authorship and Its Implications for Digital Discussions

The software continues to evolve. Version 3 of VideoPaper Builder (2005) contains a few notable features that were part of a redesign in response to user feedback to Versions 1 and 2. The most significant of these are improved menu navigation and authoring wizards, as well as a built-in html editor that allows authors to write text directly in the program (see Fig. 29.2).

For novice users, this removes the need for a separate html editor. In addition to enabling fast or slow motion replay, the new version also introduces an overlay feature that allows authors to highlight elements (e.g., gestures, facial expressions) in the video by outlining them for a specified time in a colored geometric shape (see Fig. 29.2). This adds greater precision to the commentary that authors write when referring to the video footage. Although video appears to be self-effacing, it is not uncommon for audiences to miss a particular gesture, expression, or detail that the author believes deserves greater attention. VPB3 enables authors to close caption their video footage di-

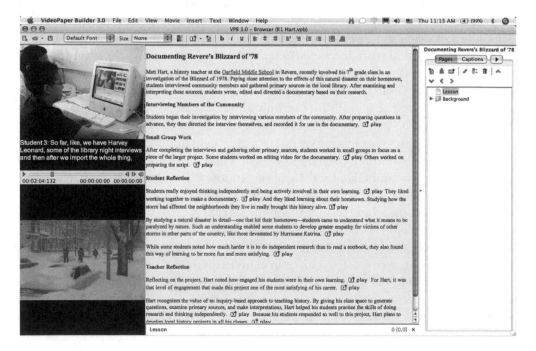

Figure 29.2.   The VideoPaper builder browser view.

rectly in the application or to import existing captioning tracks. It also offers a citation hyperlink feature to better accommodate the research community. These tools will aid authors wishing to submit to academic journals such as the *Journal for Research in Mathematics Education* (JRME), which have begun publishing VideoPapers in special monograph editions.

At the moment, VideoPaper authorship is concentrated at the document's origin. As the community of VideoPaper authors grows, however, the definition of authorship must expand to include greater consideration of the community of audiences. Dan responded to his colleagues by revising the design of his VideoPaper. Given a digital library of dozens of VideoPapers, how might readers respond? What might they want from their reading? What would readers do with the ability to respond to published VideoPapers by annotating the published text as they read it? Students who could read a VideoPaper and respond by creating their own PLAY buttons online would be able to engage in a collaborative research effort as they shared their microanalyses of video data. Teachers might access VideoPapers from online research libraries and use these as points of departure for their own classroom investigations. New media that shape teacher-researcher collaborations breathe life into the lives of practitioners and scholars.

CONCLUSION

There is a great deal to learn from the experiences cited in this chapter. When VideoPaper is used in both in-service professional development for teachers and preservice teacher education programs, it allows us to acknowledge that the very nature of the profession is marked by two parallel discourses. Each of these discourses supports teachers to develop a deeper understanding of their work. As a field of scholarship, education has a tradition of tension and division between academic discourse grounded in theoretical studies and the practitioner's discourse that thrives on the vitality of the "wild triangle," the relationship among teacher, students, and curriculum to be taught. Too often, teachers are asked to choose between these two intellectual traditions in attempting to understand their classrooms. In choosing one over the other, educators lose key insights into the complex nature of the profession. Progressive professional development programs and progressive teacher education curriculum acknowledge that teachers must learn that research must be an integral part of teaching. The teacher must also be a researcher (Meyers & Rust, 2003), probing the questions that arise as one examines the dynamics of the classroom.

VideoPaper allows the two discourses to co-exist; it allows the teacher to create a text of her classroom that, through close study and conversation with colleagues, can provide access to academic discourse and theoretical understandings that underlie the text of the video. In developing VideoPapers, Dan and Nakia worked with the discourses of both teacher and researcher, "knower and agent in both the classroom and larger educational contexts" (Cochran-Smith & Lytle, 1999). In these contexts, able to move between Academic and Practical Discourses, teachers like Dan and Nakia can have a fresh approach to incidents that intrigue them. Thus, VideoPaper helps to underscore the notion that learning to teach is not completed at the end of one's teacher education program. Improving one's craft is an intellectual process that continues throughout one's professional life span (Cochran-Smith & Lytle, 1999).

Both projects at Tufts University and at the University of Bristol demonstrate the profound insight that is possible when teachers use a medium that allows them to represent and share the vitality of their classrooms. Using such a lively text as a basis for analysis, reflection, and connection to theoretical study allows teachers to go beyond sharing "tips for teachers" and professional development as survival training. Video-Paper is a tool that represents and transforms teachers' practice while enabling research and practice to come together. VideoPaper also promises to be a tool that may encourage closer collaboration among academic researchers and educational practitioners to improve teaching and learning.

ACKNOWLEDGMENTS

"The InterActive Education project: Teaching and Learning in the Information Age" discussed in this chapter is funded by the ESRC Teaching and Learning Programme (Award No. L139251060). We would like to thank teachers Dan Sutch and Nakia Keizer.

REFERENCES

Andrasick, K. D. (1990). *Opening texts: Using writing to teach literature*. Portsmouth, NH: Heinemann.

Bartels, N. (2003). How teachers and researchers read academic articles. *Teaching and Teacher Education, 19*, 737–753.

Bruner, J. (1996). *The culture of education*. Cambridge, MA: Harvard University Press.

Carraher, D., Schliemann, A. D., & Brizuela, B. (2000, March). *Bringing out the algebraic character of arithmetic*. VideoPaper presented at the Videopapers in Mathematics Education conference, Dedham, MA.

Cogan-Drew, D. (2003). *VideoPaper as a tool for new teacher mentoring*. Grant funded by the Center for Leadership Development, Boston, MA. Retrieved October 15, 2005, from http://cogandrew.com/projects/vpmentoring/fenway/index.html

Cochran-Smith, S., & Lytle, S. L. (1999). The teacher research movement: A decade later. *Educational Researcher, 28*(7), 15–25.

Day, C. (1999). *Developing teachers: The challenge of lifelong learning*. London: Falmer Press.

Derry, S. J., Siegel, M., Stampen, J., & the STEP Research Group. (2002). The STEP system for collaborative case-based teacher education: Design, evaluation and future directions. *Proceedings of Computer Support for Collaborative Learning (CSCL) 2002* (pp. 209–216). Mahwah, NJ: Lawrence Erlbaum Associates.

Dewey, J. (1933). *How we think*. Boston: D.C. Heath.

Eraut, M. (1995). Schön shock: A case for re-framing reflection-in-action. *Teachers and Teaching: Theory and Practice, 1*(1), 9–22.

Gee, J. (1990). *Social linguistics and literacies: Ideology in discourses*. Philadelphia: Falmer Press.

Griffiths, M., & Tann, S. (1992). Using reflective practice to link personal and public theories. *Journal of Education for Teaching, 18*, 69–84.

Hawkins, D. (1974). *The informed vision: Essays on learning and human nature*. New York: Agathon.

Hoyle, E., & John, P. D. (1995). *Professional knowledge and professional practice*. London: Cassell.

John, P. (2003). Conceptions, contentions and connections: How teachers read different genres of educational research. In R. Sutherland, G. Claxton, & A. Pollard (Eds.), *Learning where world views meet* (pp. 231–244). London: Trentham Books.

McDonald, J. P. (1992). *Teaching: Making sense of an uncertain craft*. New York: Teachers College Press.

McDonough, J., & McDonough, S. (1990). What's the use of research? *English Language Teaching Journal, 44*(2), 102–109.

Meyers, E., & Rust, F. (Eds.). (2003). *Taking action with teacher research*. Portsmouth, NH: Heinemann.

Nemirovsky, R., Galvis, A., Kaplan, J., Cogan-Drew, D., & DiMattia, C. (2005). VideoPaper Builder (Version 3.0) [Computer software]. Concord, MA: Concord Consortium.

Olivero, F., Sutherland, R., & John, P. (2004). Learning lessons with ICT: Using videopapers to transform teachers' professional knowledge. *Cambridge Journal of Education, 44*(2), 179–191.

Pea, R. D. (2003, March). *Point-of-View authoring of video for learning, education and other purposes.* Paper presented at the PARC Forum, Palo Alto, CA.

Richardson, V. (1990). Significant and worthwhile change and teaching practice. *Educational Researcher, 16*(9), 13–20.

Sarason, S. B. (1999). *Teaching as a performing art*. New York: Teachers College Press.

Sherin, M. G. (2001). Developing a professional vision of classroom events. In T. Wood, B. S. Nelson, & J. Warfield (Eds.), *Beyond classroom pedagogy: Teaching elementary school mathematics* (pp. 75–93). Mahwah, NJ: Lawrence Erlbaum Associates.

Sherin, M. G. (2004). New perspectives on the role of video in teacher education. In J. Brophy (Ed.), *Using video in teacher education* (pp. 1–27). New York: Elsevier Science.

Shön, D. A. (1983). *The reflective practitioner: How professionals think in action?* New York: Basic Books.

Shön, D. A. (1987). *Educating the reflective practitioner: Toward a new design for teaching and learning in the profession.* San Francisco: Jossey-Bass.

Shulman, L. (1988). The dangers of dichotomous thinking in education. In P. P. Grimmett & G. L. Erickson (Eds.), *Reflection in teacher education* (pp. 31–38). New York: Teachers College Press.

Sutherland, R., Armstrong, V., Barnes, S., Brawn, R., Gall, M., Matthewman, S., Olivero, F., Taylor, A., Triggs, P., Wishart, J., & John, P. (2004). Transforming teaching and learning: Embedding ICT into every-day classroom practices. *Journal of Computer Assisted Learning* (Special Issue, Vol. 20, Issue 6, pp. 413–425).

Teachscape. (n.d.). *Who we are: The research.* Retrieved October 14, 2005, from http://www.teachscape.com/html/ts/public/html/research.htm

Tochon, F. (1999). *Video study groups for education, professional development, and change*. Madison, WI: Atwood.

Triggs, P., & John, P. (2004). From transaction to transformation: Information and communication technology, professional development and the formation of communities of practice. *Journal of Computer Assisted Learning, 20*(6), 399–473.

Zeuli, J. (1994). How do teachers understand research when they read it? *Teaching and Teacher Education, 10*, 39–55.

30

Fostering Community Knowledge Sharing Using Ubiquitous Records of Practice

Barry J. Fishman
University of Michigan

Standards and systemic reform are twin pillars supporting contemporary discourse on education reform in the United States. Such was the case at the close of the last century, the beginning of this one, and for the foreseeable future. The standards movement in education seeks to set forth agreed-on goals for student learning in a range of disciplines and across grade levels. Standards are created with an eye toward both practice and knowledge within a domain. For example, standards documents in the domain of science (e.g., American Association for the Advancement of Science, 1993) emphasize both the content knowledge that students should understand (e.g., Newton's Laws, the particulate nature of matter) as well as the habits of mind employed in the practice of science (e.g., inquiry, hypothesis formation, and testing). Systemic reform attempts to bring about school change by creating alignment across the components of school systems, such as administration and management, curriculum and instruction, assessment, policy, and technology, both within a single school district and between districts, states, and the federal government (Smith & O'Day, 1991). Academic standards have been a major tool for systemic reform proponents, providing target learning goals for policymakers that can then be used in high-stakes assessments designed to (in theory) measure educational progress toward standards. In practice, systemic reform efforts are often top down, beginning with the imposition of federal or state policy (such as the Bush administration's "No Child Left Behind" education policy), which can leave individual schools and teachers scrambling to figure out how to comply (Goertz, 2001). Because the "how" of systemic reform is left up to individual school districts to determine, it preserves (at least the illusion of) local control, which may be why this mode of reform is so appealing to policymakers.

There is near-universal agreement that a key to successful standards-based systemic reform is teacher professional development (e.g., Committee on Science and Mathematics Teacher Preparation, 2001). The question is, how can teachers, schools, and districts most profitably engage in professional development activities that help them achieve their systemic reform goals? Unfortunately, there is little empirical knowledge on the creation of effective professional development for teachers. As a field, we understand little about what teachers learn from professional development (Wilson & Berne, 1999), and even less about the impact of professional development on classroom practice and on students' learning (Fishman, Marx, Best, & Tal, 2003). I believe that an answer to the question of how to create effective professional development lies in the combination of several promising trends in research on teacher learning and the emerging digital and video technologies described throughout this volume. Later, I describe several of these trends and propose a vision of how technology, and in particular new digital video technology, could be applied to magnify their impact, moving us closer to achieving the core goals of education reform.

THE FUTURE IS (ALMOST) HERE

Will teaching and learning in the future look radically different compared with today? Will we still have places called "schools?" Despite frequent predictions that school as we know it is obsolete and that a golden age of self-directed learning facilitated by high-bandwidth networks is nigh, I believe that school is and will remain a societal necessity. In the future, it is likely that children will continue to congregate with their peers in controlled settings. And what will these settings be like? Larry Cuban (2001) is fond of pointing out that a teacher from the turn of the 20th century would not feel out of place in a classroom at the turn of the 21st. Although some details are likely to change, I would expect the broad outlines of schooling to remain the same in the future, with teachers organizing and supporting educational experiences for students. The aspect of classroom life where I expect to see the greatest change is in terms of student and teacher access to technology and information. Rapidly increasing "zorch" and decreasing size and cost are creating a situation where it is conceivable for everyone to have access to a personal computing device (Soloway et al., 2001), whether that be in the form of what we currently call a personal computer, in the form of a hand-held computer, or some other information and media access/creation device whose form factor we have not yet seen. Of course, knowing what to do with those devices is another matter. That's where quality teaching enters the picture.

Some readers, on seeing my prediction of little change in teaching and schools, may recoil at the thought that the current problems and insufficiencies of the U.S. education system could continue indefinitely into the future. But that is to read the words "teacher" and "teaching" with a negative valence. Reform-minded innovators who imagine a future without teachers or schools frequently describe a negative school experience as formative in their thinking. I use the terms "teacher" and "teaching" neutrally, recognizing that there are good teachers and teaching as well as bad. But even the words "good" and "bad" need to be defined within a context. I am opposed to behaviorally oriented teaching that emphasizes quiet and order over interaction and creativity, "teaching by telling," or the rote memorization of de-contextualized facts. Interestingly, so are the writers of current academic standards. The standards documents related to science education, for instance, describe learning environments that employ inquiry, communities of learners, and language that is reminiscent of that common in the learning sciences; cognitive apprenticeship, distributed cognition, and so on. In my view, the challenge for the future is not to completely reinvent our notions of teaching and schooling, but rather to figure out how to help teachers employ practices that more closely resemble the notions of constructivist teaching reflected in current national standards documents (e.g., American Association for the Advancement of Science, 1993) and learning sciences research (e.g., Bransford, Brown, & Cocking, 1999).

There are many ways to accomplish the goal of improving teaching (none of them easy). One could work to change the ways that schools of education prepare teachers, change standards for teacher certification or tenure, alter the support structure for teachers, put more money into (the right kinds of) professional development, improve administrator knowledge of standards-based teaching, alter standardized testing ... the list could go on. One of the most direct routes to improving teaching (and one highlighted in many policy arguments about systemic reform) is to focus on teacher learning through both the preparation of teachers and through in-service teacher professional development. The work of teachers is vastly more complex than most casual observers realize (Shulman, 1986). To do their jobs, teachers need to draw on a broad range of expertise that includes disciplinary knowledge (e.g., in chemistry, "What is the nature of matter?"), pedagogical knowledge (e.g., "What are the ways I can organize my students to create a productive learning environment?"), and pedagogical content knowledge (e.g., "How will the pedagogical choices I make affect my students' ability to learn about the nature of matter?"). These are the "big three" forms of knowledge that appear frequently in the literature on teaching and teacher learning, but there are also many more, including knowledge of students and their backgrounds, knowledge of the broader curriculum goals, knowledge of human social and cognitive development, and so on. In the broadest sense, the objective of teacher professional learning should be to improve teachers' facility with this broad range of knowledge and its application in classrooms.

My colleagues and I in the Center for Highly Interactive Classrooms, Curricula, and Computing in Education (hi-ce) use an approach that combines curriculum development and professional development to improve teacher practice, and ultimately, student learning. Much of this work has taken place in the context of the Center for Learning Technologies in Urban Schools (LeTUS), where we have created a set of middle-grades science curriculum materials that help teachers use inquiry-oriented instructional approaches and challenging science content rooted in standards, linked to a comprehensive program of professional development to help teachers learn how to employ these materials in their classroom contexts (Blumenfeld, Fishman, Krajcik,

Marx, & Soloway, 2000). Our work is motivated by a range of pedagogical and learning theory, but our decision to embed that theory in materials is in part motivated by Ball and Cohen's (1996) argument that because curriculum materials shape much classroom practice, it makes sense to use them as a vehicle to carry explicit messages and guidance about reform-oriented teaching and learning. Such materials, called "educative" serve as important sources of professional development for teachers.

KNOWLEDGE NETWORKS ON THE WEB (KNOW)

In LeTUS, we go beyond the creation of print-based educative curricula to employ a broad range of media designed to help teachers become proficient users of inquiry-oriented and standards-based materials. We call the vehicle for this media Knowledge Networks On the Web (KNOW, http://know.umich.edu/; Fishman, 2003), which is a web-based environment for teacher professional development. We conceptualized KNOW as a forward-looking collaborative environment for teachers working together to improve their practice with particular curriculum materials. I refer to KNOW as "forward looking" because I believe that the technology for realizing the full potential of online environments such as KNOW exists today only in part. Before looking ahead to the (hopefully) near future in terms of technologies that will fully enable our vision of an online teacher learning community, I will describe the present-day KNOW in more detail, including an overview of the learning theory that motivates its design, and describe some of the practical technology issues that bound its use.

An Overview of KNOW

KNOW exists within the broader context of LeTUS professional development. Professional development for LeTUS teachers includes week-long summer workshops, monthly face-to-face workshops, and KNOW. This range of professional development opportunities is designed to work as part of a complete package, with the various types of professional development providing different affordances for teacher learning. The online environment is helpful both for its "anytime anywhere" value and for its ability to present material that is not otherwise easy to recreate in face-to-face settings, such as video depictions of particular classroom or other settings.

The heart of KNOW is the print-based LeTUS curriculum materials, but presented as hypertext, taking advantage of the abilities that medium has for making contextual connections to related and supporting ideas and materials (Landow, 1997). KNOW contains materials related to each lesson in the curricula that go beyond what is presented in the print versions. For example, we provide links to related web sites that contain information about science content for both teachers and students. Multiple examples of student work from lessons have been made available for downloading and examination. Teachers' comments and framing/guiding questions are presented along with each student work example. We attempt to provide both "good" examples of student work to give teachers an idea of the end state of a lesson, and "bad" examples to prompt discussion about student misconceptions and common challenges in teaching various lessons. For many lessons, videos are available.

There are three major types of video available within KNOW. First are "images of practice" videos, which are shot in the classroom and edited together with teachers' interpretive commentary. These videos, as their name suggests, are primarily intended to give teachers unfamiliar with inquiry-oriented teaching some notion of what to expect and what to look out for. Next are "how to" videos, which are typically shot in a studio, and serve as visual guides for the set-up and use of scientific apparatus and computer software. Finally, we have videos that are intended to be shown to students as part of teachers' instructional practice. In many cases, these videos were created by teachers as part of workshop activities intended to help them think deeply about content (in terms of what to present in the video) and to gain experience with the technology of digital video creation. In addition, KNOW has a discussion system, again organized according to curricula, that teachers use to pose questions, ask for help, and share information beyond what is represented in the structured curriculum or multimedia materials in KNOW. Our discussion tools allow for the easy creation of Web logs, so that teachers can create journal-type narratives of their teaching and invite others to comment and discuss their version of events. The same facility allows for private journals to be kept online, which can be used as a record of how the curriculum enactment went from lesson to lesson, serving as a guide for future enactment of the same curriculum. Anything kept in a private journal may be "promoted" for inclusion in a public Web log at the election of the individual teacher. Integrated planning tools help teachers make decisions about how to modify the curriculum while maintaining fidelity to its overall goals (Lin & Fishman, 2004). Finally, a download area facilitates access to software, PDF versions of the curricula, and editable word processor versions of student worksheets and readers for teachers to customize to their own particular instructional needs.

KNOW is unique among online teacher learning environments in that it is organized with direct reference to curriculum materials, as opposed to being organized around "big ideas." By this I mean that if a teacher is working in a particular portion of the curriculum, she or he can log in to KNOW and ask to see video clips, examples of student work, or other forms of multimedia guidance, including peer-to-peer discussion forums, that are related directly to that day's teaching. By contrast, there are many high-quality online professional development environments where teachers gather to learn about, discuss, or share information related to "big ideas" such as inquiry (e.g., the Inquiry Learning Forum, http://ilf.crlt.indiana.edu/) or integrating technology into teaching (e.g., InTime, http://www.intime.uni.edu/). My belief is that, while more general sites may be valuable to teachers who are motivated to try new ideas and who possess a good amount of self-efficacy in relation to those ideas, KNOW and sites like it that are rooted in daily practice as guided by specified curriculum materials are better suited to teachers with a broad range of motivations and backgrounds. KNOW is useful for teachers precisely because it is close to their everyday practice, and can therefore provide more highly contextualized support for changing teachers' professional knowledge and practice. KNOW links powerful representations of knowledge for teaching directly to the classroom contexts in which the activities and events of teaching take place. Teacher knowledge is practical knowledge; it is not stored in terms of abstract principles, but rather in terms of how to act and react in classroom settings. Lampert and Ball (1998) describe the materials generated in the act of teaching as "records of practice," and view them as a rich vein of knowledge that can be tapped in order to create new understandings of the act of teaching, an essential component on the road to improving teaching. KNOW can be thought of as a collection of such records from the community of teachers who use the system.

As of this writing, more than 500 teachers and other education professionals belong to the KNOW community, including approximately 85 LeTUS teachers in Detroit. For those beyond Detroit, KNOW serves as a means to get access to the LeTUS curricula for use in their own schools, and also to access (via the discussion forums) teachers who have experience using these materials. In Detroit, KNOW is used by teachers both on their own (from home and from school) and as an integral part of face-to-face workshops, where professional development leaders direct participants to call up a video or student work on KNOW to serve as the focus of discussion. The LeTUS collaboration resulted in strong gains in student learning (Marx et al., 2004) and more importantly on state standardized tests, where students involved in LeTUS outperformed their within-district peers by more than 14% (Geier et al., 2004). In the LeTUS work, we were able to document a direct connection between the PD, changes in classroom practice, and student learning (Fishman et al., 2003; Kubitskey, Fishman, & Marx, 2004).

Theoretical Underpinnings for the Design of KNOW

KNOW builds on learning sciences research and theory that characterizes knowledge as constructed, situated, social, and distributed. Teachers need experience in order to tailor abstract ideas about teaching to concrete situations. To this end, KNOW is more than a collection of add-ons to curriculum, such as might be found in an annotated teachers' guide or on the CD-ROM now included in the back of so many college science texts. The material in KNOW emphasizes practical knowledge that emerges from the community of teachers who use the curriculum materials in the system. This is consistent with theory in the learning sciences that depicts knowledge as situated, with many researchers arguing for the importance of "communities of practice" for shaping knowledge, facilitating learning, and giving meaning to our actions (Brown, Collins, & Duguid, 1989). In the face of these social views of knowledge and learning, teaching is often characterized as a profession where practitioners work in extreme isolation from their peers, making the formation of communities of practice an even more urgent need in support of ongoing teacher learning (Putnam & Borko, 2000). For instance, increased collaboration and contact with colleagues experiencing similar challenges can offer teachers opportunities for examining their own thinking in relation to others' thinking, thus clarifying their own beliefs. Peer exchanges offer increased opportunities for access to information with relevance to one's current situation. There are also emotional benefits, including support for experimentation with ideas that teachers might not be familiar with, and encouragement for risk taking in the face of community or organizational norms that discourage deviation from business as usual.

A corollary to the claim that knowledge is social is that it is distributed across groups, or across community members and their materials (Pea, 1993). Certainly, teachers need to master a repertoire of understandings and practices, but professional development must balance a focus on individual competence and on socially shared, collaborative activities that can support individual proficiency. We therefore try whenever possible to structure teacher learning activities within KNOW to leverage both the knowledge of other teachers and also their supportive online presence. The aim is to develop teachers who are adaptive learners and who have skills to attain competence and information when the need arises rather than have everyone learn the same thing at the same time. In the context of professional development, a community of teacher practice must be framed with respect to a culture of learning.

Practical Limitations to the Realization of KNOW's Designed Intentions

In designing KNOW, our intention was to create an environment where teachers could easily share images and information from their own practice and gain access to the same from their peers enacting the same curriculum units. In practice, the most serious limitation to this vision is that it is still too difficult for individual teachers to create and share digital media. Consider classroom video as a prime example: Classroom video need not have elaborate production values (indeed, my experience indicates that teachers find the video to be more "realistic" when it is somewhat "raw"), but it does need to clearly depict the phenomenon of interest, and contain intelligible audio and/or a transcription of important utterances by both teacher and students. To create video with these qualities takes some skill. A single camera set up in the back of a classroom is rarely sufficient. If the viewing angle is too wide, any single teacher or student action will be hard to make out clearly, and audio is unlikely to be distinct, especially in a constructivist classroom setting where students are often working in groups with multiple simultaneous conversations going on. So, right away we have created the need for a camera operator, someone who is not otherwise either teaching or learning in the classroom, who can make sure to keep the camera focused in the proper location. That camera operator may also be able to control sound recording equipment, perhaps having the teacher wear a radio lapel microphone or select from several table-mounted microphones designed to record student conversation.

The need for an "outsider" to be present in the room in order to capture video adds a layer of both interpretive and practical complexity to the enterprise; interpretive complexity because this person brings their own perspective to what is important to capture, which may or may not be the same as the teacher's. Practical because it is costly to have another person in the classroom for the purpose of gathering video, and this limits the number of classrooms where video can be gathered to the number of trained people who can gather video. This places a severe restriction on the variety of "records of practice" that can be included in the system. Furthermore, the camera operator has to schedule their time, and it is not always possible for a teacher to know in advance when a particular day's lesson will contain something that is of value for sharing with other teachers.

A second limitation comes at the video editing phase. Currently, all videos in KNOW are edited by university personnel, and then shown to the depicted teacher for approval before being added to the system. This is in part because video editing tools, although much simpler today than they were just a few years ago, still require some specialized skills in order to be effectively used. It is also because video editing can be frustrating without relatively fast computer hardware and large amounts of digital storage. But the main limitation is time. Editing video takes time, which is a commodity most teachers lack. The simple fact of the matter is that creating videos (or other types of digital artifacts, such as student work samples) does not currently fit within the realm of important tasks a teacher must accomplish each day. For that reason above all others, the teachers who use KNOW do not create their own records of practice for sharing.

A third limitation to teachers' ability to use KNOW as envisioned in our original design has to do with the fluidity of their access to the system. In many classrooms, teacher access to technology is limited, or teachers' time to refer to online materials like KNOW in the context of a full day of teaching is constrained. Two issues arise as a result. On the one hand, teachers cannot use KNOW as an "at-hand" reference for interpreting student understanding, setting up scientific apparatus, and so forth. This is not a major limitation for teachers' use of KNOW, as working around it requires primarily that teachers look up information in advance on KNOW and make a record of the information in a format that can be handy for classroom use, such as in a paper notebook. But such translations give up the richness of the media. A second and potentially greater issue has to do with teachers' ability to keep reflective notes on their teaching either for their own use or for sharing with others. If teacher access to technology is limited, they must make their notes on enactment in a convenient form and then transcribe them into KNOW at a later time. Or alternatively, teachers may opt to wait until they have access to KNOW to make their notes. Many teachers opt to do neither, because access to KNOW is not viewed by many as a daily activity, but rather something saved for when they are engaged in the act of lesson planning. On the whole, reflective practice is an activity that requires careful scaffolding and support, and is not realized to its full potential in the current incarnation of KNOW.

NEW TECHNOLOGIES, NEW AFFORDANCES

Videocassette recorders revolutionized the home entertainment industry, giving consumers the ability to watch prerecorded programming on their own schedule. The most popular use of the VCR is for viewing rented content. A less common, but more powerful, use is to record content from broadcast television for later (sometimes called "time shifting"). Time shifting is less frequently used than is viewing prerecorded content largely because programming the VCR to record desired content is a complex and confusing task for many end-users. Recent developments, in particular personal video recorders (PVRs, e.g., TiVo), which are essentially media-centric computers that can record large quantities of video on hard disks, are revolutionizing the way that viewers record content for later viewing. One way the PVR does this is by replacing the VCR's date-and-time based programming method with one that is content centric, allowing users to name a favorite show and then have the PVR record all instances of that program. A second approach is to record live broadcast content continuously with a 30-min buffer, allowing the viewer to pause and rewind "live" television, and also to decide on the fly to record the current program for later viewing, even if that program began prior to the decision to record. My own thinking about how future video technologies are likely to enhance teachers' use of KNOW is heavily influenced by devices like TiVo. In particular, the "constant recording" features of such technologies can be used to place control of content creation into teachers' hands.

The basic idea is to place several video cameras within a classroom that are designed to capture the classroom from various angles (similar to the way video-conferencing rooms have cameras pre-positioned to video the presenter, the audience, the whiteboard, etc.). Perhaps one camera would be set up to record activity at a particular student work area, one pointed toward the front of the room, one toward the back, and so on. These cameras can be linked to a central computer that would record from all angles, all the time. The power of this system is that, as with TiVo, control can be given to the teacher to decide when to save and when to discard the video. In fact, this could be designed explicitly as an "opt-in" system, so that if a teacher did not tell the system to keep the recordings from a particular class session, it would automatically be discarded at the end of the day. This arrangement preserves control for the teacher, avoiding a "big brother" scenario. When something happens during the day that the teacher wishes to share with others through KNOW, they hit the "save" button and that video is archived for later editing. By capturing multiple views simultaneously, editing is also simplified, although if a teacher does not wish to edit the video themselves, the raw video could be shipped over high-bandwidth networks along with notes for central editors to work with, with the final product sent back to the teacher for final edits, approval, and posting to KNOW. The problem of capturing good-quality audio would remain with this approach, but might be solved through permanent mounting of high-quality microphones at strategic locations in the classroom.

Dealing with the other major issue introduced previously, teachers' ready access to information in KNOW might be addressed in the future through the use of handheld computer devices with multimedia capabilities. We have been exploring the idea of "KNOW To Go," a handheld-based application that allows teachers to extract a subset of information from KNOW, perhaps information pertaining to a single lesson, and place that information on a handheld computing device. As the processing power and storage capabilities of these small devices increases, it has become feasible to use them for viewing video (some recent devices contain the ability to capture video, albeit at low resolutions and bit rates). This capability makes it possible for teachers to store selected video clips from KNOW (perhaps a video about how to set up a complex piece of science equipment) for use during their instruction. Alternatively, handhelds could be used by teachers to record notes on their own enactment or questions they have during enactment, which could be automatically synchronized with KNOW at a later time. Teachers could use the built-in camera capabilities that are rapidly becoming commonplace in handhelds in order to capture still photos of interesting student work for sharing with others online, or for storing in their personal web journals.

As we develop these or similar tools in the future, we must keep in mind the paramount importance of usability. If teachers do not feel that the tools are a good fit with their daily work lives, they will ultimately be rejected. The introduction of new technology into classrooms and teachers' lives is a constant negotiation between existing practice and the practices we wish to see through the process of reform. In particular, it is critical to pay attention to the relationships between teachers' capabilities, the culture within the school and classroom, and the policy and administrative environment in which we hope these technologies will be used (Fishman, Marx, Blumenfeld, Krajcik, & Soloway, 2004). For example, if teachers feel accountability pressure from a particular high-stakes test, we need to show them how the new technologies can be used to help them prepare their students more effectively for that test. If teachers feel that school administrators might use information from their classrooms against them in ways that might damage their careers, special care needs to be taken in the design of the video recording technology to assure teachers that ultimate control of any content rests in their hands.

CONCLUSION

Our goal in developing KNOW was to create an online environment where teachers can share and access multiple images of inquiry-oriented teaching. These online records of practice can help teachers gain confidence with new ways of teaching that match the demands of standards-based reform. A strength of KNOW is that this information is organized for teachers according to the structure of curriculum materials designed to work in real classroom contexts. Because all teachers who use KNOW are teaching with the same curriculum materials, the possibilities for peer-group sharing and community formation are enhanced. Looking to the future, my hope is that a combination of TiVo-like digital video recording tools and handheld computers can make the capturing, annotation of, and most importantly, the sharing of records of practice easier for teachers. Such tools could facilitate the creation of ubiquitous records of practice to support teacher learning, and ultimately advance systemic reform agendas.

ACKNOWLEDGMENTS

Thanks to my colleagues at the University of Michigan: Phyllis Blumenfeld, Joe Krajcik, Ron Marx, and Elliot Soloway. I also thank the administration and teachers of the Detroit Public Schools, the students of Detroit, and the staff of the Center for Highly Interactive Highly Interactive Classrooms, Curricula, & Computing in Education (hi-ce). The KNOW research and development team includes Stein Brunvand, Johanna Craig, Jay Fogleman, Andrew Godsberg, Hsien-Ta Lin, Jon Margerum-Leys, Joy Reynolds, and Damon Warren. The research described in this chapter was funded with support from the W. K. Kellogg Foundation and the National Science Foundation under the following programs; REPP (REC-9720383, REC-9725927, REC-9876150) and USI (ESR-9453665). Thanks also to Roy Pea for his critique of a draft of this chapter. The views contained in this work are those of the author and do not necessarily represent the views of either its funders or the University of Michigan.

REFERENCES

American Association for the Advancement of Science. (1993). *Benchmarks for science literacy, Project 2061*. New York: Oxford University Press.

Ball, D. L., & Cohen, D. K. (1996). Reform by the book: What is—or might be—the role of curriculum materials in teacher learning and instructional reform? *Educational Researcher, 25*(9), 6–8.

Blumenfeld, P., Fishman, B., Krajcik, J. S., Marx, R. W., & Soloway, E. (2000). Creating usable innovations in systemic reform: Scaling-up technology-embedded project-based science in urban schools. *Educational Psychologist, 35*(3), 149–164.

Bransford, J. D., Brown, A. L., & Cocking, R. R. (Eds.). (1999). *How people learn: Brain, mind, experience, and school*. Washington, DC: National Academy Press.

Brown, J. S., Collins, A., & Duguid, P. (1989). Situated cognition and the culture of learning. *Educational Researcher, 18*(1), 32–42.

Committee on Science and Mathematics Teacher Preparation. (2001). *Educating teachers of science, mathematics, and technology: New practices for the new millennium*. Washington, DC: National Academy Press.

Cuban, L. (2001). *Oversold and underused: Computers in the classroom*. Cambridge, MA: Harvard University Press.

Fishman, B. (2003). Linking on-line video and curriculum to leverage community knowledge. In J. Brophy (Ed.), *Advances in research on teaching: Using video in teacher education* (Vol. 10, pp. 201–234). New York: Elsevier.

Fishman, B., Marx, R., Best, S., & Tal, R. (2003). Linking teacher and student learning to improve professional development in systemic reform. *Teaching and Teacher Education, 19*(6), 643–658.

Fishman, B., Marx, R., Blumenfeld, P., Krajcik, J. S., & Soloway, E. (2004). Creating a framework for research on systemic technology innovations. *The Journal of the Learning Sciences, 13*(1), 43–76.

Geier, B., Blumenfeld, P., Marx, R., Krajcik, J. S., Fishman, B., & Soloway, E. (2004). Standardized test outcomes of urban students participating in standards and project-based science curricula. In Y. B. Kafai, W. A. Sandoval, N. Enyedy, A. S. Nixon, & F. Herrera (Eds.), *Proceedings of the Sixth International Conference of the Learning Sciences* (pp. 206–213). Santa Monica, CA: Lawrence Erlbaum Associates.

Goertz, M. E. (2001). Standards-based accountability: Horse trade or horse whip? In S. H. Fuhrman (Ed.), *From the capitol to the classroom: Standards-based reform in the states: 100th Yearbook of the National Society for the Study of Education* (Part II, pp. 39–59). Chicago, IL: University of Chicago Press.

Kubitskey, B., Fishman, B., & Marx, R. (2004, April). *Impact of professional development on a teacher and her students: A case study*. Paper presented at the Annual Meeting of the American Educational Research Association, San Diego, CA.

Lampert, M., & Ball, D. L. (1998). *Teaching, multimedia, and mathematics: Investigations of real practice*. New York: Teachers College Press.

Landow, G. P. (1997). *Hypertext 2.0*. Baltimore, MD: Johns Hopkins University Press.

Lin, H.-T., & Fishman, B. (2004). Supporting the scaling of innovations: Guiding teacher adaptation of materials by making implicit structures explicit. In Y. B. Kafai, W. A. Sandoval, N. Enyedy, A. S. Nixon, & F. Herrera (Eds.), *Proceedings of the Sixth International Conference of the Learning Sciences* (p. 617). Santa Monica, CA: Lawrence Erlbaum Associates.

Marx, R. W., Blumenfeld, P., Krajcik, J. S., Fishman, B., Soloway, E., Geier, B., & Tal, R. T. (2004). Inquiry-based science in the middle grades: Assessment of learning in urban systemic reform. *Journal of Research in Science Teaching, 41*(10), 1063–1080.

Pea, R. D. (1993). Practices of distributed intelligence and designs for education. In G. Salomon (Ed.), *Distributed cognitions: Psychological and educational considerations* (pp. 47–87). New York: Cambridge University Press.

Putnam, R., & Borko, H. (2000). What do new views of knowledge and thinking have to say about research on teacher learning? *Educational Researcher, 29*(1), 4–15.

Shulman, L. S. (1986). Those who understand: Knowledge growth in teaching. *Educational Researcher, 15*(2), 4–14.

Smith, M. S., & O'Day, J. (1991). Systemic school reform. In S. H. Fuhrman & B. Malen (Eds.), *The politics of curriculum and testing* (pp. 233–267). New York: Falmer.

Soloway, E., Norris, C., Blumenfeld, P., Fishman, B., Krajcik, J. S., & Marx, R. W. (2001, June). Handheld devices are ready at hand. *Communications of the ACM, 44,* 15–20.

Wilson, S. M., & Berne, J. (1999). Teacher learning and the acquisition of professional knowledge: An examination of research on contemporary professional development. In A. Iran-Nejad & P. D. Pearson (Eds.), *Review of research in education* (pp. 173–209). Washington, DC: American Educational Research Association.

ORION™, An Online Digital Video Data Analysis Tool: Changing Our Perspectives as an Interpretive Community

Ricki Goldman
New York University

THE POINTS OF VIEWING THEORY

This chapter integrates the theoretical framework, design, and application of a tool called Orion™—a tool for enabling online video data analysis research communities. The purpose of using video data analysis tools such as Orion™ in the learning sciences is to create tools for both individual and collaborative video analysis. Wulf (1989) and Edelson, Pea, and Gomez (1996) refer to these online communities as collaboratories. As an ethnographer, I refer to these participating and virtual members who interpret videotaped events as members of video analysis cultures. In short, the central idea is that when we use online digital video tools for analysis, our perspectives continually change as we become an interpretive community.

In this chapter, I discuss Orion™ in the context of how our perspectives are altered, clarified, and enhanced for both individual researchers and research communities, thereby enhancing research in the learning sciences.

Over the last 20 years I have developed several video analysis technologies—Learning Constellations, 1989; Constellations, 1992; the Global Forest, 1995; WebConstellations, 1997; the Points of Viewing Website at http://www.pointsofviewing.com, 1998; and Orion™, 2000 at http://www.videoresearch.org.

Underlying these software applications, I have developed a theoretical framework called the *points of viewing theory* (R. Goldman, this volume; Goldman-Segall, 1995; Goldman-Segall, 1998; Goldman-Segall & Maxwell, 2002; Goldman-Segall & Reicken, 1989). This theory describes how researchers layer their viewpoints—an act that when conducted with rigor and robustness, results in conclusions that become configurationally valid accounts of what occurred when the camera was turned on.

The points of viewing theory (POVT) has at its heart the intersecting perspectives of all participants with a stake in the community. It is a theory about how the interpretive actions of participants with video data overlap and intersect. To embrace how these points of viewing converge (and diverge) leads to a deeper understanding of, not only the event and the video event, but also the actual physical and the recorded context of the topic under investigation. The theory overcomes the static, isolating, individualized approach to point of view, in favor of the dynamic tension that operates among points of view, points that generate intersecting sight lines, enabling people to catch sight of each other, as interpreters, even as they project their own point of view. In this way, the points of viewing theory underscores the importance of attending to how others project meaning on events. While attending to intersecting data of viewer and viewed, every interpretive action has the possibility of infusing meaning that creates new representations that, if carried out with sensitivity, tenderness, and humanity, resonate with the reasonable nature of members of a larger community.

LEVERS FOR CHANGE

I open this discussion by directing our attention to a book called *Changing Minds* by Howard Gardner (2004). Gardner identifies seven levers of change; reason, research, resonance, representational redescriptions, resources and rewards, real-world events, and resistances. Whereas Gardner uses these seven R-levers to identify the difficulty of changing one's mind or other people's minds, I incorporate them as levers thinking about digital video data analysis tools. For example, video analysis tools need to provide a method for interpreting videotaped Real-world events—that may seem random or inconsequential—into a logical sequence or a compelling story. These tools should provide a way to Reasonably make meaning of what has been fragmented or encapsulated into a chunk or stream of video. Gardner argues that with each new Representational Redescription, the story becomes more convincing and compelling. Video analysis tools also enable researchers to represent redescriptions, adding insights (and sight lines) as the video is viewed through intersecting viewpoints.

Resonance is perhaps the most intriguing lever. Resonance includes the ethereal aspect of how the story or argument feels. Jay Lemke (this volume) also discusses how feelings and meanings are embedded in interactive media "as we meaningfully and feelingfully traverse attentional spaces." By its nature, video, as one of the visual media forms, moves viewers to a new frame of mind where they are open to changing their viewpoints, or, as Gardner would say, to changing their minds. Using video data analysis tools, we not only see these video accounts as windows into worlds of meaning that we might never have been able to comprehend at the moment of the occurrence, we also connect with a fuller range of e-motions and emotions that are embedded in the event and our responses to it as we methodically explore the inferences in the data. We build configurations as a community by sharing interpretations with each other, and in this act of reconfiguration, we affect the nature of the content under examination. Our data is no longer just the video, but also the layers of commentary and descriptive measures we add in our re-viewings.

Individual and Collaborative Viewpoints in the Digital Commons

One particular event in my life as an educator, community activist, and video-grapher stands out for me when I consider and reconsider the use of technologies for making meaning in educational research practices. I want to explain to you how "particular events and unique occasions, an encounter here, and a battery of interpretations … produce a sense of how things go, have been going, and are likely to go" (Geertz, 1995, p. 3). as I wrote in my fieldnotes in 1986:

> It was a mild Vancouver spring day in 1973 when I first held a video camera in my right hand, a mike in my left, and a twenty-pound videotape-recording machine with thick straps crossing my shoulders. From the moment I first went into the field to interview people, I knew that video would transform how knowledge is constructed from the experiences of grassroots members of the public sector.
>
> What happens to me each time I use this technology? I look through the video camera's viewfinder as I frame and record the subject of my attention. Why, at an exact moment and not a moment before or after, do I start rolling the tape-recorder? Is it balance I seek or the lack thereof? Is it harmony among various elements or incongruity? Light, shadow, close up, pan, zoom, steady, moving? I am often overwhelmed with difficult questions: How will I arrange what I shoot in the editing room? How will I share my perspective on what I see in ways that made sense for others to understand what I am experiencing? Why is that important to others? Will videographers ever untangle how "the personal is the political," or, does the process of videotaping enable us to envision a socially equitable system still beyond the corner, out of focus?

My interest at that time was and still is, to understand how what we see is both affected and then reflected back to us using this medium. In a sense, we create the video images we see by focusing our attention on this angle rather than on another. I somehow knew that the portability of this technology would change how we would document events and tell stories. No longer would a story be told only through the filter of television broadcast station executives, producers, directors, and camerapersons. This portable video technology would continue to arm enthusiastic counterculture videographers to record events in an intimate, personal, and political manner. It would ensure that those with stories to tell could use public resources to produce and distribute their works—yes, even online, even though we didn't think about that in 1973.

These thoughts may not have been new for the earlier generations of photographers or for communities of intrepid documentary filmmakers and artists who shot with portable film cameras. But film was an expensive medium to use and develop. And, film was not the preferred expressive tool of a generation who wanted "change" and "wanted it now." We were eager to adopt a technology that was accessible, easy to use, and could be viewed right after it was shot. This defining period between the late 1960s and early 1970s, marked by the introduction of this portable video technology, continues to influence how researchers use video. Quite suddenly, stakeholders had the power to create their versions of what was important to them. And that is why I still share my cameras with the children I videotape. I don't want to only capture them in the frame; I want to share the power of storytelling.

Video has continued to move from being controlled by the privileged few to becoming a public medium for sharing viewpoints. It is no coincidence that digital video researchers are now focused on the nature of change whether it is used for changing our minds or for changing our research practices.

Hegemonic structures privileging the few at the top to tell the story about the lives of others has been changing, even in text-based stories and research. According to postmodern anthropologist Stephen Tyler (1986), ethnography has never been a singly authored text told by one observer, but a "cooperatively evolved text consisting of fragments of discourse intended to evoke in the minds of both reader and writer an emergent fantasy of a possible world of commonsense reality" (p. 125). Geertz was affected by postcolonial views of studying others with observational techniques and writer tools that distanced the observer from the observed. In 1973, he stated that the notion of thick description could be a conceptual tool for the interpretation of cultures. At that time, he was concerned that ethnography was not understood as a scientifically valid account. He devised a method of recognizing the validity of an account based on the thickness, richness, and many layers of the description. However, by 1985, Geertz shifted his focus to the importance of local knowledge rather than scientific generalization.

> It is from the far more difficult achievement of seeing ourselves amongst others, as a local example of the forms life has locally taken, a case among cases, a world among worlds, that the largeness of mind, without which objectivity is self-congratulations and tolerance a sham, comes. (Geertz, 1983, p. 16)

In *After the Fact,* Geertz concluded that "what we can construct, if we keep notes and survive, are hindsight accounts of the connectedness of things that seem to have happened: pieced-together patternings, after the fact" (1995, p. 2). By replacing his word "notes" for "video segments and annotations," we can infer that video records are also hindsight accounts of the connectedness of things. They are pieced-together patternings. The use of online video analysis technologies facilitates the connectedness of things that seem to have happened among the community of students, teachers, and researchers. Moreover, if students, teachers, and researchers work as a community to create video datasets and analyze them together, then there is a greater likelihood that the connectedness among things might actually have happened, even if they were experienced and are then interpreted from a variety of diverse perspectives. There may not be one truth to our video data, but there may be a way in which we can establish the veracity of our collective understanding through deep, rich, and, yes, thick interpretation that affords both individual and collaborative engagement.

Video data analysis technologies can also encourage participants to become critical epistemologists as reflection, critique, peer reviewing, and interpretation occur within a community of analysis. The content will no longer be a fixed fact to be digested, but rather an internalized creative subject of inquiry enhanced by a range of views that can be complementary or antagonistic to one's own mental framework. Surely this is how change occurs—through serious engagement with the dissonance and confluence one meets on the way toward interpretation and discovery of meaning.

Collaboratively Analyzing Video Data Sets (Galaxies)

For each of my video ethnographies in schools, I designed and used a video data analysis tool to interpret how students make sense of their own learning and thinking processes. From 1985 to 1988, I regularly visited the Hennigan School in Boston videotaping students and teachers about their experiences with Logo, a programming language designed by our research director Seymour Papert. I was one of Papert's doctoral students studying children's thinking and the growth of a computer culture. I was enthralled with Josh who asked me to give him the camera and he would videotape what really happens in the classroom when the MIT researchers are not there. I thought seriously about what he said.

Later, in the early 1990s at Bayside Middle School on Vancouver Island in British Columbia, my research team and I encouraged students to use video cameras and video software to make their own movies and conduct their own studies of an endangered rain forest called Clayoquot Sound. They digitized video chunks, made short movies, and shared their views about their movies with other students (Goldman-Segall, 1998). I conducted regular interviews over a period of 3 years unpacking the notion of thinking styles during these years and found that the theory of thinking styles was too limiting. The conclusions of my study recommended that learners have thinking attitudes—metaphysical, historical, ethical, and pedagogical—like dancers standing on one leg raised behind the body with the knee bent holding attitude for a few moments as a transition (Goldman-Segall, 1998, pp. 245–255). In my following digital video ethnographic study starting in 1999 (Goldman, 2004), my research team also shared digital video cameras with the students at Burnsview Junior High School—after a perfunctory introduction to filmmaking basics. Cameras were everywhere. To view video clips showing students as videographers, go to:

View video clip Making Local Technologies.mov

1. Go to http://www.videoresearch.org
2. If this step or any other does not work in this emerging software, please contact me at ricki@nyu.edu using Subject Header: Orion troubleshooting question.
3. After the introduction, click on the Orion button.

4. Enter username as *guest* with password *guest*.
5. Select the Burnsview Galaxy from the dropdown box.
6. Select Star Map from menu bar at top.
7. In Wandering Stars cluster, select the video clip called: *Making Local Technologies.mov*
8. If you need a galaxy of your own or need to use the full set of research features, please contact me at ricki@nyu.edu using Subject Header: Orion Galaxy, please.

In this video clip (embedded in the Orion software), students hold cameras as they are in constant motion, recording the class activity from many diverse positions. At least five cameras are rolling; each camera with different image resolution and quality, feature buttons, and tape formats. Each cameraperson has a different way of framing and understanding the length of a sequence. Each decides what to videotape, how to frame the subject, and how many camera functions—such as zooming—to apply in order to describe the event. The full dataset is, of course, a cacophony of filtered video images from diverse viewpoints.

The question is: How do we analyze video data in ways that deal with this complexity of multiple cameras and therefore the multiple viewpoints of the participants?

To understand events with complex datasets, imagine how knowledge artifacts are constructed with only one computer in a room of 35 students. My point is that none of us would think that successful individual and collaborative learning would be served best with one computer in the classroom. Yet, we sometimes believe that one video camcorder will be able to make sense of the complexity of what is happening in a classroom. In most video studies in educational research, researchers depend on one view from one researcher's camera, be it in the back of room or in the hands of a solitary researcher. The argument for this camera on a tripod at the back of the room is that it adds context (Erickson, this volume). And, when multiple cameras are used, they are not usually in the hands of learners. Rarely is the learner able to videotape and make meaning of his or her own learning experiences (see Granott, 1991, for exception). I continue to ask why we do not include multiple points of videotaping an event. I believe we will ask why we did not take better advantage of existing video analysis tools that were designed for sharing multiple perspectives of all the stakeholders.

Another important question is: Why do we not provide learners with the tools they need to view and reflect on their learning? Why do we not encourage learners to use video and become epistemologists, watching, critiquing, and reconfiguring the video of their learning as an essential part of reflecting on the art and science of their learning? Learning science researchers have applied video to enable teachers to improve their teaching practice using video cases in Teachscape™ (Pea, 1999) and student teachers use STEP for teacher professional development (Derry, 2005). Preservice teachers also use video as a way to anchor their practice (Petrosino & Koehler, this volume). Why do we not understand that learners can also benefit from viewing and constructing their views? Is it that we are still reluctant to work as a community with those we study? Is this the promise of more generic tools, such as Orion and DIVER (2003)?

A possible answer is that researchers, teachers, and students are often discouraged from collaborating. The popularization of Lev Vygotsky's theories (1962, 1978) by Jerome Bruner (1985) changed the solitary notion of achievement, at least among the research community in the learning sciences. Collaboration became an important element of distributed practice (Pea, 1994). Influenced by Latour (1987), we have, as a learning sciences community, agreed that knowledge is socially constructed and that collaboration is indeed how science flourishes. The lone recorder, like an instrument playing its simple melody, has been joined by a community of instruments capable of composing complex sounds yet unknown. Each sound is now layered, creating resonance and dissonance. Emergent clusters of sounds, like the images and sounds of video datasets, are understood best when interpreted together by all the members of the learning culture.

Software for Video Data Analysis

Recently, video is entering the digital commons due to the demand of regular users of the Internet. Instant Messaging and video e-mail have appeared on our desktops. Digital cameras—both still and video, cell phones, and PDAs have become pervasive in our daily lives. One of my favorite images of the 2004 Summer Olympics in Greece was the closing ceremony. Almost every athlete was holding a camera, capturing their perspective of this event. Layers and layers of images. Not the same. Different viewpoints. Yet connected to each other in webs of significance, as Max Weber would have said (Geertz, 1973, p. 5). Yet, there are only a handful of online functioning tools that cluster, annotate, navigate, search, and most importantly enable users to make meaning collaboratively from video data.

In my first video ethnography at the Hennigan School, I was curious about how to gain a deep understanding of the growth of a computer culture. With the camera, I observed, recorded, and reflected on the process of teachers, students, and fellow researchers. I also reflected on my own process. To aid the process, I designed digital video analysis tools called Learning Constellations[1] (Goldman-Segall, 1989). Other experimental tools for video analysis include VideoNoter (Roschelle, Trigg, & Pea, 1990), which became CVideo, a commercial tool that was used by researchers to analyze videotape data, VANNA, a tool that later morphed into Timelines (Harrison, Owens, & Baecker, 1994), and a stand-alone video annotation tool called MacShappa (Sanderson et al., 1994). Existing qualitative analysis tools now using video were first designed for text data analysis. Video analysis was added to NUD*IST, now called N Vivo, Interclipper, Atlas/TI, and Ethnograph.

In 1989, Learning Constellations became the first digital video random access interactive system with a robust video dataset specifically designed for educational re-

---

[1]Experimental video ethnographic analysis tools I have designed with my various research teams include StarNotes, 1988; Learning Constellations, 1989; Constellations, 1992; Global Forest, 1995; and Web Constellations, 1998. Web Constellations was awarded a Canada Excellence in TeleLearning award in 1998 for uploading, clustering and annotating any form of digital data on the web into "constellations" that could be analyzed. Orion™ is the latest version (see references).

search. It functioned on a HyperCard platform for Macintosh computer users (1989). The computer was connected to a videodisk player. A cable (designed by Mark Abate) enabled the computer to control the videodisk player. Using Learning Constellations, we could play, pause, and stop the video clip on the videodisk player. Users could view, annotate, link, and even segment video star chunks within star chunks. And, of course, they could cluster groups of stars into "constellations" and then play the cluster like a movie—a movie with some pauses as the player searched for the video on the video-disk player.

After several other tools, mentioned earlier, my research team and I built and tested Web Constellations (1998). Web Constellations was one of the first online tools for analyzing educational digital video data. It clustered links to any online digital data including images, Web sites, and video. Pea's latest innovation DIVER encourages users to "dive" into the data using a moving frame that captures what is in the frame and stores that data in the online database (Pea, this volume). Baecker focuses on "live Web casts that can be used for real-time and retrospective access to structured archives of presentations" (Baecker, this volume). Hay has taken another route, exploring how we annotate extremely large video data streams. Hay is also developing the video case tool for teachers to develop teacher cases and analyze their own data. As mentioned earlier, both Pea's Teachscape and Derry's STEP program (2005) are tools embedded in cognitive theoretical foundations that are used for teacher and preservice teacher learning. A tool called Traces captures video data and provides users with a pointer to record gestures (Stevens, this volume). The trace can be used for others to annotate and reflect on. Another way of viewing video data is the Web site, TalkBank, a video databank that houses a cluster of tools and community resources for researchers of human and animal behavior (McWhinney, this volume). Added to this list of contributions in the educational research is the range of tools that are just beginning to emerge in the public arena. And, in the learning sciences, Orion is now functioning as a research tool.

Orion, my most recent Web-based tool available online in 2000 using Internet Explorer or Safari browsers on either a PC or a Mac platform, provides users with the opportunity to cluster online video data, add comments, links, and transcripts, and conduct analysis of video clips. The most important feature, called a significance measure (Goldman-Segall, 1993), enables users to rate the keywords and simultaneously create charts that contain the significance measures of all the users of a given database.

Video Data Analysis Design Concepts in Orion™ Software

I now invite readers to join me on a short journey through a sample digital video dataset in Orion for guest exploration. I identify three design concepts as markers for this journey using Orion; Convivial constellations; Collaborative and Private Analysis; and Configurations for validity, community and commensurability.

Design Concept 1: Convivial Constellations. The first design concept recommends that video data should be flexible enough to be organized into clusters that are easy to access and use, and that are beneficial for the task at hand. They should be open to multiple interpretations. I call this concept *convivial constellations*. This concept emerges from Illich's definition of convivial tools (1973). Orion has been designed within the purview of this concept. For example, video stars—what we usually refer to as clips, segments, or chunks—can be clustered into constellations of video stars. These constellations can be interpreted differently depending on how the stars in the galaxy are arranged. The constellations metaphor encapsulates the patterning that emerges as video data are posted, layered, and analyzed. Not only can each individual analyst build multiple constellations, but multiple analysts can also construct constellations from a given video dataset.

To create your own galaxy, you would request a new galaxy by emailing ricki@nyu.edu <mailto:ricki@nyu.edu>, Subject Header: Orion Galaxy. Once you have received notice that your galaxy has been created, you can direct your browser to http://merlin.alt.ed.nyu.edu/orion and enter Orion to login to your galaxy. The administrative window is perhaps the most important first phase of setting up a galaxy. (See Fig. 31.1) We recommend that you first open the Administrator window to add new users and set up their privileges. You can set privileges for each user. You may decide that one user may only view stars and constellations, whereas another may use the full array of qualitative and quantitative annotation tools, including deleting stars from the database. You can add (and later delete if you prefer) as many other users to your own Galaxy as you wish.

In the same window, you may also want to set up the top level of descriptors, the topics you will use for coding your data. For example, in the Burnsview galaxy, the top-level descriptors are forms of participation; using and making; curricular considerations; contextual factors; and roles.

Returning to the navigation bar along the top, you will find the Help menu item. Here you can learn how to format and store on your hard drive selected video star chunks. Once stored, you can use Orion to upload them to the Orion server to be accessed by online communities. (We envision researchers using Orion on their own servers as soon as the software has been fully tested.)

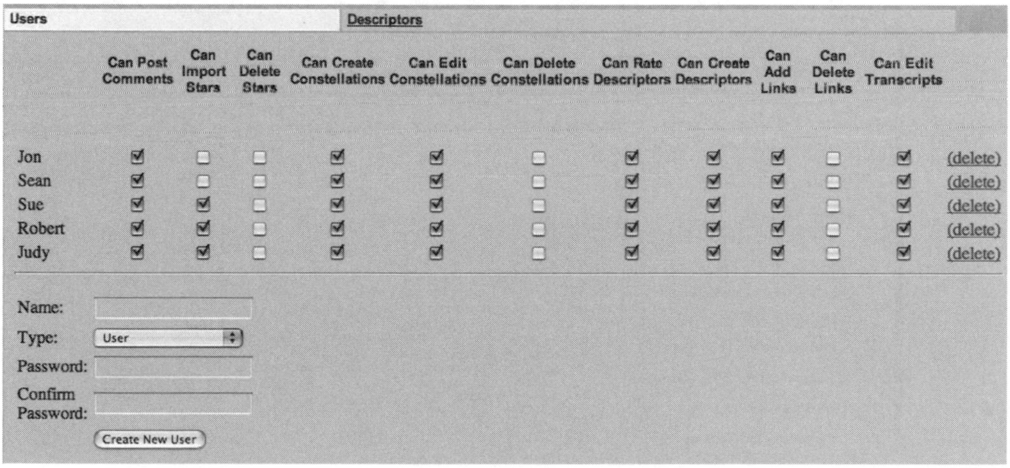

Users												Descriptors
	Can Post Comments	Can Import Stars	Can Delete Stars	Can Create Constellations	Can Edit Constellations	Can Delete Constellations	Can Rate Descriptors	Can Create Descriptors	Can Add Links	Can Delete Links	Can Edit Transcripts	
Jon	☑	☐	☐	☑	☑	☐	☑	☑	☑	☐	☑	(delete)
Sean	☑	☐	☐	☑	☑	☐	☑	☑	☑	☐	☑	(delete)
Sue	☑	☑	☐	☑	☑	☐	☑	☑	☑	☐	☑	(delete)
Robert	☑	☑	☐	☑	☑	☐	☑	☑	☑	☐	☑	(delete)
Judy	☑	☑	☐	☑	☑	☐	☑	☑	☑	☐	☑	(delete)

Name: _____
Type: [User ▼]
Password: _____
Confirm Password: _____
[Create New User]

Figure 31.1.   Administrator window: Creating users and allocating privileges.

However, for the purpose of learning more about Orion's functionality, I want you to experience an existing galaxy from a video ethnography I conduced in Burnsview middle school in Delta, British Columbia. (Simultaneously, you may also refer to an article I recently published in the *Cambridge Journal of Education,* Goldman, 2004, about the same study at Burnsview.) The video data you will meet on this journey is open to all readers to add comments, make links, and build constellations.

Login to Orion as a user in the Burnsview Galaxy once again. Select the galaxy Burnsview from the drop-down list. To view the video stars, you will want to go to the "Star Map" button on the menu bar located at the top left of the page. All uploaded video stars can be found in the "Wandering Stars" cluster. Click on a video star called Making Local Technologies.mov from the Wandering Stars cluster and the video will be downloaded on your hard drive until you exit the program. (See Fig. 31.2).

To create a constellation or cluster of video stars, go to Create Constellation item on menu bar, browse or search the existing Video Stars from the drop-down menu, and select. Notice how all the video stars appear in the Constellation window and can now be viewed from your perspective. You can code your constellation with comments, descriptors, links, and transcripts in the same way you code your video star. Easy, accessible, and beneficial for your research and perhaps, for your learning or teaching as well, as is the outcome of using the convivial constellations design concept.

Figure 31.2.   Three frames: List of stars and constellations; selected video; and analysis tools—Annotations, descriptors, links, transcripts.

Design Concept 2: Collaborative & Private Analysis.   The next design concept is ensuring that video analysis tools are designed for collaborative *or* private analysis. Both options need to be available in a video research tool. Let me continue to guide your journey:

As you have already noticed, video stars and constellations appear in the upper right-hand window. The bottom right frame has a menu with four functions. They are Comments, Descriptors, Links, and Transcript. Users can post comments, add and rate descriptors pertaining to the key themes in the specific video star, associate URL links with the star, and store transcripts or other text for the star. You can add some comments on the video you selected, transcribe and enter the transcript, add URLs, and code the data using the Descriptor function.

The most important feature in Orion is the coding of video data using descriptors. Click on the Descriptor button first under the video star window. You will see a chart with the descriptors that have been numerically coded by users that opens in your workspace below the video star. (See Fig. 31.3) Note that numerical coding is based on a 1–10 rating system. In the Burnsview Galaxy, 1 is the lowest significance to the video and 10 the highest. Now, you will want to create and add sublevel descriptors.

To add a new sublevel descriptor under your username, click on Add Descriptor. This will open a new window with the existing descriptors tree. Note that as the admin-

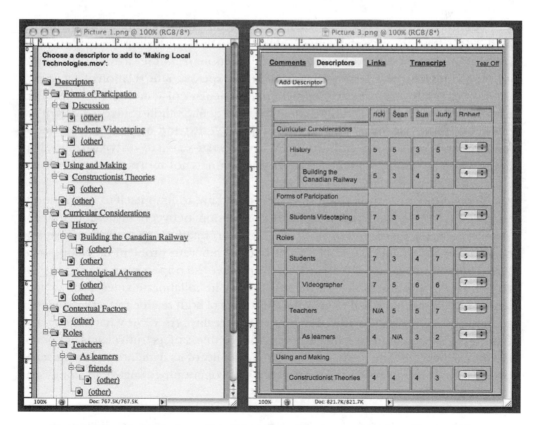

Figure 31.3.   Two windows: Tree hierarchy of descriptors; and significance rating of selected descriptions of five researchers.

istrator of this Burnsview Galaxy, I have set four top-level descriptors; forms of participation; using and making; curricular considerations; contextual factors; and roles as key areas for analysis. When you tree down the descriptors, you will find the label, *Other.* Clicking on Other lets you can add a new descriptor. As you add new subcategories to your descriptors, you will see them automatically appear in the descriptor chart in the lower frame of the main page. (We are planning to provide users with an "export to spreadsheet feature" to view data in diverse charts that can be later imported into academic papers.)

The design concept at work here is that you can create and view descriptors and associated ratings while simultaneously seeing how other members of the Galaxy coded the data, or you can decide not to have members view each other's data, thereby protecting the privacy of their coding. For example, it might be the case that you do not want users to view ratings until each user has done individual coding. (Most quantitative researchers would not want one coder to view another person's ratings.) We are currently working on a design feature that will enable the galaxy administrator to turn public coding on and off. In this way, researchers can use Orion as a private or collaborative tool. We also assume that many researchers may leave the coding open to all in some cases and closed in others, depending on what topic they are examining. After all, the concept is to provide for both private and collaborative analysis.

Design Concept 3: Configurations for Validity and Community Building.
The third design concept is that online video data analysis tools should offer the opportunity for users to build new knowledge configurations *from* the video data that build more valid and interpretive conclusions. If validity speaks to the relation of data representation to truth claims embedded within a particular epistemological stance, then validity using video most certainly changes how we define validity. Video is a very different kind of representation. Video seems to do what text and numerical data cannot. Video provides researchers with an apparently accurate visual record of what was happening when the camera was turned on. However, this is not always the case. Video can be crafted and manipulated, just like text.

The validity of a study using video does not have to limit itself to its relation to truth claims. Unpacking the traditional connection between representation and "truth" claims, the question arises: What if the interpretations and representations do not lead to truth claims but rather point out the inherent problem of making truth claims? What if, as Clifford (1986) argues, the best we can hope for is partial truths? Or, as Tyler (1986) proposes, that all we have from our collaborate video texts is "fragments of discourse intended to evoke in the minds of both reader and writer an emergent fantasy of a possible world of commonsense reality" (p. 125)? What if the validity of the video representation is based not on the fixedness of its truth claims but on the flexibility (Spiro & Jehng, 1990) and reconfigurability of its data into new representations that offer more layers of interpretations as communities engage in analysis? In 1995, I wrote that

> *Configurational validity* contributes to the expanding belief that validity in research is enriched by multiple points of view. We recognize the internal strength

(validus in Latin) of a reporting, not only by its rhetorical ability to persuade, its compelling author/ity, and its exclusive use of canon and genre, but by its ability to bend, to be resilient, and to be reconfigured into new groupings. Layers upon layers. Research also gains strength by bringing together both the discordant and the harmonious. It gains strength by providing a forum for variance and diversity. (Goldman-Segall, 1995, p. 164)

In our closing part of the journey through the basic features in Orion, you can now explore how other users clustered the video data from their perspectives. How they annotated a video star or constellation with similar or different categories, numerical codes, or commentary. Or, you can create several different constellations where you try to understand an event from different views. Or, you can create your own Galaxy of stars and constellations and upload video clips. The tool was designed as a tool to think with, an art-if-act or art-I-fact to explore meaning, intention, and inference.

CONCLUSION: WELCOME TO THE DIGITAL VIDEO COMMONS

During an event held at the media lab for our participating elementary school teacher partners in 1987, I explained that I would select several hours from the more than 100 hours of video data I had shot at the school and put digital video segments on six videodisks. These disks would be accessed and annotated using a computer running software I would design with my programmer and assistant video editor. Several years later, teachers and researchers would come into my dark red room on the third floor of the media lab to explore the Learning Constellations software and add their comments about the videotaped events. Over time, interpretations were layered and new constellations were formed. Sometimes visitors would ask: "Tell me again why anyone would want to do this kind of research?" I would answer that our research community will want to use video to study the science of learning and teaching—what we now call the learning sciences—to gain a deeper understanding of what happens at a certain time, in a certain place, when the camera is turned on and our tools offer the opportunity to build conclusions as a community. The promise of online digital video analysis tools is that learners, teachers, and researchers in the actual and the dispersed and virtual communities will construct valid accounts from a variety of points of viewing.

ACKNOWLEDGMENTS

I would like to thank John Willinsky and Anthony Petrosino for their commentaries on this chapter. My technical support team of students has demonstrated continual enthusiasm. Many thanks to Sean Orieck, Jonathan Sundy, Richard Watkins, Peter Otterstedt, and Eric Jaffe.

REFERENCES

Bruner, J. (1985). Vygotsky: A historical and conceptual perspective. In J. V. Wertsch (Ed.), *Culture, communication, and cognition: Vygotskian perspectives* (pp. 21–34). Cambridge, England: Cambridge University Press.

Derry, S. J. (2005). eSTEP as a case of theory-based web course design. In A. O'Donnell & C. Hmelo (Eds.), *Collaboration, reasoning and technology* (pp. 171–196). Mahwah, NJ: Lawrence Erlbaum Associates.

Edelson, D., Pea, R., & Gomez, L. (1996). Constructivism in the collaboratory. In B. G. Wilson (Ed.), *Constructivist learning environments: Case studies in instructional design* (pp. 151–164). Englewood Cliffs, NJ: Educational Technology Publications.

Gardner, H. (2004). *Changing minds: The art and science of changing our own and other people's minds.* Cambridge, MA: Harvard Business School Press.

Geertz, C. (1983). *Local knowledge: Further essays in interpretive ethnography.* New York: Basic Books.

Goldman, R. (2004). Video perspectivity meets wild and crazy teens: Design ethnography. *Cambridge Journal of Education, 2*(34), 147–169.

Goldman-Segall, R. (1993). Interpreting video data: Introducing a "significance measure" to layer descriptions. *Journal for Educational Multimedia and Hypermedia, 2*(3), 261–282.

Goldman-Segall, R. (1995). Configurational validity: A proposal for analyzing multimedia ethnographic narratives. *Journal for Educational Multimedia and Hypermedia, 4*(2), 163–182.

Goldman-Segall, R. (1998). *Points of viewing children's thinking: A digital ethnograper's journey.* Mahwah, NJ: Lawrence Erlbaum Associates.

Goldman-Segall, R., & Maxwell, J. W. (2002). Computers, the Internet, and new media for learning. In W. M. Reynolds & G. E. Miller (Eds.), *Handbook of psychology: Vol. 7. Educational psychology* (pp. 393–428). New York: Wiley.

Goldman-Segall, R., & Reicken, T. (1989). Thick description: A tool for designing ethnographic interactive videodisks. *SIGCHI Bulletin, 21*(2), 118–122.

Granott, N. (1991). Puzzled minds and weird creatures: Phases in the spontaneous process of knowledge construction. In I. Harel & S. Papert (Eds.), *Constructionism.* Norwood, NJ: Ablex.

Harrison, B., Owen, R., & Baecker, R. M. (1994). Timelines: An interactive system for the collection and visualization of temporal data. *Proceedings Graphics Interface '94,* 141–148.

Illich, I. (1973). *Tools for conviviality.* New York: Harper & Row.

Latour, B. (1987). *Science in action: How to follow scientists and engineers through society.* Milton Keynes: Open University Press; Cambridge: Harvard University Press.

Pea, R. D. (1994). Seeing what we build together: Distributed multimedia learning environments for transformative communications. *Journal of the Learning Sciences, 3*(3), 283–298.

Pea, R. D. (1999). New media communication forums for improving education research and practice. In E. C. Lagemann & L. S. Shulman (Eds.), *Issues in education research: Problems and possibilities* (pp. 336–370). San Francisco, CA: Jossey Bass.

Pea, R. (2003). *DIVER: Point-of-View authoring of panoramic video tours for learning, education and other purposes.* Santa Clara, CA: EOE Foundation.

Roschelle, J., Pea, R. D., & Trigg, R. (1990). VideoNoter: A tool for exploratory video analysis. Palo Alto, CA: *Institute for Research on Learning* [Tech Rep No. 90–0021].

Sanderson, P., Scott, J., Johnson, T., Mainzer, J., Watanabe, L., & James, J. (1994). MacSHAPA and the enterprise of exploratory sequential data analysis (ESDA). *International Journal of Human-Computer Studies, 41*(5), 633–681.

Spiro, R. J., & Jehng, J. C. (1990). Cognitive flexibility and hypertext: Theory and technology for the nonlinear and multidimensional traversal of complex subject matter. In D. Nix & R. J. Spiro (Eds.), *Cognition, education, and multimedia: Exploring ideas in high technology* (pp. 163–205). Hillsdale, NJ: Lawrence Erlbaum Associates.

Tyler, S. A. (1986). Postmodern ethnography: From document of the occult to occult document. In J. Clifford & G. Marcus, (Eds.), *Writing culture: The politics and poetics of ethnography* (pp. 122–140). Berkeley: University of California Press.

Wulf, W. A. (1989). The national collaboratory: A white paper. In J. Lederberg & K. Uncaphar (Eds.), *Towards a national collaboratory: Report of an invitational workshop at the Rockefeller University* (pp. 17–18). Arlington, VA: National Science Foundation.

Integrated Temporal Multimedia Data (ITMD) Research System

Kenneth E. Hay
Indiana University

Beaumie Kim
Wheeling Jesuit University

This chapter is an introduction to an innovative Internet-based model of multi-media-based research. Our Integrated Temporal Multimedia Data (ITMD) Research System incorporates new research practices and emerging digital technologies into a system that we believe will dramatically enhance the ways social scientists record and analyze complex social events and share their conclusions. The ITMD system and the new methods it allows, promises to transform social science research in four profound, interconnected ways:

- Capturing events by recording a theoretically unlimited number of temporal (i.e., time-coded) digital data streams (audio, video, computer screens, biometric data, etc.) and integrating and synchronizing these streams into flexibly blended, temporally based representations;
- Supporting a theoretically unlimited number and type of analyses with both existing and new analysis tools and blending analytical representations with temporal event representations;
- Enabling real-time and asynchronous collaboration among teams of distributed researchers in recording, analyzing, and sharing of large-scale temporal multimedia data sets; and

- Supporting the distillation of high-bandwidth, multistream multimedia data into research products that can be disseminated through Web-based, DVD/CD-based, and traditional means.

TRANSFORMATIONAL RESEARCH TECHNOLOGY

Progress in all sciences is greatly affected by tools. Bruno Latour and Steve Woolgar (1979) showed that tools are not only a source of new insights, but constitute part of the community of scientists. Most new tools comfortably fit into the normal practices of science. However, on rare but important occasions, a new technology such as the telescope changes the very nature of science and the practices of a scientific community. The computer is one such "transformative" technology. It catalyzed a new branch of science called computational science, leveraging two attributes of the computer to conduct "mathematical experiments" and to visualize findings graphically. Atmospheric scientists can now develop mathematical models of tornadoes and run billions of calculations, display the results with interactive 3-D visualizations, and share them among collaborating scientists and the general public via the Web. Increased computer efficiency and new affordances have yielded similar fundamental changes in many disciplines.

History suggests that event-recording technologies have had similar effects. The transformation in the world cultures when event-telling stories were first transcribed into written words profoundly affected our collective understanding of events. Guttenberg conquered the core problem in mass dissemination of print-based knowledge; photography provided historians visual records unbiased by the creative filters of artists; and audio recordings gave linguists new opportunities to analyze spoken language. Most recently, video camcorders have aided social scientists in the in-depth study of human events. Each of these technological developments advanced scientists' capabilities to re-examine and reconsider events, enabling increasingly deeper analysis. Like advances in computation, advances in event recording technology have begun to fundamentally reshape behavioral and social science research (Pea, 1999). The goal of the ITMD Research System is to dramatically enhance the emerging synergy between computation technology and event recording technology in order to transform social science and behavioral research.

INTEGRATED TEMPORAL MULTIMEDIA DATA (ITMD) IN SOCIAL SCIENCE RESEARCH

We stand at the threshold of a transformation in event recording technology that will significantly impact social science research. This stage is marked by dramatic increases in event recording using what we call temporal multimedia data (TMD). TMD refers to any concurrent temporal digital data collected on an event (video, audio, computer screen, pulse rates, etc.). Our ITMD system will enable researchers to record, analyze, collaborate, and disseminate findings on complex events using TMD. ITMD builds on the emergence of ubiquitous consumer-level digital video. Coupled with developments in compression technologies, high capacity storage, and network bandwidth, digital video should soon be as easy to capture, index, post, share, and integrate with other media as text on a Web page. This will have profound effects. ITMD promises to revolutionize social science research through four transformations.

1. Multiple Stream Recording. Video enables researchers to re-examine and reflect on events many times to conduct deep analysis, reducing the possibility of premature inferences and conclusions. However, video records an event from a single dominant point of view. This currently forces video-based researchers to make compromises in their methods. Some researchers study single events that artificially simplify conditions (i.e., a clinical interview or lab-based task); cognitive science has been justifiably criticized for over-reliance on such oversimplification (i.e., Lave, 1988). Other researchers examine authentic, naturally occurring events but must focus on events that have primarily only one point of view (i.e., a teacher-led classroom discussion). This reduces artificially both the complexities in the event itself (i.e., students off camera) and the kinds of events that can be studied (i.e., eliminating group-based learning). Understanding complexity seems likely to yield fundamental insights in the social sciences. Reflecting on a decade of largely unproductive video-based study of classroom practices, Ball and Lampert (1999, p. 373) implored researchers to develop "better understandings of ... complexity, not efforts to eradicate that complexity." Although multiple video and audio channels are available in tape-based research, daunting practical logistics limit their use. With our ITMD prototype, we have already demonstrated the capability to record a theoretically unlimited TMD number of event streams and view them as single threaded data representations. This can be extended to recording network-based collaborative events that are impossible for video cameras to capture.

2. Multiple Analysis. Social scientists typically use proxy (rather than primary) variables to analyze events. Proxy variables—indicators that approximate or predict a true variable or event—are used because the events themselves are considered too difficult or impractical to gather or too complex analyze (e.g., written tests of knowledge v. actual performance; recollections of stress during a treatment v. direct physical evidence of stress such as GSR). However, proxy variables inherently reflect a particular idiosyncratic methodological or disciplinary perspective; we collect, analyze and interpret proxy indicators of events, not the events themselves. By realistically capturing an event, video has the potential to overcome many of the limitations presented by proxy variables. Due to the overwhelming difficulty in manipulating and analyzing video relative to text (Reid, 1992), however, the reuse of video is rare. In addition to the 3–8 hours needed to transcribe a typical video, the source video itself is no longer integral to the data set; rather, proxy variables replace the primary video event data. Few source video clips are ever seen or analyzed again by the researcher or other members of the research community. The potential of the rich, initial video data is compromised; the utility of tens of thousands of videotapes is minimized by banishment to inaccessible archives. By mak-

ing multistream TMD as "ready at hand" as proxy variables, our single-click prototype allows researchers to collect, view, thread, and access primary TMD largely unfettered by initial biases. Source data can be indexed and "viewed" through different, complementary, or competing perspectives without compromising the source data. We have already prototyped several indexing strategies including hierarchical indexes and 3-D visualizations. Our data display interfaces can thread different proxy information together with the primary TMD in a blended representation, creating the potential to challenge or build on different analyses of the same event (i.e., displaying their's and other researchers' codes simultaneously as they watch a multistream display of the event).

3. Collaborative Analysis. Currently, researchers who digitize video usually do so after the research is concluded, in order to distribute findings. This is extremely inefficient. ITMD employs real-time digitizing, compressing, storing, and streaming that enable the creation of what Pea (1999) calls "new media communications forums," which offer immediate and pervasive synchronous collaboration between geographically dispersed researchers. At its highest level, distributed collaborative researchers could observe live events through streaming net casts. Collaborative researchers could selectively attend to any data streams (listen to selected students, look at computers, select specific cameras, etc.) of interest. In the nearer term, event recordings could be rapidly stored on media servers for immediate postevent analysis by a theoretically unlimited number of dispersed researchers. This will be possible because aggregation-level metadata (i.e., data about all TMD involved with a particular group at a particular time) will be used to dynamically construct Web interfaces that flexibly blend the data streams. Future versions of ITMD will support code development, instrument validation, collaborative video interpretation, visualization sharing, and tentative hypothesis testing. Collectively, this creates a potential for intensively iterative research across distributed research communities.

4. Research to Practice. The final transformational change concerns what the Vice President of the American Council of Learned Councils, calls a "context of disconnection" between social science research and practice (in Willinsky, 2001). This has been echoed in education by the National Research Council (1999). The groundbreaking potential of primary multimedia data to bridge this void is highlighted by several projects recently funded by the National Science Foundation (e.g., Beardsley, Cogan-Drew, & Olivera, this volume; Carraher & Nemirovsky, 2001; Miller, Brady, Stigler, & Perry, 2000; Stigler, Gallimore, & Hiebert, 2000). But even these leading-edge projects use digitized video to disseminate and publicize findings at the conclusion of their efforts. Each attempts to leverage video's compelling nature to model exemplary teaching practices. Carraher and Miller's (Miller et al., 2000) project is also pursuing "community-building tools supporting conversations in person-to-person and Web-based settings." By using primary temporal data from initial collection through ultimate dissemination, the pro-

posed system will support further transformational change in this area. ITMD will dramatically reduce the "friction" involved in creating Web-based reports, professional development material, evaluation exemplars, assessment benchmarks, and the like.

ANALYSIS OF TECHNOLOGICAL AND METHODOLOGICAL FRICTION IN SOCIAL SCIENCE RESEARCH

Friction is the central empirical research construct we use to characterize the current state of video research and the commensurate potential for ITMD. We borrow this term from information-age analysis by economist T. G. Lewis. Lewis (1997) advances the notion that the information age portends a "friction-free economy," that among other things promises zero production and distribution costs. We have taken Lewis' notion that fundamental economic principles must be rethought in the "new economy" to document the reduced friction of the proposed ITMD tools and practices.

We define friction in video-based research as the additional time it takes to do research beyond the time it takes for the actual event to occur and be reviewed. Some of this time or friction is necessarily due to the interpretive nature of social science research. However, the vast majority of friction is unnecessary and can, we believe, be eliminated with the ITMD model's technologies and methodologies. In order to analyze friction, we have identified three points on a continuum of video-based research. One end is anchored by the tape-based "entry level" system (tape-based recording, distributing and viewing, paper database, and word processing) that continues to be the mainstay of video-based research. At the center are current "best practices" such as those that researchers are developing around current, but not Web-enabled, technologies (tape-based recording, digitized CD distributing and viewing, and digital database). With real-time digital recording, compressing, chunking, and coding, and Web-based viewing and analysis, the proposed ITMD system anchors the other end of the continuum.

To analyze friction, we benchmarked a test case based on an ongoing study of science learning in computer-rich inquiry-based learning focused on small group interactions and knowledge construction. This case requires recording 30 one-hour events with two cameras and distributing copies to four remote collaboration sites. It should be noted that the coding conducted here employed a relatively straightforward chunking and coding framework. The development and refinement of codes within a framework, more subtle coding frameworks, and coding frameworks with a larger number of codes would require added postevent reflection time and was not accounted for in this friction analysis. The recordings are chunked into appropriately coded segments and 200 comparisons are made of five coded segments. Table 32.1 shows the results of our friction analysis for recording, distributing, chunking and coding, and accessing in each of the three models of research practice. Whereas the current "best practices" system reduced entry-level friction by nearly one-half, ITMD further reduces it by one full order of magnitude. Much of the reduction is in eliminating duplicating and distributing tapes or CDs, and allowing real-time chunking and coding during recording.

TABLE 32.1
Friction Analysis for One Video-Based Research Case
with Three Different Levels of Tools

	Entry-Level	*Best-Practices*	*ITMD*
RECORDING Record 30 one-hour events with two cameras and create backups. Excludes minutes to set up and make recording.	4,200	9,000	660
DISTRIBUTING Copy recordings for 5 teams and deliver to 4 other sites.	21,960	6,600	0
CHUNKING & CODING Create temporal metadata tags that identify segments sufficiently to locate clip for subsequent analysis.	13,500	10,800	1,800
ACCESSING Locate 5 clips from metadata and compare them by viewing each 2 times. Do this 200 times. Excludes viewing time.	12,800	2,940	500
TOTAL MINS. FRICTION TOTAL HOURS FRICTION	52,460 874	29,340 489	2960 49

Approach and Design Principles

We believe that the development of an effective model of technology and practice will occur among social scientists rather than computer scientists. Reflecting their unique orientation, engineers and inventors seldom realize the full potential of a new medium. Rather, it is the users—those creating, manipulating, and studying content in the new medium—who define the medium and its genres (Oren, 1990). Many important initial advances in video analysis were due to their use (e.g., Goldman-Segall, 1992, 1998) rather than their invention. Although leading computer scientists will undoubtedly have a significant impact of the range of options digital video researcher have available to them, the transformational power of ITMD follows from its support of ongoing efforts of social science researchers. The transformational power of the system has been guided by three ITMD design principles: 1) TMD Start to Finish—All data must be in time-coded digital formats from capture to reporting stage, which will enable all benefits of digital data through the entire research process; 2) Minimize Friction—All systems, interfaces, and procedures must be analyzed to minimize friction; and 3) Networked Systems—All data, displays, analysis tools, and reporting mechanisms must be Web based. We have exploited language and platform independence solutions to be as independent as possible.

Levels of ITMD Research System

Three levels of ITMD Research System include General, Internal Subsystems, and External Subsystems. Figure 32.1 depicts the flow of the internal and external subsystems' components.

ITMD Research System

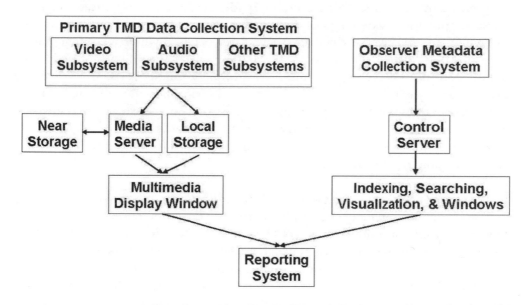

Figure 32.1.   ITMD research system flow.

General Level—Data and Metadata Standards.   We have developed a set of standards based on a thorough survey of technology and social scientists' needs. They had to be extensible, allow for independent audio, video, and capture streams, use a common time code standard that allows streams to be integrated and synchronized, and use compression techniques that capture and store streams of typical length. We have identified three levels of standards: Baseline metadata concerns context and biographical information about the participants of the event, such as those needed for secondary analysis; Aggregation metadata concerns information that links multiple streams to an event (i.e., three students work together at a computer for 3 min, where aggregation metadata links each student's audio, video, and capture streams for those 3 min). Analysis metadata concerns higher level information about the actual research assessments of the primary TMD. This includes high level information about research genre and data partitions, and the resultant codes, labels, and numeric assessments of the primary TMD. This metadata will support the original researchers' analyses, comparison of research findings between researchers, as well as verification of findings by other researchers and building knowledge across consecutive analyses by multiple researchers.

Internal Subsystem Level—Temporal Multimedia Data Collection. Collecting multiple TMD streams is logistically challenging. ITMD requires tools and practices that drastically reduce TMD collection friction (see Fig. 32.2). Unlike their predecessors, the range of commercially available digital VCRs, camcorders, and audio recorders digitize and compress in real time using downloadable standard formats. The data collection system in our prototype system can simultaneously capture and compress 12 tracks of video and 24 tracks of audio in real time. A significant challenge will be supporting a wide range of these heterogeneous hardware and software systems while meeting diverse social science research needs. An interview might only require an MPEG digital camcorder, whereas researchers recording more complex, computer-supported events need multistream multiformat collections capacity. We have explored the utility of wireless cameras and microphones and portable digital audio recorders for high-movement research environments. A key fiction dimension concerns the merits of controlling and coordinating all data streams through one centralized computer, compared to reliance on decentralized media recording elements such as portable digital audio recorders.

Internal Subsystem Level—Metadata Collection.   In tape-based research, labels and shelving can meet organizational needs. ITMD requires effective collection of metadata about the multiple primary TMD streams. Our prototype system has demonstrated the collection of aggregation and analysis-level metadata (see Fig. 32.3). We will provide facilities for the computer-based generation of some metadata, but at a routine level (e.g., random sampling of events). Although researchers (e.g., Jones & Jones, 2004; Zhao, Chellappa, Phillips, & Rosenfeld, 2003) are making great strides in

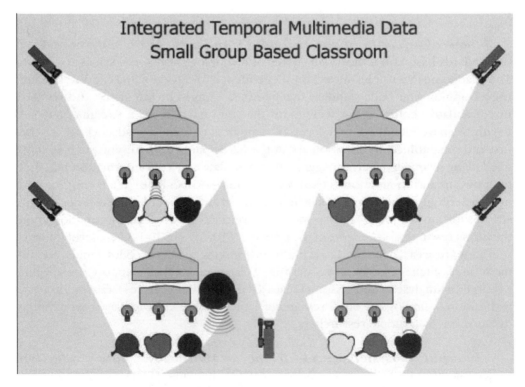

Figure 32.2.   Illustration of TMD collection set-up.

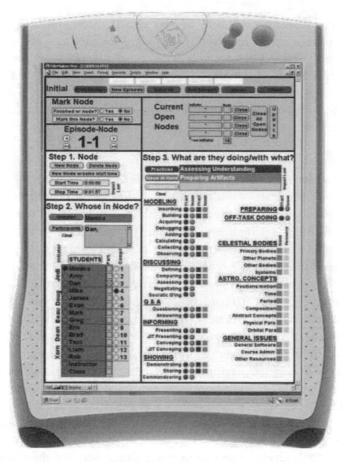

Figure 32.3. Observer metadata collection system prototype.

computer-based interpretation of video or still images, such techniques are not yet sophisticated enough to recognize event dynamics of interest to social scientists or they only provide a first pass analysis that helps eliminating video that is not of interest. We will focus our efforts on enhancing human interpretation efficiency.

For Baseline metadata (i.e,. date, teacher, school, students, etc.) and Aggregation metadata (i.e., Students 1, 5, & 7 worked together at computer 3 from 09:45:33 to 09:46:55), we have prototyped a touch-screen tablet computer running an SQL database enabling one-touch, on-site, real-time metadata collection. Our Web-based TMD display tools will then enable researchers to do postevent metadata coding or refinements of real-time coding. For analysis metadata (i.e., the theoretical codes, labels, numerical assessments, etc.), we have developed tools for researchers to work with on both an on-site tablet computer and/or postevent Web-based TMD display tools. On-site recording requires a sufficiently developed metadata framework (i.e., where coding categories have been refined through piloting and intercoder reliability and are sufficiently small in number to realistically enable real-time coding). More intensive analyses will be conducted with Web-based TMD displays. Whereas specific baseline and aggregation metadata collection will be standardized across research projects, analysis metadata collection needs to be specified for each project. Current analysis software seriously hampers researchers in this regard (Stanley & Temple, 1995), forcing them to "subvert" software structure to meet their needs (Fielding & Lee, 1998). We have developed a suite of tools with a significant range of units of analysis, analytical techniques, and epistemological options to meet the specific demands of different research perspectives and a general technology framework for readily sharing that data.

Internal Subsystem Level—TMD and Metadata Storage.   ITMD research will generate huge amounts of multimedia data that, until recently, would have overwhelmed storage capacity. Storage costs are plummeting at 60% annually (see Pea & Hoffert, this volume for a detailed discussion), while compression standards like MPEG-2 and RealMedia have slashed storage and bandwidth requirements. Nonetheless, architectures and strategies that support TMD's unique storage and distribution requirements are keys to ITMD's success. We have developed a model with distributed TMD servers and a central ITMD control server that minimizes network traffic and creates a natural researcher portal. The distributed servers reside on the users' LANs and store all the TMD for a project or set of projects, whereas the control server houses the metadata files. The control server also functions as an application service provider, serving tools for indexing, displaying, visualizing, searching, collaborating, and reporting findings. This approach is supported by empirical studies of researchers showing "clusters of users" with similar approaches and supporting technologies (Fielding & Lee, 1998). We have also developed software what will enable the researcher to point either to video stored on the media server or video stored on a computer's local hard drive where the TMD access in almost instantaneous. Because the researchers' work, the metadata, is stored on the control sever, it can later be pointed back to the media server.

Internal Subsystem Level—Display Interface for TMD and Analysis Metadata.   Displaying and controlling multiple coordinated, integrated, and synchronized streams of TMD is technically challenging. Both stand-alone (i.e., MPEG-4) and Web-based standards (SMIL) have been explored and are reviewed by Pea and Hoffert (this volume); we have developed and tested a SMIL prototype (see Fig. 32.4). SMIL files are small because they are used to coordinate temporal and display elements of a multistream Web page. They can control all streams via a VCR-styled controller, and can present analysis metadata in a temporal fashion within a rolling text window that is treated as another temporal data stream.

We hope to greatly extend our initial success and preliminary research in our prototype system and explore these and other emergent media integration standards and languages to meet researcher display needs. Although SMIL creates an ideal environment for multistream TMD display, data display needs will vary widely. Analysis of one student working on a computer requires synchronizing a relatively small video window of the student and a large window of the computer screen with a single audio stream. In contrast, secondary analysis of three scientists using networked computers

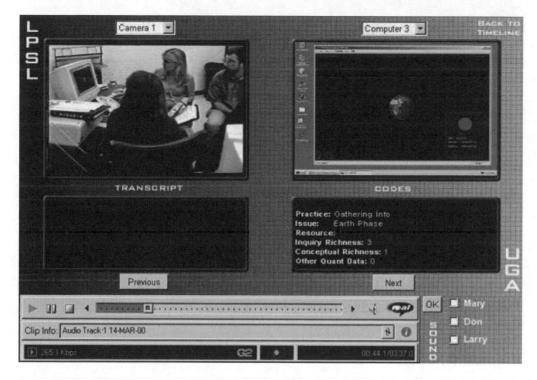

Figure 32.4.   TMD and metadata display interface prototype.

with one of the three in a remote location would require three displays of screen capture streams, two video streams, three audio streams, and a text display of the previous analysis of the event. ITMD requires a flexible display creation interface, akin to database management software's "forms" creation facilities that allows researcher to customize their particular format of their display.

   Internal Subsystem Level—Data Indexing, Visualization, and Analysis.
Interpreting and analyzing TMD requires huge data sets to be organized conceptually. We have prototyped a set of tools that integrate general needs for sharing with the specific needs of researchers. Web-based, hyperlinked indexing facilities will use the aggregation metadata to organize an event into major subevents with one-click access to TMD display. This structure will be used for viewing TMD and will be integrated into a larger depository system for sharing. Analysis metadata related to the hierarchies within a particular analysis can be integrated into this structure, giving researchers ready access to their unique data partitioning scheme.

   In the hard sciences, 3-D data visualization gives meaning to huge volumes of primary data. ITMD does the same thing-and more. By hyperlinking the 3-D elements back to the primary TMD, researchers that identify patterns in the visualization can click on elements in that pattern and immediately access the underlying primary TMD, giving added meaning to the visualization. We call this *event intimacy*. Our prototype uses VRML to visualize Aggregation and Analysis metadata and allows 3-D objects to be hot-linked to our SMIL-Data Display interfaces, keeping the primary TMD close at hand (see Fig. 32.5). This overcomes pervasive concern among social science event researchers about "distance from the data" (Richards, 1998). Event intimacy in our prototype is further enhanced by powerful navigational techniques for exploring the data space, visually organizing the data, and exploring patterns in the data in powerful ways. As explained by pioneers Keller and Keller (1993), data visualization "consists of exploration, analysis, and then presentation," and therefore based on the researcher's particular needs. Our goal is most certainly not developing all the possible ways to visualize such data, but rather developing a set of foundational specifications and operationalizing those specifications in ITMD.

Data analysis will come through relatively unconstrained exploration of 3-D visualization patterns and through powerful SQL search engines that can locate patterns using hierarchical indexing and 3-D search strategies. Our prototype uses a Web-based Boolean search engine to find event nodes in the index structure and visualization. Once located, events in the display are highlighted so patterns can be found and verified in other data. Simplified navigation of 3-D space in our prototype enables the exploration of event content.

External Subsystem Level—Collaboration

ITMD supports immediate, persistent, and enduring research collaboration among communities of scientists centered on primary TMD. To support this type of

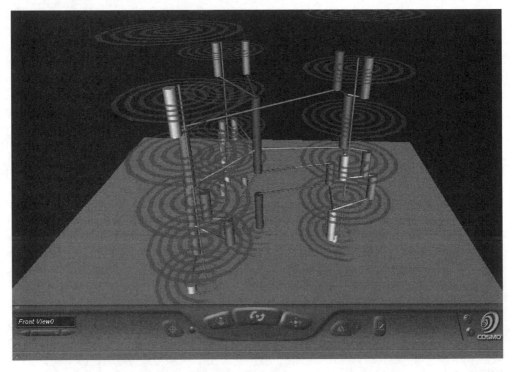

Figure 32.5.   Metadata visualization window.

collaboration, we cannot realistically produce a different interface with unique functionality for each type of user. However, we can create malleable collaboration spaces that users can readily adapt. Researcher can build and use data presentation templates that enable them to determine how many streams of video to display and their positions and sizes in the Web page. These templates can be shared among research teams. Our distributed TMD storage strategy enables researchers to work in low bandwidth situations and can provide an added level of security. The ability to e-mail collaborators a node/clip has been developed and used in a limited way. However, due to the drive of our immediate research needs, we have not developed broader collaborative support within ITMD. Some of these features have been developed in a related effort. In a partnership with the college board, we have developed a system to support teachers building video cases of their own teaching and utilizing several collaborative features. These include attaching discussion forums to video data, a system of accessing rights and invitations, and sharing data between teachers.

External Subsystem Level—Distribution of Research Products. A goal of ITMD is connecting research TMD to communities that derive value from that research. Minimizing the gap between "info-rich" and "info-poor" with a "distributed information space" (Logoze & Fielding, 1998) should narrow the infamous schism between social science researchers and their constituents. Much of this is accomplished by ITMD's inherently Web-based nature that simplifies distribution of findings and illustrations of practices (see Fig. 32.6). But we also need further collaboration constituent communities using the toolkits available to researchers. Eventual success hinges on adequately supporting a variety of users from a variety of disciplines. Within educational research, this variety was well articulated in guiding principles defined by (Willinsky, 2001). Peer or secondary researchers will interact with the full database through XML-based search facilities either to collaborate on the analysis or to perform further analyses; peer research community is less interested in seeing all the data, but wants to review it to search for confirming or conflicting evidence; publications-based community needs exemplary TMD available to the readers with varied access needs (Web, CD, or print); policymakers and stakeholders are farthest from the original researchers and mostly want summary data or lived experience of the event that is compact—a *60 Minutes*-style report.

Several tools exist to support digital video dissemination. Beardsley, Cogan-Drew, and Olivera (this volume) are advancing "video papers" that combine text, images, and video. To date, this work has not addressed fundamental usability and user interface issues. Video supports a wide range of researchers, constituent needs, and tasks (e.g., see the range of video interfaces described in Davenport, 1998). A single type distribution mechanism of annotated video presentation is insufficient for the range of user needs. We are currently developing these mechanisms in ITMD.

General Level—Ethical, Privacy, and Human Subjects Issues. Ethical treatment and privacy is of paramount concern to the social scientists. Web-based video and collaborative research poses challenges in terms of standards of privacy and anonymity. This includes the exemplified definition of "levels of possible harm" that

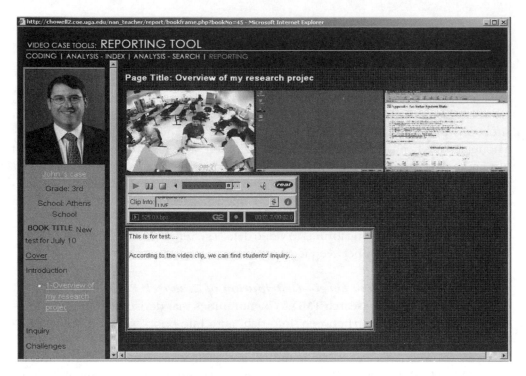

Figure 32.6.   ITMD Research reporting tool prototype.

would be matched to standards for informed consent, and consideration of online multimedia informed-consent tools. In the future, there needs to be community involvement in developing a publication that would allow ITMD research and similar work to begin from the highest ethical grounds.

CONCLUSION

ITMD aims to reconfigure the "exchange economy" of scholarly production, but this will not be accomplished solely by exploring Internet distribution strategies or making incremental "improvements" in fundamentally flawed approaches. We challenge the enterprise itself to provide tools that not only simplify but transform social science research, development, and dissemination processes. We need to provide different products that address diverse requirements, not just to distribute traditional scholarly products more widely. This cannot be achieved by simply retrofitting the products of traditional research methods; we must fundamentally change how "research is conceived, conducted, authored, and critically responded to by it audience" (Pea, 1999). ITMD has the potential to cause social scientists to reassess existing methods and approaches and to reexamine core assumptions about the nature of data and the goals of data analysis.

Just as computational science has shaken the hard sciences to the very core, ITMD will push social science researchers to consider "at a more fundamental level about what it means to do research" (Eisner, 1997, p. 5). ITMD will enable researchers to capture and analyze events through multistream event capturing. Because this represents the actual event rather than a proxy or approximation, multiple analyses can be made without some current, but certainly not all, initial confounding bias. Collaboration among primary researcher teams, researchers conducting secondary analyses, and consumers of research can be immediate and pervasive throughout research projects. The products of research will be used to bridge the chasm from research to practice.

The ITMD past and future is ambitious but we as a field are up to the task. We have developed a viable proof of concept and are prepared to address the field's goal to develop new knowledge, models, concepts, and design principles that incorporate an understanding of social and technical systems that will impact social change and transformation.

ACKNOWLEDGMENTS

This material is based upon work supported by the National Science Foundation under grant 0101941. Kenneth E. Hay is an Associate Professor in the Learning Sciences Program of the Department of Counseling and Educational Psychology at Indiana University. Beaumie Kim is an Assistant Professor, Learning Sciences Lab/Learning Sciences & Technologies, National Institute of Education, Nanyang Technological University. Correspondence regarding this article can be sent to Kenneth E. Hay, Department of Counseling and Educational Psychology, School of Education, Indiana University, 201 N. Rose Ave., Bloomington, IN 47405-1006. E-mail addresses for the authors are kehay@indiana.edu ; mail to:kehay@indiana.edu ; and beaumie.kim@nie.edu.sg

REFERENCES

Ball, D. L., & Lampert, M. (1999). Multiples of evidence, time, and perspective: Revising the study of teaching and learning. In E. C. Lagemann & L. S. Shulman (Eds.), *Issues in education research: Problems and possibilities* (pp. 371–398). San Francisco, CA: Jossey-Bass.

Carraher, D., & Nemirovsky, R. (2001). *Bridging research and practice.* Retrieved October 1, 2005, from http://brp.terc.edu/

Davenport, G. (1998). Curious learning, cultural bias, and the learning curve. *IEEE Multimedia, 5*(2), 14–19.

Eisner, E. (1997). The promises and perils of alternative forms of data representation. *Educational Researcher, 26*(6), 4–10.

Fielding, N. G., & Lee, R. M. (1998). *Computer analysis and qualitative research.* London: Sage.

Goldman-Segall, R. (1992). Collaborative virtual communities: Using learning constellations, a multimedia ethnographic research tool. In E. Barrett (Ed.), *Sociomedia: Multimedia, hypermedia, and the social construction of knowledge* (pp. 257–296). Cambridge, MA: MIT Press.

Goldman-Segall, R. (1998). *Points of viewing children's thinking:A a digital ethnographer's journey.* Mahwah, NJ: Lawrence Erlbaum Associates.

Jones, V., & Jones, M. (2004). Robust real-time face detection. *International Journal of Computer Vision, 57*(2), 137–154.

Keller, P., & Keller, M. (1993). *Visual cues: Practical data visualization.* Los Alamitos, CA: IEEE Press.

Latour, B., & Woolgar, S. (1979). *Laboratory life: The social construction of scientific facts*. Los Angeles, CA: Sage.

Lave, J. (1988). *Cognition in practice*. Cambridge, England: Cambridge University Press.

Lewis, T. G. (1997). *The friction-free economy: Marketing strategies for a wired world*. New York: Harper Business.

Logoze, C., & Fielding, D. (1998). Defining collections in distributed digital libraries. *D-Lib Magazine, 4*(11). Retrieved October 1, 2005, from http://www.dlib.org/

Miller, K. F., Brady, D. J., Stigler, J. W., & Perry, M. (2000). *Representing and learning from classroom processes* (NSF Grant REC-0089293). Champaign, IL: University of Illinois.

National Research Council. (1999). *Improving student learning: A strategic plan for educational research and its utilization*. Washington, DC: National Academy Press.

Oren, T. I. (1990). A paradigm for artificial intelligence in software engineering. In T. I. Oren (Ed.), *Advances in artificial intelligence in software engineering* (Vol. 1, pp. 1–55). Greenwich: JAI Press.

Pea, R. (1999). New media communications forums for improving education research and practice. In E. C. Lagemann & L. S. Shulman (Eds.), *Issues in education research: Problems and possibilities* (pp. 336–370). San Francisco, CA: Jossey-Bass.

Reid, A. O. J. (1992). Computer management strategies for text data. In D. F. Crabtree & W. L. Miller (Eds.), *Doing qualitative research* (Vol. 3, pp. 125–145). London: Sage.

Richards, L. (1998). *Nud*Ist introductory handbook*. Melbourne, Australia: Qualitative Solutions and Research Pty.

Stanley, L., & Temple, B. (1995). Doing the business? Evaluating software packages to aid the analysis of qualitative data sets. In R. G. Burgess (Ed.), *Computing and qualitative research* (Vol. 5, pp. 169–193). Greenwich, CT: JAI Press.

Stigler, J. W., Gallimore, R., & Hiebert, J. (2000). Using video surveys to compare classrooms and teaching across cultures: Examples and lessons from the TIMMS video studies. *Educational Psychologist, 35*(2), 87–100.

Willinsky, J. (2001). The strategic educational research program and the public value of research. *Educational Researcher, 30*(1), 5–14.

Zhao, W., Chellappa, R., Phillips, P. J., & Rosenfeld, A. (2003, December). Face recognition: A literature survey. *ACM Computing Surveys (CSUR), 35*(4), 399–458.

33

A Transcript-Video Database for Collaborative Commentary in the Learning Sciences

Brian MacWhinney
CMU

Researchers in the learning sciences have acquired great sophistication in the use of video recordings as ways of understanding instructional interactions. Conference and lecture presentations on instructional processes are now typically grounded on the analysis of authentic video material. For example, at the 2002 meeting of the American Educational Research Association (AERA), there were 44 scientific panels and symposia that relied on analysis of classroom video. Video is also used in teacher training programs (Derry, in press; Pea, 1999) and materials illustrating proposed nationwide educational standards (Daro, Hampton, & Reznick, 2004). The field also enjoys a great range of high-quality tools for the analysis of video interactions. Systems such as N-Vivo (www.qrsinternational.com), DIVER, TransAna (www.transana.org), ATLAS.ti (www.atlasti.com), Elan (www.mpi.nl/tools/elan.html), MacShapa (Sanderson & Fisher, 1994), CLAN (childes.psy.cmu.edu), VideoNoter/C-Video (Roschelle, Pea, & Trigg, 1990), Ethnograph (www.qualisresearch.com), Anvil (www.dfki.de/~kipp), Orion, aka Constellations (<http://merlin.alt.ed.nyu.edu>; Goldman, this volume; Goldman-Segall & Riecken, 1989), ePresence (Baecker, Fono, & Wolf, this volume), Informedia (Wactlar, Christel, Gong, & Hauptmann, 1999), and VideoPaper (Beardsley, Cogan-Drew, & Olivero, this volume) are allowing researchers to produce large quantities of well-analyzed video interactions.

Despite the high quality of video analysis methodology, the large quantity of data being produced, and the centrality of video to the scientific study of learning and instruction, there has not yet been a community-wide acceptance of the importance of a shared database of instructional interactions. There has been extensive discussion of the formation of collaboratories for the study of instructional interactions (Baecker, Fono, & Wolf, this volume; Edelson, Pea, & Gomez, 1996). However, without a general method for sharing data across projects, collaboratories are limited to data sets collected from single projects (Abowd, Harvel, & Brotherton, 2000). However, many of the most interesting questions in learning and instruction involve comparison between alternative teaching frameworks and situations. This type of diversity in the database can best be achieved by having data from many different laboratories and groups channeled into a uniform, but distributed database.

Development of a shared database is a crucial next step in the maturation of the learning sciences as a scientific discipline. In genetics, projects such as the Human Genome Project (www.ornl.gov/hgmis), GenMapp (www.genmapp.org), or Protein Map (Aisenman & Berman, 2000) are now storing all published genetic sequences in forms that are open to analysis and data mining through the Web. In fact, gene sequences are not accepted for publication until they have been entered in these systems. In paleontology, museums worldwide preserve fossils whose specific physical structure, radiological dating, and stratificational location are crucial to our reconstruction of the history of life and of the earth. Electronic records and scans based on this evidence are now being made available electronically (www.ucmp.berkeley.edu/pdn/) for deeper analysis and data mining. Internet databases are now fundamental to progress in astronomy (van Buren, Curtis, Nichols, & Brundage, 1995), physics (Caspar et al., 1998), economics, medicine, history, political science, experimental psychology, linguistics, and other sciences.

To address this need, the TalkBank Project has begun an effort to construct a shared database for the learning sciences. TalkBank (http://talkbank.org) is an international collaborative effort that has been building a Web-accessible database for spoken language interactions. All of the video and audio media in TalkBank are fully transcribed and each transcribed utterance is linked directly to the corresponding segment of the media. The media and transcripts can be downloaded from the Web. Users can also open a browser window, scroll through transcripts, play back the corresponding audio or video, and insert commentary regarding their analyses. The current TalkBank database has large collections of data in the areas of child language (CHILDES), aphasia (AphasiaBank), second language learning (SLABank), bilingualism (LIDES), formal meetings, and spontaneous conversational interactions (CABank and MOVIN).

A shared database for the learning sciences will have some interesting features unique to this area. It will be important to develop a taxonomy of educationally relevant activities, events, and interaction types that can serve as metadata for coding and retrieval. It will also be important to supplement video records with additional ethnographic materials such as diaries, notebooks, drawings, and class records. However, the most powerful feature of a shared database in the learning sciences will certainly be its availability to collaborative commentary. The idea of scientific collaboratories has been developed and discussed elsewhere in this volume. With the context of collaboratories, projects such as Orion (Goldman, this volume), DIVER (Pea, in press), and WebCast (Baecker, Fono, & Wolf, this volume) have shown how a group of educational researchers can work together to analyze interactions and evaluate competing interpretations. However, for the process of collaborative commentary to work as a general model for the learning sciences, it must be linked to a commitment to the process of data sharing. What is unique about the TalkBank Project is not its emphasis on collaborative commentary, but rather its emphasis on data sharing. However, the greatest value for scientific progress arises when data sharing is joined with collaborative commentary.

DATA SHARING

The goal of TalkBank is to support data sharing and direct, communitywide access to naturalistic recordings and transcripts of human and animal communication. The concept emerged from two ongoing initiatives that had already proven important to their respective user communities. The first is the Linguistic Data Consortium (http://ldc.upenn.edu) that has published some 288 large corpora over the past decade. The second is the CHILDES system (http://childes.psy.cmu.edu) that has constructed a database of 150 corpora of parent–child interactions in 20 languages. The data-sharing model for the new TalkBank project is based on the model from the CHILDES project (MacWhinney, 2000).

Having reviewed best practice in 12 very different research areas studying communicative interaction, Talkbank has identified these seven shared needs:

1. Guidelines for ethical sharing of data.
2. Metadata and infrastructure for identifying available data.
3. Common, well-specified formats for text, audio and video
4. Tools for time aligned transcription and annotation.
5. A common interchange format for annotations.
6. Network based infrastructure to support efficient (real time) collaboration.
7. Dissemination of shared data, tools, standards, and best practices to the research community.

DATA ON INSTRUCTIONAL INTERACTIONS

Materials in our initial collection of video studies of instructional interactions include:

1. Studies of problem-based learning in medical school education from Tim Koschmann and Curtis LeBaron (Koschmann, 1999).
2. Discussions of the interpretations of graphs from a special issue of the *Journal of the Learning Sciences* (2002), edited by Anna Sfard and Kay McClain.
3. Recording of students engaged in group problem solving in algebra from James Greeno and Carla Van Sande.

4. Materials from the TIMMS study (Stigler, Gallimore, & Hiebert, 2000) comparing math and science instruction in Australia, Czechoslovakia, Hong Kong, Japan, and the Netherlands.
5. College lectures on research methods in psychology from Brian MacWhinney.
6. A lecture on map reading to a sixth-grade class contributed by Wolf-Michael Roth.
7. A discussion of a unit on camels in a fifth-grade class contributed by Rosalind Horowitz.
8. A comparison of classroom, business, and meeting contexts contributed by Reed Stevens.
9. Teachers' discussion of the basis of gravity from Beth Warren.
10. Science museum visit materials from Jrene Rahm and Kevin Crowley.
11. Dyadic tutorial sessions on the f-ratio in the analysis of variance from Carl Frederiksen.
12. Dyadic tutorial sessions on psychology research methods from Natalie Person and Arthur Graesser.
13. Dyadic tutorial sessions on how to play a video game from Nikolinka Collier.
14. Videos of seventh-grade children in Dresden, Germany learning English, French, and Czech from Angelika Kubanek-German.

In addition, the related CHILDES (Child Language Data Exchange System) database (http://childes.psy.cmu.edu) contains several major video studies of more informal learning in the home in English, German, Japanese, Spanish, Cantonese, and Thai.

In an early attempt to promote collaborative commentary, TalkBank promoted the creation of a CD-ROM for a special issue in the *Journal of the Learning Sciences* (Sfard & McClain, 2002). This CD-ROM contains articles commenting on two lessons on graphs in a seventh-grade classroom. The PDF files for these articles contain links that replay the relevant video. In addition, there is a demonstration transcript that serves as a compendium of commentary on particular analyses.

We also hope to provide streaming video access to three large longitudinal classroom corpora. One corpus, from Carolyn Maher and her associates at Rutgers, contains 3,000 hours recorded over a span of 12 years tracking the math learning of a group of 15 students. Another, from Rich Lehrer and Carmen Curtis, records a year's worth of integrative geometry lessons from a third-grade classroom. The third, from Juliet Langman, compares alternative formats for bilingual classrooms.

Related Corpora

The TalkBank database contains a wide range of materials outside of this particular initial corpus for learning contexts. Among the major datasets in these other areas are:

1. CallFriend phone conversations in English, Spanish, and Japanese, provided by the LDC and transcribed by TalkBank.

2. The Santa Barbara Corpus of Spoken American English (SBCSAE), including conversations between college friends, lectures, and meetings.
3. Recorded phone calls from the Nixon Whitehouse, transcribed by Gail Jefferson.
4. European political television programs and other components of the MOVIN database from Johannes Wagner.
5. Informal interview materials from a special issue of the *Journal of Communication*.
6. A complete collection of all the oral arguments of the last 30 years of the Supreme Court of the United States (SCOTUS). This enormous dataset is currently being formatted into the TalkBank schema and linked to the digitized media.
7. Over 12 corpora from second language learners and bilinguals.
8. A new database of 12 corpora from aphasic speakers.

Ethics for Data-Sharing

Public sharing of data over the Web brings with it a variety of challenges regarding participant rights and professional ethics. These issues have been an ongoing topic of discussion within the TalkBank communities. The current result of this process is a set of ethical and practical guidelines adopted for all TalkBank data sets, described at http://talkbank.org/share and available for use beyond the TalkBank project. The centerpiece of this approach is the idea that participants can opt to provide releases for the use of their data at any one of eight different levels. The lowest level of protection allows for full Web access to transcripts and video with no attempt at anonymity. Higher levels of protection add anonymity in transcripts and media, password protection, and finally, no access but only archiving for the future. The choice of an appropriate level for a given data set is decided first by the human subjects review process at each institution and then by the participants themselves. In addition, TalkBank discourages any use of the data that is critical of the performance or motives of individuals recorded in the interactions. Groups that require further privacy and respect considerations include indigenous groups, speakers of endangered languages, clinical subjects, subjects in psychiatric treatment, and classroom teachers.

Infrastructure for Annotation

Researchers working with video records from learning contexts need support for a wide range of transcription, editing, and analysis functions. The TalkBank project has supported the development of a variety of annotation tools, including the AG ToolKit (agtk.sourceforge.net; Maeda, Bird, Ma, & Lee, 2002), TransAna (www.transana.org), and Transcriber (www.ldc.upenn.edu/mirror/Transcriber/). However, the annotation tool that is most fully compatible with other TalkBank programs and that provides best integration with the database is CLAN (childes.psy.cmu.edu/clan). CLAN allows the transcriber to link directly to audio and video material, to rewind the material, to use CA transcription, and to analyze files using a variety of search programs. By inserting metadata fields in CLAN, the contents of all TalkBank files can be published over the Web in conformity with the standards of the OLAC (Open Language Archives Community) project (www.linguistlist.org/olac).

Interchange Formats for Data Annotation

The greatest challenge facing TalkBank has been the need to bring hundreds of corpora created in diverse ways into conformity with a common standard. The first step in this direction involved the specification of a proper XML schema for the CHAT transcription system. The system involves three major steps. In the first step, the ANTLR parser generator creates a parse tree that is converted to a JAXB tree that is then serialized into XML. JAXB is Sun's data binding framework that generates Java code for specialized DOM construction, validation, and serialization. In the second step, XSLT outputs CHAT. In the third step, a modified version of Unix DIFF compares the original CHAT with the converted CHAT and reports differences for correction. Once a corpus passes through this process with no errors, it is included in TalkBank. CHAT versions are zipped so that users can download complete data sets and the XML versions are shipped to the server (http://xml.talkbank.org) to support online transcript and media browsing.

Browsable Transcripts

Once the transcripts are in the TalkBank XML format, they can easily be rendered as HTML pages. The Java WebStart program called TBViewer then allows users to view transcripts over the Web and to play the underlying audio or video by clicking on utterances. After clicking on the first utterance, the user can decide to either stop playback or allow the viewer to play back the whole transcript by highlighting each utterance as it is played. The transcript can be scrolled and paged to allow for full control of playback over the Web. The playback application is installed by clicking on a link at the TalkBank page that then runs Java WebStart.

The playback facility relies on the use of hinted video and audio from Apple's QuickTime Streaming Server. For audio-only corpora, the QuickTime window displays the utterance and plays the sound without any video.

Collaborative Commentary

We can define collaborative commentary as the process by which a research community engages in the interpretive annotation of electronic records. The goal of this process is the evaluation of competing theoretical claims. To achieve clear connections between data and theory, commentators need to link their comments and other evidentiary materials to specific segments of either transcripts or electronic media. Now that researchers have access to browsable corpora representing various learning contexts, we can begin to think about how to implement the process of collaborative commentary.

To illustrate the goals of this process, let me use some of my own explorations through the database as the example. I was interested in exploring evidence for the neo-Vygotskyan claim (Nelson, 1998) that word meanings are shaped through communicative interactions. While browsing through online media at the CHILDES childes.psy.cmu.edu site, I located several instances of videos of mother–child book reading in the McMillan and Rollins corpora. In these interactions, mothers help children turn the pages and name the animals or objects in the pictures. In some cases, children call the pictures by the wrong name. I wanted to see whether mothers would use these errors as opportunities to provide corrective positive feedback (MacWhinney, 2005b). For example, if the child calls a bear a "doggie," the mother should respond, "no, that's a bear, not a doggie." Then, the child might engage in self-correction by saying "a bear."

To initiate the process of collaborative commentary, I then write up a short summary of my analysis. In reality, I have not yet conducted this step, although an initial attempt can be found in MacWhinney (2005a). I would then want to make this analysis available in three ways. First, I would like to post my claim to some discipline-based commentary space on the Web. Second, I would want to make sure that others who view the relevant segments from the Julie and Rollins corpora could see that I have provided detailed interpretive commentary regarding certain specific segments. Finally, I hope to receive feedback from other researchers regarding my interpretations and arguments.

To further illustrate how this process can work, let us consider a second example from the area of mathematical learning. In this fictive example, Harriet Keck is a developmental psychologist specializing in children's concepts of number. She and her colleague Robin Clark are both interested in understanding how children solve problems such as $3 + 4 = ?$ Keck believes that children solve the problem in an internal mental model and then read out the solution to their fingers. Clark believes that children use their fingers to form external representations of the addends and then count their fingers visually. Keck's model predicts that children will count directly across the fingers, whereas Clark's model suggests that children will begin with placing one addend on each hand separately.

To explore this issue, Dr. Keck uses metadata search tools to find video cases in TalkBank format involving "4-year-old children AND counting." Exploring these videos using the TalkBank viewer, she finds that 70% of their gestures support her theory, whereas only 30% are in line with Clark's account.

Over the next several weeks, Keck and her colleagues use CLAN to link each case of finger counting to comments that also point to a brief report summarizing her conclusions. Not surprisingly, Clark disagrees with Keck's conclusions and responds by reinterpreting the same video cases that Keck has just analyzed. His analysis points to several counterexamples that do not fit Keck's theory. He also argues for including trials that have no overt finger counting in the denominator. Keck, in turn, responds to Clark's criticism by asserting that the gestures he has coded are inadvertent hand movements and revises her paper to anticipate his objection. Keck and students submit their revised paper, including the video data, to the online edition of cognitive development. One of the reviewers has a question on whether the authors have properly categorized a set of gestures from one of the videos. Keck responds with a close analysis of the gesture in question using a fine-grained analysis of the actual hand movements. The reviewer is convinced by her response and the paper is published with links to the video data and analysis.

Although this scenario may seem a bit futuristic, it is not very different conceptually from forms of collaborative commentary we have already produced. One example is a special issue of the *Journal of the Learning Sciences* (Sfard & McClain, 2002) that focuses on learning about graphs and numerical distributions in a seventh-grade classroom. The difference is that in this new framework, analyses will be directly linked to the data, rather than hidden within PDFs. Moreover, in this new framework, analyses will be directly accessible from browsers.

Infrastructure for Collaborative Commentary

The TBViewer is now being configured to support this process. In the browser window, pencil icons next to utterances represent the commentary field. When the user clicks on this field, a separate commentary window opens up. This commentary window allows the researcher to create the following elements:

1. A brief summary of the claim or analysis relevant to the current utterance or utterance sequence.
2. Typing of the claims and analysis into specific categories.
3. Explanations of the evidentiary role of the texts and media being referenced.
4. Links to other texts or claims that are relevant to the current claim.
5. Links to external Web content, including material (HTML, PDF, Word) that presents the proposed analysis more fully.
6. Embedded HTML code.

Once this material has been entered into the commentary field, it can be redisplayed through the TBViewer facility. The QuickTime window echoes the comments that have been entered in the commentary database as streaming playback progresses. If a given segment has more attached commentary than can be displayed in the reserved segment of the QuickTime window, the window will have a final line listing the number and size of the comments that cannot be displayed.

Naked Media

Because TalkBank transcripts are subject to ongoing modifications, reference to line numbers is not stable. A more reliable method links commentary to time points in the media. The idea of linking commentary to media is also in accord with the theoretical emphasis in the annotation graph framework of Bird and Liberman (2001). Thinking of a database in this way also opens up a more general possibility for multimedia databases that we will refer to as "naked media." Consider the case of a large database of classroom video data contributed to TalkBank by Rich Lehrer from the geometry lessons of Carmen Curtis. This database consists of 200 hours of classroom video with no accompanying transcripts. It would take perhaps a full year to transcribe all of these sessions. On the other hand, the video can be prepared for streaming Web access in about a month. Once the naked video is posted on the Web, it can be a target for collaborative commentary. In cases of this type, collaborative commentary can operate effectively even without accompanying transcripts.

CONCLUSION

The TalkBank project has provided openly accessible databases for the study of spoken language interactions. We are now implementing support for collaborative commentary targeted to these databases. Construction of these new methods will open up many exciting new lines of investigation for each of the several disciplines studying human communication.

ACKNOWLEDGMENTS

The discussion of the process of collaborative commentary and the example of the dialog between Keck and Clark was developed in collaboration with Roy Pea. Prabhu Raghunathan has programmed the TBViewer. Leonid Spektor has programmed CLAN. Franklin Chen has programmed XML support for TalkBank. Kelley Sacco has developed and maintained the TalkBank database.

From 1999 to 2005, TalkBank received support from NSF Grant BCS-9978056 to Carnegie Mellon University. TalkBank senior investigators included Steven Bird, Ken Dauber, Jerry Goldman, Mark Liberman, Brian MacWhinney, Roy Pea, and Howard Wactlar.

REFERENCES

Abowd, G. D., Harvel, L. D., & Brotherton, J. A. (2000). Building a digital library of captured educational experiences. *Kyoto International Conference on Digital Libraries*, 395–402.

Aisenman, R. A., & Berman, R. (2000). Rethinking lexical analysis. In M. Aparici, N. Argerich, J. Perera, E. Rosado, & L. Tolchinsky (Eds.), *Developing literacy across genres, modalities, and language* (Vol. 3, pp. 187–196). Barcelona: University of Barcelona.

Bird, S., & Liberman, M. (2001). A formal framework for linguistic annotation. *Speech Communication, 33*, 23–60.

Caspar, T. A., Meyer, W. M., Moiler, J. M., Henline, P., Keither, K., McHarg, B., et al. (1998). Collaboratory operation in magnetic fusion research. *ACM Interactions, 5*, 56–64.

Daro, P., Hampton, S., & Reznick, L. (2004). *Speaking and listening for preschool through 3rd grade: Book discussion and narratives*. Pittsburgh: LRDC and the National Center for Education and the Economy.

Derry, S. (in press). STEP as a case of theory-based Web course design. In A. O'Donnell & C. Hmelo (Eds.), *Collaboration, reasoning, and technology*. Mahwah, NJ: Lawrence Erlbaum Associates.

Edelson, D., Pea, R., & Gomez, L. (1996). Constructivism in the collaboratory. In B. G. Wilson (Ed.), *Constructivist learning environments: Case studies in instructional design* (pp. 151–164). Englewood Cliffs, NJ: Educational Technology Publications.

Goldman-Segall, R., & Reicken, T. (1989). Thick description: A tool for designing ethnographic interactive videodisks. *SIGCHI Bulletin, 21*, 118–122.

Koschmann, T. (1999). Special issue: Meaning making. *Discourse Processes, 27*(2), 98–167.

MacWhinney, B. (2000). *The CHILDES project: Tools for analyzing talk*. Mahwah, NJ: Lawrence Erlbaum Associates.

MacWhinney, B. (2005a). Can our experiments illuminate reality? In L. Gershkoff-Stowe & D. Rakison (Eds.), *Building object categories in developmental time* (pp. 301–308). Mahwah, NJ: Lawrence Erlbaum Associates.

MacWhinney, B. (2005b). Item-based constructions and the logical problem. *ACL 2005*, 46–54.

Maeda, K., Bird, S., Ma, X., & Lee, H. (2002). *Creating annotation tools with the annotation graph toolkit*. Paper presented at the Proceedings of the Third International Conference on Language Resources and Evaluation, European Language Resources Association, Paris France.

Nelson, K. (1998). *Language in cognitive development: The emergence of the mediated mind*. New York: Cambridge University Press.

Pea, R. (in press). Video-as-data and digital video manipulation techniques for transforming learning sciences research, education and other cultural practices. In J. Weiss, J. Nolan, & P. Trionas (Eds.), *International handbook of virtual learning environments*. Amsterdam: Kluwer.

Pea, R. D. (1999). New media communication forums for improving education research and practice. In E. C. Lagemann & L. S. Shulman (Eds.), *Issues in education research: Problems and possibilities* (pp. 336–370). San Francisco, CA: Jossey Bass.

Roschelle, J., Pea, R. D., & Trigg, R. (1990). *VideoNoter: A tool for exploratory video analysis* (Technical). Palo Alto, CA: Institute for Research on Learning.

Sanderson, P. M., & Fisher, C. (1994). Exploratory sequential data analysis: Foundations. *Human-Computer Interaction, 9*, 251–317.

Sfard, A., & McClain, K. (2002). Analyzing tools: Perspective on the role of designed artifacts in mathematics learning [Special issue]. *Journal of the Learning Sciences, 11*, 153–388.

Stigler, J., Gallimore, R., & Hiebert, J. (2000). Using video surveys to compare classrooms and teaching across cultures: Examples and lessons from the TIMSS video studies. *Educational Psychologist, 35*(2), 87–100.

van Buren, D., Curtis, P., Nichols, D. A., & Brundage, M. (1995). The AstroVR collaboratory: An on-line multi-user environment for research in astrophysics. In H. E. Payne & J. J. E. Hayes (Eds.), *Astronomical data analysis software and systems IV, ASP conference series*. Astronomical Society of the Pacific.

Wactlar, H. D., Christel, M. G., Gong, Y., & Hauptmann, A. G. (1999, February). Lessons learned from building a terabyte digital video library. *IEEE Computer*, 66–73.

Capturing Ideas in Digital Things: A New Twist on the Old Problem of Inert Knowledge

Reed Stevens
University of Washington

This chapter proposes a new solution to an old problem. Whitehead's famous formulation of the problem of inert knowledge in his 1929 collection, *The Aims of Education*, is well known, but Whitehead actually developed the idea quite a bit earlier in a talk to British mathematicians and mathematics educators in 1916. The concern hung around for Whitehead for at least 13 years and it is still hanging around today.

> In training a child to activity of thought, above all things we must beware of what I will call "inert ideas"—that is to say, ideas that are merely received into the mind without being utilized, or tested, or thrown into fresh combinations... Except at rare intervals of intellectual ferment, education in the past has been radically infected with inert ideas. (Whitehead, 1929, p. 1)

In a less scholastic but still trenchant analysis, Father Guido Sarducci explored one implication of the inert knowledge problem when he proposed a new educational venture, "the five minute university."

> I find that education, I think it donta matter where you go to school, Italy, America, Brazil, it's alla the same, it's all just memorization. It don't matter how long you can remember anything, just so you can parrot it back for the test. And I gotta this idea for a school I would like to start, something called the five minute

---

[1]My thanks to colleagues Ed Lazowska and John Bransford for sharing Sarducci's routine with me.

university. And the idea is that in five minutes, you learn what the average college graduate remembers five years after he or she is outofa school ....[1] (*Gilda Live,* Warner Studios, 1993)

For those of us who are making a career of trying to create substantive change in educational practice, this is certainly gallows humor. Sarducci's view echoes some stocktaking by Jan Hawkins who, shortly before her untimely death, named two primary disappointments for the field that has come to be known as the learning sciences (Hawkins, 1997). The first disappointment related to what Hawkins saw, in 1997, as an unmet goal of creating conditions for many more students in school to be thinkers, interpreters, makers, and builders: in other words, an alternative image to students as receptacles of inert ideas. Her second disappointment was that educational technology had not met its transformative promise, having typically done little more than to play out surface variations on the old knowledge transmission theme.

Why, we may ask, are schools so frequently factories of inert ideas? There are many possible explanations and many of these are simplistic like deficit models of students and teachers. A more complex analysis looks to the social organizational conditions under which education is conducted. As sociologist Howard Becker once noted, it may be the very structural organization of schooling that produces its failure (Becker, 1972). In other words, this particular type of failure—that schools are inert idea factories—may be built into the very infrastructure of schools as they typically are organized. How could that be? As historians of education remind us, schools were set up under a transmission or factory model of learning and over decades, a complex set of interlocking practices and devices have formed an infrastructure that supports that model. In that model, teachers transmit, students repetitively practice, and then face decontextualized knowledge display rituals. The typical outcome makes the five minutes university worthy of serious consideration.

A NEW TWIST ON AN OLD PROBLEM: THE VIDEO TRACES DIGITAL ANNOTATION MEDIUM FOR CAPTURING AND CIRCULATING IDEAS

This chapter argues that if the learning sciences are to make a lasting contribution to educating people in a substantially new way, then we need to build collectively a socio-technical infrastructure for learning that is quite different from the one that was assembled to fit the factory model of education. This is no easy task, because infrastructures are often invisible to those that rely on them and are hard to change piecemeal (Becker, 1995; Bowker & Star, 2000). In addition, there are powerful political forces now working tirelessly to marginalize, if not extinguish, educational practices that treat the having, making, revising, and sharing of ideas a centerpiece of education. By all measures, there seems a clear goal on the part of these forces to turn the clock back to the factory model of education. These challenges are the reason I stress the collective nature of building an alternative infrastructure. This chapter reports on one possible element of such an infrastructure—the *Video Traces* digital annotation medium. Video Traces is a medium for people to represent their ideas in a pretty natural way and to circulate them to others.

The images of learning and knowledge work that Video Traces is designed to support involve helping people become active, critical makers, revisers, and users—of ideas, tools, objects, images, and all the other categories of nonhumans that we humans care about. To do so, we need to make it straightforward and natural for learners to generate ideas, to critically reflect on them, to revise them, and to get feedback on these ideas from many others. We need to be able to put different ideas about common objects into interaction (cf. Goldman, this volume; Pea, this volume).

Ideas, for the purposes of this chapter, are not treated primarily as mental objects. This does not, of course, discredit the view that ideas also take mental form, and for the long-term systematic goal of understanding and supporting learning and cognitive activity, we need to understand the coordination of all "media" in which ideas are embodied (Hutchins, 1995). However, as a corrective to the largely mentalist conception of ideas that is the legacy of cognitivism, my use of the term in this chapter refers to ideas given material form—in voice or matter. In my view, it is little more than feckless reductionism to say that a thought I have with my eyes closed and mouth shut is more of an "idea" than a drawing I make and hand across the table to a colleague. At least with regard to learning from and with others (and most likely also ourselves), the kinds of ideas that need more of our attention are those that people embody and share in public, material ways (at least until we discover that learning from others takes place through mental telepathy). From here on, my use of "ideas" should be understood in this public, embodied sense.

Like the problem of inert ideas, the images of learning that I am affirming here are not so new; Dewey substantially articulated them almost a century ago in opposition to transmission and factory models of education. And these ideas remain current under various banners, including project-based learning and constructionism (Barron et. al., 1998; Harel & Papert, 1991; Kafai & Resnick, 1996; Stevens, 2000a). What is proposed as potentially new in this chapter is a representational medium called *Video Traces* (Stevens & Hall, 1997) that fits the image of the alternative infrastructure for education I have in mind. Video Traces, in short, is a digital annotation medium with which people capture and circulate ideas in a particular digital form.

In what follows, I describe some of the key conceptual ideas embodied in the Video Traces medium, drawing on examples of its use, between 2001 and 2005, in a variety of communities. The chapter is more suggestive than exhaustive about the uses of Video Traces, seeking to give the reader a sense of the range of uses to which people have and can put the medium to use. I try to show how the design of the medium addresses some of the nagging issues I see within the learning sciences, including and beyond the inert ideas problem. With some irony, it is worth noting that this medium (text) does not lend itself particularly well to representing a medium (Video Traces) that was designed explicitly to overcome some of the limitations of static text as a representational medium for learning, teaching, and exchanging ideas. The chapter is therefore best understood as an invitation to readers to explore the Video Traces medium itself in light of the arguments and descriptions presented here. Finally, the chapter continues to work through my evolving perspective on what has become an important goal for learning sciences generally—to describe how research on learning and the design of learning technologies inform each other (Stevens, 2000a, 2000b, 2006).

The basic digital object that users create and circulate with the Video Traces medium is called a *trace*. The word "trace" was chosen to represent both something re-

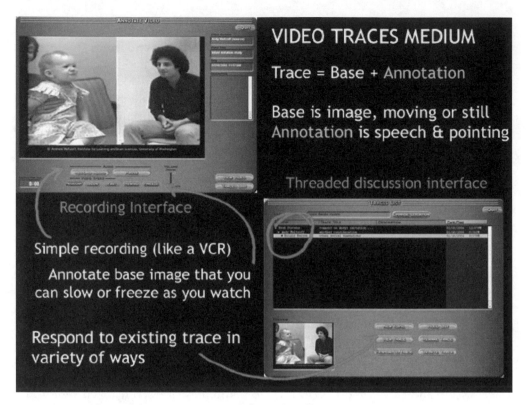

Figure 34.1.   An overview representation of Video Traces.

maining from a prior experience and to acknowledge something lost in transformation. The basic conceptual formula by which traces are composed can be expressed as
base + annotation = trace. The base is an image, either still or moving. New base images can be directly recorded into the medium with a standard video camera attached
by a cable to the computer or images can be imported from other sources. Once a trace
is placed in the threaded discussion interface, other users can respond to it or begin a
new thread. Responses can take two basic forms, (1) making a new base and a new annotation, or (2) making a new annotation over the same base in the trace being responded to.

Annotations are composed of both speech and pointing gestures that are recorded together and layered over the underlying image as it is viewed. Users can also
draw on still images, either when the base is a still image or when a frame of a video is
frozen on during annotation recording. The basic layered form of a trace involves temporally coordinated speech and pointing that references a visible scene or object (the
base). The form of a trace is built directly from insights from interaction analysis researchers who have found that these coordinated embodied resources enable people
to work and learn together across a wide range of settings (e.g., Goodwin, 1994, 1996,
2000, 2003a, 2003b; Stevens & Hall, 1998). Pointing is especially critical to face-to-face
interaction, supporting what linguists call *deixis* (Hanks, 1996). Words like "here,"

"this," "that," and "there"—which are extremely natural and common in everyday speech—make sense in use, because pointing orients a hearer/viewer to a particular referent (Kita, 2003). In the Traces medium, pointing gestures (in the shape of a pointing finger) are recorded as the mouse-controlled cursor is moved over an image during recording.[2] In general, the design of the medium has been guided by the goal of translating features of face-to-face interaction into a representational medium as culturally transparent as possible (Lave & Wenger, 1991; Wenger, 1990). Part of designing for cultural transparency has been to use now-familiar interface conventions from common cultural artifacts like VCRs (cf. Stevens & Hall, 1997).

As a representational medium, Video Traces materializes a conversational metaphor. "Turns" (traces) are composed of coordinated speech (annotation) and visible scenes or objects (base). These turns can be linked to prior ones to constitute a threaded "conversation" and users can point to common objects (i.e., common base images). As objects, traces are primarily anchored by people's actual voices,[3] so the prosodic properties of speech are available as resources for interpreting the visual scenes being annotated. Users report also that the presence of other people's actual voices enhance the sense of personal connection they experience using Video Traces.

There are important differences between conversations in the Traces medium and face-to-face interaction; these differences are both unavoidable and by design. Turns in Video Traces are sequenced asynchronously rather than synchronously as in face-to-face interaction. Unavoidably, joint reference involving two or more sets of hands and eyes "on" the same unfolding scene or object at the same moment is not possible, but reference to *common objects*—in the form of the same visual base object—across turns is nonetheless the norm in trace conversations. Also, the asynchronous sequencing of turns in Video Traces does not allow for the immediate "next turn repair" (Sacks, Schegloff, & Jefferson, 1974) recognized to be so critical to joint sense making in face-to-face interaction (Clark & Brennan, 1991). Yet, although the Traces medium loses some important features from face-to-face interaction, other affordances are gained.

By design, Traces is an asynchronous medium, and asynchrony has its virtues. In particular, it affords conversations to be distributed in time and across space. It affords "turns" to be composed with deliberation while retaining the natural expressive modalities of looking, speaking, and pointing. Another important difference between trace conversations and face conversations is that traces are preserved durably rather than lost as ephemera in a passing moment. This feature makes these "conversations" more like the kinds that readers have with texts when they inscribe in the margins of books and articles, and more like the ways that designers inscribe draw-

---

[2]Various studies of human–computer interaction have shown that pointing is so integral a part of human activity that computer users typically repurpose the mouse and cursor to point. We took advantage of this finding in our design.

[3]Attention to "voice" has gained prominence in education and discourse studies over many years, but the term is usually used metaphorically as when produced texts are interpreted as displaying a writer's voice. In the Traces, medium, the human voice itself anchors the medium.

ings and other images.[4] A trace therefore is a durable object that is both immutable and mobile in the sense of these terms meant in Latour's analysis of scientific representations (Latour, 1990). Latour argued that representations with these properties made a decisive difference for the accumulation of collective knowledge in science. As I argue later in the chapter, traces might play the role of "immutable mobiles" for education just as graphs and similar representations have played for science.

Another designed feature of the Traces medium that distinguishes it from ordinary conversation is that users can alter the time structure or visual focus of the underlying image prior to or during annotation. Users can "slow" and "freeze" (when images are moving) or "zoom" and "scroll" (when images are still). These features of the medium complement the pointing function in allowing trace makers to guide attention to specific aspects of the visual image by re-presenting the image on a different time scale or from a different perspective than is possible in real time (cf. Goodwin, 1994; Stevens & Hall, 1997).[5] Importantly, these features that alter the time structure of the base image can be manipulated in the real-time production of the trace, giving the user significant flexibility over how the annotation is layered over the base image. For instance, in the example trace made by the rowing coach (Fig. 34.2), she allows the base video to move at regular speed through certain segments, slows it at others, and freezes it at still others while continuing to narrate the action. A final means of guiding attention is one that, although not designed, has emerged in relation to the time-structure altering features. Often, when a user is about to alter the time structure or visual focus of the base image, she will comment on the action (e.g., "now I am going to freeze it so you can see …"), a practice that prospectively orients the trace viewer to the shift and thereby to signal the sequentially imminent relevance of what follows.[6] Taken together, these features enhance a trace maker's ability to *discipline the perception* (Stevens & Hall, 1998) of her or his future recipients.

I refer to traces as idea objects because they carry ideas, bits of everyday knowledge and inquiry in Dewey's sense of the term. Dewey used the term *inquiry* to refer to a line of investigation initiated by a felt problematic (cf. Dewey, 1934; Stevens & Hall, 1997).[7] A trace can as easily be about how to make turkey gravy[7] as about how to interpret a bubble chamber diagram. At a certain basic level of coordinated speech and action, many "ordinary" and disciplined forms of apprenticeship and knowledge work are achieved interactionally through the same basic resources of speaking, looking,

---

[4]Architects and engineers call this "red-lining," and it is a common annotation practice in both these fields (Stevens, 2000b).

[5]When a base video image includes relevant audio, users can move sliders to control the respective volumes of the base video and the annotations.

[6]These speech practices perform a similar function within trace conversations that affective loading and prospective indexicals do in face-to-face conversations (Goodwin, 1996).

[7]A couple years ago, a woman I sat next to on an airplane ride asked me about my work. I described *Video Traces* and she said, "I know what I would use it for" and proceeded to explain she would have sent a trace of herself making turkey gravy to her daughter, who had just made her first Thanksgiving dinner for her family. She explained that her daughter had called just before dinner with questions about how to do this and that she had trouble explaining it and wished she "had been there to show her."

Figure 34.2.   A still from a trace made by a university rowing coach for one of her beginning rowers. The directional line follows the recording path of the pointing gesture tracking along the left arm and back to the oar. These actions were recorded in temporal synchrony with sequential parts of her verbal recording: "The main problem here is that your hands are really low and your blade is really high above the water here."

and pointing (Stevens & Hall, 1998), a similarity that is the basis for the root technical form of traces as media objects. A trace is made when users layer speech and pointing gestures over embodied action or visual images.

Uses of the Medium

Just as everyday knowledge comes in many forms, so also do traces that people make. Traces are good for capturing, among other things, interpretations, instructed actions, noticings, questions, and assessments. The medium has been used by people in a wide variety of situations and settings, including coaches and athletes, choreographers and dancers, conducting professors and their students, nurses and patients, and future teachers and the network of people who are responsible for their induction into the teaching profession.

The following nine vignettes describe a number of ways that people have made use of Video Traces for their own learning, teaching, and knowledge construction work. They are drawn from a range of communities that used the Traces medium between 2001 and 2005.[8]

> A teacher education student annotates a video of a student's problem-solving activity with whom she is working during her first field placement in an elementary school. She was puzzled by something the student was consistently doing, so she recorded his activity and made a trace. She used this trace to solicit feedback on this activity, framed by her specific question, from her field mentor and a university teacher education faculty member. Both made response traces that led her to try something new with the student, the outcome of which she captured in another trace. The conversation continued.

> A graduate student in a university's Instrumental Conducting Program studies a video of a recent conducting performance with the wind ensemble. He is drawn to a particular musical response that has been elicited by a problematic conducting gesture. Through the use of the drawing function of Video Traces, the student is able to represent a potential alternative for the ineffective gesture. The supervising professor then reviews the trace and makes minor modifications.

> A college student in a critical film studies course annotates a clip from a Hollywood film to present her analysis of an underlying theme in the film. Her interpretation details the way camera edits and perspective reveal this theme. She shares this trace with her instructor who uses it as a performance assessment of her ability to achieve her own interpretation through the critical techniques explored in the course. The trace is used as a bridging representation into the writing of an academic paper.

> An elderly person who has just had surgery demonstrates a prescribed exercise and points out, through speech and gesture, what parts of the exercise she does not understand. A nurse responds to the trace by annotating the same video of the woman doing the exercise, answering her question.

> An expert in sustainable building practices takes a group of architecture students to a Native American community to build new housing and to teach students a specific type of sustainable building practice known as straw bale construction. During the day, part of the professor's expert labor is video-recorded. Later he annotates the footage with *Video Traces,* explaining key features of what he is doing and why. The trace goes into an archive, with many possible uses, including helping students who can't participate at the field site learn about this unique building practice and forming an archive for future indigenous builders through a seemingly more natural representational medium (Video Traces) than technical architectural drawings.

---

[8] Because *Video Traces* is a medium that has been taken up in a number of communities beyond my resources for studying them all, many of the examples described here, although real, cannot be represented visually because they were not collected by me or with human subjects approval to share them for publication. The example that is displayed in detail later, from dance composition, was collected with human subjects approval that allows published reproduction.

In a science museum, a boy and his father record their interaction with an interactive exhibit. They then record a conversation, expressing different thoughts about "what's going on" and pose an experiment. Another visitor who visits the display can now respond to their trace in a variety of ways, including doing the experiment, offering a theoretical viewpoint, and trying to replicate the "data" from the father and son's prior trace.

An interaction analyst annotates a videotaped experimental interaction between a developmental psychologist and a baby. The scholar of interaction looks for and sees different things in the video than does the developmental psychologist. This forms the basis for further grounded conversation across the two scholars' disciplinary perspectives and is partially responsible for the conception of a new interdisciplinary research study drawing on both traditions.

A graduate student in a university choreography program annotates an early performance of one of her dances. Through the annotation, she sees things that she wants to change and uses the speech and the pointing tool to simulate moving the dancers around in space. This sets her up to make substantive revisions to the dance. She also shares the trace with her choreography professor, who makes a trace over the performance and suggests further revisions.

Video Traces is set up in the corner of a gallery in a contemporary art museum filled with famous modern paintings. Visitors make traces over digitized images of paintings that hang in the gallery. They offer their interpretations of the paint-

Figure 34.3.   Video still from trace made by choreography professor (part 1).

ings to which other visitors and experts can respond, thus creating a distributed conversation. As with the traces made in science museums, these traces also offer a glimpse into how visitors think and respond to exhibits, information much desired by museum educators and exhibit designers (Stevens & Toro-Martell, 2003).

The following example shows the temporal organization of a single trace (excerpted) to give readers a more moment-to-moment sense of how the medium is used. (As I explained earlier, the representational medium of static text and image is a somewhat strained one for displaying objects made in Video Traces). Among the things this trace shows are how an instructor coordinates speech with the moving image, alters the time structure of the base video to highlight particular features, and uses the pointing tool to indexically ground speech to action all in service of providing a detailed interpretation and critique of a student's work. This trace was made by a university dance professor, responding to a composition by one of her students.

I like the way this starts.

(3 sec)

It's a good set up of the bodies in space.

(5 sec)

You also establish a nice relationship to the music.

(12 sec)

Figure 34.4.   Video still from trace made by choreography professor (part 2).

Figure 34.5. Video still from trace made by choreography professor (part 3).

I think that's a good choice, having them facing different directions.

And then here I wonder if you could, I'm going to freeze it for a second,

for that, those little hops forward, well they are actually hops sideways with some people are coming down and some people going directly side—, if you could have those hops a little bit longer or a little bit more, you could do those, that one hop two hops then quickly hop, hop, hop so that we can see that that you are kind of pulling apart this little fivesome, just so it registers that there are different facings. ok, I'm going to start.

(2 sec)

I love that wiggle.

I could do a lot more of the wiggle and just pull it out of the context.

(7 sec)

I'm not sure what that run weave was, like do you know what that is at that moment, how that relates to those gestures, because we start relating movement to those gestures because its, you've established a pretty clear message. All of this movement stuff from that woman that weaves to now to here, I think you could repeat a lot of the material so we see each portion of this much longer.

This excerpt of a trace made by a choreography professor displays a number of common features of how the Video Traces medium is used. An analysis of the temporal

Figure 34.6.   Video still from trace made by choreography professor (part 4).

structure of the trace shows that the instructor has organized the temporal relations between base and annotation in two different ways. One way involves commenting on a just completed subsegment of the composition. For example, when she says, "I think that's a good choice, having them facing different directions," she has just watched and was silent for 12 seconds before referring to a particular "that" that she evaluates as a "good choice." At other times, base and annotation are related in the more common way they are related among the broader population of users—she comments directly on the unfolding action, a practice that is afforded by prior viewings of the base video.

Another common feature of use represented in this trace is displayed by the abundance of deictic terms like "this," "that," "here," and "there." As described earlier, deictic terms and the indexical grounding they enable are supported by two features of the medium: the temporal coordination of speech over base images and the actual pointing or drawing on base images that is recorded with cursor movement. In this trace, the trace maker has "pointed" with her speech but not the explicit pointing tool, when she identifies a temporal segment of the performance with a synchronized phrase "to now to here." More typically, users take advantage of both the temporal coordination of speech with the represented action in the base image and explicit pointing or drawing.

Traces are made by users that are both evaluative and more strictly interpretative. When traces are made to evaluate as often the case when conversations transpire between students and teachers, they often have the quality exemplified by the dance instructor in being evaluative but constructively so; both speech and gestural action are used to move from evaluation to the projection of possible alternatives (e.g., "I could do a lot more of the wiggle and just pull it out of the context" and "you could repeat a lot of the material"). The traces medium seems to support this sort of constructive evaluation and consideration of alternatives, perhaps because the medium allows one to move from a basic evaluation to the embodiment of an alternative with one's gesturing or drawing hand and a co-occuring verbalized narrative.

DISCUSSION

To this point in the chapter, I have described in detail how findings and concepts from microethnographic research, otherwise known as research on situated practice, have informed the design and use of Video Traces. This articulates the relationship between design and research in one direction, describing a familiar directionality from "basic" research to design. One of the reasons that this is important to articulate, at least in this case, is to combat the often overheard view that research in the situated practice tradition has little to contribute to informing actual design. The view seems to stand on the belief that documenting specific practices in detail prohibits informing a design that is useful to more than the specific persons or community who were studied. Video Traces is a strong counterexample to this unexamined view in that this medium has been useful to a range of communities, almost all beyond those studied in prior microethnographic research. Combining a comparative perspective involving studying practice across a range of settings with the situated, microethnographic approach to specific settings has allowed for a very robust generalization about key features of face-to-face interaction that we have, in turn, embodied in our design.

Another issue I want to take up in this discussion also involves the relationship between design and research, but runs the question in a somewhat different direction. Having described how basic research on interaction shaped the features of Video Traces, I want to now turn to some of the core conceptual issues and challenges in the learning sciences and describe how uses and potential uses of Video Traces may be responsive to these issues and challenges.

Ideas—Things Just in the Mind or Also in the World?

One of the achievements of the cognitive revolution was to return to education a concern for ideas, entities that had, in a prior academic generation, been all but banished by behaviorists. However, like most intellectual movements that have the run of the table for too long, cognitivism (like behaviorism before it) has had its debilitating effects. Chief among these effects has been the monocular tendency to see ideas in almost exclusively mental terms—as concepts, categories, schemata, and the like. These mentalistic conceptions leave little room for understanding and taking advantage of the material character of ideas as they are embodied in activities and things. In this light, traces may be seen as hybrid idea objects that smudge the hard boundary between object and mind in how they layer ideas, concepts, and abstractions onto concrete images and events.

The materiality of representations matters (Hall, 1990). We have found that trace-making resembles everyday representation in face-to-face interaction in many ways but differs in consequential ways, because of their specific material form. Traces differ because they are durable, portable,[9] and replayable. Traces also retain an indexical relation between the object (the base image or video) and descriptions and inscriptions of this object; this indexical link is lost with the typical "occasioned" representation like hand-drawn maps (Psathas, 1979). The durability of traces supports distributed, asynchronous conversation. This is a feature of considerable potential usefulness for those learners for whom expertise is hard to locate in one's proximal environment.

One of the emergent features that the durability of traces affords is the possibility of an *incidental archive*. As users make traces for current purposes, the traces are retained and can be re-purposed and reorganized for later, different uses (Stevens, 2002a). For example, the rowing coach's traces that were made to provide specific guidance to her current group of beginning rowers can be taken as a collection to provide a succinct initial representation of teaching expertise in rowing. Taken as a set, the traces she made for these specific rowers provide a coherent representation of her disciplined perception that might be of use to a beginning rowing coach. This use would be possible as an incidental outcome of having made this set of traces for current purposes, thus the notion of an incidental archive.

Making Perspectives and Knowledge Visible

Models of apprenticeship, cognitive or otherwise, emphasize that expert knowledge must be made visible to learners.[10] Models differ on whether this visibility is achieved by the natural organization of collective activity (Hutchins, 1995; Lave & Wenger, 1991; Rogoff et al., 2003) or by explicit pedagogical intent (Collins, Brown, & Newman, 1989), but this feature nonetheless characterizes all images of apprenticeship. The Traces medium has been an exemplary one for experts to make their knowledge visible; in the terms I have used elsewhere, the medium allows experts to display their disciplined perception, their way of seeing and interpreting objects and events in the world through everyday resources of speech, pointing, and inscription.

What is less explicit in apprenticeship models of learning but no less important if we are to clearly step away from transmission models of education is the fact that novice knowledge needs as well to be made visible through representational practice, a point most explicitly articulated by social-constructivist models of learning. One often overlooked feature of most explicitly designed learning environments (like schools) is that most of the powerful representational media (chalkboards, grade sheets, overheads, computer projection, etc.) are designed for and controlled by the teachers

---

[9]Video Traces began as a desktop application. As this chapter goes to press, we are in the process of releasing a Web-based version thus dramatically enhancing the possibilities for distributed conversation across space and time.

[10]I am using the terms expert and novice because they are familiar. In my view, these terms are limited by being binary and suggesting that expertise and noviceness reside within individuals rather than being distributed properties. They therefore should be understood here as stand-ins for my actual view that recognizes contextualized, often contingent positionings of reciprocal roles of learner and teacher, novice and expert.

rather than by the learners. And most of the representations that learners make are discarded. What Traces provides, in this regard, is a symmetric common medium for both teachers and learners to represent how they see and interpret the world in the form of durable objects (traces).

One possible implication of Video Traces being such a common symmetric medium is that, in the context of an extended traces conversation, expertise is both decentered and reauthorized. It is decentered in the sense that all knowledge sharing need not pass through the recognized expert or teacher. Anyone inside a community of users (or outside, if granted access) can make a trace and take an expert position. As a result, recognized expertise is potentially reauthorized in the sense that expertise can be seen as authoritative rather than arbitrary. A trace made by an expert need not appeal to authority but can display it; it can show a new way of seeing, by providing an interpretation or asking a question that is compelling, subtle, useful, insightful, or disciplined.[11]

Immutable Mobiles for Education: Toward an Educational Infrastructure of Idea Creation and Circulation

Sociologist of science Bruno Latour has made the compelling argument that science became science as we know it—progressive and powerful—when it created a representational infrastructure for the circulation of a certain category of representations called immutable mobiles. In Latour's argument, the immutable mobiles of science are typically graphs and charts of numerical data that summarize, that "draw together" various data from distant locations into a single "center of calculation" to compose synoptic representations that supports claims about other localities (Latour, 1990).

What I would like to propose here, as a final provocation, is that traces might function as immutable mobiles for education but with a twist on the idea. If immutable mobiles allowed science to become Western society's authoritative question answering institution, I see Video Traces as supporting a more diverse role; of supporting idea circulation and refinement, question asking as much as question answering. What would be mobilized and immutable would be current versions of peoples' ideas in the form of traces and trace conversations.

For traces to perform the role I am sketching for them here, the unmet goal I began this chapter with needs to be addressed—building an alternative infrastructure for education. The transmission model's infrastructure is firmly entrenched and if current sociopolitical forces succeed, this infrastructure will become more efficient and more far reaching. Despite this, I want to resist dark and deterministic predictions, because I see the situation through a particular analogy. If the transmission infrastructure is analogous to a particular world-dominating computer operating system (you know the one I am talking about), it is not the only operating system and other operating systems and their affiliated artifacts continue to enroll allies of their own. This is the image I have for the alternative infrastructure for education; as smaller, better designed, and better fitted to people's learning needs and desires.

---

[11]See Stevens and Toro-Martell, 2003 for a more extended argument about how Video Traces can reauthorize expertise in museum settings.

This alternative infrastructure probably won't win market share, but it might win over educators who care most about learning.

I have proposed traces as idea objects that flow through a circulation system of education (rather than test scores and the like). This circulation system remains unbuilt. I have few illusions that such an infrastructure can be built in short order, but I do believe that as a research and design field, we need to be widely focused on what such an alternative infrastructure would look like, one that allows learners and teachers to represent their ideas in more natural ways, to bring them into contact with others, and to see them grow, change, and find expression in all the places that learners work, learn, and live. My hope is that through this collective endeavor and with media objects like traces, we might someday find ourselves with an educational system different from the one that Whitehead and Dewey were worrying about nearly a century ago, one in which ideas are not inert but in motion, in contact, and in use.

ACKNOWLEDGMENTS

My thanks to Amit Saxena and Tom Satwicz and for assistance in preparation of this chapter.

REFERENCES

Barron, B., Schwartz, D. L., Vye, N. J., Moore, A., Petrosino, A., Zech, L., Bransford, J. D., & The Cognition and Technology Group at Vanderbilt. (1998). Doing with understanding: Lessons from research on problem- and project-based learning. *The Journal of the Learning Sciences, 7*(3/4), 271–311.

Becker, H. (1972) School is a lousy place to learn anything in. *American Behavioral Scientist, 16*, 85–105.

Becker, H. S. (1995). The power of inertia. *Qualitative Sociology, 18*, 301–309.

Bowker, G., & Star, S. (1999). *Sorting things out: Classification and its consequences.* Cambridge, MA: MIT Press.

Clark, H. H., & Brennan, S. A. (1991). Grounding in communication. In L. B. Resnick, J. M. Levine, & S. D. Teasley (Eds.), *Perspectives on socially shared cognition* (pp. 127–149). Washington, DC: APA Books.

Collins, A., Brown, J. S., & Newman, S. E. (1989). Cognitive apprenticeship: Teaching the craft of reading, writing and mathematics. In L. B. Resnick (Ed.), *Knowing, learning and instruction: Essays in honor of Robert Glaser* (pp. 453–494). Hillsdale, NJ: Lawrence Erlbaum Associates.

Dewey, J. (1934). *The logic of inquiry.* New York: Holt.

Goodwin, C. (1994). Professional vision. *American Anthropologist, 96*(3), 606–633.

Goodwin, C. (1996). Transparent vision. In E. Ochs, E. A. Schegloff, & S. Thompson (Eds.), *Interaction and grammar* (pp. 370–404). Cambridge, England: Cambridge University Press.

Goodwin, C. (2000). Practices of seeing: Visual analysis: An ethnomethodological approach. In T. van Leeuwen & C. Jewitt (Eds.), *Handbook of visual analysis* (pp. 157–182). London: Sage.

Goodwin, C. (2003a). Pointing as situated practice. In S. Kita (Ed.), *Pointing: Where language, culture and cognition meet* (pp. 217–241). Mahwah, NJ: Lawrence Erlbaum Associates.

Goodwin, C. (2003b). The semiotic body in its environment. In J. Coupland & R. Gwyn (Eds.), *Discourses of the body* (pp. 19–42). New York: Palgrave.

Hall R. (1990). *Making mathematics on paper: Constructing representations of stories about related linear functions.* Unpublished doctoral dissertation, Department of Information and Computer Science, University of California, Irvine.

Hanks, W. F. (1996). *Language and communicative practices.* Boulder, CO: Westview Press.

Harel, I., & Papert, S. (Eds.). (1991). *Constructionism.* Norwood, NJ: Ablex.

Hutchins, E. (1995). *Cognition in the wild.* Cambridge, MA: MIT Press.

Kafai, Y. B., & Resnick, M. (Eds.). (1996). *Constructionism in practice: Designing, thinking, and learning in a digital world.* Hillsdale, NJ: Lawrence Erlbaum & Associates.

Kita, S. (Ed.). (2003). *Pointing: Where language, culture and cognition meet.* Mahwah, NJ: Lawrence Erlbaum Associates.

Latour, B. (1990). Drawing things together. In M. Lynch & S. Woolgar (Eds.), *Representation in scientific practice.* Cambridge, MA: MIT Press.

Lave, J., & Wenger, E. (1991). *Situated learning: Legitimate peripheral participation.* Cambridge, England: Cambridge University Press.

Psathas, G. (1979). Organisational features of direction maps. In G. Psathas (Ed.), *Everyday language: Studies in ethnomethodology* (pp. 203–225). New York: Irvington.

Rogoff, B., Paradise, R., Mejia Arauz, R., Correa-Chavez, M., & Angelillo, C. (2003). Firsthand learning through intent participation. *Annual Review of Psychology, 54,* 175–203.

Sacks, H., Schegloff, E. A., & Jefferson, G. (1974). A simplest systematics for the organization of turn-taking for conversation. *Language, 50,* 696–735.

Stevens, R. (2000a). Who counts what as math: Emergent and assigned mathematical problems in a project-based classroom. In J. Boaler (Ed.), *Multiple perspectives on mathematics education* (pp. 105–144). New York: Elsevier.

Stevens, R. (2000b). Divisions of labor in school and in the workplace: Comparing computer and paper-supported activities across settings. *Journal of the Learning Sciences, 9*(4), 373–401.

Stevens, R. (2002b). Keeping it complex in an era of big education. In T. Koschmann, N. Miyake, & R. Hall. (Eds.), *Computer supported collaborative learning II: Continuing the conversation.* Mahwah, NJ: Lawrence Erlbaum & Associates.

Stevens, R. (2002a). *Video traces*: An Innovative digital medium supporting interpretation, exchange, and knowledge construction. *Innovations in Learning Colloquium Series.* Stanford University.

Stevens, R., & Hall, R. (1997). Seeing tornado: How *Video Traces* mediate visitor understandings of (natural?) spectacles in a science museum, *Science Education, 18*(6), 735–748.

Stevens, R., & Hall, R. (1998). Disciplined perception: Learning to see in technoscience. In M. Lampert & M. L. Blunk (Eds.), *Talking mathematics in school: Studies of teaching and learning* (pp. 107–149). New York: Cambridge University Press.

Stevens, R., & Toro-Martell, S. (2003). Leaving a trace: Supporting museum visitor interpretation and interaction with digital media annotation systems. *Journal of Museum Education, 28*(2), 25–31.

Warner Studios. (1993). "The Five Minute University" from *Gilda Live 1980* [video recording]. Burbank, CA: Warner Studios.

Wenger, E. (1990). *Toward a theory of cultural transparency: Elements of a social discourse of the visible and the invisible.* Unpublished doctoral dissertation. University of California at Irvine.

Whitehead, A. (1929). *The aims of education and other essays.* New York: Macmillan.

Educational Research Visualizations: Creating New Research Warrants

Raul A. Zaritsky
University of Illinois at Urbana-Champaign

After a decade of conducting classroom-based research, the Children's Math World's (CMW) Project led by Karen Fuson has developed for mass distribution an innovative and comprehensive K–6 elementary math curriculum along with a professional development companion, a DVD-based workshop-in-a-box, both to be published commercially (Fuson & Zaritsky, 2005). Many of the innovations in Fuson's CMW curriculum are based on the creation of new accessible algorithms that allow students to utilize visual methods to derive solutions and explain problems to their class. The CMW curriculum also includes a new pedagogy, a new clinical practice for the teacher, and new forms of peer learning for the students. In CMW classrooms, mathematics is a student-guided enterprise. As the students become the leaders the teacher's role moves to providing scaffolding and explanations as needed. In bilingual classes or with students needing special attention, teachers find more time for individual instruction.

The CMW classroom represents a dramatic change in social structure which cannot be described via journal articles alone. Our project needed a new visual form to dessimate the CMW innovations and the solution was video cases created from filming in many CMW classrooms over 3 years. To design a scientific form for a media-based research report, we turned to the field of scientific visualization to provide effective models for visualizing complex research results. Our final database is an almost 6-hour, DVD, in what we consider an Educational Research Visualization (ERV) form produced to broadcast standards and based on the research claims of Scientific Visualization to provide authentic representations of the nature of the CMW world.

Our ERV form uses DVD technology rather than current web video technologies in order to take advantage of the large scale adoption of DVD players in U.S. homes and to provide the needed higher image quality necessary for viewers to follow the mathematics. As an extensive resource of classroom cases, this ERV form supports both early research claims of efficacy and provides a media for the diffusion of the CMW innovation by providing communication of the specific clinical details based on the major CMW results. In construction we maximized communication of the understood variables needed to support increased adoptions (Rogers, 2003; Zaritsky, Kelly, Flowers, Rogers, & O'Neill, 2003).

The complete professional development package contains case-based segments that show how to create a community of math talk and segments that show how to use the mathematic innovations such as various accessible algorithms. Teacher interviews are included to provide personal details of adoptions and reflections on the development of their math community in K–6. Our ERV includes many focused segments; however, we also included many largely unedited classroom segments. These longer video cases show the integration of CMW innovations and mathematical methods into a student-led community and counter some concerns that the video case materials are biased and unrepresentative.

Although projects such as Fuson's CMW are often at the discovery or early trial phase of the research cycle, even after a decade of research in the classroom, so we need to create these visual clinical forms of scientific warrants for our report in this earlier phase. Such early phase warrants are required by this project as well as by many other educational research projects, and improve on the information in research reports provided to funding agencies. As visual warrants supporting research claims far earlier, they allow a scientific basis for adoptions of educational innovations in response to immediate societal demands. Educators should not, indeed will not, wait for a 20-year research cycle.

Phase-appropriate warrants must be capable of communicating early results to differing stakeholders. Each group makes a separate balance of confirmation and immediate need; thus, each needs to be informed in sufficient detail so that adoption possibilities are communicated to teachers, policymakers, and commercial interests. Creating an effective visualization method that provides such warrants required (a) borrowing benchmark validation from engineering; (b) building a media form of a scientific visualization method for education; and (c) using the filming methods of visual ethnography.

RESEARCH METHOD

In order to create our 'existence proofs' demonstrating the effectiveness of CMW results such as the innovations in accessible algorithms, it was necessary to film in detail and in close enough with multiple cameras to follow the mathematic solutions used by the students to follow the classroom interactions. Figure 35.1 provides an image typical of our filming and segment construction. As can be seen in these sample video stills, our ethnographic filmmaking team was able to film very close up, and with the use of two video cameras and a digital still camera, we could follow both the students working at the board, the interactions with the class, and teacher and student desk work very close up.

Figure 35.1.   Example of the case-based classroom footage. A graphical circle over the video helps the viewer note that the "borrowed 1" is placed on the "equals" line.

In this figure, a student is adding a two-digit number. However, rather than in the typical algorithm, the extra 10 created by adding 6 and 8 is placed on the equals line rather than above the 10s column. Fuson's research has shown that such transformations in the visualizations of the problem are easier for students. In this particular math problem, the student first created a proof drawing in order to make the problem clearer. Proof drawings exemplify the principles of scientific visualization that underlies the Children's Math World's research. Figure 35.1 shows that eight 1s and two 1s are combined with two more 1s from the 6 to make one additional 10. Four 1s are left and, thus, 4 is the answer in the 1s column to 8 plus 6. The additional 10 is written graphically by students on the "equals" line rather than above it, as is more familiar. We used a simple graphic, a circle, as shown in the photo, to highlight for viewers/teachers that the additional 10 has been placed on the equals line.

The Design Criteria and Validity Testing Methodology for the CMW's ERV

The ERV was designed using an intiterative design process; in the alpha phase, with major segments of about 10 to 12 min in length, organized around the project's major results of the significant cognitive components and pedagogy of this curriculum. Using benchmark testing of iterative versions of the ERV we could determine whether the specific construction of cases and DVD features was effective with teachers. In the alpha phase our benchmark testing focused on the major features of the ERV case length, choice of specific video chips from the classrooms, and the usefulness of DVD separate audio commentary from math experts and teachers via the other tracks of the DVD. We recorded focus groups viewing the DVD and recorded individual viewings and sorted results to find "time on task" versus "time distracted from task."

Initial testing resulted in design adjustments to our construction of cases: repeats of similar structures were reduced and cases were shortened. Also, when constructions used more elaborate graphical grammars (via digital video effects) to indicate change in place, time, and grade, these were eliminated as they produced cognitive overload, cluttering the teachers' ability to focus on the salient features in the case, especially identifying the class grade. Our results uncovered general classes of unrelated malfunctions; a type of close ups, uncertain words, certain clips, or grammatical elements could all distract the teachers. In focus group testing of later versions, we discovered that teachers not involved with the CMW curricula project also required more explicit and detailed explanations to understand the CMW math. Thus, explanatory mathematic section were constructed for inclusion in the final version.

The final ERV workshop-in-a-box has four major parts: The first one is explanatory; the second one devoted to the major classroom components that demonstrate the main engines of this innovation; the last is composed of on-camera teacher interviews. In the explanatory section of the DVD, we borrowed the pedagogical form of seminal ethnographic filmmaking by using the form of *The Axe Fight* (Asch & Chagnon, 1975). These main sections provide footage of student work. Then, via slow motion, narration, and simple graphics over the student work, we explain the mathematics used in the curriculum. Third, we return to show the student work in real time so that viewers may apply their acquired understanding to following the CMW mathematics in a recap. A fourth component is made up of longer classroom segments that show the full contexts of the use of exemplars and other aspects of classroom interactions not included in the shorter cases.

The following figures are examples of the graphics we created in order to explain the new accessible algorithms that have been shown by Fuson's team to improve student work. In this explanatory case, we narrate the process of addition providing our explanations of the advantages for the student of placing the new 10 on the equals line.

Projects Appropriate to the Methods of an Educational Research Visualization

The effort and expense of creating an ERV is more effective and the filming is less disruptive when used in classroom projects where the presence of visitors is common and considered helpful by both teachers and students. Further, a project well developed in its research and implementation can base filming on a principled preselection of core features of the innovation and results. In the Children's Math World's Project, the preselection was based on variables including the mathematical domain specific to developmental stages, the capturing of a broad cross section of students and teachers from various cultural backgrounds, and the theoretical formulations of essential features of the innovation.

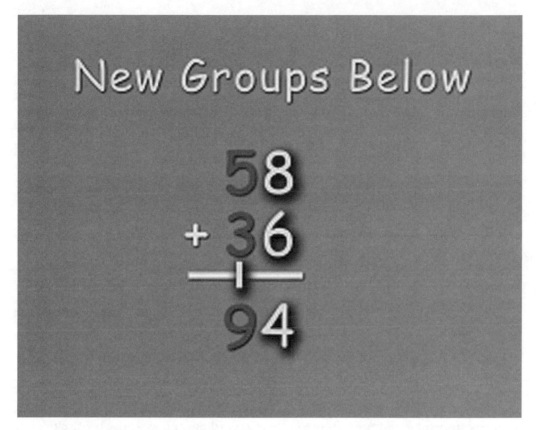

Figure 35.2. In the math explanation section, simple student graphics are redrawn, and simple animations of colors and drop shadow clarify the accessible algorithm called "New Group Below."

Methods for recording in the classroom are many. Often, projects use a single camera from the back of the class to provide a record; or they use a hidden camera in the ceiling in specially designed classrooms. However, in order to build our ERV, we needed eye-level filming from close in, thus our clinical methods were modified from documentary filmmaking. These essential methods were developed for recording real-world events and constructing a document to present a view of the world; these technologies and techniques date from the earliest documentary film; from Russia it is seen in the 1920s work of Dziga Vertov who created films he called *Kino-Eye* or *Kino Pravda* (literally cinema truth). The technology and methods required by Vertov for Kino-Eye are also requirements of documentary filmmakers today. His methods specified:

1. Rapid means of transport.
2. Highly sensitive film stock.
3. Light, handheld film cameras.
4. Equally light lighting equipment.
5. A crew of super-swift cinema reporters. (Vertov & Michelson, 1984)

Vertov's truth claim was that film has a truth in the service of a polemical argument supporting "scientific socialism." In the 1960s, French ethnographic filmmakers created a less polemical and less intrusive method of filmmaking with "direct cinema" with the creation of Jean Rouch's *Chronique d'un été (Chronicles of a Summer;* Rouch, 1961b); however, their publicity person translated Vertov's more polemical notion of Kino- Pravda (cinema truth) into French (Brault, 2003) and *Cinema Vertité* became the intellectual standard for documentary films. The epistemological mischief created by this misnomer "Verité" claiming that a film record is more than a representation bedevils filmmaking and journalism to this day. The direct cinema methods were in fact proposed in 1961 by ethnographic filmmaker Jean Rouch, and are relevant for education; he required that the ethnographer be the camera person; he argued against outsiders as filmmakers: "… I am violently opposed to film crews. My reasons are several. The sound engineer must fully understand the language of the people he is recording. It is thus indispensable that the filmmaker belong to the ethnic group being filmed and that he also be trained in the minutiae of his job" (Rouch, 1961a, online). Our filmmaking utilized Rouch's criteria for a team able to create direct cinema.

Our ERV case construction constructs a scientific visualization representation of the CMW classroom and its innovations in mathematics and is based in part on the authors' prior experience in the creation of video cases for academic reports about the state of the art of computer graphics as can be seen in tens of hours of the *Siggraph Video Review*, a video publication of the Academic Society of Computer Graphics. The form was further developed in an early set of theoretical papers by this author and colleagues (Herr & Zaritsky, 1987, 1989). In this ERV, we wanted to balance filming bias against our need to capture effective close-in recordings and we chose being close but not seen as disruptive by being part of the class rather than applying the creed of direct cinema to be "a fly on the wall." From 90-plus hours of footage shot over 3 years, an epistemic backbone was created that compresses and focuses on the major research results. The major sections of the ERV were based on the following initial criteria: The research team identified the highest priority issues and then reviewed and selected from the raw classroom video footage the clips that portrayed the fullest examples of the innovations.

In early versions, our editing developed many advanced grammatical forms of media construction between and within scenes, grammatical forms with promising theoretical foundations. However, as we were taking this project to scale, we were obligated to validate all such features for inclusion in the released national product. Benchmark testing discovered that many of the theoretically promising features were in fact malfunctioning. Thus, we removed features not validated and utilized only the validated functional affordances in our final ERV construction.

Our ability to make a claim for the representational validity of these math innovations is enhanced by results of focus groups within the community of CMW teachers who report that the video cases present a realistic sample of practice and outcomes. As our DVD form provides a large database of video materials for review by teachers and other researchers, this claim of the ERV that it represented a correct picture of the CMW classrooms is partially validated by our clinical community—without the need for a statistical method.

Our recording technology included using two broadcast-quality DV-25 camcorders and multiple wireless and boom microphones, plus a high-resolution still camera. The results always met professional standards for broadcast TV with full-screen images. Using a digital camera throughout the filming created even more editing choices as these still images may be postprocessed into moving video with inexpensive software (see www.stagetools.com) thus creating third and forth camera angles of smooth zooms and pans around the classroom. One may see the degree of zooming that can be done with a still image and this inexpensive editing tool. Thus, with digital still images above 6 mega pixels, one has enough resolution to zoom from a shot of the whole class into the math problem on the board at full video resolution. By using postprocessing tools, a digital still camera should be considered as another important image capturing device when filming in a classroom.

Using postprocessing in case constructions is an important feature noted in other systems. The feature was crucial here and is more complete in systems such as DIVER™

Figure 35.3.   Shown is the range of postprocessing zoom researchers can use and still maintain a video resolution image.

(Pea, in press). The DIVER™ system for recording allows such processing with all the live video.

What Constitutes a Case?

Although there are numerous theories that informed our case construction, in practice, we discovered via iterative design cycles and testing that educational theories were only generally suggestive at the large grain level. For example, in describing how text-based cases should be constructed for teachers, Shulman suggested that text-based case studies should contain

> a plot with some dramatic tension that must be resolved, several embedded problems that can be framed and analyzed from many perspectives, thoughts and feelings of the author as the narrative unfolds, and reflective comments and questions about events that occurred. (Shulman, 1992, 2001)

Although we implemented these features in creating our video cases (and we also note that these suggestions are identical to the plot forms for most film scripts), such goals and theories were not specific enough (of small enough grain size) to inform our actual practice of creating educationally appropriate filming, editing, or DVD production.

Another example of useful theory comes from cognitive flexibility theory (Spiro, 1988; Spiro, Zaritsky, & Feltovich, 2001); it informed our case recording and case construction by suggesting that we include a large number of cases. This theory states, in general, that in very messy domains, one needs to present a large population of cases in order that learners may determine the common underlying features of the family. In our ERV versions, we do provide a large database of video cases of the CMW innovation rather than just a few specific cases. Thus cognitive flexibility theory was helpful as a guiding notion in building a large database of cases within the ERV construction. However, we could not practically validate that our collection of cases constitutes the most effective family of cases. With our benchmarks, we could only be assured that our many case repetitions of various student solutions were not too "boring" to focus groups. Yet to be discovered are more practical benchmarks, for such design-based projects, to test if deeper notions of the family of solutions are constructed from the provided video cases.

In general, our goals for the ERV were to report the central CMW research results; all of the major significant and effective innovations in mathematics teaching. These main results are structured around five essential CMW classroom components with their new approaches to mathematics within a new classroom culture of mathematics talk. Our pre-selection method assured that we could construct these main case segments; this provided the first filter for selecting when and where to film. Thereafter, four other criteria were added: (a) Show a broad and varied group of students and teachers to demonstrate that the innovations were implemented in inner city classrooms including bilingual classrooms with a broad cross section of students; (b) create a pace and length that would engage without boring any of the intended audiences and yet provide enough length and details common to videos used in professional developmental workshops; (c) create a musicality in which a monotone in construction is avoided, and instead, there is an emotion arc in each segment, when possible; and (d) select demonstrations of classroom exemplars directly tied to the main concepts in the grade by grade student development of math skills.

Innovative features we built and tested included; (a) advanced media grammars to mark a change in focus, a change in time, or a change in classrooms; (b) titles and other graphics using techniques such as rotoscoping that highlighted or enchased key points such as the graphical solutions of students via zooming into the video scene or graphically outlining the student's solution methods and various types of graphics over video to indicate the grade level of each classroom; (c) two alternative angles via video cameras (DV-25) plus high-resolution still cameras allowed an edited construction where the viewer could select from either camera's point of view; and (d) the use of separate audio channels to provide expert commentary on both the mathematics being viewed and the nature of classroom management.

RESULTS

Our most significant outcome was the discovery that a large number of theoretically promising media design features and media grammars grossly malfunctioned, for various reasons, with general teacher audiences, while media grammars based on common filmmaking methods could be shown in test groups to function as expected. Our benchmark testing methods were adapted from engineering design (Ulrich & Eppinger, 2000) and are discussed in detail in earlier work (Zaritsky et al., 2003). Given the possibilities to develop new media forms, and some promising ideas for possible new media grammars (Spiro et al., 2001), we hoped to validate the usefulness of innovative grammar form as well as other innovative affordances including the use of the audio commentary tracks and the multiple video tracks available with DVD media. Validation of features was essential for each prototype ERV to create a reasonable prediction that all features would function for the broad national audience.

The following is a brief summary of some surprising malfunctioning components and some of our solutions used in constructing our final commercial version:

Promising, but Unexpected Malfunctioning Features

1. Media grammars: Complex media grammars between and in the cases decreased the ability of teachers to say where they were in the cases, to identify the grade level, or to say what they were seeing even though they had been provided this information through narration, graphics, and video effects. These more complex media grammars were replaced with standard film grammars (cuts and dissolves) and silent film type intertitles when we needed graphics to indicate grade level. Very slow pan and scan images were used when providing detailed information via narration (common in all history documentaries). No advanced grammatical innovations were used in the final version.

2. Innovative video graphics or computer graphics to explain the math domain: Complex graphics created via 3-D software were found to distract teachers. Instead, simple re-creations of the images from the board in off-the-shelf graphic packages were used in the final version.

3. Multiple angles: The DVD feature of being able to switch between two points of view was a feature unknown to most teachers. When built into the menu structure, only further complexity ensued. However, when a picture-in-picture image contained both camera views, experienced DVD owners were able to switch between the camera angles. This feature was not used in the final version because it was still too complex.

4. Multiple audio commentary track: Besides the sync sound of the case, two additional audio commentary channels were created. One included the teachers in the video discussing their goals and classroom management issues while the viewer watched them in their classrooms. A second audio commentary channel was created in which math educators discussed the domain-specific features of the new curriculum. As only some of the teachers who owned DVDs knew how to use this feature, it was excluded from the final product. It remains a highly promising feature for a time when switching between audio channels is simplified by hardware manufacturers and by PC software and when more viewers are familiar with this potential.

DISCUSSION: BUILDING ON THE HISTORIES OF SCIENTIFIC VISUALIZATION AND ETHNOGRAPHIC FILM

As a scientific method of representing a research project's major results, scientific visualization provides the closest fit to the scientific needs in education for methods; further, an Educational Research Visualization can provide early phase research warrants by representing the essential clinical results of a classroom innovation. Although our ERV does not provide predictive probabilities, it does demonstrate the clinical development of the CMW across K–6. As in a medical grand rounds presentation, such clinical details allow the community to understand the interaction of various and highly situated variables. Modern image construction for meaning making in science dates from the medical detective work and visual analysis used in 1848 London to stop a devastating outbreak of cholera. Dr. John Snow in part created both the new field of epidemiology and an early effective form of scientific visualization (Vinten-Johansen, 2003). To find the cause and a solution to the outbreak, Dr. Snow plotted mortalities on a map of the city and then, using overlaid maps of possible variables such as elevation and water sources, he discovered the fit. Visually he found that the deaths were all clustered around the area whose water came from the Broad Street pumping station. His visual presentations also convinced the Parliament to close that pumping station of polluted Thames water and as the story goes, ended the epidemic (Tufte, 1990, 2001). With increased computational power, scientific visualization has become a major engine of meaning making in science.

In the social sciences, anthropology developed a "visual" subdiscipline furthered by technological inventions, in the 1960s, of the Éclair portable synchronous film cameras and the synchronous portable Nagra sound recorders. Prior to this point, to record synchronous sound required that the camera and the recorder be tied together thus restricting the freedom of filming and the intimacy of the location sound. Modern documentary and ethnographic films are based on the use of these inventions. These cameras and recorders borrowed the recently invented transistors and quartz circuits from the first quartz watch to lock their motors independently. The result of this technological innovation was that a new generation of lightweight, synchronous filmmaking equipment was developed for ethnographic projects. Math educators interested in the problem of understanding rate would be delighted with the result seen in *Under the Men's Tree*, from 1974, an ethnographic film in which African men discussed how it was impossible for their friend who is leaving tomorrow to make it to the city before their friends who left yesterday (MacDougall & MacDougall, 1974). (Solution hint: One group was going by car.)

Another technology convergence occurred recently between the digital methodologies of recording/filmmaking and the methodologies of scientific visualization. New digital video cameras record an hour on very small cassettes; computer technologies allow massive storage of the video; postprocessing software and editing systems provide full professional ability to construct scenes; and elaborate 3-D effects including morphing between images are available on the desktop. Educational Research Visualization methods are based on this new mix of digital video recording and high-resolution digital still photography with desktop editing. As a set of technologies, these can be moved into any learning location; however, the filming methods are very close in and would be disruptive if employed in classrooms that do not regularly have researchers join the class. Educational Research Visualization methods require new educational researchers trained in the skills of ethnographic filmmaking.

Needless to say, a nonconvergence for educational research projects is the practice found in education when using this new technology with graduate students unskilled in filmmaking. Almost universally, sufficiently clear and readable images and understandable audio are not obtained. As we have often argued, the solution is for education, like anthropology, to develop its own form of visual methodologists.

Having found that new media grammars suggested by theories in education and psychology malfunctioned in our tests, our ERV form returned to the cinema grammars developed in over a century as seen in the earlier films and writings of Dziga Vertov and Sergei Mikhailovich Eisenstein, from 1920s Russia (Eisenstein & Leyda, 1948). Thus, when a complex grammar failed, we returned to the common film solution. However, for our explanatory sections, we successfully borrowed the pedagogical form from an innovative ethnographic film *The Axe Fight* (Asch & Chagnon, 1975) created by Timothy Asch based on his anthropological field work with the Yanomami. Showing the footage to his classes, he found the footage itself left them confused so he created for the final film version of *The Axe Fight* an new effective pedagogical form; providing what we might call "just-in-time domain knowledge" so that the students would understand what they were seeing in *The Axe Fight* (Asch & Chagnon, 1975; Elder, 2001). Asch, during the first section of the film, presents a fight between villagers and visitors; this section is notable for what looks like extreme violence and for confusion, as the viewer cannot understand what triggered the fight or ended it. Having presented the fight in detail, Asch then stops the ethnographic footage and switches to charts and narrations to explain the kinship relationships of the village. The explanation answers the question of what triggered the fight and the last segment replays the fight scene and allows viewers to apply their new anthropological knowledge to understanding what they are seeing in the fight. The result was a very successful pedagogical structure that this author borrowed immediately on seeing an early screening of Asch's work (Zaritsky, Kamii, & DeVries, 1976). In our ERV, we used Asch's pedagogy when constructing our explanatory sections.

CONCLUSION

The construction of our Educational Research Visualization on DVD is part of a convergence of scientific visualization methods and ethnographic methods. While providing exemplar new media affordances as suggested by theory and laboratory experiments were initial research goals for our project; however, when the project needed to create a professional development product for mass distribution we included few grammars and advanced features in the final DVD as their effectiveness could not be validated. Rather than suggested grammars, we returned to the literature of film for our media grammars. The results suggest that with each specific media creation and technology, educational researchers must use an empirical design-based process with increasingly demanding benchmarks.

Our iterative testing of versions with complex media grammars seem to provide us with results that diverge from other researchers who have promoted image sound and media innovations. However, in part, we suggest that all multimedia forms are not the same. We suggest that when using a higher resolution video picture and image size, different outcomes should appear as in our results. Compared to multimedia, our images provide 30 degrees of visual angle on television screens; this is a significant increase in size and detail. What we call multimedia forms incorrectly lump DVD with Web-based video and CD-ROM multimedia when in fact multimedia forms provide only approximately 2.5 to 7 degrees of viewing angle (at normal distances from the screen of approximately an arms length). Wide-screen feature films or home theater provides approximately 70 plus degrees of viewing angle; virtual reality technology requires even higher viewing angles in order to create a cognitively immersive effect. For a multimedia CD-ROM or Web-based delivered image to provide similar visual acuity to DVD, one would have to place one's nose on the space bar of your laptop. The results would not be pretty!

For the viewer then, DVD technology provides more accessible clinical detail in both sound and image quality. Clusters of distraction that we report are consistent with this increased clinical detail. We suggest that cognitively, our DVD media then only contains some features in common with multimedia forms and our failures with new grammars may not conflict with research using 3–5 degrees of viewing angle and new grammar forms. As we report, during our design iterations, we returned to standard film grammars and likely our success with these grammars is due to image-size-appropriate grammars found successful by many millions of filmgoers every week. If one remembers the allegory of Plato's Cave, we suggest this comparison can be stated metaphorically by considering multimedia forms as the allegory of Plato's Box!

In summary, education needs to create new visual warrants for an early research innovations and creating an exemplar was the primary goal of this ERV research as the form for the research report of Fuson's CMW mathematics curriculum research. The long research cycle in educational innovations involving math and science with technology is well documented. For PLATO, Logo, and other exemplars, it has taken about 15 years. SimCalc (Roschelle, Kaput, & Stroup, 2000) is a recent example where moving from its early phase work has taken 10 to 12 years prior to the current large-scale, Texas-based clinical trial with IERI funding. The WorldWatcher™ software tools for middle school atmospheric science began in 1992 (Edelson, Gordin, & Pea, 1999), and 12 years later it is a product, without at-scale trials. In summarizing the typical research cycle in mathematics education, Schoenfeld stated for an NSF presentation that, "It may take 25 to 30 years from the first basic research to large-scale implementation of ideas grounded in the research" (2003).

Given the clinically situated nature of these innovations, their practical value and the possibility of a 20-year research cycle, we argue that visualization methods such as our exemplar Educational Research Visualization method can suffice to provide authentic new scientific warrants supporting the claims of these educational innovations earlier and also, they can create needed effective communication media for the diffusion of these classroom innovations.

ACKNOWLEDGMENT

This material is based on work supported by the National Science Foundation under Grant No. 222599. Any opinions, findings, and conclusions or recommendations expressed in this material are those of the author and do not necessarily reflect the views of the National Science Foundation.

REFERENCES

Asch, T., & Chagnon, N. (1975). *The axe fight* [Documentary]. From *The Yanomamo Series*.

Edelson, D. C., Gordin, D. N., & Pea, R. D. (1999). Addressing the challenges of inquiry-based learning through technology and curriculum design. *Journal of the Learning Sciences, 8*(3/4), 391–450.

Eisenstein, S., & Leyda, J. (1948). *The film sense* [New ed.]. Faber & Faber.

Elder, S. (2001). Images of Asch. *Visual Anthropology Review, 17*(2).

Fuson, K., & Zaritsky, R. A. (2005). *Children's math worlds* (DVD multi audio and video tracks, interactive). Evanston, Il: NSF Grant.

Herr, L., & Zaritsky, R. (1987). Using video resources for scientific visualization reports. In T. A. DeFanti, M. D. Brown, & B. H. E. McCormick (Eds.), *Visualization in scientific computing* (Vol. 21). IEEE.

Herr, L., & Zaritsky, R. A. (Writers/Producers). (1989). *Visualization in scientific computing* [Video tape]. (Available from SIGGRAPH Video Review, ACM/SIGGRAPH).

MacDougall, D., & MacDougall, J. (Writer). (1974). *Under mens' tree*: U.C. Extension.

Pea, R. (in press). Video-as-data and digital video manipulation techniques for transforming learning sciences research, education and other cultural practices. In J. Weiss, J. Nolan, & P. Trifonas (Eds.), *International handbook of virtual learning environments*. Dordrecht: Kluwer.

Rogers, E. M. (2003). *Diffusion of innovations* (5th ed.). New York: Free Press.

Roschelle, J., Kaput, J., & Stroup, W. (2000). Simcalc: Accelerating student engagement with the mathematics of change. In M. J. Jacobsen & R. B. Kozma (Eds.), *Innovations in science and mathematics education: Advanced designs for technologies of learning* (pp. 47–75). Mahwah, NJ: Lawrence Erlbaum Associates.

Rouch, J. (1961a). Retrieved from http://pro.imdb.com/name/nm0745541/personal

Rouch, J. (1961b). *Chronique d'un été* [Chronicle of a summer]. Retrieved October, 2005, from http://pro.imdb.com/title/tt0054745/

Schoenfeld, A. (2003). *Expanding research opportunities at EHR.* Presented October 7, 2003 in Washington, DC: National Science Foundation.

Shulman, J. H. (1992). *Case methods in teacher education.* New York:

Shulman, L. S. (2001). Appreciating good teaching: A conversation with Lee Shulman. *Educational Leadership, 58*(5), 6–11.

Spiro, R., J., Zaritsky, R., & Feltovich, P. (2001). *Defining the video case: What we know and how we (use video to) know it. Learning technology developments with video case-based learning utilizing DVD technologies.* Paper presented at the AERA.

Spiro, R. J. (1988). *Cognitive flexibility theory advanced knowledge acquisition in ill-structured domains.* Champaign, IL: University of Illinois at Urbana-Champaign.

U.S. Dept. of Education Office of Educational Research and Improvement Educational Resources Information Center.

Tufte, E. R. (1990). *Envisioning information.* Cheshire, CN: Graphics Press.

Tufte, E. R. (2001). *The visual display of quantitative information* (2nd ed.). Cheshire, CN: Graphics Press.

Ulrich, K. T., & Eppinger, S. D. (2000). *Product design and development* (2nd ed.). Boston: Irwin/McGraw-Hill.

Vertov, D., & Michelson, A. (1984). *Kino-eye: The writings of Dziga Vertov.* Berkeley, CA: University of California Press.

Vinten-Johansen, P. (2003). *Cholera, chloroform, and the science of medicine: A life of john snow.* Oxford, NY: Oxford University Press.

Zaritsky, R., Kelly, A. E., Flowers, W., Rogers, E., & O'Neill, P. (2003). Clinical design sciences: A view from sister design efforts. *Educational Researcher, 32*(1), 32–34.

Zaritsky, R. A., Kamii, C., & DeVries, R. (Writer). (1976). *Playing with rollers* [Film]. (Available from University of Illinois Chicago)

Author Index

Note: *f* indicates figure; *n* indicates footnote.

Subject Index

Note: *f* indicates figure; *n* indicates footnote; *t* indicates table.